W0051367

Biology of Turbellaria and some Related Flatworms

Developments in Hydrobiology 108

Series editor
H. J. Dumont

Biology of Turbellaria and some Related Flatworms

Proceedings of the Seventh International Symposium on the Biology of the Turbellaria, held at Åbo/Turku, Finland, 17–22 June 1993

Edited by

Lester R. G. Cannon

Reprinted from Hydrobiologia, vol. 305 (1995)

Springer Science+Business Media, B.V.

Library of Congress Cataloging-in-Publication Data

International Symposium on the Biology of the Turbellaria (7th : 1993 : Turku, Finland)
Biology of Turbellaria and some related flatworms : proceedings of the Seventh International Symposium on the Biology of the Turbellaria, held at Åbo/Turku, Finland, 17-22 June 1993 / edited by Lester R.G. Cannon.
p. cm. -- (Developments in hydrobiology ; v. 108)
"Reprinted from Hydrobiologia, vol. 305 (1995)."
Includes bibliographical references and index.
ISBN 978-94-010-4025-9 ISBN 978-94-011-0045-8 (eBook)
DOI 10.1007/978-94-011-0045-8
1. Turbellaria--Congresses. 2. Platyhelminthes--Congresses.
I. Cannon, L. R. G. (Lester Robert Glen), 1940- . II. Title.
III. Series: Developments in hydrobiology ; 108.
QL391.P7I56 1993
595.1'3--dc20 95-14932

ISBN 978-94-010-4025-9

Printed on acid-free paper

Originally published by Kluwer Academic Publishers in 1995
Softcover reprint of the hardcover 1st edition 1995

Contents

ECOLOGY AND BEHAVIOUR

ANATOMY AND ULTRASTRUCTURE

Preface

"What's in a name?" muses Juliet, "that which we call a rose, by any other name would smell as sweet". It is perhaps fitting that at a time when the serious question was posed as to what we call our animals – is it Plathelminthes or Platyhelminthes? – we should have held our conference in the town of Åbo, if you are a Swedish speaking Finn, or Turku if your native tongue is Finnish. We were all in the same place, however, at our host institution the Swedish language University, Åbo Akademi.

Provided there is no ambiguity most of us are happy to accommodate to different names of things in our own language (Australian, American and British English all have their own idiosyncracies) and especially between languages. Of course, there can be subtle differences, so tolerence is needed to ensure that confusion is minimal. So in everyday life we cope with synonyms with little difficulty. There are times, however, when names are very important. As a taxonomist and Museum worker I am daily reminded of name changes and the need to strive for accuracy and consistency in naming organisms. My discipline demands it, but so does my computer. Computers are quite literal and unless we make sure they are aware of synonyms and aliases they recognise words as quite different unless they are exactly the same. A search for Plathelminthes may turn up nothing on Platyhelminthes – a serious error for they are one and the same.

To me as an English speaker, Platyhelminthes is correct and I have adopted this for all the papers in this volume, all published in English, except for those where the authors whose native tongue is not English prefer Plathelminthes. In these cases their choice of word remains, but I have added Platyhelminthes to the key words.

Words, of course, can also change their meaning. About ten years ago 'Turbellaria' lost its status as representing the **Class Turbellaria** to become a group name for the mainly free-living flatworms: the taxon set to disappear under the thrust of modern phylogenetic studies. These studies continue apace with increasing use of molecular techniques. The relative simplicity of turbellarians makes them attractive candidates for such studies, and many others as was outlined in this symposium, which probe the fundamental structure and function of metazoans.

Of course, not everyone agrees about words, or taxa, or indeed interpretation of data. The discussion, opinion, interchange of ideas all contribute to our search for truth and these conferences provide an exciting opportunity for us to get together. It seems incongrous that we have no name for our group who meet regularly and talk of turbellarian biology. We have no formal structure, no society, no journal, no newsletter. Nevertheless these regular Proceedings volumes belie that we are so amorphous. Perhaps the time is ripe for us to evolve into a more cohesive group.

LESTER CANNON

Organizers of the Seventh International Symposium on the Biology of the Turbellaria

Convener
Lecturer Maria Reuter
Department of Biology, Åbo Akademi University, SF-20520 Åbo, Finland

Local Organizing Committee
Dr Margaretha Gustafsson
Dr Paula Lindroos
Department of Biology, Åbo Akademi University, SF-20520 Åbo, Finland

International Organizing Committee
Prof. Ernest Schockaert
Limburgs Universitair Centrum, B-3590 Diepenbeek, Belgium

Prof. Jan Hendelberg
Dept. of Zoology, University of Gothenburg, S-413 90 Gothenburg, Sweden

Acknowledgments

Financial support for the 7th International Symposium of the Biology of the Turbellaria came from
The Academy of Finland
Ministry of Education, Finland
The Council of the Åbo Academy Foundation
Åbo Akademi University

Editorial review and assistance of the following is gratefully acknowledged:

Adlard, R.
Anderson, D.
Arthington, A.
Baguna, J.
Barker, S.
Best, J.
Beveridge, I.
Blair, D.
Boyer, B.
Burt, M.
Cribb, T.
Curini-Galletti, M.
Dittman, S.
Dobson, C.
Domrow, R.
Fairweather, I.
Ghiradelli, E.
Goggin, L.
Halton, D.
Healy, J.
Jennings, J.
Joffe, B.
Jondelius, U.
Jones, M.
King, M.
Koopowitz, H.
Lanfranchi, A.
Lawn, I.
Martens, P.
Mead, R.
Morita, M.
Moritz, C.
Newman, L.
Ogren, R.
Orii, H.
Palmberg, I.
Peter, R.
Reuter, M.
Rieger, R.
Rohde, K.
Saló, E.
Schockaert, E.
Sluys, R.
Smith, J.
Teshirogi, W.
Tyler, S.
Wallace, C.
Watson, N.
Webb, R.
Whittington, I.
Williams, J.
Winsor, L.

1. R. Sluys
2. M. Curini-Galletti
3. S. Morita
4. M. Morita
5. I. Böckerman
6. G. de Clerk
7. B. Sopott-Ehlers
8. A. Best
9. A. Takai
10. E. Kotikova
11. M. Reuter
12. K. Niewiadomska
13. B. Boyer
14. M. Querci-Gremigni
15. A. Falleni
16. E. Ghiradelli
17. M. Kawakatsu
18. E. Schockaert
19. M. Silveira
20. L. Cannon
21. P. Swiderski
22. M. Gustafsson
23. V. Fagerholm
24. A. Streng
25. I. Elo
26. M. Takai
27. W. Schürmann
28. J. Ogren
29. J. Hauser
30. R. Peter
31. B. Grabda-Kazubska
32. J. Moens
33. V. Gremigni
34. O. Raikova
35. E. Kornakova
36. A. Rogosin
37. I. Drobysheva
38. S. Ishida
39. B. Joffe
40. P. Lucchesi
41. C. Cannon
42. D. Halton
43. A. Swiderski
44. S. Ishii
45. A. Hendelberg
46. K. Lundin
47. P. Boaden
48. R. Ogren
49. U. Ehlers
50. K. Rohde
51. H. Koopowitz
52. J. Baguñà
53. M. Wahlberg
54. A. West
55. M. Grahn
56. S. Tamura
57. I. Hori
58. I. Oki
59. H. Kasai
60. T. Shinozawa
61. Y. Mamkaev
62. P. Panula
63. J. Hendelberg
64. T. Sakurai
65. J. Best
66. U. Jondelius
67. M. Burt
68. J. Smith III
69. A. Kumar Aditya
70. M. Sarperi
71. H. Orii
72. P. Lindroos
73. G. Bylund
74. G. Malmberg
75. A. Czubaj
76. R. Rieger
77. I. Fairweather
78. C. Vreys
79. D. Bueno
80. P. Martens
81. K. Kuznedelov
82. L. Beukeboom
83. O. Timoshkin
84. N. Michiels
85. K. Karlstedt
86. K. Eriksson
87. E. Saló
88. J. Holleman
89. Y. Shirazawa

List of Participants and Contributors

Aditya, Ajit Kumar
Dept. of Zoology
Visva-Bharati University
Santiniketan-731 235
West Bengal
India

Agata, Kiyokazu
Dept. of Life Science
Himeji Institute of Technology
678-12 Harima Science Park City,
Hyogo
Japan

Alberti, A.
Istituto de Zoologia
Università di Sassari
Italy

Aragao, Pedro Henrique A.
Laboratório de Microscopia
Electrónica
Instituto de Física
Universidade de São Paulo
05508-900 São Paulo
Brazil

Ax, Peter
II. Zoologisches Institut und
Museum der Universität
Berliner Str. 28
3400 Göttingen
Germany

Bae, Tony
Dept. of Ecology and Evolutionary
Biology
University of California
Irvine CA 92717
USA

Baguñà, Jaume
Departamento de Genètica
Universitad de Barcelona
Diagonal 645
08071 Barcelona
Spain

Bandyopadhyay, M.P.
Dept. of Zoology
Visva-Bharati University
Santiniketan - 731 235
West Bengal
India

Batlle, E.
Departamento de Genètica
Universitad de Barcelona
Diagonal 645
08071 Barcelona
Spain

Baverstock, P.R.
School of Resource Science and
Management
University of New England
Lismore N.S.W.2480
Australia

Bayascas-Ramirez, J.R.
Departamento de Genètica
Universitad de Barcelona
Diagonal 645
08071 Barcelona
Spain

Bedini, Celina
Dipartimento de Scienze del
Comportamento Animale e
dell'Uomo
Università di Pisa
Via. A. Volta 6
56126 Pisa
Italy

Best, Jay Boyd
Dept. of Environmental Health
Colorado State University
Fort Collins, CO 80526,
USA

Birstein, V.A.
N.K.Koltzov Institute of
Developmental Biology RAN
117334 Moscow
Russia

Boaden, Patrick
The Queen's University of Belfast
Marine Biology Station
Portaferry BT 1 PF
Northern Ireland
UK

Boyer, Barbara
Department of Biology
Union College
Schenectady, NY 12308
USA

Bueno, David
Departamento de Genètica
Universitad de Barcelona
Diagonal 645
08071 Barcelona
Spain

Burt, Michael D.B.
Dept. of Biology
University of New Brunswick
Fredericton N.B.E3B 6E1
Canada

Böckerman, Inger
Dept. of Biology
Åbo Akademi University
BioCity, Artillerigatan 6
FI-20521 Åbo
Finland

Cannon, Christine
International Food Institute of
Queensland
19 Hercules St. Hamilton
Queensland 4007
Australia

Cannon, Lester R.G.
Queensland Museum
Box 3300, South Brisbane
Queensland 4101
Australia

Canovai, R.
D.C.D.S.L. sez. Entomologia agraria
Via San Michele 2
I-56124 Pisa
Università di Pisa
Italy

Carranza, S.
Departamento de Genètica
Universitad de Barcelona
Diagonal 645
08071 Barcelona
Spain

Charchenko, V.V.
Limnonlogical Institute
Russian Academy of Sciences
664 033 Irkutsk
Russia

Chomicz, Lidia
Dept. of General Biology and
Parasitology
Medical Academy
02-004 Warzawa
Chatubinskiego 5
Poland

Corominas, M.
Departamento de Genètica
Universitad de Barcelona
Diagonal 645
08071 Barcelona
Spain

Cumming, Meg S.
School of Biological Sciences
University of Manchester
Manchester M13 9PL
UK

Curini-Galletti, Marco
Istituto di Zoologia
Università di Sassari
Via Muroni 25
I-07100 Sassari
Italy

Czubaj, Andrzej
Dept. Cytology
University of Warzawa
Krakowskie Przedmiescie 26/28
00-927/1 Warszawa
Poland

De Clerk, Gert
Dept. SBG
Limburgs Universitair Centrum
Universitaire Campus
B-3590 Diepenbeek
Belgium

Denegri, Guillermo Maria
Cátedra de Parasitología y Enf.
Parasitarias-Facultad de
Ciencias Veterinarias – 60 y 118
1.900 La Plata
Argentina

Drobysheva, Irina
Zoological Institute
Russian Academy of Sciences
199034 St Petersburg
Russia

Dyganova, Rosa
Dept. of Invertebrate Zoology
Kazan State University
420008 Kazan
USSR

Ehlers, Ulrich
II. Zoologisches Institut und
Museum der Universität
Berliner Str. 28
3400 Göttingen
Germany

Elo, Iina
Dept. of Biology
Åbo Akademi University
BioCity, Artillerigatan 6,
FI-20521 Åbo
Finland

Elvin, Mark
Dept. of Ecology and Evolutionary
Biology
University of California
Irvine, CA 92717
USA

Eriksson, Krister
Dept. of Biology
Åbo Akademi University
BioCity, Artillerigatan 6
FI-20521 Åbo
Finland

Espinosa, Lluis
Departamento de Genètica
Universitad de Barcelona
Diagonal 645
08071 Barcelona
Spain

Ezaki, M.
Dept. of Biological and Chemical
Engineering
Faculty of Engineering
Gunma University
Japan

Fairweather, Ian
School of Biology & Biochemistry
The Queen's University of Belfast
Belfast BT7 INN
Northern Ireland
UK

Falleni, Alessandra
Dipartimento di Biomedicina
Sperimentale
Sezione di Biologia e Genetica
Università di Pisa
Via A. Volta 4
56126 Pisa
Italy

Fujino, H.
Dept. of Biological and Chemical
Engineering
Faculty of Engineering
Gunma University
Japan

Galleni, Lodovico
D.C.D.S.L. sez. Entomologia
Agraria
Via San Michele 2
I-56124 Pisa
Italy

Garcia-Fernandez, Jordi
Departamento de Genètica
Universitad de Barcelona
Diagonal 645
08071 Barcelona
Spain

Gevaerts, H.
Dept. SBG
Limburgs Universitair Centrum
B-3590 Diepenbeek
Belgium

Ghosh, P.N.
Dept. of Zoology
Visva-Bharati University
Santiniketan – 731235
West Bengal
India

Grabda-Kazubska, Bozena
W. Stefan'ski Institute of
Parasitology
Polish Academy of Sciences
ul, Pasteura 3, S.p 153
00-973 Warszawa
Poland

Grahn, Malin
Dept. of Biology
Åbo Akademi University
BioCity, Artillerigatan 6
FI-20521 Åbo
Finland

Gremigni, Vittorio
Dipartimento di Biomedicina
Sperimentale
Sezione di Biologia e Genetica
Università di Pisa
Via A. Volta 4
56100 Pisa
Italy

Grintsov, V.
Institute of Biology of the Southern
Seas
Ukrainian Academy of Sciences
2 Nakhimov Str.
335011 Sevastopol
Ukraine

Gustafsson, Margaretha
Dept. of Biology
Åbo Akademi University
BioCity, Artillerigatan 6
FI-20521 Åbo
Finland

Halton, David
School of Biology & Biochemistry
The Queen's University of Belfast
Belfast BT7 1NN
Northern Ireland
UK

Hauser, Josef
Instituto de Pesquisa de Planária
Universidade do Vale do Rio dos
Sinos – UNISINOS
Praca Tiradentes, 35 – CP 275
93010-020 São Leopoldo
Brazil

Hefford, C.
Dept. of Microbiology and
Infectious Diseases
Flinders Medical Centre
Bedford Park, S.A. 2480
Australia

Hendelberg, Jan
Dept. of Zoology
University of Göteborg
Box 25059
S-413 90 Göteborg
Sweden

Hoek, Robert M.
Research Institute Neurosciences
Vrije Universiteit
De Boelelaan 1087
1081 HV Amsterdam
The Netherlands

Holleman, John
8320 El Matador Dr
Gilroy, CA 95020
USA

Hori, Isao
Dept. of Biology
Kanazawa Medical University
Uchinada-machi
Ishikawa-ken 920-02
Japan

Hydén, A.
Dept. of Biology
Åbo Akademi University
BioCity, Artillerigatan 6
FI-20521 Åbo
Finland

Iida, M.
Dept. of Biology
Hirosaki University
Hirosaki 036
Japan

Inoue, K.
Institution of Internal Secretion
Gunma University
Maebashi
Japan

Ishida, Sachiko
Dept. of Biology
Hirosaki University
Hirosaki 036
Japan

Ishii, Saburo
Central Research Laboratory
Fukushima Medical College
1 Hikariga-oka
960-12 Fukushima
Japan

Joffe, Boris I.
Zoological Institute,
Russian Academy of Sciences
199034 St Petersburg V-034
Russia

Jondelius, Ulf
Dept. of Zoology
University of Göteborg
P.O. Box 25059
S-413 90 Göteborg
Sweden

Jones, Hugh D.
Department of Environmental
Biology
University of Manchester
M13 9PL Manchester
UK

de Jong Brink, Marijke
Research Institute Neurosciences
Vrije Universiteit
De Boelelaan 1087
1081 HV Amsterdam
The Netherlands

Johnson, A.M.
Dept. of Microbiology and
Infectious Diseases
Flinders Medical Centre
Bedford Park, S.A. 5042
Australia

Karlstedt, Kaj
Dept. of Biology
Åbo Akademi University
BioCity, Artillerigatan 6
FI-20521 Åbo
Finland

Kawakatsu, Masaharu
Biological Laboratory
Fuji Women's College
Kita-16,
Nishi-2, Kita-ku,
Sapporo (Hokkaido) 001
Japan

Kawarada, H.
Dept. of Biological and Chemical
Engineering
Faculty of Engineering
Gunma University
Maebashi
Japan

Koopowitz, Harold
Dept. of Ecology and Evolutionary
Biology
University of California
Irvine, CA 92717
USA

Kornakova, Elena
Zoological Institute
Russian Academy of Sciences
199034 St Petersburg
Russia

Kotikova, Elena A.
Zoological Institute
Russian Academy of Sciences
199034 St Petersburg
Russia

Kreshchenko, N.D.
Institute of Cell Biophysics
Russian Academy of Sciences
Moscow Region 142292
Russia

Kuwahara, T.
Dept. of Biology
Faculty of Science
Hirosaki University
Hirosaki 036
Japan

Kuznedelov, Konstantin
Dept. of Chemistry and
Biochemistry of Nucleic Acids
Limnological Institute of Irkutsk
Ulan-Batorskaya 3
664033 Irkutsk
Russia

Lanfranchi, Alberto
Dipartimento di Scienze del
Comportamento Animale e
Dell'Uomo
Via Volta 6
56126 Pisa
Italy

Lucchesi, Paolo
Dipartimento di Biomedicina
Sperimentale
Via Volta 4
56126 Pisa
Italy

Lundin, Kennet
Dept.of Zoology
University of Göteborg
Box 25059
S-413 90 Göteborg
Sweden

Makino, Naoya
Dept. of Biology
Tokyo Medical College
6-1-1, Shinjuku
Shinjuku-ku
Tokyo 160
Japan

Malmberg, Göran
Narturhistoriska Riksmuseet och
Zoologiska Institutionen
Stockholm University
S-106 91 Stockholm
Sweden

Mamkaev, Yuri V.
Zoological Institute
Russian Academy of Sciences
199034 St Petersburg
Russia

Markevich, I.
Zoological Institute
Russian Academy of Sciences
199034 St Petersburg
Russia

Martens, Paul
Dept. SBG
Limburgs Universitair Centrum
B-3590 Diepenbeek
Belgium

Maule, Aaron G.
School of Clinical Medicine
The Queen's University of Belfast
Belfast BT7 1NN
Northern Ireland
UK

Michiels, Nico K.
Max-Planck-Institut für
Verhaltensphysiologie
D-8130 Seewiesen
Germany

Moens, J.B.
Dept. SBG
Limburgs Universitair Centrum
B-3590 Diepenbeek
Belgium

Morita, Michio
Dept. of Anatomy and Neurobiology
Colorado State University
Fort Collins, CO 80523
USA

Mukherjee, D.
Dept. of Zoology
Visva-Bharati University
Santiniketan-731 235
India

Muñoz-Mármol, Ana Maria
PZA Antonio Lópezno 3-4
08003 Barcelona
Spain

Murina, Galena-Vanzetti
Dept. of Mariculture
Institute of the Biology of Southern
Seas
Ukrainian Academy of Sciences
Nahimov St. 2
335011 Sevastopol
Ukraine

Naidenova, Nonna
Institute of Biology of the Southern
Seas
Ukrainian Academy of Sciences
Nahimov St. 2
335011 Sevastopol
Ukraine

Nakamura M.
Dept. of Biology
Faculty of Science
Hirosaki University
Hirosaki 036
Japan

Newman, Leslie
Queensland Museum
Box 3300, South Brisbane
Queensland 4101
Australia

Niewiadomska, Katarzyna
W. Stefan'ski Institute of
Parasitology
Polish Academy of Sciences
ul Pasteura 3 S.p. 153
00-973 Warszawa
Poland

Nomura, Takahiro
Dept. of Molecular Biology
Cancer Research Institute
Kanazawa University
Kanazawa 920
Japan

Ogren, Robert E.
Dept. of Biology
Wilkes University
Wilkes-Barre, PA 18766
USA

Oki, Iwashiro
Osaka Environmental Project
Authority
MFC7F
1-2-15 Uchihon-machi
Chuoch-ku
Osaka 540
Japan

Orii, Hidefumi
Dept. of Life Science
Himeji Institute of Technology
Harima Science Park City,
Akou, Hyogo 678-12
Japan

Paatero, Gun
Department of Biology
Åbo Akademi University
Biocity, Artillerigatan 6
FI-20521 Åbo
Finland

Pala, M.
Istituto di Zoologia
Università di Sassari
Via Muroni 25
I-07100 Sassari
Italy

Palmberg, Irmeli
Dept. of Teacher Education
Åbo Akademi University
Box 311
65101 Vasa
Finland

Panula, Pertti
Dept. of Biology
Åbo Akademi University
BioCity, Artillerigatan 6
FI-20521 Åbo
Finland

Peter, Roland
Institute of Genetics and General
Biology
University of Salzburg
Hellbrunnerstrasse 34
A-5020 Salzburg
Austria

Ponce de León, Rodrigo
Departamento de Biología Animal
Facultad de Ciencias
Universidad de Alcala de Henares
28871 Alcal de Henares (Madrid)
Spain

Porfirjeva, Nina A.
Dept. of Invertebrate Zoology
Kazan State University
Lenina 18
Kazan 420008
Russia

Privitera, I.
D.C.D.S.L. sez. Entomologia agraria
Università di Pisa
Italy

Raikova, Olga
Zoological Institute
Russian Academy of Sciences
199034 St Petersburg
Russia

Reiter, Dietmar
Institut für Zoologie
Innrain 109
A-6020 Innsbruck
Austria

Reuter, Maria
Dept. of Biology
Åbo Akademi University
BioCity, Artillerigatan 6
FI-20521 Åbo
Finland

Rieger, R.M.
Institut für Zoologie
Universität Innsbruck
Technikerstrasse 25
A-6020 Innsbruck
Austria

Riutort, Marta
Departamento de Genètica
Universitad de Barcelona
Diagonal 645
08071 Barcelona
Spain

Rogosin, Alexander
Biomonitoring Laboratory of Ilmen
Academician Reservation
456301 Miass
Cheljabinsk District 456301
Russia

Rohde, Klaus
Dept. of Zoology
University of New England
Armidale, N.S,W. 2351
Australia

Romero, Rafael
Departamento de Genètica
Universitad de Barcelona
Diagonal 645
08071 Barcelona
Spain

Ryoyama, Kazuo
Dept. Experimental Therapy
Cancer Research Institute
Kanazawa University
Kanazawa 920
Japan

Saheki, Toshihiko
Faculty of Engeneering
Gunma University
1-5-1 Tenijn-cha, Kiryu
376 Gunma
Japan

Sakharov, Dmitri
Institute of Developmental Biology
Russian Academy of Sciences
26 Vavilov str.
117808 Moscow
Russia

Sakurai, Taka
Division of C
Central Research Laboratory
Fukushima Medical College
1 Hikariga-Oka
Fukushima 960-12
Japan

Saló, Emili
Departamento de Genètica
Universitad de Barcelona
Diagonal 645
08071 Barcelona
Spain

Salvenmoser, Willi
Institut für Zoologie
Universität Innsbruck
Technickerstrasse 25
A-6020 Innsbruck
Austria

Sato, Y
Dept. of Biology
Hirosaki University
Hirosaki 036
Japan

Schockaert, Ernest R.
Dept. SBG
Limburgs Universitair Centrum
B-3590 Diepenbeek
Belgium

Schürmann, Wolfgang
Institute of Genetics and General Biology
University of Salzburg
Hellbrunnerstrasse 34
A-5020 Salzburg
Austria

Serras, F.
Departamento de Genètica
Universitad de Barcelona
Diagonal 645
08071 Barcelona
Spain

Sewell, Kim B.
Departments of Parasitology and Anatomy
University of Queensland
Queensland 4072
Australia

Skakurova, Natalia
Dept. of Invertebrate Zoology
Kazan State University
Lenina 18
420008 Kazan
Tatarstan
Russia

Shaw, Chris
School of Clinical Medicine
The Queen's University of Belfast
Belfast BT7 1NN
Northern Ireland
UK

Sheiman, Inna
Institute of Cell Biophysics
Russian Academy of Sciences
Moscow District
142292 Pushchino
Russia

Shinozawa, Takao
Dept. of Biological and Chemical Engeneering
Faculty of Engineering
Gunma University
Kiryu, Gunma 376
Japan

Shiozaki, Syuichi
Dept. of Biological and Chemical Engineering
Gunma University
Kiryu Gunma 376
Japan

Shirasawa, Yasuko
Dept. of Biology
Tokyo Medical College
6-1-1 Shinjuku-ku
160 Tokyo
Japan

Silveira, Marina
Laboratório de Microscopia Eletrônica
Instituto de Física
Universidade de São Paulo
Caixa Postal 20516
05508-900 São Paulo
Brazil

Sluys, Ronald
Institute of Taxonomic Zoology
University of Amsterdam
Box 4766
1009 AT Amsterdam
The Netherlands

Smith III, Julian
Dept. of Biology
Winthrop University
Rock Hill, SC 29733
USA

Solonchenko, Alla
Institute of Biology of the Southern Seas
Ukrainian Academy of Sciences
2 Nakhimov Pr.
335001 Sevastopol
Ukraine

Sopott-Ehlers, Beate
II. Zoologisches Institut und Museum der Universität
Berliner Str. 28
3400 Göttingen
Germany

Soriano, M.A.
Departamento de Genètica
Universitat de Barcelona
Diagonal 645
0871 Barcelona
Spain

Solis Soto, Manuel
Research Institute Neurosciences
Vrije Universiteit
De Boelelaan 1087
1081 HV Amsterdam
The Netherlands

Stocchino, G.
Istituto di Zoologia
Università di Sassari
Via Muroni 25
I-07100 Sassari
Italy

Swiderski, Peter
Laboratory of Comparative
Anatomy and Physiology
University of Geneva
Sciences III 30 quai Ernest-
Ansermet
CH-1211 Geneva 4
Switzerland

Takahashi, H.
Dept. of Biology
Hirosaki University
Hirosaki 036
Japan

Takai, Masayuki
Biology Laboratory
Saga Medical School
5-1-1 Nabeshima
Saga 849
Japan

Takezaki, K.
Dept. of Biological and Chemical
Engineering
Gunma University
Kiryu
Japan

Tamura, Sachiko
Osaka Prefectural Institute of Public
Health
1-3-69 Nakamichi,
Higashinari-ku
537 Osaka
Japan

Tanaka, H.
Dept. of Pharmacology
Gunma University
Maebashi
Japan

Teshirogi, Wataru
Dept. of Biology
Hirosaki University
036 Hirosaki
Japan

Thollesson, Mikael
Dept. of Zoology
University of Göteborg
Box 25059
S-413 90 Göteborg
Sweden

Timoshkin, Oleg
Dept. of Hydrobiology and
Systematics of Aquatic Invertebrates
Limnological Institute of Irkutsk
Siberian Division of the Russian
Academy
P.O. Box 4199
664033 Irkutsk
Russia

Tiras, Kharlampi
Institute of Cell Biophysics
Russian Academy of Sciences
Moscow District
142292 Pushchino
Russia

Tkatchuk, Lidia
Institute of Biology of South Seas
Ukranian Academy of Sciences
2, Nahimov str
335011 Sevastopol
Ukraine

Tokynova, R.P. Ryma
Dept. of Invertebrate Zoology
Kazan State University
Lenina ul. 18
420008 Kazan
Russia

Troitsky, Aleksey
Institute of Physico-Chemical
Biology
Moskow State University
119899 Moscow
Russia

Wahlberg, Monika
Dept. of Biology
Åbo Akademi University
BioCity, Artillerigatan 6
FI-20521 Åbo
Finland

Valiejo Roman, K.M.
A.N. Belozersky Institute of Psysico-
Chemical Biology
Moscow State University
119899 Moscow
Russia

Vreys, Carla
Dept. SBG
Limburgs Universitair Centrum
B-3590 Diepenbeek
Belgium

Watanabe, Kenji
Laboratory of Regeneration Biology
Department of Life Science
Himeji Institute of Technology
678-12 Harima Science Park City
Hyougo 678-12
Japan

Watson, Nikki Anne
Dept. of Zoology
University of New England
Armidale NSW 2351
Australia

Wirth, Quirino Joao
Instituto de Pesquisa de Planárias-
UNISINOS
Praca Tiradentes, 35
Caixa Postal, 275
93010-020 São Leopoldo
Brazil

Hydrobiologia **305**: 1–2, 1995.
L.R.G. Cannon (ed.), Biology of Turbellaria and some Related Flatworms.

Plathelminthes or Platyhelminthes?

Ulrich Ehlers & Beate Sopott-Ehlers
II. Zoologisches Institut und Museum der Universität Göttingen, Berliner Str. 28, D-37073 Göttingen, Germany

Key words: Plathelminthes, Platyhelminthes, taxonomy

Today, we have two names for the same taxon: 'Platyhelminthes' is often used in English texts whereas 'Plathelminthes' is used in texts of other languages, to our knowledge in most or even all the languages of the European Continent. Is one scientific term the correct one and the other the incorrect one that should be abandoned?

The International Code of Zoological Nomenclature (IRZN, 1985) does not cover names for categories higher than family-group names. Hence, there are no mandatory provisions with respect to a 'correct' spelling of the names 'Platyhelminthes' or 'Plathelminthes'. Do the provisions of the Code suggest an analogy we could use?

According to the Code, the valid name of a taxon is the oldest available name applied to it: 'Priority is the basic principle of zoological nomenclature' (see preamble of IRZN). Which name was first applied for the taxon with the flatworms?

For flatworms, several scientific names of different spellings were used during the second half of the last century: Platyelmia (introduced by Vogt, 1851); Platodes (introduced by Leuckart, 1854); Platyelminthes (introduced by Gegenbaur, 1859); Plathelminthes (introduced by Schneider, 1873); Platyhelmia (introduced by Koch, 1876); Platyhelminthes (introduced by Claus, 1887); Plathelmintha (introduced by Vaillant, 1890) and Platodaria and Platodinia (introduced by Haeckel, 1896).

Only two spellings, 'Plathelminthes', 1873, and 'Platyhelminthes', 1887, have been consistently used in original systematic articles, in review articles and in books written in different languages. Other synonyms have been used rarely in our century and can be considered *nomina oblita* (IRZN, p. 260). According to priority *and* usage, Plathelminthes would be the valid name. In analogy to this conclusion, Ricci (1983) stated that the name 'Rotifera', 1798, must be chosen for the rotifers, whereas 'Rotatoria', 1832, is a junior synonym of this designation.

But it was argued that 'Plathelminthes' is etymologically incorrect. Hyman (1951), for example, wrote: 'The etymologically correct form of the phylum name is Platyhelminthes, from the Greek *platys*, flat, and *helminthes*, worms; the spellings Platyelminthes, Plathelminthes, Platyelmia ... are not acceptable'. But in analogy, Article 32(c) of the IRZN provides: '... incorrect transliteration or latinization ... are not to be considered inadvertent errors'. Mayr (1969, p. 355) made clear comments on this topic: 'One of the curses of nomenclature has always been the zoologists' habit of "correcting" names proposed by other authors. Article 32 states clearly that the original spelling must be retained unless some exceptional condition exists ... incorrect transliteration, improper latinization ... are not to be considered inadvertent errors'. So one could argue that 'Plathelminthes' is the valid form of the name, although it is an incorrect spelling from the etymological point of view.

Accordingly, there are convincing reasons to accept 'Plathelminthes' as the correct name of the taxon. That is what we prefer to do, although the junior synonym 'Platyhelminthes' is predominant in papers and books written in English.

On the other hand, based on a discussion during this International Symposium, the participants recommend that equal rank should be given to both versions, Plathelminthes and Platyhelminthes, and both should be regarded as valid names for the taxon in question.

Acknowledgments

We express our sincere thanks to Prof. O. Kraus, president of ICZN, for helpful comments on an earlier draft of this paper.

References

Claus, C., 1887. Lehrbuch der Zoologie. 4. Auflage. N. G. Elwertsche Verlagsbuchhandlung, Marburg, Leipzig.

Gegenbaur, C., 1859. Grundzüge der Vergleichenden Anatomie. W. Engelmann, Leipzig.

Haeckel, E., 1896. Systematische Phylogenie. Entwurf eines natürlichen Systems der Organismen auf Grund ihrer Stammesgeschichte. Zweiter Theil: Systematische Phylogenie der wirbellosen Thiere (Invertebrata). G. Reimer, Berlin.

Hyman, L. H., 1951. The Invertebrates. Platyhelminthes and Rhynchocoela. McGraw-Hill Book Co., N.Y., Toronto, London.

International Code of Zoological Nomenclature, 1985. Third edition. International Trust for Zoological Nomenclature, London.

Koch, G. v., 1876. Grundriss der Zoologie. Verlag von Herman Dabis, Jena.

Leuckart, R., 1854. Bericht über die Leistungen in der Naturgeschichte der niederen Thiere während der Jahre 1848–1853. Archiv für Naturgeschichte Berlin 20, 2: 289–473.

Mayr, E., 1969. Principles of Systematic Zoology. McGraw-Hill Book Co., N.Y.

Ricci, C., 1983. Rotifera or Rotatoria? Hydrobiologia 104: 1–2.

Schneider, A., 1873. Untersuchungen über Plathelminthen. Jahresber. Oberhess. Ges. Natur- und Heilkunde Giessen 14: 1–76.

Vaillant, L., 1890. Histoire naturelle des Annelés marins et d'eau douce. Tome Troisième, second partie. Librairie Encyclopédique de Roret, Paris.

Vogt, C., 1851. Zoologische Briefe, Naturgeschichte der lebenden und untergegangenen Thiere. Literarische Anstalt, Frankfurt a. M.

Hydrobiologia **305**: 3–9, 1995.
L.R.G. Cannon (ed.), Biology of Turbellaria and some Related Flatworms.

Two peculiar new genera of Typhloplanoida from the Western Indian Ocean

Gert G. De Clerck & Ernest R. Schockaert
Research Group Zoology, Dept. S.B.G. Limburgs Universitair Centrum, B-3590 Diepenbeek, Belgium

Key words: Turbellaria, Typhloplanoida, Trigonostomidae, Promesostomidae, Western Indian Ocean

Abstract

Three new species and two new genera of Typhloplanoida from the Western Indian Ocean are described. *Feanora brevicirrus* gen. et sp. n. is a member of the Trigonostomidae von Graff, 1905 *sensu* Den Hartog, 1964. It has a cirrus, paired ovovitellaria and an undifferentiated afferent duct. An excentric bursa opens in the fecundatorium and a large ovoid vesicle enters the afferent duct in its distal part. *Gaziella pileola* gen. et sp. n. and *G. lacertosa* sp.n. are two new members of the Promesostomidae Den Hartog, 1964. They have a long cirrus, ovovitellaria and a long muscular female duct with two excentrically placed sperm-receiving vesicles. Their most striking feature is the proboscis-like structure. *Feanora* and *Gaziella* are provisionally considered as genera *incertae sedis* within the resp. families.

List of abbriviations

b:	bursa
cga:	common genital atrium
cgs:	caudal glandular system
ci:	cirrus
cob:	copulatory bulb
fad:	female afferent duct
fdd:	female deferent duct
gp:	gonopore
ovv:	ovovitellaria
ph:	pharynx
rh:	rhabdites
sr:	seminal receptable
sv:	seminal vesicle
te:	testes

Introduction

The turbellarian fauna from the Western Indian Ocean is poorly known. So far, 17 species have been mentioned from the Somalian coast (Schockaert, 1971;1982; Schockaert & Martens, 1985; 1987), and 12 species of Kalyptorhynchia from the Kenyan coast (Jouk & De Vocht, 1989). In the framework of a survey on Indian Ocean marine turbellaria, the Kenyan coast in the neigbourhood of Mombasa was sampled in September 1991 and October 1992, while different coasts of the Seychelles were sampled during the Netherlands Indian Ocean Expedition in December 1992 (Leg E: 'Oceanic Reefs'). Three new typhloplanoid species and two new genera are presented in this contribution.

Material and methods

The animals were extracted from the sediment using the $MgCl_2$ decantation method (Boaden, 1963), studied alive and mounted in lactophenol. The remaining available specimens were fixed in hot Bouin's and/or Stieve's solution (Romeis, 1968) and serially sectioned. The 5 μm sections were stained with Heidenhain's iron hematoxylin using eosin as counterstain. All material, including types, is deposited in the zoological collection of the Limburgs Universitair Centrum.

Descriptions

Feanora brevicirrus gen. et spec. nov. *(Figs 1, 2, 6 and 7)*

Type locality: McKenzie point, Mombasa (Kenya), in fine sand from mangrove-area at low tide; middle to high midlittoral (25.09.91).

Other localities. McKenzie point, fine sand near mangrove area; high midlittoral (21.10.92) and mixture of fine and coarse sand with shell debris in a rocky pool; middle midlittoral (26.10.92).

Material: two animals studied live, six mounted (one of them designated as holotype) and two individuals serially sectioned.

Derivation of name: the genus name is a derivation of a mythological personage from Tolkien's 'Silmarillion'; the species name refers to the short cirrus.

The animals are slender, whitish-transparent and 1–1.2 mm in whole mounts. Eyes are lacking. The epidermis is cellular with basophilic rhabdites (3.8 μm): Acidophilic adenal rhabdites (9.6 μm) are organised in two long tracts. The pharynx rosulatus (diameter: 96-131 μm) is situated at about 30% from the front, ciliated at the rim and with some long cilia in the middle of its lumen.

Paired testes lie ventrally just in front of the pharynx. Vasa deferentia are slightly swollen before entering the paired muscular seminal vesicles. These merge before entering the elongated copulatory bulb that is surrounded by strong internal circular and weak external longitudinal muscles. A constriction can be observed at about one third from the distal end. Ejaculatory duct runs centrally through the bulb and is surrounded by basophilic glands. Cirrus (19–31 μm long and 10–14 μm maximum width; respectively: 27 and 12 μm in the holotype) is to be found distally from the constriction of the bulbus. Spines are densely packed in the cirrus at rest; they are more apart in the extruded cirrus and here it can be seen that the bases of the hooks form a honeycomb-like lace. In the extruded cirrus the spines are 6 μm long. The cirrus at rest protrudes into the male antrum and is surrounded with a little cap.

Paired ovovitellaria lie dorsally and extend from above the testes to the level of the seminal vesicles: germinative zone is ventral at the caudal end. Epithelium lining the short oviducts is high and vacuolated. Both

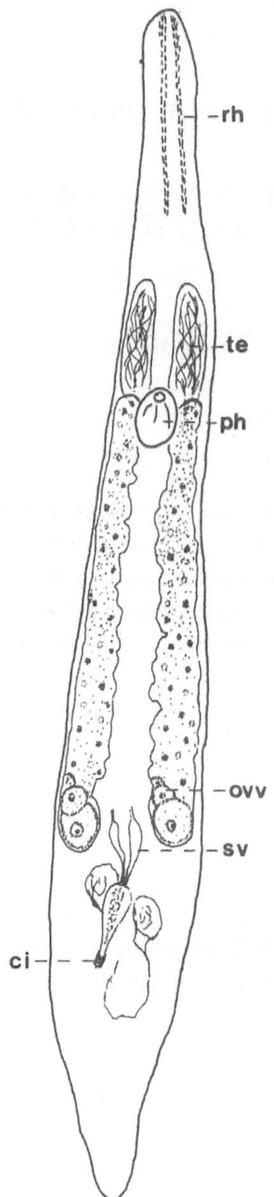

Fig. 1. *Feanora brevicirrus* gen. et sp. n. Slightly squeezed live specimen (free-hand drawing).

oviducts open into the fecundatorium (ootype) which is connected to the common genital atrium by the deferent and the afferent duct. The deferent duct is surrounded by internal circular and external longitudinal muscles. The afferent duct, broad from beginning to end, is filled with a granular material over almost its entire length, except near the fecundatorium where a wide lumen with spermatozoa can be seen. The basement membrane is slightly thickned and remnants of the lining

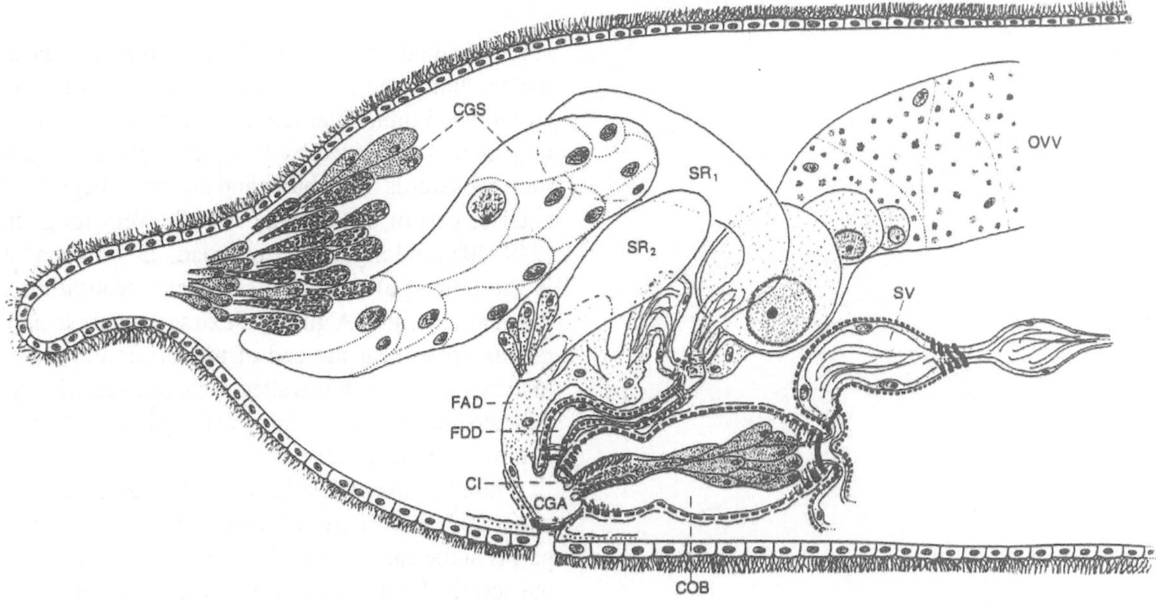

Fig. 2. *Feanora brevicirrus* gen. et sp. n. Saggital reconstruction of the genital organs (camera lucida).

epithelium with some nuclei can only be found near the atrium. There is a sphincter at the junction with the fecundatorium. Dorsally and about halfway along the afferent duct a large ovoid vesicle opens into it. It has a low, slightly basophilic epithelium and is filled with the same granular material as the afferent duct. No sperm were observed, however. Basophilic glands are present at the junction with the afferent duct. A second, very large vesicle opens dorsally into the fecundatorium. It is filled with a basophilic nucleated tissue in which sperm can be seen in a variable number of vacuole-like cavities as often seen in a bursa of the *resorbiens* type. A sphincter guards the junction with the fecundatorium. Neither the vesicles nor the afferent duct are surrounded by muscles.

Three types of glands constitute the caudal gland system: (i) glands with a coarse eosinophilic rhabdite-like secretion, (ii) glands with a coarse basophilic secretion and (iii) a tight aggregation of very large cells, situated between the caudal gland system and the bursal apparatus, filled with a slightly basophilic secretion.

Gaziella pileola gen. et spec. nov. *(Figs 3, 4, 9, 11 and 12)*

Type locality: Gazi Bay (Kenya), detritus rich, very fine sand from a shallow channel in a mangrove depressoin (02.10.1991).

Other localities: McKenzie point, Mombasa (Kenya), fine sand from exposed area; middle to high mid-littoral (05.09.1991).

Material: two animals studied live, four mounted, one designated as holotype; three specimens fixed and sectioned.

Derivation of name: the genus name is derived from the type locality while the species name refers to the cap like structure at the end of the cirrus.

Eyeless, slender, opaque animals are about 1.5 mm (mounted). Live worms move about slowly. Epidermis is cellular with few rhabdites (4 μm). Adenal rhabdites (9 μm) are organised in long tracts and converge to a small frontal invagination, 13 μm deep, forming a small proboscis-like structure, not obvious in the living animal. Epidermis of this proboscis is devoid of nuclei, but is clearly ciliated. Distal parts of the rhabdite tracts are surrounded by a very delicate septum. From the base of this septum four retractor muscles extend to the body wall behind the brain. Thin retractor fibers are also present at the base of the invagination itself. The pharynx rosulatus, (diameter: 156–164 μm) lies about 65% from the enterior end. Small granules (reduced cilia or microvilli?) can be observed on its distal rim; there are no cilia in the lumen.

Paired testes lie ventrally just in front of the pharynx. Vasa deferentia open separately into the elongate

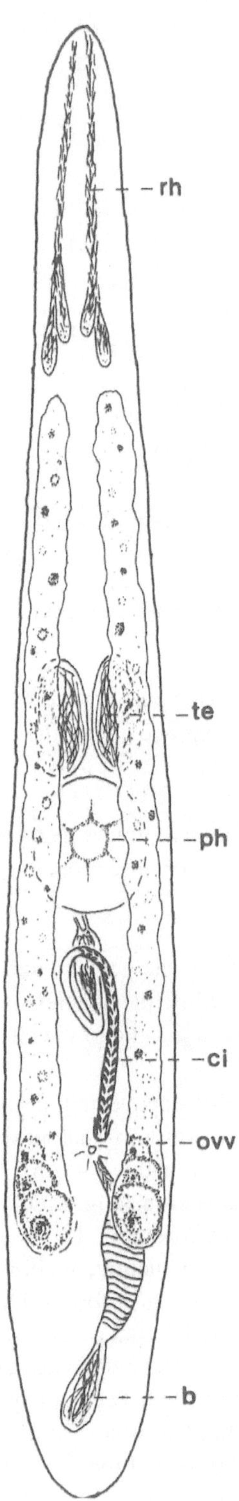

Fig. 3. *Gaziella pileola* gen. et sp. n. Slightly squeezed live specimen (free-hand drawing).

single seminal vesicle which is incorporated in the copulatory bulb; the latter is surrounded by thick inner circular and thin, outer longitudinal muscles. Ejaculatory duct, surrounded by basophilic glands, makes a loop backwards in the bulb; and the cirrus begins in the anterior part of the copulatory bulb. Very long cirrus (249–305 μm long; 7–10 μm broad; 255 μm and 8 μm in the holotype) ends in a small cap protruding into the genital atrium. A girdle of extracapsular basophilic glands opens near the end of the cirrus into the copulatory bulb. The thin walled male antrum with some delicate muscles, opens into the front part of the small, common genital atrium.

Paired ovovitellaria lie dorsally and extend from about 20% to 80% from the anterior end. Germinative part is at the caudal end. Short ovovitelline ducts enter the terminal part of the long, wide, sinuous female duct that is surrounded by weak circular muscles. Its distal part is narrowed where the epithelium is unclear (ciliated or pseudociliated?). Copulatory bursa with sperm opens ventrocaudally into the female duct near the oviducts and is surrounded by inner longitudinal muscles and outer, strong circular muscles. The seminal receptacle opens dorsocaudally into the female duct at the same place and also contains sperm as does the bursa.

Gaziella lacertosa sp. n. (Figs 5, 8 and 10).

Type locality: same locatity as for *Gaziella pileola* (02.10.91).

Other localities: Gazi Bay (Kenya): in front of 'Fisherman's hut' in detritus-rich very fine sand of an intertidal sand flat in the mangrove (24.10.1992). Kanamai (Kenya): mixture of coarse shell debris and very fine sand from a shallow intertidal pool; middle to high mid-littoral (20.10.1992). Grande Anse (Beolière), W. coast of Mahé (Seychelles): mangrove area, detritus-rich, fine sand near *Avicennia* sp.; high mid-littoral (19.12.1992).

Material: several animals studied alive, four of them mounted (one designated as holotype); four individuals sectioned.

Derivation of the name: the name refers to the highly muscular female duct.

Similar to *G. pileola* except the copulatory organ differs in that (1) the seminal vesicle is small and nearly

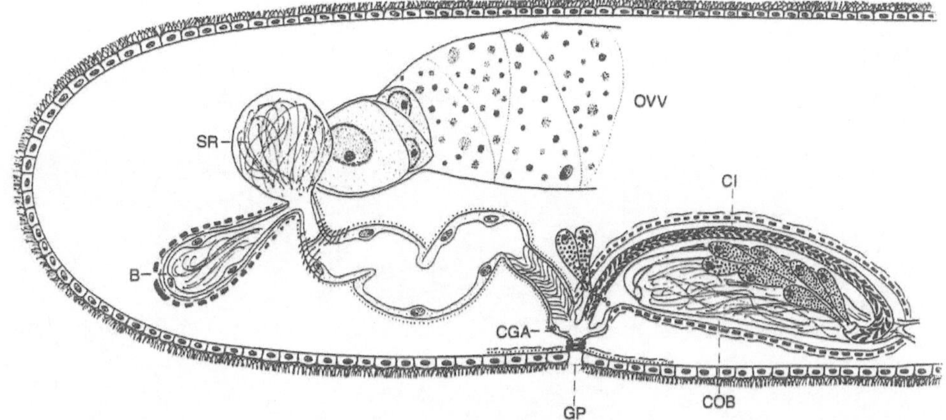

Fig. 4. *Gaziella pileola* gen. et sp. n. Saggital reconstruction of the genital organs (camera lucida).

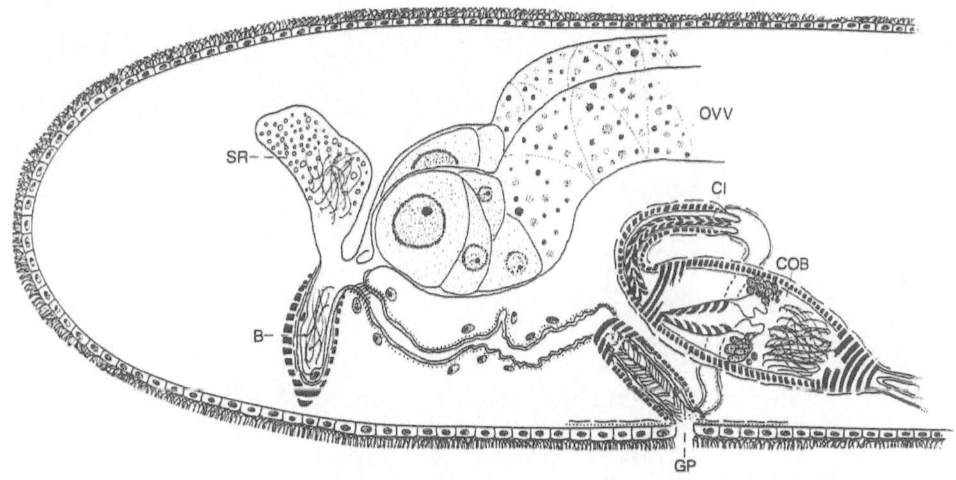

Fig. 5. *Gaziella lacertosa* sp. n. Saggital reconstruction of the genital organs (camera lucida).

spherical and only occupies the anterior half of the copulatory bulb, (2) the ejaculatory duct is short and does not make a loop and (3) the cirrus is short with a length varying between 48–78 μm (holotype: 67 μm). The narrow distal part of the female duct is highly muscular.

Discussion

With its afferent and deferent female ducts *Feanora* clearly belongs to the Trigonostomidae Graff, 1905 (*sensu* Den Hartog, 1964). However, it deviates from the known trigonostomids in a number of respects and cannot be assigned to one of the three subfamilies, illustrating the inconsistency of the current taxonomy of the Trigonostomidae. The excentrically placed large bursa opening in the fecundatorium is clearly a first autapomorphy for the genus and the large vesicle opening dorsally into the afferent duct is a second one. The afferent duct itself does not show further differentiation whatsoever and comparing it with the hypothetical evolutionary steps depicted by Luther (1943: p. 38–39 and Fig. 23; 1950: p. 36–37) *Feanora* represents an even more 'primitive' condition than *Austrotorhynchus bifidus* (McIntosh, 1874). This condition can be hypothesised as the plesiomorph situation of the afferent duct in the Trigonostomidae. A copulatory organ consisting of a cirrus enclosed in a cirrus bulb is unique among the Trigonostomidae. From ingroup comparison this might be interpreted as an apomorphy; from outgroup comparison however the conclusion is not so evident: a cirrus or a stylet can be found in most other typhloplanoid taxa and in nearly all Platyhelminth orders. A discussion on this matter is clearly beyond the scope of the present contribution.

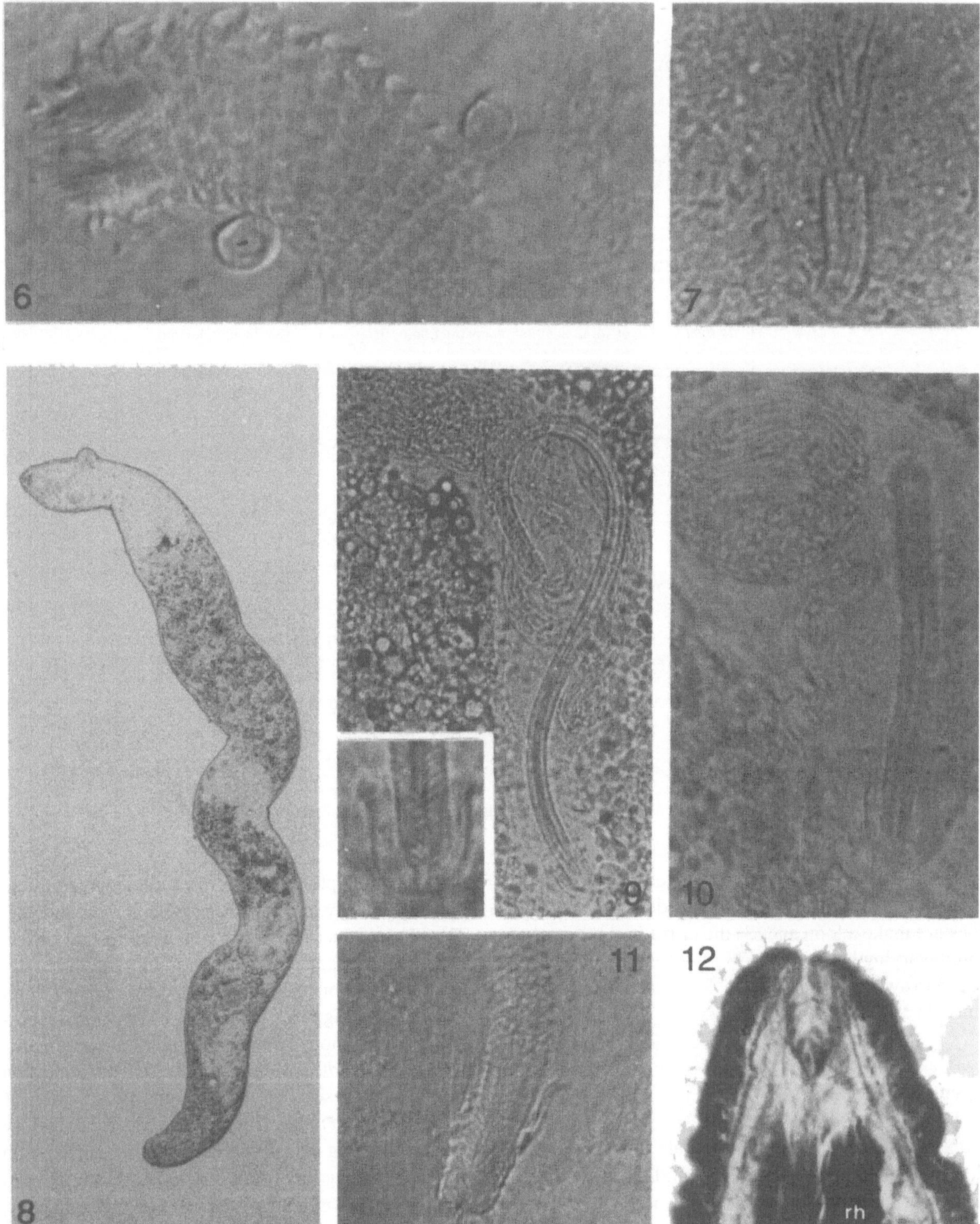

Plate 1. Fig. 6 *Feanora brevicirrus* gen. et sp. n. Everted cirrus! (100×; Nomarsky prism). Fig. 7 *Feanora brevicirrus* gen. et sp. n. Cirrus in live specimen (40×). Fig. 8 *Gaziella lacertosa* sp. n. Slightly squeezed specimen (4×). Fig. 9 *Gaziella pileola* gen. et sp. n. Male copulatory organ (25×); inset: End of cirrus (40×). Fig 10 *Gaziella lacertosa* sp. n. Male copulatory organ (40×). Fig 11 *Gaziella pileola* gen. et sp. n. Distal end of everted cirrus (100×; Nomarsky prism) Fig 12 *Gaziella pileola* gen. et sp. n. Frontal invagination (sectioned material; 100×)

Due to its paired ovovitellaria and the single connection of the female system with the atrium, *Gaziella* belongs to the Promesostomidae Den Hartog, 1964. A (short) cirrus and a long female duct can also be found in *Moevenbergia una* Armonies & Heller, 1987 but in this species four excentrically placed sperm-receiving organs are present. The very long cirrus, the single seminal vesicle, and the proboscis like structure justify the erection of a new genus. A proboscis-like structure is also present in *Pararhynchella* Ehlers, 1981, but the organisation of the genital organs of *Gaziella* deviates highly from that in Ehlers' genus. The pharynx and the structure of the proboscis on the other hand are clearly not of the same type as found in the Kytorynchidae Rieger, 1974.

Since the 'preliminary' revision of the Trigonostomidae Graff, 1905 (*sensu* Luther, 1943) by Den Hartog (1964) the system he proposed appeared stable. However, with the discovery of many new organisation types various characters appear to be distributed in a 'mozaic-like' fashion in the 'Typhloplanoida'. Without being exhaustive we can mention: ovovitellaria versus separate vitellaria and ovaria, paired versus unpaired ovaries, an atrial bursa present or absent, short undiferrentiated female duct(s) versus long female duct(s) with various differentiations, a cirrus versus a stylet (different types!), various types of pharynges, a proboscis-like structure and various types of rhabdites. These problems have been mentioned earlier e.g. by Karling & Mack-Fira (1973), Ehlers (1974), Ehlers & Ehlers (1981), Armonies & Hellwig (1987).

For the time being we prefer to place both, *Feanora* and *Gaziella* as genera *incertae sedis* resp. in the Trigonostomidae and Promesostomidae (*sensu* Den Hartog, 1964) until the phylogenetic relationships among the 'Typhloplanoida' are clarified.

Acknowledgments

Part of the work was carried out at the Kenyan Marine and Fisheries Research Institute (K.M.F.R.I.) Mombasa, Kenya. We would like to express our gratitude to Dir. Dr E. Okemwa for his hospitality. Thanks are also due to Mrs Y. Vermeulen and Dr E. Martens for their assistance during the stays. This research was supported by F.K.F.O project nr. 2.009.92 (Fund for Fundamental Scientific Research of the Belgian National Science Foundation). The junior author wishes to thank Dr J. van der Land for the given opportunity to take part in the Netherlands Indian Ocean Expedition and the financial support from S.O.Z. The help of Mr M. Withofs, Ms. N. Steffanie and Mrs L. Grosemans was very much appreciated.

References

Armonies, W. & M. Hellwig, 1987. Neue Plathelminthes aus dem Brackwasser der Insel Sylt (Nordsee). Microfauna Mar. 3: 249–260.

Boaden, P. J. S., 1963. The interstitial fauna of some North Wales beaches. J. mar. biol. Ass. U.K. 43: 79–96.

Den Hartog, C., 1964. A preliminary revision of the proxenetes group (Trigonostomidae, Turbellaria) I. Proc. k. ned. Akad. Wet. 67: 371–382.

Ehlers, U, 1974. Interstitielle Typhloplanoida (Turbellaria) aus dem Litoral der Nordseeinsel Sylt. Mikrofauna des Meeresbodens 49: 1–102.

Ehlers, U. & B. Ehlers, 1981. Interstitielle Fauna von Galapagos: XXVII: Byrsophlebidae, Promesostomidae, Brinkmanniellinae, Kytorhynchidae (Turbellaria, Typhloplanoida). Mikrofauna des Meeresbodens 83: 83-115.

Karling, T. G. & V. Mack-Fira, 1973. Zur Morphologie und Systematik der Gattung *Paramesostoma* Attems (Turbellaria Typhloplanoida). Sarsia 52: 155–170.

Jouk, P. E. H. & A. De Vocht., 1989. Kalyptorhynchia (Plathelminthes Rhabdocoela) from the Kenyan coast, with description of four new species. Trop. Zool. 2: 145–157.

Luther, A., 1943. Untersuchungen an Rhabdocoelen Turbellarien. V. úber einige Reprásentanten der Familie Proxenetidae. Acta zool. fenn. 38: 1–101. .

Luther, A., 1950. Untersuchungen an Rhabdocoelen Turbellarien IX. Zur Kenntnis einiger Typhloplanoida. X. Über Astrotorhynchus bifidus (M'Int.) Acta zool. fenn. 60: 1–42.

Romeis, B., 1968. Microscopische Technik. R. Oldenbourg Verlag, München.

Schockaert, E. R., 1971. Turbellaria from Somalia I. Kalyptorhynchia (Part 1). Monit. zool. ital. (N.S.) Suppl.4: 101–122.

Schockaert, E. R., 1982. Turbellaria from Somalia II. Kalyptorhynchia (Part 2). Monit. zool. ital. (N.S.) Suppl. 17: 81–96.

Schockaert, E. & P. M. Martens, 1985. Turbellaria from Somalia. III. Lecithoepitheliata and Typhloplanoida. Monit. zool. ital. (N.S.) Suppl. 20: 27–41.

Schockaert, E. R. & P. M. Martens, 1987. Turbellaria from Somalia. IV. The genus Pseudomonocelis Meixner, 1943. Monit. zool. ital. (N.S.) Suppl.22: 101–115.

Hydrobiologia **305**: 11–14, 1995.
L.R.G. Cannon (ed.), *Biology of Turbellaria and some Related Flatworms.*

Phylogenetic relationships within the *Archiloa* genus complex (Proseriata, Monocelididae)

Paul M. Martens[1] & Marco C. Curini-Galletti[2]
[1]*Department SBG, Research group Zoology, Limburgs Universitair Centrum, B-3590 Diepenbeek, Belgium*
[2]*Istituto di Zoologia, Università di Sassari, I-07100 Sassari, Italy*

Key words: Platyhelminthes, Proseriata, Monocelididae, phylogeny

Abstract

Of the seven genera which we have recognised within the *Archiloa* genus complex *sensu* Karling (1966) the cosmopolitan genus *Archilina* is the most 'primitive' and is characterised only by plesiomorphic characters, and has to be considered paraphyletic. All other species of the *Archiloa* genus complex are hypothesized to be derived from *Archilina*-like ancestors through different evolutionary lineages. One lineage led to the genera *Archiloa*, *Inaloa*, *Archilopsis* and *Monocelopsis*, taxa found in the Atlantic and the Mediterranean. These genera are monophyletic and their relationships are analyzed. The genera *Mesoda* (Brazil) and *Tajikina* (Northern Pacific) can be considered as two other separate lineages. Similarly, within what we now consider as the genus *Archilina* different lineages can be recognized in different regions.

Introduction

There is no general consensus on the systematics of the *Archiloa* genus complex. Karling (1966) synonymized the genera *Monocelopsis*, *Archilopsis*, *Archilina*, *Mesoda*, and *Pistrix* with *Archiloa*, on the basis of the copulatory organ of the duplex type and the presence of a vagina. He did not give weight to the details of location and morphology of vagina, bursa and copulatory organ. His proposal, however, has not been generally accepted (cf. Sopott, 1972; Curini-Galletti *et al.*, 1989; Martens *et al.*, 1989a, b). Based on a revision of the *Archiloa* genus complex, we considered most of the synonymized genera as valid. The detailed taxonomy of the group, with the description of numerous new species, as well as full karyological data, is dealt with in other papers (Martens *et al.*, 1989a; Martens & Curini-Galletti, 1994 or in preparation). Here we present the presumed phylogenetic relationships within Karling's *Archiloa* group.

Phylogenetical analysis

From a thorough revision of the Monocelididae (we have studied nearly all known species and more than 30 new species) we were not able to recognise an apomorphic feature for the *Archiloa* genus complex. Many species within this complex show characters which are widely distributed within the family and can be considered as plesiomorphic:

- insunken nuclei, many vitelline follicles beside and posterior to the testes (characters which are present in nearly all Monocelididae)
- a simple organisation of the copulatory organs with three genital pores of which the first is an external vagina connected with a prepenial bursa which is a part of the female duct (this character is also present in the genera *Monocelis*, *Minona*, *Duplominona*, *Pseudomonocelis*, *Preminona*)
- a copulatory organ of the duplex type (present in all species of the complex), in many of them the organisation of this organ is very simple: a rather short and straight cirrus and the bulb is completely filled with the prostate vesicle and seminal vesicle (see also Fig. 1 in Martens & Curini-Galletti, 1994).

This character state is also present in the genera *Promonotus*, *Duplominona* and *Duploperaclistus*.

All these plesiomorphic characters are in fact also the characters of the genus *Archilina*.

Examination of specimens of *Archiloa rivularis* de Beauchamp, 1910, the type species of the genus, revealed that the copulatory organ is provided with an accessory cirrus, a structure already described as a 'ligament du penis' by de Beauchamp (1910). An accessory cirrus is also present in *A. petiti* Ax, 1956 and *A. westbladi* Ax, 1954. In these three species there is an internal vagina with a prepenial bursa which is not directly connected with the female duct (see Fig. 1 in Martens & Curini-Galletti, 1994). Both the presence of the accessory cirrus and the internal vagina, without a direct connection with the female duct, are not present in any other of the Monocelididae species and are synapomorphies for these three species (4, 5 in Fig. 1) and justify the genus *Archiloa*. The species known so far are exclusively found in Western Europe and the Mediterranean, in brackish conditions; *A. rivularis* occurs in fresh water, and is only known from a small creek in the Pyrenee.

Monocelis cirrifera described by Meixner (1943) has a very long cirrus of about 180 μm. We have studied specimens from the type locality. The morphology of the copulatory organ, the bursa and the vagina are very similar to those of *Monocelis scalopura* Marcus, 1949 (which has a cirrus of 600 μm long), and a new species from the Mediterranean with a cirrus up to 1.5 mm. The presence of an extremely long cirrus and of a very long vagina are synapomorphies for these species (6, 7 in Fig. 1) which we have put together in the new genus *Inaloa* (see Martens & Curini-Galletti, 1994).

The genera *Archiloa* and *Inaloa* share some characteristics. All species, that could be studied alive, have 4 chromosomes in the haploid set and show similarities in the morphology of the copulatory organ which is clearly different from the plesiomorphic condition. The spherical prostatic vesicle is separated from the seminal vesicle by a muscular diaphragm; both vesicles are not in touch with the bulb wall and do not fill the whole bulb (see Fig. 1 in Martens & Curini-Galletti, 1994). The morphology of the copulatory bulb is a synapomorphy for the two genera (3 in Fig. 1).

Ax (1951) introduced the genus *Monocelopsis* for a peculiar species, *M. otoplanoides*, which moves and superficially looks like an otoplanid. This small species is characterised by three genital pores (plesiomorphy), a pre- and post penial bursa (plesiomorphy), and 4 pairs of vitelline follicles, all posterior to the testes, only one pair of which is in front of the pharynx. Sopott (1972) described *Mesoda septentrionalis* from the North Sea. She assigned the species to the genus *Mesoda* Marcus 1949, on the basis of the intraepithelial nuclei of the epidermis. *Mesoda* was only known from a single South American species. However, Ax's and Sopott's species have both the same general construction of the copulatory organ, vagina and bursa, the same otoplanoid habitus, and agree also in number and localization of the vitellaria; and both species have five chromosomes in their haploid set, with chromosomes less heterobrachial than in the genus *Archilopsis* (see below) (for a description of these karyotypes, see Martens *et al.*, 1989a and Curini-Galletti *et al.*, 1989). We postulate the number and localization of vitellaria, the otoplanoid habitus and the chromosome morphology as synapomorphies for the two species (8, 9, 10 in Fig. 1) which diagnose the genus *Monocelopsis*. Both species occur in Western and Northern Europe, in mid and high mediolittoral; *M. septentrionalis* prefers coarser sediments than *M. otoplanoides*.

The revision by Martens *et al.* (1989a) of *Archilopsis unipunctata* revealed the existence of a species-complex including four different species; one species in agreement with Fabricius' description (1826) (*A. unipunctata*), one conferming *A. spinosa* described by Jensen (1878), and two new species (*A. marifuga* and *A. arenaria*). Two synapomorphies in these four species justify the conservation of the genus *Archilopsis* Meixner, 1938: backwards oriented vaginal duct, connected postpenially with the femal duct, and the presence of two prostate channels (11, 12 in Fig. 1). For phylogenetic relationships within the genus, see Martens *et al.* (1989a). The four species are all characterised by the presence of five chromosomes in their haploid sets. The genus *Archilopsis* is amphiatlantic in distribution, and strictly boreal (from North-Western Europe to Eastern Canada).

The karyological evolution of the *Archiloa* complex has been studied by Curini-Galletti *et al.* (1989). It involves step-wise processes of fission from the basic set of the Monocelididae with $n=3$ to sets with $n=4$ (one fission) and $n=5$ (two fissions). Fission leading from $n=3$ to $n=4$ is here considered as a synapomorphy for the genera *Archiloa*, *Inaloa*, *Monocelopsis* and *Archilopsis* (1 in Fig. 1). Fission leading from $n=4$ to $n=5$ constitutes a synapomorphy for the genera *Monocelopsis* and *Archilopsis* (2 in Fig. 1).

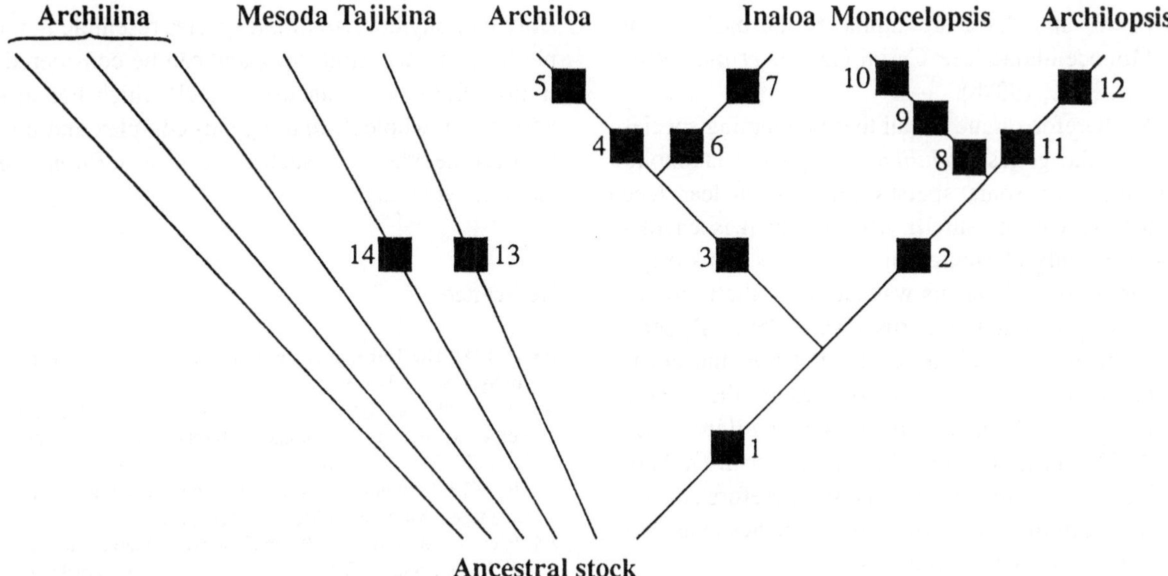

Fig. 1. Cladogram for the *Archiloa* genus complex. Numbers refer to the hypothesized apomorphies. 1. $n=4$: fission of the first metacentric chromosome of the basic set. 2. $n=5$: fission of the second metacentric chromosome of the basic set. 3. Spherical seminal vesicle and prostatic vesicle separated by a muscular diaphragm, both vesicles are more or less not in touch with the bulb wall. 4. Accessory cirrus. 5. Internal vagina ending in the prepenial bursa without direct connection with the female duct. 6. Extremely long cirrus. 7. Extremely long vagina. 8. Four pairs of vitelline follicles, only one pair of which is in front of the pharynx. 9. Otoplanoid habitus. 10. Karyotype nearly symmetrical. 11. Copulatory organ with prostatic channels. 12. Vaginal duct connected postpenially with the female duct. 13. Vagina interna. 14. Intraepithelial nuclei.

Two other species, described as *Archiloa juliae* by Tajika (1982) and *Archiloa tajikai* by Ax & Armonies (1990), both from the Northern Pacific have an internal vagina. An internal vagina, but not connected with the female duct, is also present in *Archiloa* as defined above. However, in these two species the seminal and the prostate vesicle fill the whole bulb, and the female duct is connected with the vagina. In addition, the geographic distribution of these species is different and *Archiloa juliae* has only two chromosomes in the haploid set. We consider the presence of an internal vagina as a parallelism. These species cannot conveniently be assigned to any of the other genera. We therefore proposed the new genus *Tajikina* Martens & Curini-Galletti, 1994, with the internal vagina as the provisional autapomorphy for this genus (13 in Fig. 1).

The phylogenetic relationships among the remaining species of the complex are not clear. These species mostly show plesiomorphic characters as mentioned above. Two species described by Marcus – *Mesoda gabriellae* Marcus, 1949 and *Pistrix thelura* Marcus, 1951 – show the general plesiomorphic features of female and male organs. The existence of two different genera is not justifiable; the only character which can be considered an apomorphy are the intraepithelial nuclei of the epidermis. Furthermore both species occur in the same geographical area (Southern Brazil). We have thus provisionally placed these species into one single genus, *Mesoda*, with intraepithelial nuclei as its presumed derived character (14 in Fig. 1).

The remaining species are *Archilina endostyla* Ax, 1959, *Archiloa vanderlandi* Martens & Curini-Galletti, 1989, *Archiloa subtilis* Karling, Mac-Fira and Dörjes, 1972, *Archiloa papillosa* Ax & Ax, 1977, *Archiloa duplaculeata* Ax & Armonies, 1990. In addition to these, numerous new species have been found in the Mediterranean and in tropical areas (some of which are described in Martens & Curini-Galletti, 1994, Curini-Galletti & Martens, in press). All these species show a very similar organisation of the genital organs (as far as we could understand from available descriptions and our own observations), *i.e.* a rather short external vagina, opening in the female channel which forms at that place a bursa (prepenial bursa) provided with some glands; in front of this bursa the female duct can be provided with an epithelium of the resorbiens type; the copulatory bulb is mostly compact with no specialised structure of the prostatic glands. All these characters are found also in other Monocelididae not belonging to the *Archiloa* group and represent the plesiomorphic condition. The karyotype of *A. endostyla* and of most of the new species is known: the chromosome number ($n=3$) and shape are similar to the basic set of the Monocelididae (see Curini-Galletti *et al.*, 1989; Martens *et al.*, 1989b).

We therefore argue for all these remaining species to be in the genus *Archilina*, though the taxonomic position of some species remains unclear (see Martens & Curini-Galletti, 1994), which is characterised by only plesiomorphic characters. This originally monospecific genus was based on the presence of a stylet within the cirrus (Ax, 1959). Presently, in the Monocelididae, a stylet within the cirrus is known to occur in some species of the genera *Duplominona*, *Duploperaclistus* and *Archilopsis* (see Ax, 1977; Martens 1983; Martens & Curini-Galletti, 1989; Martens *et al.*, 1989a), and we therefore consider this within the genus *Archilina*, a species character only known for *Archilina endostyla*.

Because the genus *Archilina* is presently based only on plesiomorphic characters, it may be a non-monophyletic taxon. On the other hand morphological details of the *Archilina* species, especially the cirrus morphology, have revealed the existence of different evolutionary lines in different geographical areas (Martens & Curini-Galletti, 1994).

Conclusions

Within the *Archiloa* genus complex we can recognize seven valid genera, including the possibly non-monophyletic *Archilina*. Phylogenetic relationships between four of these genera can be resolved.

The ancestor(s) of this genus complex was cosmopolitan in distribution and in different areas it evolved along different evolutionary pathways. In tropical and subtropical areas the evolution of the group has produced different lineages but mostly with small morphological and karyological modifications. In contrast, in temperate and boreal areas the morphological and karyological evolution was much more pronounced, resulting in clear evolutionary lines with different taxa which all can be characterised by autapomorphies. We recognize an evolutionary line in the Atlantic and the Mediterranean which includes the genera *Archiloa*, *Inaloa*, *Archilopsis* and *Monocelopsis*. These taxa together are monophyletic and their relationships can be resolved (Fig. 1). The genus *Mesoda* (Brazil) and *Tajikina* (Northern Pacific) can be considered as two other, evolutionary lines.

From this we argue that the recent *Archilina* is possibly paraphyletic. *Archilina* species resemble in morphology the ancestral stock and can be considered as living relicts of the ancestral stock which lies at the base of the whole *Archiloa* genus complex and possibly of some other taxa, such as *Boreocelis*, *Promonotus* and *Paramonotus*.

References

Ax, P., 1951. Die Turbellarien des Eulitorals der Kieler Bucht. Zool. Jb. Syst. 80: 277–378.

Ax, P., 1959. Zur Systematik Ökologie und Tiergeographie der Turbellarienfauna in den ponto-kaspischen Brackwassermeeren. Zool. Jb. Syst. 87: 43–184.

Ax, P., 1977. Problems of speciation in the interstitial fauna of the Galapagos. Mikrofauna Meeresboden 61: 29–43.

Ax, P. & W. Armonies, 1990. Brackish water Plathelminthes from Alaska as evidence for the existence of a boreal brackish water community with circumpolar distribution. Microfauna Marina 6: 7–109.

Beauchamp, P. de, 1910. Archiloa rivularis n.g. n.sp. Turbellarié Alloeocoele d'eau douce. Bull. Soc. zool. Fr. 35: 211–219.

Curini-Galletti, M., I. Puccinelli & P. M. Martens, 1989. Karyometrical analysis of ten species of the subfamily Monocelidinae (Proseriata, Platyhelminthes) with remarks on the karyological evolution of the Monocelididae. Genetica 78: 169–178

Fabricius, O., 1826. Fortsaettelse of Nye Zoologiske Bidrag VI. Nogle lidet bekjendte og tildels nye Flad-Orme (Planarier). Kongl. Danske Vidensk. Selskabs. Naturv. og Mathem. Afhandl. 2: 13–35.

Jensen, O., 1878. Turbellarier ved Norges vestkyst., Bergen, 97 pp.

Karling, T. G., 1966. Marine turbellaria from the Pacific Coast of North America IV. Coelogynoporidae and Monocelididae. Ark. Zool. 18: 493–528.

Martens, P. M., 1983. Three new species of Minoninae (Turbellaria, Proseriata, Monocelididae) from the North Sea, with remarks on the taxonomy of the subfamily. Zool. Scr. 12: 153–160.

Martens, P. M. & M. C. Curini-Galletti, 1989. Monocelididae and Archimonocelididae (Platyhelminthes, Proseriata) form South Sulawesi (Indonesia) and Northern Australia with biogeographical remarks. Trop. Zool. 2: 175–205.

Martens, P. M. & M. C. Curini-Galletti, 1994. Revision of the Archiloa species complex, with description of seven new *Archilina* species from the Mediterranean (Proseriata, Platyhelminthes). Bijdragen tot de Dierkunde, 64: (in press).

Martens, P. M., M. C. Curini-Galletti & I. Puccinelli, 1989a. On the morphology and karyology of the genus Archilopsis (Meixner) (Platyhelminthes, Proseriata). Hydrobiologia 175: 237–256.

Martens, P. M., M. C. Curini-Galletti & P. Van Oostveldt, 1989b. Polyploidy in Proseriata (Plathelminthes) and its phylogenetical implications. Evolution 43: 900–907.

Meixner, J., 1943. Über die Umbildung einer Turbellarienart nach Einwanderung aus dem Meere ins Süsswasser. Int. Revue ges. Hydrobiol. Hydrogr. 43: 458–468.

Sopott, B., 1972. Systematik und Ökologie von Proseriaten (Turbellaria) der deutschen Nordseeküste. Mikrofauna Meeresboden 13: 1–72.

Tajika, K.-I., 1982. Marine Turbellarien aus Hokkaido, Japan IX. Monocelididae (Proseriata). Bull. Lib. and Sci. Course, Sch. Med. Nihon Univ. 10: 9–34.

Hydrobiologia **305**: 15–19, 1995.
L.R.G. Cannon (ed.), Biology of Turbellaria and some Related Flatworms.

An elaboration of the evolutionary morphological basis for the systematics of the Plathelminthes

Yurij V. Mamkaev
Zoological Institute, Russian Academy of Sciences, 199034 St. Petersburg, Russia

Key words: Plathelminthes, Platyhelminthes, Turbellaria, evolutionary morphology, systematics

Abstract

Compared with Hennig's phylogenetical systematics which has as its aim the retracing of genealogical relations between taxonomic groups, evolutionary morphological systematics is equally justified. Classifications of basic plans, morphological types, and morphofunctional systems of organisms serve as the foundation of evolutionary morphological systems. They are constructed on the basis of thorough understanding and further iteration of morphological transformation in phylogenetical branches based on the constructional pecularities of the morphofunctional systems. The evolutionary morphological approach in systematics is important especially for elaborating macrosystems dealing with vastly divergant groups where it is impossible to trace their real genealogy. The general principles of evolutionary morphological systematics are considered. A variant of the classification system of the Plathelminthes is suggested.

Introduction

According to Remane (1955, 1956) establishing morphological types is indispensable for elaborating natural evolutionary system of organisms and classifying morphological types forms the basis for taxonomic ranking, i.e. the natural system was considered by him as an evolutionary classification of morphological types. He noted some types are so remote from each other that they may be connected only with either single bridges of homology or intermediate forms (Remane, 1955). So Remane considered morphological classification as preceding the revelation of phylogenetical relations between more remote groups.

This evolutionary morphological approach is quite different from the present phylogenetic systematics which seeks for dichotomous classifications of sister groups. The phylogenetic (cladistic) systematics of Hennig (1966) has as its aim retracing genealogical relationships between taxonomic groups. It is only the features marking genealogically related forms that are significant for phylogenetic systematics. It does not deal with any estimation of morphological transformations. By comparison, evolutionary morphological classifications are built by thorough analysis of morphological transformations in phylogenetic branches.

Comparing the two approaches in systematics, it must be emphasized that both are applicable but in different situations. Each has its specific aim. The aim of evolutionary morphological systematics is to estimate a degree of morphological proximity 'morphological conformity' between different groups of species. This approach is of particular importance for taxonomic ranking of groups.

Systematicians usually emphasize their aspirations to elaborate 'natural systems'. But what are natural system? Genuine natural systems are only those which really do exist in nature. They are characterized by the active interactions between their elements. So, in nature we observe the following hierarchy of systems: cells, tissues, organs, systems of organs, organisms, populations, communities, then to ecosystems and on up to the whole biosphere. Classification systems are man made. They are purely logical, gnoseological constructs. Thus, side by side with phylogenetic systems we can elaborate evolutionary morphological ones.

The evolutionary morphological system

The evolutionary morphological approach in systematics is important especially for elaborating macrosystems dealing with vastly divergent groups because tracing their real genealogy is impossible. The phylum Plathelminthes represents a set of such groups. Undoubtedly, between such taxa as the Gnathostomulida, Acoelomorpha, Catenulida, Macrostomida, Polycladida and Neoophora there were very many intermediate, now extinct groups. So at present it is impossible to reveal their true genetical relationships. If we refuse to recognise the phylum Plathelminthes and class Turbellaria as entities, what do we do with the aggregation of such different groups as those cited above? They can not be drawn nearer to any other groups outside this set of forms. I propose the evolutionary morphological approach gives the only basis for elaborating plathelminth systematics.

The main tast of evolutionary morphological systematics is to give a better notion of the basic plan used for establishing a high taxon - the phylum. The basic plan is a type of organism's construction formed within the limits of a group, characterized by architectonic peculiarities. These pecularities consist in the constructional specificity of the morphofunctional systems typical for a given organisation and, further, in the uniqueness of the complex formed by the subsystems of the organism based on intersystem relationships and distinguished by a particular disposition of the parts - by promorphological (symmetrical) charactericts.

The problem of the basic plan is quite complicated. The number of basic plans is constantly increasing, accordingly the number of phyla and even kingdoms is growing excessively. Incidentally, phylogenetic systematics promotes this because the following principle begins to work: if some forms are not derived monophyletically they represent special basic plans, special phyla. In this connection it is expedient, to my mind, to consider as a special basic plan only such a type of organism's construction which is present not in a single but in a few rather prominent morphological types. The basic plan should be considered as the general scheme of construction. This is a constructional basis for different morphological types - a certain principle of organization perspective for the development of a set of morphological types. If only one form of the previous diversity has remained it does not represent the basic plan of the former group; it can even give an erroneous impression of the group. So it is better to add such a form to the 'mother' basic plan (basic plan of the ancestral phylum).

As the above definition shows, the basic plan (a type of organism's construction... formed *within* the limits of a group) is considered here not as a prototype but as a typical scheme of organisation - a certain morphological centre of a group. Any evolutionary formation of a new basic plan appears to be accompanied by the phenomena of initial morphological diversity, parallelisms, repetitions and mosaic evolution, all of which speak against the idea of a 'prototype' (Mamkaev, 1985, 1987, 1991). Indeed, there is a basic plan of an ancestral group and its complicated development and transformation into a new basic plan. It should be emphasized that a wide range of forms is usually incorporated in this process of transformation. It seems quite evident that any group characterized by a new morphological pattern takes its origin from a large set of forms (from a 'multi-source complex', 'large basis').

Taking into account all the above considerations, we can suggest that the phylum should be defined as a taxon of the highest level which is characterized by a special basic plan, or of interconnected basic plans, showing a certain continuity of forms, and at the same time clearly separated from other basic plans. The classification scheme for such a taxon appears to be as follows: (1) diversity of groups presenting the formation of a morphological center (groups at the beginning of a given evolutionary path including forms showing other evolutionary tendencies can be ascribed to this stage), (2) a morphological center representing a given basic plan in its formed state and in its characteristic morphological types, (3) the most advanced groups whose representatives show in their organisation certain features of the next basic plan.

In accordance with the demands of strict monophyly the majority of groups of the first stages should be excluded from any such highest taxon. But then what are we to do with such groups? Add them to the nearest lower phylum? But then they will unjustifiably enlarge its limits, making its diagnosis worse. Do we leave them as independent taxa? But then it is necessary to suggest a new category of taxa ('intermediate', 'splinter'). The latter decision is possible, to my mind, but it will require an essential change of the canons of systematics. So the suggested variant appears to be more preferable for the present, reflecting better the real chain of events which has been fully demonstrated for the evolution of certain groups (e.g. for the processes of tetrapodization and mammalization). I have tried to reveal such processes for the plathelminths (Mamkaev, 1985, 1986, 1987). As a result the following evolutionary morphological variant of the plathelminths system can be suggested (Mamkaev, 1991).

An evolutionary morphological classification of the Plathelminthes

Phlylum Plathelminthes
- Superclass Turbellariomorphae
  - Class Turbellaria
    - Subclass Archoophora
      - Superorder Acoelomorpha
        - Order Nemertodermatida
        - Order Acoela
      - Superorder Notandropora
        - Order Catanulida
      - Superorder Macrostomomorphae
        - Order Macrostomida
        - Order Haplopharyngida
      - Superorder Polycladida
        - Order Polyclada
    - Subclass Neoophora
      - Superorder Prolecithophora
        - Order Combinata
        - Order Separata
        - Order Lecithoepitheliata
        - Order Gnosonesimida
      - Superorder Seriata
        - Order Proseriata
        - Order Tricladida
      - Superorder Rhabdocoelida
        - Order Rhabdocoela
        - Order Kalyptorhynchia
        - Order Temnocehalida
  - Class Udonellida
  - Class Monogenoidea
  - Class Gyrocotyloidea
  - Class Amphilinida
  - Class Cestoda
  - Class Aspidogastrea
  - Class Trematoda
- Superclass Gnathostomularia
  - Class Gnathostomulida
- Superclass Xenoturbellaria
  - Class Xenoturbellida

The features showing a level of system organization and evolutionary tendencies have been taken into consideration (first of all: the polymerization of the kinetids, the compartmentalization, delimitation and ordering of histological and anatomical systems, the development of the epitheliomusculary organs). The latter can form rather complicated and stable morphofunctional complexes - moduli (Markevich, 1988, 1990) which are characteristic of a number of groups. So it is possible to distinguish modular morphological types, modular groups (such as the Ghathostomulida, Kalyptorhynchia, Udonellida, Monogenoidea).

For the Plathelminthes as the lowest recent Bilateria the following main peculiarities are characteristic:

1. blind-ending digestive system without anus,
2. soft integument without cuticle,
3. uncentralized distributive system without special uniting cavity,
4. hermaphroditism.

As an example of this scheme of the plathelminths' basic plan, Ax's picture can be used (Ax, 1961). However it is not a prototype but an image of the typical organization - of the morphological centre of the Plathelminthes. All characteristic systems of the plathelminths are seen on the picture in their fully-developed state.

Plathelminthes within the Bilateria

Comparing the given basic plan with those of neighbouring phyla, we can see essential differences. The plathelminths include the quite distinctive basic group (Turbellaria) showing the wide spectrum of formative evolution which affects the fundamental systems of multicellular organism, missing states being fortunately supplied by the gnathostomulids and *Xenoturbella*. It is the formation of the given organization that can be demonstrated in this case. The basic plans of the other Scolecida and Nemertini are already presented only in their fully-developed state - without any basic groups.

The evolutionary morphological condition in the plathelminths can be presented as follows: the basic ('mother') group together with its derivative 'daughter' ones (Turbellariomorpha), plus the Gnathostomulida and Xenoturbellida (?) which quite early digressed from the plathelminths' way of evolution and show peculiar evolutionary tendencies. If we withdraw the Gnathostomulida and Xenoturbellida from the Plathelminthes where can we find a place for them within the Bilateria? As to the former taxon, in phylogenetic systematics it is the adelphotaxon of the plathelminthes (Ax, 1987). However, in accordance with the principles of evolutionary morphological systematics it can certainly be included in the phylum. Indeed, its morphological type fully corresponds to the plathelmints' basic plan. The gnathostomulids can be characterized as monociliated plathelminths ('Monociliaria') which, owing to their protonephridia, an essentially two-layered organization, a triangular mouth, and cases of regular disposition of the epidermal cells, show evolutionary tendencies to the Nemathelminthes (especially Gastrotricha). But the main specificity of the group consists in the presence of a strongly pronounced modulus (pharynx having complicated jaw apparatus). Primarily it determines the morphological diversity of the group.

As to the mysterious *Xenoturbella* (hermaphroditic, but without any copulatory organs, wholly epithelized, and having an intraepidermal nervous system), at present it also could be placed (however rather conditionally) among the lower plathelminths. At any rate, owing to the ciliary rootlets drawn together, it is possible to put this form in the beginning of the morphological row leading to a unique rootlet apparatus of the Acoelomorpha with the anterior and posterior rootlets merged together (Raikova, 1989). At the same time, the high multiseriated epidermis having a quite well developed thick basal lamina and a basiepithelial nervous system appears to be a peculiar characteristic of deuterostomian evolution (Reisinger, 1960). From the evidence of a quite unique statocyst (Ehlers, 1991), I should like to suggest the name Mastigocystia for this superclass.

Relationships within the Plathelminthes

From their morphological diversity (first of all, owing to the peculiarities of the body's stratification) the archoophoran turbellarians are easily subdivided into four groups which could be considered superorders.

The neoophoran turbellarians are subdivided into three flocks of formative evolution which can also be considered superorders. Among them the Prolecithophora Combinata is remarkable: it forms the basic group because of its morphological diversity including its primitive diffuse gonads. The plagiostomids, prorhynchids, and gnosonesimids show quite stable morphological types and can be considered derivative groups. The relationships of the Lecithoepitheliata Prorhynchidae to the Prolecithophora has been quite cogently demonstrated by Karling (1940) and Timoshkin (1991 a,b). Thus, a superorder including four orders appears to be a suitable status for this flock of forms.

Similarly, the Proseriata is a mother group of the Tricladida. The kalyptorhynchians have such a pronounced modulus (proboscis) which has given an outstanding spectrum of formative evolution, that this appears to present sufficient grounds for the foundation of an independent order.

Discussing the suggested system is difficult within the limits of a short communication; this was done earlier (Mamkaev, 1991). The main aim of the paper is to show how it is possible to approach the problem from an evolutionary morphological point of view.

Acknowledgments

I am cordially thankful to my Finnish colleagues, especially Dr Maria Reuter and Dr Margaretha Gustafsson, for the opportunity to take part in the VII ISBT and for their constant trouble. I am also grateful to reviewers for their constructive criticism. This work was supported by a grant from Russian Biodiversity Foundation.

References

Ax, P., 1961. Verwandtschaftsbeziehungen und Phylogenie der Turbellarien. Ergebnisse der Biologie 24: 1–68.

Ax, P., 1987. The phylogenetic system (The systematization of organisms on the basis of their phylogenesis). J. Wiley & Sons, Chichester, N.Y., 340 pp.

Ehlers, U., 1991. Comparative morphology of statocysts in the Plathelminthes and Xenoturbellida. Hydrobiologia 227 (Dev. Hydrobiol. 69): 263–271.

Hennig, W., 1966. Phylogenetic systematics. University of Illinois Press, Urbana, Chicago, London, 263 pp.

Karling, T. G., 1940. Zur Morphologie und Systematik der Alloeocoela Cumulata und Rhabdocoela Lecithophora (Turbellaria). Acta Zool. fenn. 26: 1–120.

Mamkaev, Yu. V., 1985. The principle of morphological radiation and its significance for evolutionary morphology. In: J. Mlikovsky, V. J. A. Novak (eds), Evolution and Morphogenesis. Academia, Praha: 421–427.

Mamkaev, Yu. V., 1986. Initial morphological diversity as a criterion in deciphering turbellarian phylogeny. Hydrobiologia 132/Dev. Hydrobiol. 32: 31–33.

Mamkaev, Yu. V., 1987. Resnichnye chervi i metodologicheskie printsipy evolyutsionnoj morfologii. Tr. Zool. Inst. Akad. Nauk USSR 167: 4–33. (Turbellaria and methodological principles of evolutionary morphology. In Russian, English summary).

Mamkaev, Yu. V., 1991. O morfologicheskikh osnovakh sistemy ploskikh chervej. Tr. Zool. Inst. Akad. Nauk USSR 241: 3–25. (On morphological principles of the plathelminthes system. In Russian, English summary).

Markevich, G. I., 1988. Zakonomernosti morfologicheskoj evolutsii slozhnykh skleritno-myshechnykh sistem kolovratok. Problemy Makroevolutsii, S. Severtsov (ed.), Nauka, Moskow: 81–82. (Regularities of the morphological evolution of the complicate sclerit-muscle systems of the Rotatoria. In Russian)

Markevich, G. I., 1990. Istoricheskaya rekonstruktsiya filogeneza kolovratok kak osnova postroeniya ikh makrosistemy. Kolovratki. Materialy Tret'ego Vsesoyuznogo simpoziuma. L. A. Kutikova (ed.), Zool. Inst. Akad. Nauk USSR: 140–156. (Historical reconstruction of the phylogenesis of the Rotatoria as basis for constructing their macrosystem. In Russian.)

Raikova, O. I., 1989. O gomologii kornevykh nitej resnichnogo apparata Acoela i drugikh Turbellaria. DAN SSSR 308, 6: 1506–1509. (On homology of the rootlet apparatus of the Acoela and other Turbellaria. In Russian.)

Reisinger, E., 1960. Was ist Xenoturbella? Zeitschr. wiss. Zool. 164: 188–198.

Remane, A., 1955. Morphologie als Homologienforshung. Zool. Anz. 18. Supplementband: 159–183.

Remane, A., 1956. Die Grundlagen des Naturlichen Systems, der Vergleichenden Anatomie und der Phylogenetik. Akademische Verlagsgesellschaft, Geest & Portik K.-9, Leipzig, 364 S.

Timoshkin, O. A., 1991a. Resnichnye chervi oz. Bajkal. I. Turbellaria Prorhynchidae. Morfologya, sistematika i filogeniya Lecithoepitheliata. Morfologya i evolutsia bespozvonochnykh. A. A. Linevich and E. L. Afanasieva (eds), Nauka, Novosibirsk: 63–185.

Timoshkin, O. A., 1991b. Turbellaria Lecithoepitheliata: morphology, systematics, phylogeny. Hydrobiologia 227 (Dev. Hydrobiol. 69): 323–332.

Hydrobiologia **305**: 21–26, 1995.
L.R.G. Cannon (ed.), Biology of Turbellaria and some Related Flatworms.

The basic organization of the Plathelminthes

Ulrich Ehlers
II. Zoologisches Institut und Museum der Universität Göttingen, Berliner Strasse 28, D-37073 Göttingen, Germany

Key words: basic organization, stem species, phylogenetics, Plathelminthes, Platyhelminthes

Abstract

The basic organization of the Plathelminthes is summarized. Special attention is given to epidermal structures, musculature, extracellular matrices, nervous system and sensory structures, digestive system, totipotent stem cells, protonephridia, reproductive system and life cycle. The latest common ancestor of the Plathelminthes lacked any 'parenchymal' cells and tissues; Plathelminthes do not represent 'Parenchymia'. Discussions concerning the relationships of the Plathelminthes with other Metazoa must be based on the characteristics of the plathelminth stem species.

Introduction

In many discussions on metazoan evolution the Plathelminthes take a central position when considering the origin and basic systematics of the bilaterian or so-called 'tripoblastic' animals. Discussions with a clear phylogenetic approach can be found already in textbooks on invertebrate zoology (Brusca & Brusca, 1990; Meglitsch & Schram, 1991). In these two books, but also in other publications, the 'basic anatomy' or the 'general plathelminth anatomy' corresponds with that of a triclad, often based on a *Dugesia* species.

Of course, triclads represent well-known representatives of the free-living flatworms. But a triclad like any other known species of the Plathelminthes does not represent the basic organization of the Plathelminthes. All extant species and all well-known supraspecific monophyletic entities like the Tricladida possess a mosaic of plesiomorphic and apomorphic characteristics. Several of the plesiomorphies were inherited from the latest common ancestor of the Plathelminthes. Other plesiomorphies were not yet present in this common ancestor but have evolved later representing autapomorphies of subordinated monophyletic entities (cf. Ehlers, 1985). For example, the heterocellular female gonad of triclads does not belong to the 'basic anatomy' of the Plathelminthes but is an apomorphous feature of the Neoophora and a plesiomorphy for the Tricladida.

Comparisons of the Plathelminthes with any other taxon of the Metazoa must be based exclusively on those characteristics that have been present in the latest common ancestor of the Plathelminthes. Characteristics that have evolved secondarily within the Plathelminthes are without significance for such comparisons.

On the basis of our present knowledge, we can hypothesize several distinct characteristics, mostly morphological ones, for the stem species of the Plathelminthes. Part of these characteristics have been mentioned already by Ehlers (1984, 1985) and Ax (1984, 1987) and will not be discussed here in detail again.

The stem species of the Plathelminthes

The most recent common ancestor of the Plathelminthes had a body length of about one millimetre like nearly all extant species of free-living flatworms. Body sizes of a centimetre and more have been evolved secondarily within a few subordinate taxa such as the Polycladida, Tricladida or parasitic groups.

This small ancestor had a body circular in transverse section (Fig. 1), not a dorso-ventrally flattened

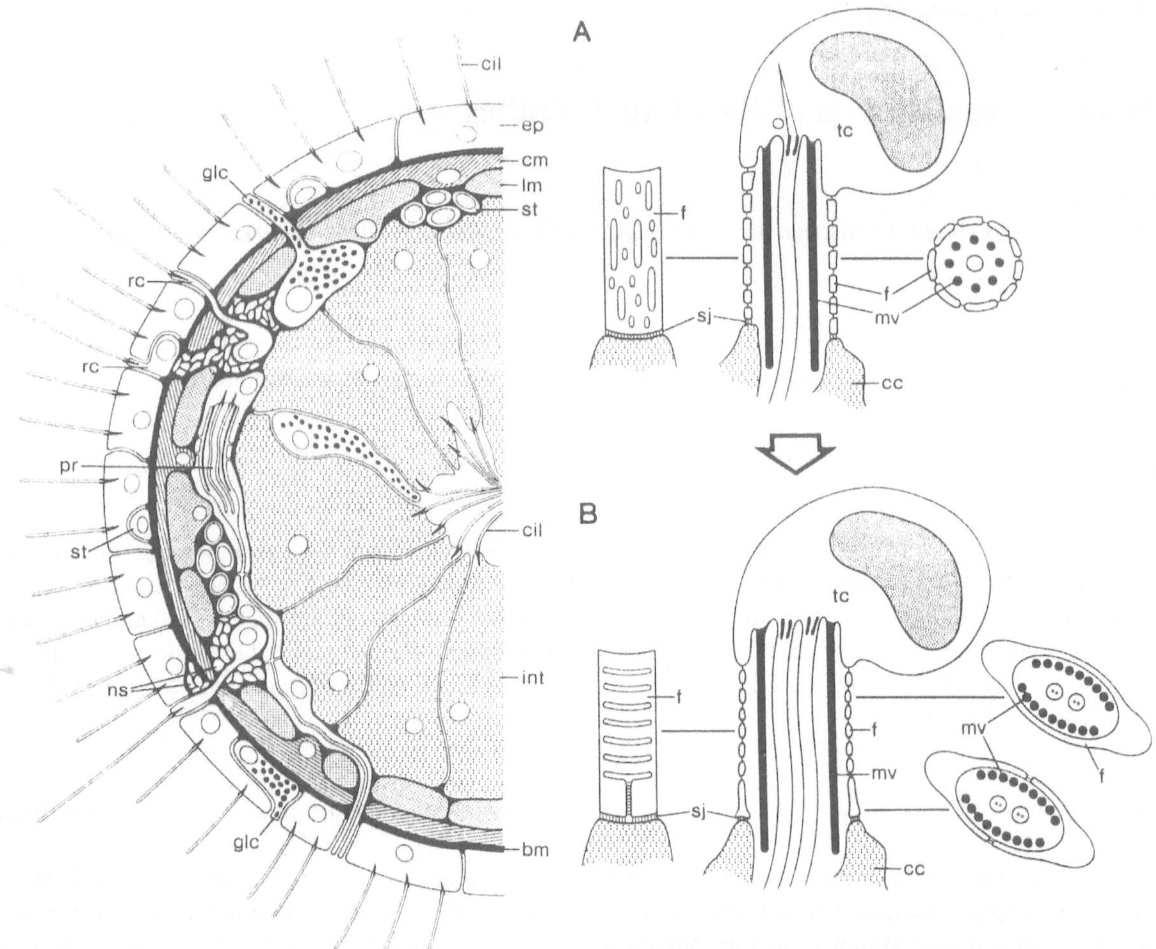

Fig. 1. Basic organization of the Plathelminthes (one-half of cross section; reproductive system not drawn). Abbreviations: *bm*, basement membrane and extracellular matrix (ECM); *cil*, cilium; *cm*, circular muscle fiber; *ep*, multiciliary epidermis cell; *glc*, intra- and subepidermal gland cells; *int*, ciliated intestine; *lm*, longitudinal muscle fiber; *ns*, intra- and subepidermal nerve cells; *pr*, protonephridium with biciliary terminal cell; *rc*, monociliary intra- and subepidermal receptor cells; *st*, intra- and subepidermal stem cells.

Fig. 2. Basic organization of the protonephridial terminal cell with the filter of the Bilateria (A) and of the Plathelminthes (B) (longitudinal and cross sections). Bilateria (fig. from Bartolomaeus & Ax, 1992): the monociliary terminal cell (*tc*) shows a single cilium surrounded by 8 microvilli (*mv*) and a cytoplasmic tube with slit-like perforations serving as a filter (*f*); septate junctions (*sj*) connect the slashed tube and the adjacent canal cell (*cc*). Plathelminthes: the biciliary terminal cell (*tc*) shows two cilia surrounded by about 16 microvilli (*mv*) and a cytoplasmic filter (*f*); septate junctions (*sj*) connect the slashed tube with the adjacent canal cell (*cc*).

body. The organization of the ancestor was bilaterally symmetrical.

The monolayered cellular epidermis consisted of multiciliary cells. Most probably a uniform ciliation as known for the Catenulida can be hypothesized for the stem species of the Plathelminthes (Ehlers, 1986). An external strengthened cuticle was lacking, but a so-called glycocalyx between epidermal microvilli was present. A basement membrane underlay the layer of epidermal cells and functioned in establishing and maintaining polarity in the peripheral epidermal cell layer in such a way that the nuclei of the epidermal cells took an intraepithelial position. The lack of a basement membrane in the Acoelomorpha and the occurrence of so-called 'insunk' perikarya of epidermis cells in several plathelminth groups are derived features which have been evolved secondarily within the Plathelminthes. In all probability, the epidermal cilia displayed a rostral and a caudal rootlet, but no accessory centriole, a situation known from the Catenulida and from distinct taxa of the Macrostomida. The interconnected rootlet-system of the Acoelomorpha is an evolutionary novelty of this taxon. The existence of a fine terminal intracellular fibril layer (a terminal web) also belongs to the basic organization of the Plathelminthes (Rieger *et al.*, 1991b) as is also the case with the presence of distinct intercellur junctions (zonulae adhaerentes, septate junctions and gap junctions).

Several other epidermal features known from all taxa of the free-living flatworms can be hypothesized for the latest common ancestor, e.g. the existence of gland cells and of sensory cells. Within several taxa, e.g. Catenulida, Acoelomorpha, *Haplopharynx*, Macrostomida or Polycladida, there exist intraepidermal gland and sensory cells as well as gland and sensory cells with a subepidermal position. Intraepidermal as well as subepidermal gland and sensory cells can be hypothesized for the stem species of the Plathelminthes (Fig. 1). All the dendritic processes of sensory cells and the discharging gland necks took positions between epidermal cells; penetration of epidermal cells by gland necks and sensory processes was not present but has been evolved secondarily within the Neoophora (Sopott-Ehlers, 1984). Only monociliated epidermal sensory receptors occurred in the stem species (cf. Ehlers, 1992a); multiciliated receptors are evolutionary novelties that arose secondarily within the Plathelminthes. Duo-gland adhesive systems as described by Tyler (1976) and glands producing secretions in the form of lamellated rhabdites (cf. Smith *et al.*, 1982) are known to exist in free-living Rhabditophora; these differentiations were still lacking in the common ancestor of the Plathelminthes.

A thin subepidermal body-wall musculature consisting of outer circular and inner longitudinal fibers can be hypothesized for the plathelminth stem species (Fig. 1). Diagonal muscle fibers may have been absent (cf. Rieger *et al.*, 1991a: 127). The muscle cells were surrounded by an extracellular matrix (Fig. 3), a feature common for nearly all plathelminths (except the Acoelomorpha) and inherited from the common ancestor of the Bilateria. There were no epitheliomuscular cells in the plathelminth stem species. Distinct molecules are known to constitute the subepidermal matrices of at least several flatworms: collagens, laminin, fibronectin and glycosaminoglycans (e.g. Hori, 1992; Lindross & Still, 1988; Pascolini *et al.*, 1989; Pedersen, 1991). Because these constituents also occur in other Metazoa (e.g. Exposito & Garrone, 1990; Pedersen, 1991), they can be hypothesized to have been present in the common ancestor of the Plathelminthes.

The plathelminth stem species possessed a nervous system that was localized both basiepidermally and subepidermally (Fig. 1). Longitudinal nerve cords might have been present besides plexus-like peripheral arrangements of the neurones. Most probably, accumulations of neurons formed a weakly lobed ganglionic to ring-form cerebrum in the anterior end of the stem species (Joffe, 1991; Raikova, 1991; Rieger *et al.*, 1991b). As discussed by Reuter & Gustafsson (1989), peptidergic and sensory cells can be hypothesized for the stem species whereas more specialized ganglion cells and glia cells have been evolved secondarily within the Plathelminthes (cf. Reuter, 1991). Neuromediators (e.g. a number of neuropeptides, monoamines, acetylcholine, gamma aminobytyric acid etc.) and different growth factors (EGF, ECFr, bFGF) are now known for several flatworms and belong to the basic characteristics of the Plathelminthes (Gustafsson & Eriksson 1992; Gustafsson *et al.*, 1993; Halton *et al.*, 1990; Palmberg, 1991; Reuter & Kuusisto, 1992; Reuter & Eriksson, 1991).

There exist pigment-cup ocelli and statocysts in free-living flatworms (cf. Sopott-Ehlers, 1991; Ehlers, 1991). These well-known sensory organs cannot be hypothesized to represent basic features of the Plathelminthes.

Most probably, a ventral simple mouth opening marked the connection between epidermis and gastrodermis in the stem species (cf. Doe, 1981; Ehlers, 1985). An anus was lacking. The sac-like intestine consisted of a monolayer of ciliated gastrodermal cells (phagocytes) and of glandular cells (Fig. 1). The absence of ciliation or of gland cells and the existence of a branched gut or of digestive syncytia are novelties that have evolved within the Plathelminthes.

A remarkable characteristic of the Plathelminthes is the existence of totipotent undifferentiated cells (stem cells, neoblasts) which give rise to all differentiated cell types (Auladell *et al.*, 1993). Differentiated somatic cells do not proliferate. Populations of subepidermally located stem cells can be hypothesized as a basic characteristic of the Plathelminthes (Fig. 1). Within the Catenulida, undifferentiated cells also exist in a basiepidermal position (cf. Drobysheva, 1991; Ehlers, 1992b). Such intraepidermal stem cells may represent a characteristic of the stem species as well (Fig. 1).

One pair of protonephridia is hypothesized for the basic organization of the Plathelminthes (Fig. 1) as discussed in detail by Ehlers (1985). The situation to be found within the Catenulida with an unpaired system is an evolutionary novelty of this taxon. Nearly all other plathelminths (except the Acoelomorpha, see below) possess protonephridia with few to many branches;

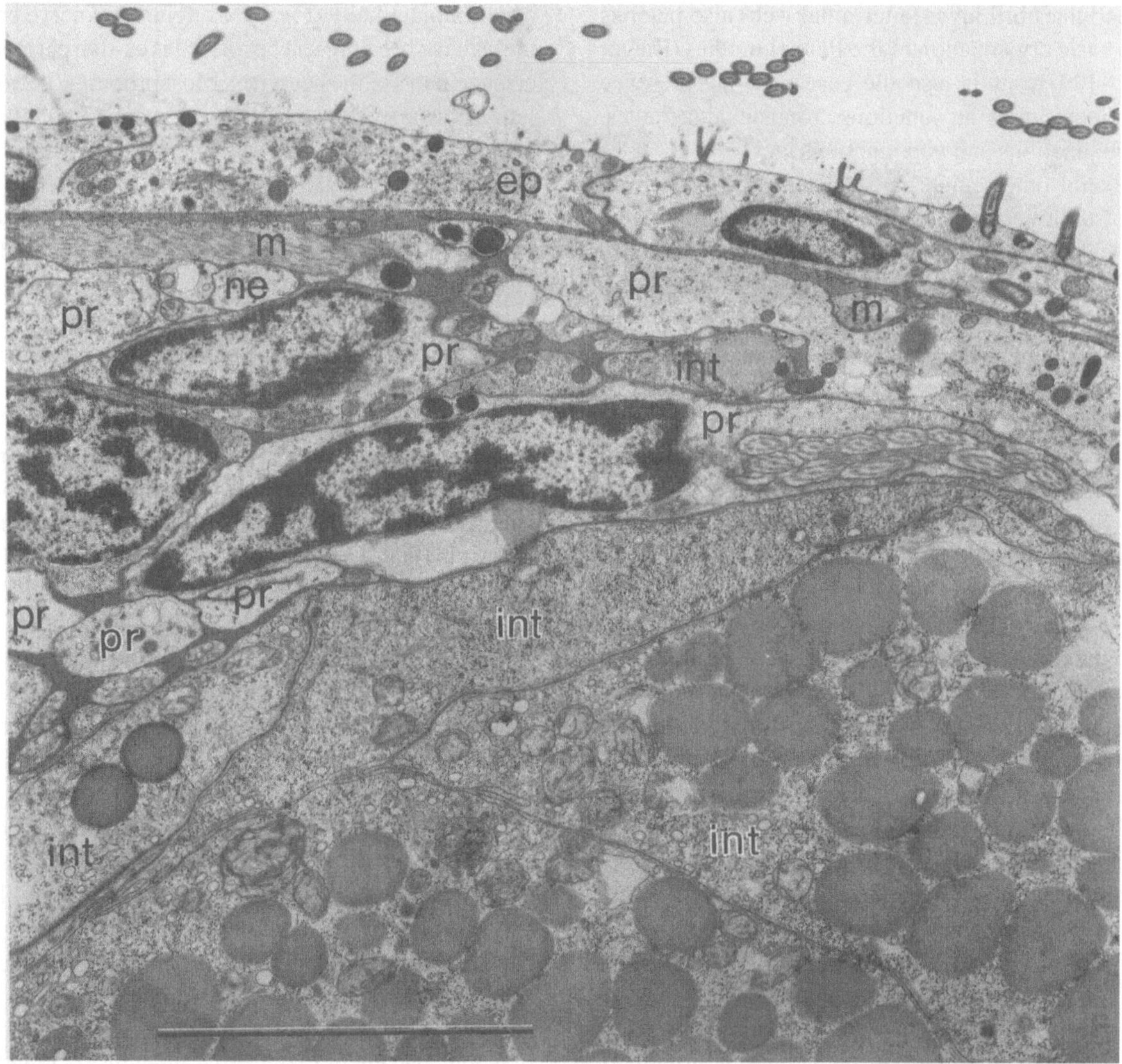

Fig. 3. Lack of 'parenchymal' cells in *Archimonocelis oostendensis* Martens & Schockaert, 1981. The body space between epidermis (*ep*) and intestine (*int*) is filled with muscle cells (*m*), nerve cells (*ne*), protonephridial cells (*pr*) and extracellular matrix. Scale bar represents 5 μm.

such branches may have been present already in the stem species. Both protonephridia of the most recent common ancestor of the Plathelminthes consisted of at least three cell types inherited from the stem species of the Bilateria (cf. Bartolomaeus & Ax, 1992): terminal cells, canal cells and specialized canal cells, the nephroporus cells. In the ancestor of the Bilateria and of the Gnathostomulida (the sister group of the Plathelminthes) as well, the terminal cell was monociliated (cf. Ax 1984, 1987). Eight microvilli surrounded the cilium; the tube-like and partly slashed distal cytoplasm of the cell formed the filter and surrounded the cilium and the microvilli (Fig. 2A). The filtration barrier consisted of a fine extracellular matrix (ECM) covering the fenestrations of the cytoplasmic tube. Within the Plathelminthes, slightly modified terminal cells exist in the Catenulida (Fig. 2B). Here, the cells are biciliated and 15–21 (often 16) microvilli can be found in each terminal cell. That means the number of cilia and microvilli of a terminal cell has been doubled with respect to the cell of the Gnathostomulida and the Bilateria, respectively. A biciliary terminal cell is hypothesized to represent a basic feature of the Plathelminthes. This condition has been modified secondarily in taxa of the Rhabditophora by increasing the number of cilia or by the evolution of a weir, in which a terminal cell and the adjacent canal cell together form a filtration area. Most probably, the lack of protonephridia in the Acoelomorpha is correlated with the (secondarily achieved) absence of a subepidermal ECM, which is an essential part of a filtration structure.

Many representatives of the free-living Plathelminthes lack any 'parenchymal' or 'mesenchymal' cells as can be seen in Fig. 3. The absence of such cells is a plesiomorphic state and can be attributed to the common ancestor of the Plathelminthes (Fig. 1). Within the flatworms, several types of 'parenchymal' cells like fixed 'parenchymal' cells or reticular cells in triclads (e.g. Hori, 1991; Morita, 1991) or chordoid cells, pigment cells etc. (see also Rieger *et al.*, 1991b) have evolved secondarily a number of times in different taxa as discussed by Ehlers (1985). Besides such 'new' cell types, already existing 'old' somatic cell types like muscle cells can become enlarged giving the impression of a 'parenchymal' nature at the light microscopical level (unpubl. findings). Because of the primary absence of parenchymal cells and tissues, the Plathelminthes should not be called 'Parenchymia'.

Without doubt, the stem species of the Plathelminthes was hermaphroditic with ovaries and testes, most probably in the form of germ layers. The homocellular female gonad gave rise to oocytes which were inseminated and fertilized within the organism. The eggs were entolecithal. Discharging female canals (oviducts) and so-called shell glands were not yet present; the eggs lacked a 'hard' sclerotic shell (cf. Gremigni & Falleni, 1992). The mature spermatozoa, which were filiform and monociliated, were released through a male genital pore; most probably, a copulatory organ was already existent. Cleavage was of the spiralian type (cf. Baguñà & Boyer, 1990). The ciliated hatchling was a young adult, meaning that the Plathelminthes have a direct development. Vegetative (asexual) reproduction, most probably in the form of paratomy, also belongs to the basic characteristics of the Plathelminthes (cf. Rieger *et al.*, 1991b; Ehlers, 1992a).

In accord with our present knowledge, the small common ancestor was holobenthic like the majority of the existing free-living flatworms (cf. Boaden, 1989). This does not exclude the possibility of pelagic dispersion, but did not represent a 'pelago-benthic life cycle'.

Conclusions

The hypothetical latest common ancestor of the Plathelminthes can be characterized by many characteristics, mostly morphological. Of course, there is still dispute about the basic phylogeny of the Plathelminthes and there exist competing hypotheses of relationships, but most of the characteristics mentioned above are not affected by these current contrasting discussions.

The hypothesized basic characteristics are those relevant data on which discussions concerning the relationships of the Plathelminthes with other Metazoa must be based. All other features which have evolved secondarily within the Plathelminthes must be left out of consideration. To give two examples: (1) Lymphatic systems, which are present in a few parasitic taxa but do not belong to the basic organization of the Neodermata, cannot be interpreted as homologous structures of 'coelomic circulatory systems' in other Bilateria (Ruppert & Carle, 1983). (2) Protonephridial weirs with two rows of cytoplasmic rods are present in distinct taxa (*Macrostomum*, Proseriata, Neodermata) but cannot be hypothesized as a basic feature of the Plathelminthes. Therefore, any idea of 'close relationships' between the Plathelminthes and the Rotifera cannot be based on such a characteristic (Clément & Fournier, 1981).

In the past few years, distinct molecular data (ribosomal RNA nucleotide sequences) have become increasingly available for the plathelminths. Of course, it is not yet possible to hypothesize the ancestral 5S rRNA or 18S rRNA (and 28S rRNA) sequences for the latest common ancestor of the living Plathelminthes. Nevertheless, molecular data, especially from the longer rRNA sequences, are important for analyses of relationships between Plathelminthes and other taxa. But a proposed phylogeny must be based on all (!) the available information. That includes molecular and morphological data but not selected data. We can have alternative diagrams of relationships, but not different morphological and molecular phylogenies whether there seems to be consensus or not between morphological and molecular data.

References

Auladell, C., J. Garcia-Valero & J. Baguñà, 1993. Ultrastructural localization of RNA in the chromatoid bodies of undifferentiated cells (neoblasts) in planarians by the RNase-gold complex technique. J. Morph. 216: 319–326.

Ax, P., 1984. Das phylogenetische System. Systematisierung der lebenden Natur aufgrund ihrer Phylogenese. G. Fischer, Stuttgart, N. Y., 349 pp.

Ax, P., 1987. The phylogenetic system. The systematization of organisms on the basis of their phylogenies. J. Wiley & Sons, Chichester, N.Y., Brisbane, Toronto, Singapore, 340 pp.

Baguñà, J. & B. C. Boyer, 1990. Descriptive and experimental embryology of the Turbellaria: present knowledge, open questions and future trends. In H.-J. Marthy (ed.), Experimental embryology in aquatic plants and animals. Plenum Press, N.Y.: 95–128.

Bartolomaeus, Th. & P. Ax, 1992. Protonephridia and metanephridia – their relation within the Bilateria. Z. zool. Syst. Evolut. forsch. 30: 21–45.

Boaden, P. J. S., 1989. Meiofauna and the origins of the Metazoa. Zool. J. Linnean Soc. 96: 217–227.

Brusca, R. C. & G. J. Brusca, 1990. Invertebrates. Sinauer Ass., Sunderland, Mass., 922 pp.

Clément, P. & A. Fournier, 1981. Un appareil sécréteur primitif: les protonéphridies (Plathelminthes et Némathelminthes). Bull. Soc. zool. Fr. 106: 55–67.

Doe, D. A., 1981. Comparative ultrastructure of the pharynx simplex in Turbellaria. Zoomorphology 97: 133–193.

Drobysheva, I. M., 1991. Cambiality of epidermis in the Turbellaria. Proc. zool. Inst. Sankt–Petersburg 241: 53–87 (in Russian).

Ehlers, U., 1984. Phylogenetisches System der Plathelminthes. Verh. naturwiss. Ver. Hamburg (NF) 27: 291–294.

Ehlers, U., 1985. Das phylogenetische System der Plathelminthes. G. Fischer, Stuttgart, N.Y., 317 pp.

Ehlers, U., 1986. Comments on a phylogenetic system of the Plathelminthes. Hydrobiologia 132: 1–12.

Ehlers, U., 1991. Comparative morphology of statocysts in the Plathelminthes and the Xenoturbellida. Hydrobiologia 227 (Dev. Hydrobiol. 69): 263–271.

Ehlers, U., 1992a. On the fine structure of Paratomella rubra Rieger & Ott (Acoela) and the position of the taxon Paratomella Dörjes in a phylogenetic system of the Acoelomorpha (Plathelminthes). Microfauna Marina 7: 265–293.

Ehlers, U., 1992b. No mitosis of differentiated epidermal cells in the Plathelminthes: mitosis of intraepidermal stem cells in Rhynchoscolex simplex Leidy, 1851 (Catenulida). Microfauna Marina 7: 311–321.

Exposito, J.-Y. & R. Garrone, 1990. Characterization of a fibrillar collagen in sponges reveals the early evolutionary appearance of two collagen gene families. Proc. nat. Acad. Sci. USA 87: 6669–6673.

Gremigni, V. & A. Falleni, 1992. Mechanisms of shell-granule and yolk production in oocytes and vitellocytes of Platyhelminthes-Turbellaria. Anim. Biol. 1: 29–37.

Gustafsson, M. K. S. & K. Eriksson, 1992. Never ending growth and a growth factor. I. Immunocytochemical evidence for the presence of basic fibroblast growth factor in a tapeworm. Growth Factors 7: 327–334.

Gustafsson, M. K. S., D. Nässel & A. Kuusisto, 1993. Immunocytochemical evidence for the presence of substance P-like peptide in Diphyllobothrium dendriticum. Parasitology 106: 83-89.

Halton, D. W., I. Fairweather, C. Shaw & C. F. Johnston, 1990. Regulatory peptides in parasitic platyhelminths. Parasitology Today 6: 284–290.

Hori, I., 1991. Role of fixed parenchyma cells in blastema formation of the planarian Dugesia japonica. Int. J. Dev. Biol. 35: 101–108.

Hori, I., 1992. Localization of laminin in the subepidermal basal lamina of the planarian Dugesia japonica. Biol. Bull. 183: 78–83.

Joffe, B. I., 1991. On the number and spatial distribution of the catecholamine-containing (GA-positive) neurons in some higher and lower turbellarian – a comparison. Hydrobiologia 227 (Dev. Hydrobiol. 69): 201–208.

Lindroos, P. & M. J. Still, 1988. Extracellular matrix components in Polycelis nigra (Turbellaria, Tricladida). Fortschritte der Zoologie/Progress in Zoology 36: 157–162.

Meglitsch, P. A. & F. R. Schram, 1991. Invertebrate zoology (3rd ed.). Oxford University Press, N.Y., Oxford, 623 pp.

Morita, M., 1991. Phagocytic response of planarian reticular cells to heat-killed bacteria. Hydrobiologia 227 (Dev. Hydrobiol. 69): 193–199.

Palmberg, I., 1991. Differentiation during asexual reproduction and regeneration in a microturbellarian. Hydrobiologia 227 (Dev. Hydrobiol. 69): 1–10.

Pascolini, R., M. Camatini, R. Maci, A. Colombo & F. Panara, 1989. Immunoelectron microscopic localization of a fibronectin-like molecule in Dugesia lugubris s.l.. Tissue & Cell 21: 589–604.

Pedersen, K. J., 1991. Invited review: structure and composition of basement membranes and other basal matrix systems in selected Invertebrates. Acta zool. (Stockh.) 72: 181–201.

Raikova, O. I., 1991. Fine structural organisation in the nervous system and ciliary receptors of acoelan turbellarians. In D. A. Sakharov & W. Winlow (eds), Simpler nervous systems (Studies in Neuroscience 13). Manchester University Press, Manchester, N.Y.: 37–50.

Reuter, M., 1991. Are there differences between proseriates and lower flatworms in ultrastructure of the nervous system? Hydrobiologia 227 (Dev. Hydrobiol. 69): 221–227.

Reuter, M. & K. Eriksson, 1991. Catecholamines demonstrated by glyoxylic-acid-induced fluorescence and HPLC in some microturbellarians. Hydrobiologia 227 (Dev. Hydrobiol. 69): 209–219.

Reuter, M. & M. Gustafsson, 1989. 'Neuroendocrine cells' in flatworms – progenitors to metazoan neurons? Arch. Histol. Cytol. 52 Suppl.: 253–263.

Reuter, M. & A. Kuusisto, 1992. Growth factors in asexually reproducing Catenulida and Macrostomida (Plathelminthes)? Zoomorphology 112: 155–166.

Rieger, R., W. Salvenmoser, A. Legniti, S. Reindl, H. Adam, P. Simonsberger & S. Tyler, 1991a. Organization and differentiation of the body-wall musculature in Macrostomum (Turbellaria, Macrostomida). Hydrobiologia 227 (Dev. Hydrobiol. 69): 119–129.

Rieger, R. M., S. Tyler, J. P. S. Smith III & G. E. Rieger, 1991b. Platyhelminthes: Turbellaria. In F. W. Harrison & B. J. Bogitsh (eds), Microscopic anatomy of invertebrates, vol. 3 Platyhelminthes and Nemertinea. Wiley-Liss, N.Y.: 7–140.

Ruppert, E. E. & K. J. Carle, 1983. Morphology of metazoan circulatory systems. Zoomorphology 103: 193–208.

Smith III, J., S. Tyler, M. B. Thomas & R. M. Rieger, 1982. The morphology of turbellarian rhabdites: phylogenetic implications. Trans. Am. Microsc. Soc. 101: 209–228.

Sopott-Ehlers, B., 1984. Epidermale Collar-Receptoren der Nematoplanidae und Polystyliphoridae (Plathelminthes, Unguiphora). Zoomorphology 104: 226–230.

Sopott-Ehlers, B., 1991. Comparative morphology of photoreceptors in free-living plathelminths – a survey. Hydrobiologia 227 (Dev. Hydrobiol. 69): 231–239.

Tyler, S., 1976. Comparative ultrastructure of adhesive systems in the Turbellaria. Zoomorphologie 84: 1–76.

Hydrobiologia **305**: 27–35, 1995.
L.R.G. Cannon (ed.), *Biology of Turbellaria and some Related Flatworms.*

Aspects of the phylogeny of Platyhelminthes based on 18S ribosomal DNA and protonephridial ultrastructure

K. Rohde[1], A. M. Johnson[2], P. R. Baverstock[3,4] & N. A. Watson[1]
[1]*Department of Zoology, University of New England, Armidale, N.S.W. 2351, Australia*
[2]*Department of Microbiology, University of Technology, Sydney, Westbourne Street, Gore Hill, N.S.W., 2065, Australia*
[3]*School of Resource Science and Management, University of New England, Northern Rivers, Lismore N.S.W. 2480, Australia*
[4] *Current address: Faculty of Resource Science, Southern Cross University, Lismore, N.S.W. 2480, Australia*

Key words: Platyhelminthes, phylogeny, DNA, ultrastructure, protonephridia

Abstract

DNA studies of 23 taxa (20 platyhelminths, 1 nemertean, *Homo* and *Artemia*) and electron-microscopic studies of the protonephridia of many platyhelminths (supported by some additional ultrastructural data) have led to the following conclusions: the Neodermata are monophyletic; Temnocephalida and Dalyelliida form one clade and are not the 'primitive' sister group of the Neodermata; Gyrocotylidea, Amphilinidea and Eucestoda form one monophylum; Pterastericolidae and Umagillidae are dalyelliids and not the sister group of the Neodermata; and Proseriata are unlikely to be closely related with the Tricladida. A large taxon consisting of the Proseriata and some other 'turbellarians' may represent the sister group of the Neodermata.

Introduction

The Platyhelminthes have particular significance in the animal kingdom, as the likely sister group (jointly with the Gnathostomulida) of all the 'higher' metazoans (Ax, 1984). Their phylogeny has recently received much attention (Ax, 1984; Brooks *et al.*, 1985; Ehlers, 1985; Rohde, 1990; Baverstock *et al.*, 1991; Rohde *et al.*, 1993). However, authors do not agree on the relationships of many taxa. Thus, Ivanov (1991) does not accept the monophyly of the Neodermata (major groups of parasitic platyhelminths) suggested first by Ehlers (1985); some authors have suggested that different parasitic taxa have originated from different turbellarian taxa (Cannon, 1986); the ultrastructure of the protonephridia has been claimed as evidence that dalyelliid rhabdocoels are not the sister group of the Neodermata as usually assumed, for example by Ehlers (1985) (Rohde, 1990, 1991); a taxon Endoaxonemata, comprising the last common ancestor of the Neodermata, Pterastericolidae and some other 'rhabdocoel' families, has been proposed (Jondelius & Tholleson, 1993); and even the monophyly of the Platyhelminthes has been doubted (Smith *et al.*, 1986). Concerning some smaller taxa, the position of the Aspidogastrea is not clear (Blair, 1993), the Gyrocotylidea are sometimes considered to be close relatives of the cestodes and sometimes of the monogeneans (discussion in Rohde, 1990), and concerning the Monogenea, recent ultrastructural studies have shown that *Anoplodiscus* combines characters of typical Monogenea and other flatworms (Rohde *et al.*, 1992b).

In this paper, we try to answer the following questions:

1. Is the Neodermata monophyletic?
2. Are the Aspidogastrea and Digenea sister groups, or is the Aspidogastrea the sister group of all other Neodermata?
3. Do the Gyrocotylidea, Amphilinidea and Eucestoda form a monophylum or are the Gyrocotylidea aberrant Monogenea? (As suggested by Llewellyn, 1986.)
4. Is the Monogenea (Monopisthocotylea plus Polyopisthocotylea) paraphyletic? (A possibility sug-

gested by the findings of Baverstock *et al.*, 1991; Justine, 1991a, b; and Blair, 1993).

5. What is the taxonomic position of *Anoplodiscus*, a genus combining aberrant, 'turbellarian'-like spermiogenesis and aberrant, cestode-like protonephridia with monogenean characters (Watson & Rohde, 1992; Rohde *et al.*, 1992b)?
6. Is the Temnocephalidae the sister group of the Neodermata? (As suggested by Brooks *et al.*, 1985.)
7. Is the Dalyellioida the sister group of the Neodermata?
8. Are the Pterastericolidae and Umagillidae sister groups of the Neodermata? (See Cannon, 1986; Jondelius & Thollesson, 1993.)
9. Is the Proseriata the sister group of the Neodermata? (See Rohde, 1991.)
10. Do Lecithoepitheliata and Rhabdocoela form a monophylum? (As suggested by similar flame bulbs, see Rohde, 1991.)

Material and methods

In order to test some of the conflicting hypotheses, we obtained partial sequences (approximately 550 base pairs) of the 5′ region of the small subunit ribosomal RNA gene (18S) amplified by PCR of the enoplan nemertean *Prostoma* sp., and of 13 free-living and parasitic platyhelminths. Sequences of the same species and of a further seven platyhelminths obtained by reverse transcription of RNA were used to test the same hypotheses (Table 1). Sequences of the first 13 species were aligned with each other along with homologous regions of *Homo* and *Artemia* as outgroups, using the program CLUSTAL V (Higgins *et al.*, 1991) run on a Sun SPAR c Server computer. Sequences of the additional seven species were aligned with each other and with the other 16 taxa by eye, since data were not suitable for alignment with CLUSTAL. The data were analyzed using PAUP (Swofford, 1990) in three parts: 1) 16 taxa for all nucleotides, 2) 16 taxa for truncated data (hypervariable regions excluded), 3) 23 taxa for truncated data. (For sequences and details of methods see Rohde *et al.*, 1993).

Flame bulbs of protonephridia were reconstructed from ultrathin sections, many of them serial or semiserial (for details see Rohde *et al.*, 1992a, b; Watson *et al.*, 1993; Rohde & Watson, 1993a, b).

Table 1. Species used for the analysis. Species amplified with PCR indicated by an asterisk. Sources of segments from other authors in parentheses.

Species
Acanthobothrium heterodonti - Eucestoda Tetraphyllidea (Baverstock *et al.*, 1991)
Anoplodiscus cirrusspiralis * - Monogenea Monopisthocotylea
Artemia - Crustacea (Nelles *et al.*, 1984)
Artioposthia sp.* - Tricladida Terricola
Austramphilina elongata - Amphilinidea (Baverstock *et al.*, 1991)
Calicophoron calicophorum - Trematoda Digenea (Baverstock *et al.*, 1991)
Coelogynoporidae gen. sp.* - Proseriata
Diclidophora merlangi - Monogenea Polyopisthocotylea (Baverstock *et al.*, 1991)
Dictyocotyle coeliaca - Monogenea Monopisthocotylea (Baverstock *et al.*, 1991)
Dugesia tigrina - Tricladida Paludicola (Baverstock *et al.*, 1991)
Gyratrix sp.* - Rhabdocoela Kalyptorhynchia
Gyrocotyle rugosa - Gyrocotylidea (Baverstock *et al.*, 1991)
Homo - (McCallum & Maden, 1985)
Lobatostoma manteri - Trematoda Aspidogastrea (Blair, 1993)
Notoplana australis - Polycladida (Baverstock *et al.*, 1991)
Opisthorchis sp. - Trematoda Digenea (Korbsrisate *et al.*, 1991)
Prorhynchus sp.* - Lecithoepitheliata
Prostoma sp.* - Nemertini Enopla
Pterastericola australis * - Rhabdocoela Dalyelliida
Schistosoma mansoni - Trematoda Digenea (Ali *et al.*, 1991)
Stenostomum sp. * - Catenulida
Syndisyrinx punicea * - Rhabdocoela Dalyelliida
Temnocephala sp. * - Rhabdocoela Temnocephalida

Results and discussion

DNA

Fifty percent Majority-rule consensus of the 17 and 6 trees, respectively, within two steps of the most parsimonious ones, for 16 taxa all data (Fig. 1) and truncated data (Fig. 2) are illustrated. Both trees strongly support the monophyly of the Neodermata (Monogenea: *Anoplodiscus* and *Diclidophora*; Trematoda: *Opisthorchis*, *Lobatostoma* and *Schistosoma*) and the Trematoda within them, as well as that of the Dalyellioida (Dalyelliida: *Syndisyrinx* and *Pterastericola*; Temnocephalida: *Temnocephala*). There is no support for the monophyly of the Monogenea (Monopisthocotylea: *Anoplodiscus*; Polyopisthocotylea: *Diclidophora*), and for the Rhabdocoela (Kalyptorhynchia:

Gyratrix; Dalyellioida: *Syndisyrinx*, *Temnocephala*, *Pterastericola*). The triclad *Artioposthia* appears as the sister group of the Dalyellioida, the catenulid *Stenostomum* and the nemertean *Prostoma* form one clade, the lecithoepitheliate *Prorhynchus* appears as the sister group of all the other Platyhelminthes (excluding *Stenostomum*), and the proseriate (Coelogynoporidae gen.sp.) appears as the sister group of the Rhabdocoela plus Tricladida (Fig. 1) or forms a trichotomy with the Neodermata and Rhabdocoela plus Tricladida (Fig. 2). In the 50% Majority – rule consensus tree of the five equally most parsimonious trees (23 taxa) (Fig. 3) relationships are similar, but, in addition, monophyly of the Tricladida (Paludicola and Terricola), of the Monogenea Monopisthocotylea, and of a clade consisting of the Cestoda (Eucestoda, Amphilinidea and Gyrocotylidea) and Monogenea Polyopisthocotylea is supported. The polyclad appears to be closely linked with the nemertean and catenulid, also shown by the 50% Majority-rule consensus of the 1639 trees within two steps of the most parsimonious trees (not illustrated).

Bootstrapping data for 16 taxa (all nucleotides) 50 times gave strong support for the monophyly of the Neodermata (although the position of the monopisthocotylean *Anoplodiscus* was not resolved) and of the Dalyellioida. Bootstrapping data for 16 taxa (truncated data) led to a 50% Majority-rule consensus tree in which the nemertean *Prostoma* plus *Stenostomum* (70% of all trees) appeared as the sister group of *Prorhynchus* (60%), *Prorhynchus* as the sister group of all others (54%), and the Dalyellioida (53%) and the Trematoda plus *Diclidophora* (64%) appeared as monophyla, but the position of most taxa remained unresolved. Bootstrapping data for 23 taxa 50 times led to a 50% Majority-rule consensus tree in which the two triclads (100% of trees), the catenulid and nemertean (35%), all Rhabdocoela (52%), Dalyellioida (63%), the two monopisthocotylean monogeneans (96%), the trematodes *Opisthorchis* and *Calicophoron* (86%), the amphilinid *Austramphilina* and the eucestode *Acanthobothrium* (60%), and the cestodes (Eucestoda, Amphilinidea and Gyrocotylidea) (54%) appeared as monophyla.

In view of the very large number of possible trees for 16 and 23 taxa and since our aim was to test various specific hypotheses on the phylogeny of the Platyhelminthes, we compared most parsimonious trees with trees that were constrained in various ways, using PAUP on the three data sets (16 taxa all data, 16 taxa truncated data, 23 taxa truncated data).

This approach led to the following conclusions: Trematoda Digenea and Aspidogastrea are monophyletic (0 additional steps required for all three data sets, *i.e.* very strongly supported); Amphilinidea, Gyrocotylidea and Eucestoda are monophyletic (O additional steps required, tested only for 23 taxa); Aspidogastrea and Digenea are sister groups (1 extra step, 23 taxa only); Lecithoepitheliata and Catenulida are monophyletic (1 extra step, 23 taxa only); Aspidogastrea and all the other Neodermata are probably not sister groups, although this possibility cannot be entirely ruled out (4 extra steps, 23 taxa only); it is doubtful that Monogenea Polyopisthocotylea and Monopisthocotylea form one clade (3 extra steps, 23 taxa only); it is unlikely that the following pairs of taxa are sister groups: Gyrocotylidea and Monogenea Monopisthocotylea (7 extra steps, 23 taxa only); Gyrocotylidea and Monogenea (5 extra steps, 23 taxa only); Proseriata and Neodermata (15 and 9 extra steps, 16 taxa only); Pterastericolidae and Neodermata (19 and 25 extra steps, 16 taxa only); Neodermata and Rhabdocoela (8, 7 and 3 extra steps); Neodermata and Kalyptorhynchia (6, 6 and 4 extra steps); Temnocephalidae and Neodermata (12 extra steps, 23 taxa only). Gyrocotylidea and Monogenea may be monophyletic (3 extra steps, 23 taxa only). It is unlikely that the following taxa are monophyletic: Temnocephalida and Neodermata (10 extra steps, 23 taxa only), Lecithoepitheliata and Rhabdocoela (7 and 8 extra steps, 16 taxa only).

Transitions, usually, occur far more frequently than tranversions and transversions may therefore be more useful for establishing higher branching relationships. We analyzed data weighting transversions 10 times transitions. In such weighted analyses for 16 taxa all nucleotides, 16 taxa truncated data, 23 taxa truncated data, the Platyhelminthes did not appear as a monophylum, and the position of many taxa remained unresolved. However, Aspidogastrea plus Digenea, Dalyelliida (in some analyses all Rhabdocoela), all Neodermata (except for *Anoplodiscus* whose position was not resolved in some analyses) and the cestodes (*Austramphilina*, *Acanthobothrium* and *Gyrocotyle*) appeared as monophyla, supporting the conclusions reached by the unweighted analyses.

Flame bulbs

The types of flame bulbs found in the Platyhelminthes are illustrated in Fig. 4. Three main types can be dis-

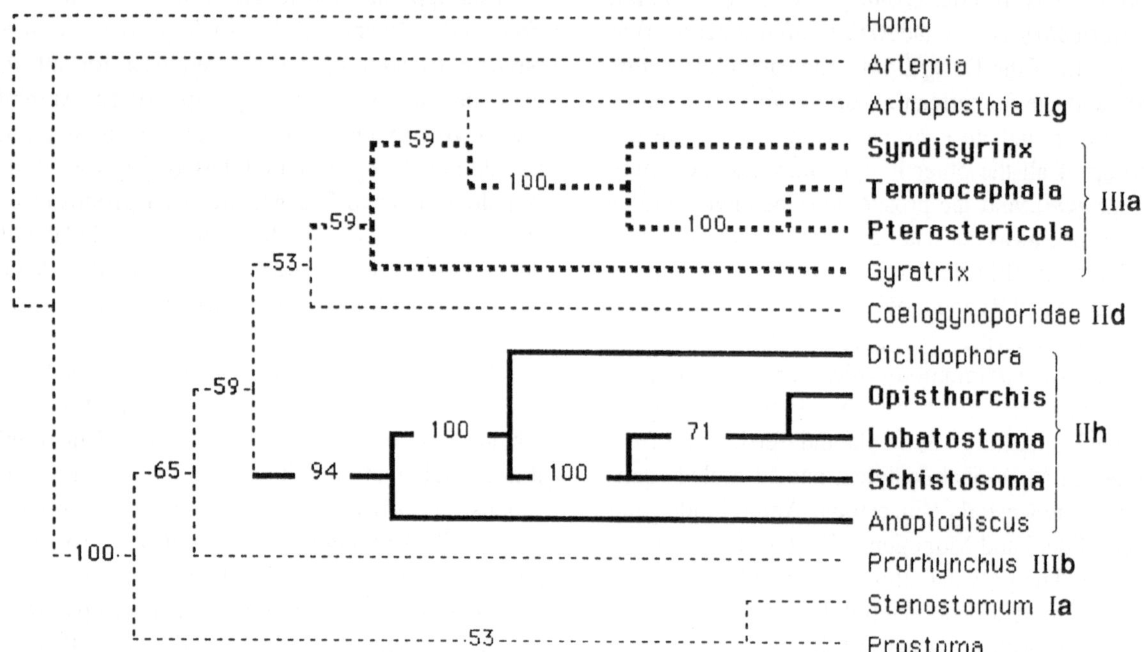

Fig. 1. 16 taxa, all data. 50% Majority – rule consensus of 17 trees with a length of ≤807 steps (most parsimonious trees 805 steps). Numbers indicate proportion (%) of trees supporting a clade. Bold unbroken lines: Neodermata (bold print: Trematoda Aspidogastrea (*Lobatostoma*) plus Digenea); bold broken lines: Rhabdocoela (bold print: Dalyellioida). Roman numerals and letters refer to types of flame bulb ultrastructure. Based on Rohde *et al.* (1993), with additional data on flame bulb ultrastructure.

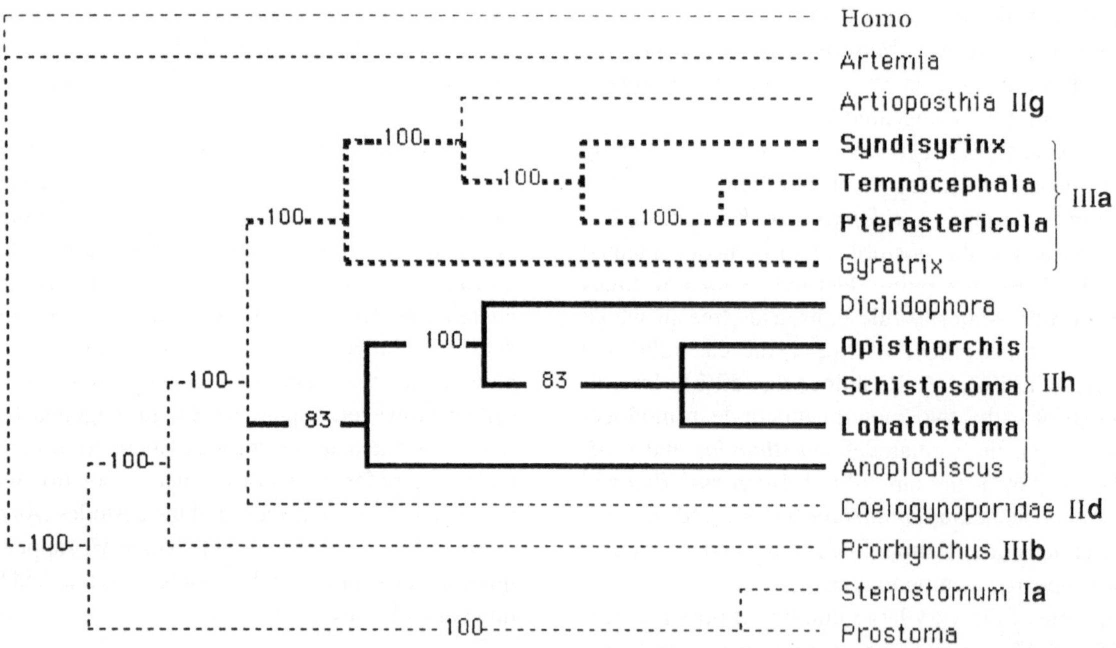

Fig. 2. 16 taxa, truncated data. 50% Majority – rule consensus of 6 trees with a length of <653 steps (most parsimonious trees 651 steps). Numbers indicate proportion (%) of trees supporting a clade. Bold unbroken lines: Neodermata (bold print: Trematoda Aspidogastrea (*Lobatostoma*) plus Digenea); bold broken lines: Rhabdocoela (bold print: Dalyellioida). Roman numerals and letters refer to types of flame bulbs (see Fig. 4). Based on Rohde *et al.* (1993), with additional data on flame bulb ultrastructure.

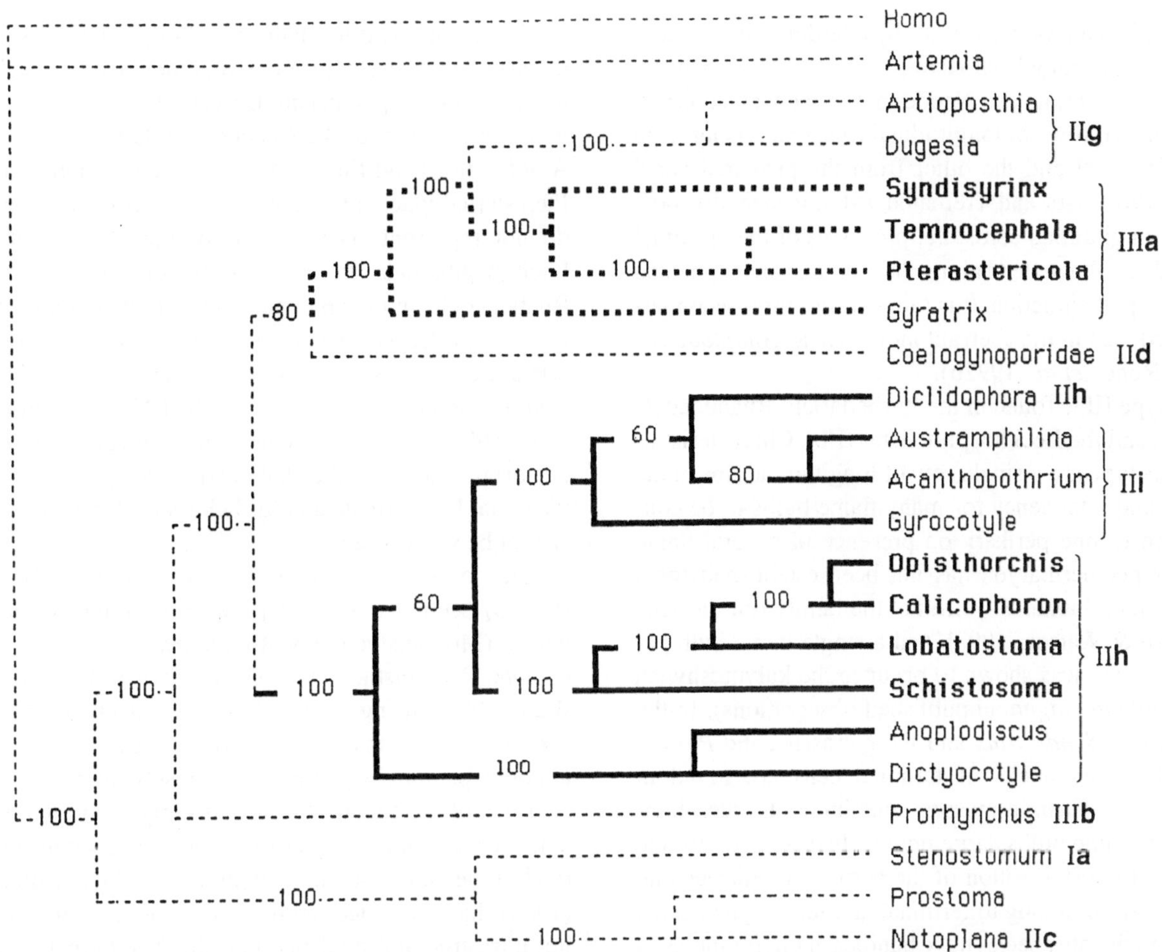

Fig. 3. 23 taxa, truncated data. 50% Majority – rule consensus of the 5 equally most parsimonious trees. Numbers indicate proportion (%) of trees supporting a clade. Bold unbroken lines: Neodermata (bold print: Trematoda Aspidogastrea (*Lobatostoma*) plus Digenea); bold broken lines: Rhabdocoela (bold print: Dalyellioida). Roman numerals and letters refer to types of flame bulbs (see Fig. 4). Based on Rohde *et al.* (1993), with additional data on flame bulb ultrastructure.

tinguished. Type I is restricted to the Catenulida. It has two cilia and the weir consists of longitudinal ribs in the genus *Retronectes*, and of interdigitating transverse ribs alternatingly coming from the 'left' and 'right' in *Catenula*, *Suomina* and *Stenostomum*, all formed by the terminal cell. In the last three genera, which have been examined most recently, there also are internal leptotriches (Rohde & Watson, 1993a, 1994) and not a second row of longitudinal ribs as suggested by earlier investigations (for references see Rohde & Watson, 1993a), and cilia have two cross-striated rootlets, one vertical and one bent 'forward', extending for some distance in the cytoplasmic columns along the weir. Canal cells do not contribute to the weir. Type II is found in most platyhelminth taxa and shows many modifications. The flame consists of many cilia, and, the proximal canal cell is close to the flame, but does not always contribute to the formation of the weir. For example, in the Macrostomida, the weir is formed by ribs arising from the terminal cell in *Microstomum*, and by two rows of ribs, one from the terminal and the other from the proximal canal cell, in *Macrostomum*. In a third macrostomid genus, *Paromalostomum*, weirs of three terminal cells, each consisting of a single row of ribs, face a common lumen (for references see Rohde, 1991). In Götte's larva of the polyclad *Stylochus*, only the terminal cell contributes to the weir (Watson *et al.*, 1992). In the Prolecithophora, *Urastoma* (possibly not a prolecithophoran) has a wall of the flame bulb consisting of rods coiled around each other, some continuous with the terminal and some with the proximal canal cell (Rohde *et al.*, 1990), and *Archimonotresis* has a weir with a single row of ribs, continuous with the terminal cell (Ehlers, 1989). In the

Tricladida the weir is formed by interdigitating branches of the terminal cell (Ishii, 1980; Rohde & Watson, 1992). Proseriata and Neodermata have weirs consisting of two rows of longitudinal ribs, one arising from the terminal and the other from the proximal canal cell. Proseriata and Trematoda/Monogenea also possess longitudinal cords, *i.e.* processes of the proximal canal cell that extend along the weir and are connected by a septate junction. It is (almost certainly) secondarily reduced in the aberrant monogenean *Anoplodiscus* (see Rohde *et al.*, 1992b).

Type III is found in the 'turbellarian' Rhabdocoela (IIIa) and the Lecithoepitheliata (IIIb). Characteristics of this type are a single row of longitudinal ribs in the weir and a tendency for many flame bulbs to be connected to one perikary on presence of several flame bulbs per perikaryon has not been established for a small meiofaunal dalyelliid of the family Luridae (see Rohde & Watson, 1993b and a single flame bulb per perikanyon was shown to occur in the kalyptoshynch *Baltoplana magna*, unpublished observations). In the umagillid *Syndisyrinx* and the pterastericolid *Pterastericola*, the nuclei of terminal cells are found in bulky cytoplasmic masses containing, besides bundles of flame bulbs, large liquid-filled lacunae (Rohde *et al.*, 1992a). Position of the perikarya indicates that they indeed belong to terminal and not to canal cells. Although both species have bundles of microtubules as found in other rhabdocoels, the weir is not well developed and, in *Syndisyrinx*, may even be rudimentary. Flame bulbs of the Lecithoepitheliata, like those of rhabdocoels, have a single row of longitudinal ribs, but the ribs lack microtubules and they have a characteristic, almost triangular shape in cross-section (Rohde & Watson, 1991).

Of particular interest, the fecampiid *Kronborgia isopodicola*, usually included in the Rhabdocoela, has a rudimentary weir formed by a terminal and a canal cell resembling that of the Neodermata (Watson *et al.*, 1992). For this reason, and because of the lack of any other synapomorphies shared with the rhabdocoels, it cannot be included in that group.

Conclusions

In conclusion, concerning evidence from DNA studies, we have found convincing evidence for the following: the Neodermata are monophyletic; the Temnocephalida and Dalyelliida form one clade; the Trematoda Aspidogastrea and Digenea form one clade (although we do not entirely rule out the possibility that the Aspidogastrea is the sister group of all other Neodermata); neither Temnocephalida nor Dalyelliida are the 'primitive' sister group of the Neodermata; Gyrocotylidea, Amphilinidea and Eucestoda form one monophylum; Pterastericolidae and Umagillidae are dalyelliids and do not represent sister groups of the Neodermata; Lecithoepitheliata are not closely related with the Rhabdocoela, although unpublished DNA findings by us, using a larger data set, suggest otherwise; Proseriata are unlikely to be closely related with the Tricladida. (Recent studies using 18S rRNA by Riutort *et al.* (1993) led to the conclusion that trematodes have not originated from the rhabdocoels, supporting our view that Neodermata and 'turbellarian' Rhabdocoela cannot be sister groups.)

The position of *Diclidophora* in our analysis (Fig. 3) seems to suggest polyphyly of the Monogenea (Monopistocotylea *Anoplodiscus* and *Dictyocotyle*, Polyopisthocotylea *Diclidophora*). However, Blair (1993), using longer DNA sequences of fewer taxa, has shown that *Dictyocotyle* and *Diclidophora*, although paraphyletic, are indeed closely related.

Most of the conclusions are strongly supported by evidence from ultrastructural studies of protonephridia (and some other organs and tissues). Thus, Ehlers (1985) has provided convincing evidence including ultrastructural evidence that the Neodermata are monophyletic. Temnocephalida, Pterastericolidae and Umagillidae have type IIIa flame bulbs (a synapomorphy of the Rhabdocoela) and, since such flame bulbs are unlikely to be plesiomorphic (Rohde, 1991) cannot be the 'primitive' sister group of the Neodermata. Ultrastructure of spermiogenesis of *Pterastericola* also differs fundamentally from that of Neodermata (Watson *et al.*, 1993). Aspidogastrea and Digenea share a type of flame bulb (type IIh) plesiomorphic for the Neodermata which, although not sufficient evidence for a monophyletic relationship, does at least not contradict it. Gyrocotylidea, Amphilinidea and Eucestoda have type IIi flame bulbs, a clear synapomorphy of the cestodes. Monogenea Polyopisthocotylea and Monopisthocotylea share type IIh flame bulbs (probably plesiomorphic for the Neodermata and therefore insufficient to characterize a monophylum), but both also have a unique larva, the oncomiracidium, and two pairs of eyes, suggesting that they are monophyletic. Proseriata and Tricladida have different flame bulbs (types IId and IIg, respectively). Ultrastructural and DNA evidence, therefore, does not support monophyly of a taxon Seriata, consisting of the Tricladida and

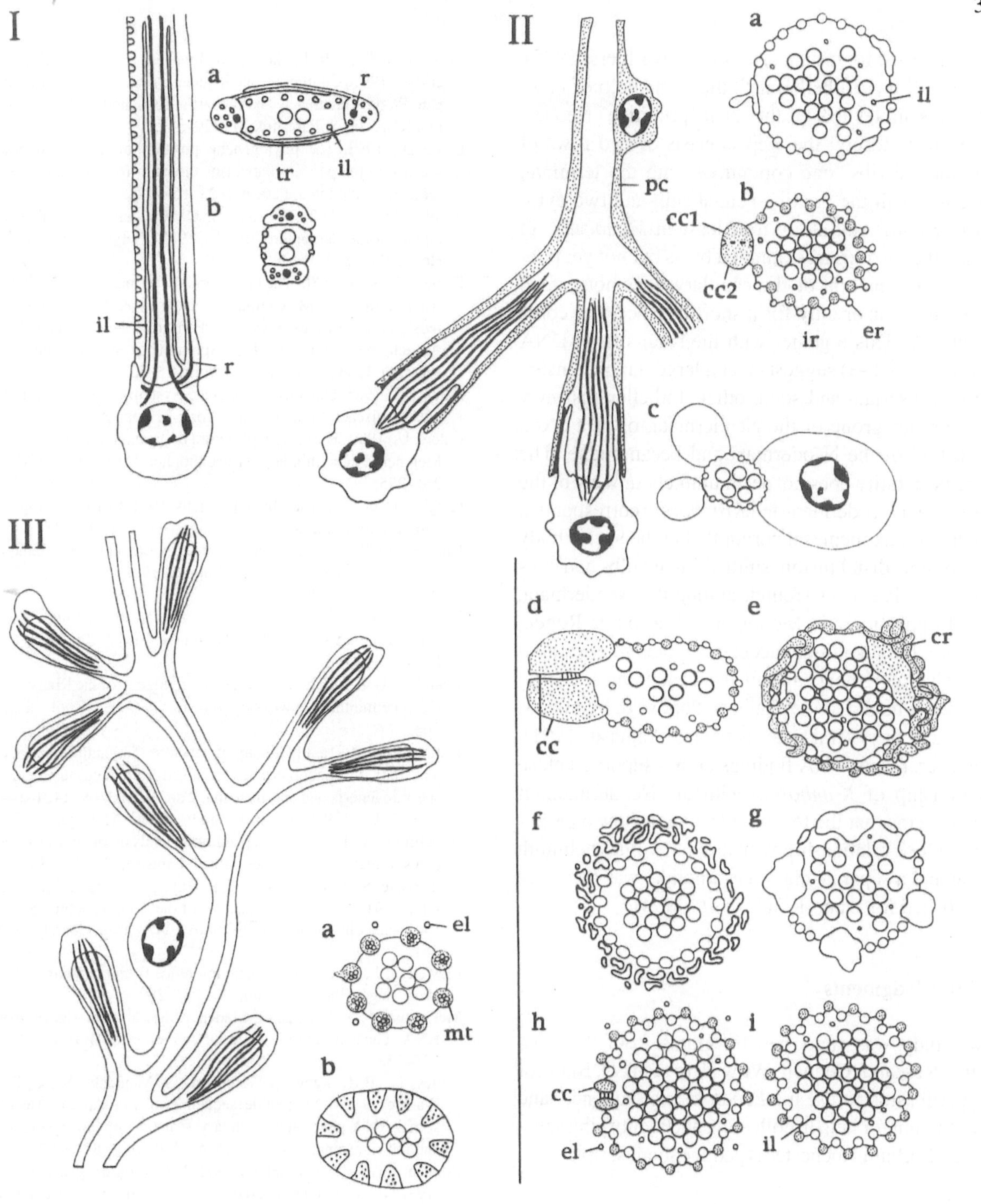

Fig. 4. Diagrams of a representative longitudinal section of each of the three main types of flame bulb (terminal part of the protonephridial system), and of cross-sections of various kinds within each main type. I. Flame bulb type I. Catenulida: *Catenula* and *Stenostomum* (Ia), *Retronectes* (Ib). II. Flame bulb type II, Macrostomida: *Microstomum* (IIa), *Macrostomum* (IIb), Polycladida: Götte's larva of *Stylochus* (IIc), Proseriata: *Monocelis* (IId), Prolecithophora: *Urastoma* (IIe), *Archimonotresis* (IIf), Tricladida (IIg), Neodermata: Trematoda/Monogenea (IIh), Gyrocotylidae/Amphilinidea/Eucestoda (IIi). III. Flame bulb type III. 'Turbellarian' Rhabdocoela (IIIa), Lecithoepitheliata (IIIb). Based on Rohde (1991) with additional data on Catenulida, Polycladida, and Proseriata (modified from Rohde, 1994). cc – cytoplasmic cord, cr – coiled rod, cc_1 – 1st canal cell, cc_2 – 2nd canal cell, el – external leptotrich, er – external rib, il – internal leptotrich, ir – internal rib, mt – microtubule, pc – proximal canal cell, r – rootlet of cilium, tr - transverse rib.

Proseriata, as, for instance, proposed by Ehlers (1985). Among all the 'turbellarians', the flame bulb of Proseriata resembles that of the most 'primitive' Neodermata most closely: the weir consists of two rows of longitudinal ribs, one continuous with the terminal, the other with the proximal canal cell, and two cytoplasmic cords run along it at least in *Monocelis*; in some other Proseriata examined by us but not yet published, the junction of the capillary does not extend along the weir or only for a short distance, *i.e.* cords are absent. This together with the results from DNA studies (Figs 1–3) suggests that a larger taxon consisting of Proseriata and some other 'turbellarians' may be the sister group of the Neodermata, or of a taxon consisting of the Neodermata and Fecampiidae. The latter possibility does not seem unlikely in view of the finding that Neodermata and *Kronborgia* correspond in having two axonemes incorporated in the sperm body by proximo-distal fusion, similar flame bulbs and similar 'reflective' eyes (found, among the Neodermata, only in polystomatid Monogenea, Watson & Rohde, 1993a,b, but also in the acoelan *Convoluta convoluta* which has a pigment cup cell also containing a 'central vacuole' with reflective platelets, Popova & Mamkaev, 1985, and in several other acoelans, Yamasu, 1991). However, recent DNA findings do not support a close relationship of *Kronborgia* with the Neodermata, it rather seems that the fecampiids do not show a particularly close relationship with any other platyhelminth taxon and a separate class, Fecampiida, was therefore established for them (Rohde *et al.*, 1994).

Acknowledgments

This study was supported by grants from the Australian Research Council. We wish to thank D. Blair for his permission to use his *Lobatostoma* sequence, and S. Dittmann for help in collecting and identifying specimens. Barbara Rochester typed the manuscript.

References

Ali, P. O., A. J. Simpson, R. Allen, A. P. Waters, C. J. Humphries, D. A. Johnston & D. Rollinson, 1991. Sequence of a small subunit rRNA gene of *Schistosoma mansoni* and its use in phylogenetic analysis. Mol. Biochem. Parasitol. 46: 201–208.

Ax, P., 1984. Das phylogenetische System. Systematisierung der lebenden Natur aufgrund ihrer Phylogenese. Gustav Fischer Verlag, Stuttgart, New York, 349 pp.

Baverstock, P. R., R. Fielke, A. M. Johnson, R. A. Bray & I. Beveridge, 1991. Conflicting phylogenetic hypotheses for the parasitic Platyhelminthes tested by partial sequencing of 18S ribosomal RNA. Int. J. Parasitol. 21: 329–339.

Blair, D., 1993. The phylogenetic position of the Aspid obothsea within the parasite flatworms inferred from ribosomal RNA sequence data. Int. J. Parasitol. 23: 169–178.

Brooks, D. R., R. T. O'Grady & D. R. Glen, 1985. The phylogeny of the Cercomeria Brooks, 1982 (Platyhelminthes). Proc. Helminthol. Soc. Wash. 52: 1–20.

Cannon, L. R. G., 1986. The Pterastericolidae: parasitic turbellarians from starfish. In M. Cremin, C. Dobson & D. E. Moorhouse (eds), Parasite Lives. Papers on Parasites, their Hosts and their Associations to honour J. F. A. Sprent. University of Queensland Press, St. Lucia, London, New York: 15–32.

Ehlers, U., 1985. Das Phylogenetische System der Plathlelminthes. Gustav Fischer, Stuttgart, New York, 317 pp.

Ehlers, U., 1989. The protonephridium of *Archimonotresis limophila* Meixner (Plathelminthes, Prolecithophora). Microfauna Mar. 5: 261–275.

Higgins, D. G., A. J. Bleasby & R. Fuchs, 1991. Clustal V: improved software for multiple sequence alignment. Cabios 5: 151.

Ishii, S., 1980. The ultrastructure of the protonephridial flame cell of the freshwater planarian *Bdellocephala brunnea*. Cell Tissue Res. 206: 441–449.

Ivanov, A. V., 1991. O sovremennoi sisteme Plathelminthes. (The state-of-the-art system of Plathelminthes) Zool. Zh. 70: 5–19 (In Russian).

Jondelius, U. & M. Thollesson, 1993. Phylogeny of the Rhabdocoela (Platyhelminthes): a working hypothesis. Can. J. Zool. 71: 298–308.

Justine, J.-L., 1991a. Phylogeny of parasitic Platyhelminthes: a critical study of synapomorphies proposed from the ultrastructure of spermiogenesis and spermatozoa. Can. J. Zool. 69: 1421–1440.

Justine, J.-L., 1991b. Cladistic study in the Monogenea (Platyhelminthes), based upon a parsimony analysis of spermiogenetic and spermatozoal characters. Int. J. Parasitol. 21: 821–838.

Korbsrisate, S., S. Mongkolsuk, J. R. Haynes, D. England & S. Sirisinha, 1991. Nucleotide sequences of the small subunit ribosomal RNA-encoding gene from *Opisthorchis viverrini*. Gene 105: 259–261.

Llewellyn, J., 1986. Phylogenetic inference from platyhelminth life-cycle stages. Int. J. Parasitol. 17: 281–289.

McCallum, F. S. & B. E. H. Maden, 1985. Human 18s ribosomal RNA sequence inferred from DNA sequence. Biochem. J. 232: 725–733.

Nelles, L., B.-L. Fang, G. Volckaert, A. Vandenberghe & R. De Wachter, 1984. Nucleotide sequence of a crustacean 18s ribosomal RNA gene and secondary structure of eukaryotic small subunit ribosomal RNAs. Nucl. Acid. Res. 12: 8749–8768.

Popova, N. V. & Y. V. Mamkaev, 1985. Ultrastruktura glas *Convoluta convoluta* (Turbellaria, Acoela) i ich primitivnie osobennosti. (Ultrastructure of the eyes of *Convoluta convoluta* (Turbellaria, Acoela) and its primitive features). Dokl. Akad. Nauk SSSR 283: 756–759 (In Russian).

Riutort, M., K. G. Field, R. A. Raff & J. Bagũna, 1993. 18S rRNA sequences and phylogeny of Platyhelminthes. Biochem. Syst. Ecol. 21: 71–77.

Rohde, K., 1990. Phylogeny of Platyhelminthes, with special reference to parasitic groups. Int. J. Parasitol. 20: 979–1007.

Rohde, K., 1991. The evolution of protonephridia of the Platyhelminthes. Hydrobiologia 227 (Dev. Hydrobiol. 69): 315–321.

Rohde, K., 1994. The minor groups of parasitic Platyhelminthes. Adv. Parasitol. 33: 145–234.

Rohde, K. & N. Watson, 1991. Ultrastructure of the flame bulbs and protonephridial capillaries of *Prorhynchus* (Lecithoepitheliata, Prorhynchidae, Turbellaria). Zool. Scr. 20: 99–106.

Rohde, K. & N. A. Watson, 1992. Ultrastructure of flame bulbs and protonephridial capillaries of *Artioposthia* sp. (Platyhelminthes, Tricladida, Geoplanidae). Acta zool. (Stockholm) 73: 231–236.

Rohde, K & N. A. Watson, 1993a. Ultrastructure of the protonephridial system of regenerating *Stenostomum* sp. (Platyhelminthes, Catenulida). Zoomorphology 113: 61–67.

Rohde, K. & N. A. Watson, 1993b. Ultrastructure of epidermis and protonephridium of an undescribed species of Luridae (Platyhelminthes, Rhabdocoela). Aust. J. Zool. 41: 415–421.

Rohde, K. & N. A. Watson, 1994. Ultrastructure of the protonephridial system of *Suomina* sp. and *Catenula* sp. (Platyhelminthes, Catenulida). J. Submicrosc. Cytol. Pathol.

Rohde, K., N. A. Watson & U. Jondelius, 1992a. Ultrastructure of the protonephridia of *Syndisyrinx punicea* (Hickman, 1956) (Rhadocoela, Umagillidae) and *Pterastericola pellucida* (Jondelius, 1989) (Rhabdocoela, Pterastericolidae). Aust. J. Zool. 40: 385–399.

Rohde, K., N. A. Watson & F. A. Roubal, 1992b. Ultrastructure of the protonephridial system of *Anoplodiscus cirrusspiralis* (Monogenea, Monopisthocotylea). Int. J. Parasitol. 22: 443–457.

Rohde, K., N. Noury-Sraîri, N. Watson, J.-L. Justine & L. Euzet, 1990. Ultrastructure of the flame bulbs of *Urastoma cyprinae* (Platyhelminthes, 'Prolecithophora', Urastomidae). Acta zool. (Stockholm) 71: 211–216.

Rohde, K., C. Hefford, J. Ellis, A. M. Johnson, P. R. Baverstock, N. A. Watson & S. Dittmann, 1993. Contributions to the phylogeny of Platyhelminthes based on partial sequencing of 18S ribosomal DNA. Int. J. Parasitol. 23: 705–721.

Rohde, K., K. Luton, P. R. Baverstock & A. M. Johnson, 1994. The phylogenetic relationships of *Kronborgia* (Platyhelminthes, Fecampiida) based on comparison of 18S ribosomal DNA sequences. Int. J. Parasitol. 24: 651–669.

Smith, J. P. S., III, S. Tyler & R. M. Rieger, 1986. Is the Turbellaria polyphyletic? Hydrobiologia 132 (Dev. Hydrobiol. 32): 13–21.

Swofford, D. L., 1990. PAUP: Phylogenetic Analysis Using Parsimony, version 3.0. Computer program distributed by the Illinois Natural History Survey, Champaign, Illinois, 162 pp.

Watson, N. A. & K. Rohde, 1992. Ultrastructure of sperm and spermatogenesis of *Anoplodiscus cirrusspiralis* (Monogenea, Monopisthocotylea). Ann. Parasitol. Hum. Comp. 67: 131–140.

Watson, N. A. & K. Rohde, 1993a. Ultrastructural evidence for an adelphotaxon (sister group) to the Neodermata (Platyhelminthes). Int. J. Parasitol. 23: 285–289.

Watson, N. A. & K. Rohde, 1993b. Ultrastructure of sperm and spermiogenesis of *Kronborgia isopodicola* (Platyhelminthes, Fecampiidae). Int. J. Parasitol. 23: 737–744.

Watson, N. A., K. Rohde & U. Jondelius, 1993. Ultrastructure of sperm and spermiogenesis of *Pterastericola astropectinis* (Platyhelminthes, Rhabdocoela, Pterastericolidae). Parasitol. Res. 79: 322–328.

Watson, N., K. Rohde & J. B. Williams, 1992. Ultrastructure of the protonephridial system of larval *Kronborgia isopodicola* (Platyhelminthes, Fecampiidae). J. Submicrosc. Cytol. Pathol. 24: 43–49.

Yamasu, T., 1991. Fine structure and function of ocelli and sagittocysts of acoel flatworms. Hydrobiologia 227 (Dev. Hydrobiol. 69): 273–282.

Hydrobiologia **305**: 37–43, 1995.
L.R.G. Cannon (ed.), *Biology of Turbellaria and some Related Flatworms.*

5S rRNA sequences of 12 species of flatworms: implications for the phylogeny of the Platyhelminthes

B. I. Joffe[1], K. M. Valiejo Roman[2], V. Ya. Birstein[3] & A. V. Troitsky[2]
[1] *Zoological Institute RAN, St-Petersburg 199034, Russia*
[2] *A. N. Belozersky Institute of Physico-Chemical Biology, Moscow State University, Moscow 119899, Russia*
[3] *N. K. Koltzov Institute of Developmental Biology RAN, Moscow 117334, Russia*

Key words: 5S rRNA, Platyhelminthes, Turbellaria, parasitic flatworms, phylogeny

Abstract

5S rRNAs from 12 species of free living and parasitic platyhelminthes were sequenced. In the phylogenetic analysis, attention was focused on the statistical estimates of the trees corresponding to existing phylogenetic hypotheses. The available 5S rRNA data agree well with widely accepted views on the relationships between the Acoela, Polycladida, Tricladida, and Neorhabdocoela; our analysis of the published 18S rRNA sequences also demonstrated good correspondence between these views and molecular data. With available 5S rRNA data the hypothesis that the dalyellioid turbellarians is the sister group of the Neodermata is less convincing than the hypotheses proposing the Neodermata as the sister group of the Neorhabdocoela, or of the Seriata, or of the branch uniting them. A relatively low rate of base replacement in parasitic flatworms, probably, accounts for the uncertain position of the Neodermata, while a relatively high rate in planarians may explain a relatively too early divergence of the Tricladida in several published phylogenetic trees constructed from various rRNA data.

Introduction

The phylogeny of the Platyhelminthes is a promising field for application of molecular methods because, while some problems have already found reasonable solutions and may serve as a 'control' (see Ehlers, 1985; Rieger *et al.*, 1990; Rohde, 1990), many others remain unsolved. By now, portions of 18S rRNA have been sequenced from various platyhelminthes (Baverstock *et al.*, 1990; Ruitort *et al.*, 1992 a,b). Yet, additional sequence data are still necessary, in particular, from other molecules. With the 12 new sequences presented here, a total of 14 sequences of 5S rRNA may now contribute to this goal of elucidating platyhelminth phylogeny.

The shortness of the 5S rRNA limits its phylogenetic usage (Hendriks *et al.*, 1986; Hori & Osava, 1987). Therefore, we focused our attention not on inferring phylogenetic trees, but on the statistical estimation of the existing phylogenetic hypothesis, *i.e.*, on their compatibility with 5S rRNA sequence data. This aspect of the problem is important irrespective of the length of the analyzed sequences. Indeed, a tree computed from molecular data is a statistical estimate, and one must know whether there are statistically significant reasons to prefer it (at least, under the assumptions of the used tree constructing method) to some other topology, e.g. one supported by morphological evidence.

Material and methods

The species studied, their systematic positions and collection sites are listed in Table 1. The total RNA was isolated by the hot phenol extraction procedure with 4M guanidinium isothiocyanate; 5S rRNA was purified by electrophoresis on 8% polyacrylamide gel and sequenced by the Peattie's chemical method after 3′-end labelling with (5′-32P)pCp as described earlier (see Troitsky *et al.*, 1991 for the protocol used).

The sequences used for rooting the platyhelminth tree were taken from the 'Berlin RNA databank' (Specht *et al.*, 1990) with the exception of 5S rRNA of *Echinorhynchus gadi* which was sequenced by

Table 1. Systematic positions and sites of collection of the platyhelminth species used in this study

'Turbellaria'	
Archoophora	
Acoela	
Convoluta convoluta (Abildgard, 1806)	(1)
Polycladida	
Notoplana humulis (Stimpson, 1857)	(2)
**Planocera reticulata* (Stimpson, 1855)	
Neoophora	
Tricladida	
Fam. Dugesiidae	
**Dugesia (Dugesia) japonica* (Ichikava et Kawakatsu, 1964)	
Dugesia (Girardia) tigrina (Girard, 1850)	(3)
Fam. Planariidae	
Planaria torva O. F. Muller, 1773	(4)
Fam. Dendrocoelidae	
Dendrocoeluim lacteum (O. F. Muller, 1773)	(4)
Neorhabdocoela	
Typhloplanoida	
Bothromesostoma esseni (Braun, 1885)	(5)
Kalyptorhynchia	
Macrorhynchus crocea (O. Fabricius, 1826)	(1)
Dalyellioida	
Graffilla graffi Mitin, 1970	(6)
Pseudograffilla arenicola (Meixner, 1938)	(7)
Monogenea	
Discocotyle sagittata (Leuckart, 1842)	(8)
Cestoda	
Khavia sinensis Hsu, 1935	(9)
Trematoda	
Fasciola hepatica Linne, 1758	(10)

* Species sequenced by other authors. For the species sequenced by us, the numbers in parentheses indicate the collection sites, as well as the hosts for the parasitic species. (1) Dalnezelenetskaya Inlet, Barentz Sea; (2) Biological station 'Vostok', Sea of Japan; (3) laboratory stock; (4) a pond near St-Petersburg; (5) a pool in Borok (Yaroslavl District, Russia); (6) from the marine gastropod, *Neptunea despecta*, Biological station 'Kartesh', White Sea; (7) Biological station 'Kartesh', White Sea; (8) from the fish *Coregonus albula*, Ladoga lake; (9) from the fish *Cyprinus carpio*, fish-farm in Estonia; (10) from cows from a slaughter house.

us and will be published elsewhere. All the platyhelminth sequences wholly fitted to the alignment universally used for 5S rRNA (see Specht *et al.*, 1990). 3′-terminal uracil residues were not used in the analysis because variation in number of 3′-terminal uracils (2 or 3) was observed in practically all studied species. Phylogenetic analysis was done using PHYLIP 3.3 package (Felsenstein, 1990). Maximum likelihood (ML) method was always implemented with empirical estimation of base frequencies and a transition/transversion ratio = 2. Distances were calculated using the 2-parameter Kimura's model, also with a transition/transversion ratio = 2. Lower probability for fixation of transversions is suggested by a very high level of conservation of the secondary structure of rRNAs, which is always affected by transversions, while the influence of transitions is moderated by noncanonical pairing (GU). The value of 2 (rather than, e.g., 1.5 or 3) was empirically chosen based on a series of preliminary experiments with various rRNA data sets. Computation of the ML trees was repeated three times with different species input order (one starting with the root sequence and two random ones) for each species set with global rearrangements. Being mainly interested in the statistical estimates of given tree topologies, we omited bootstrapping; moreover, with the ML method, branch lengths which are not significantly positive themselves indicate the places where changes in the tree are most probable (Felsenstein, 1990).

Results and discussion

Twelve 5S rRNA sequences of the species studied are presented in Fig. 1 together with the sequences from two species studied earlier (Ohama *et al.*, 1986; Hori *et al.*, 1988).

Approach to the phylogenetic analysis

The first question we had to answer was whether the tree(s) constructed from 5S rRNA data reproduced the relations between the main groups of turbellarians represented in our material (Fig. 2A) which had been convincingly established from morphological evidence (Ehlers, 1985; Rieger *et al.*, 1990; yet, see Rohde, 1990 and present volume for different views — these views are not discussed below because no tree topology suitable for statistic estimation has been published).

In view of the fact that changes in the set of species are the most important test for the consistency of the tree topology (see, e.g. Wainwright *et al.*, 1993), we constructed the trees for several sets of species. Each of them included all the 14 platyhelminth sequences and differed from the others in the single outgroup (rooting) sequence. This explicitly showed the effect of each outgroup on the tree topology and provided a

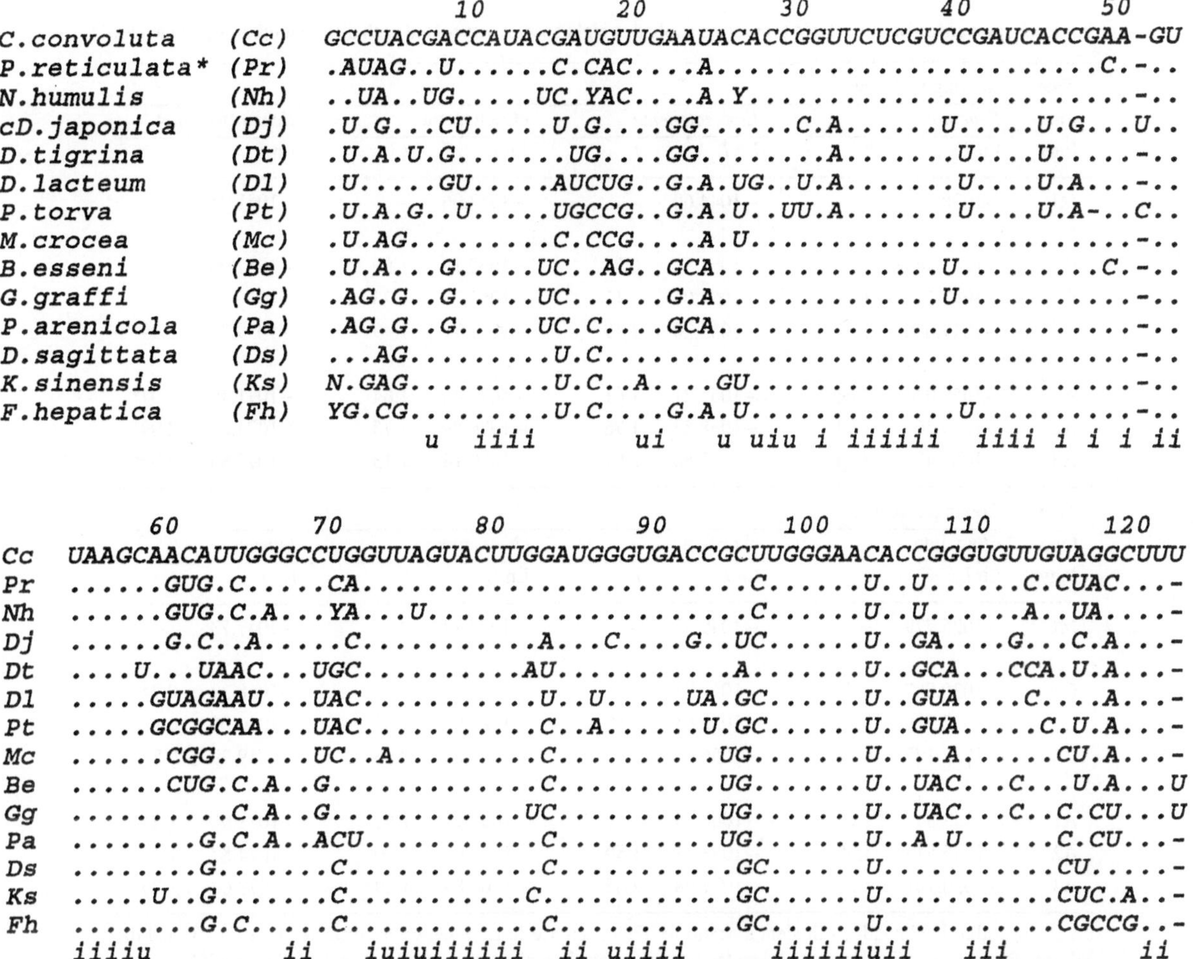

10 20 30 40 50
C.convoluta (Cc) GCCUACGACCAUACGAUGUUGAAUACACCGGUUCUCGUCCGAUCACCGAA-GU
P.reticulata* (Pr) .AUAG..U......C.CAC....A.........................C.-..
N.humulis (Nh) ..UA..UG....UC.YAC....A.Y.........................-..
cD.japonica (Dj) .U.G...CU.....U.G...GG......C.A......U.....U.G...U..
D.tigrina (Dt) .U.A.U.G.......UG...GG........A.......U....U.....-..
D.lacteum (Dl) .U....GU....AUCUG..G.A.UG..U.A.......U...U.A...-..
P.torva (Pt) .U.A.G..U....UGCCG..G.A.U..UU.A.......U...U.A-..C..
M.crocea (Mc) .U.AG.........C.CCG...A.U......................-..
B.esseni (Be) .U.A...G....UC..AG..GCA..............U.........C.-..
G.graffi (Gg) .AG.G..G....UC.....G.A..............U...........-..
P.arenicola (Pa) .AG.G..G....UC.C...GCA...........................-..
D.sagittata (Ds) ...AG.........U.C................................-..
K.sinensis (Ks) N.GAG.........U.C..A...GU.........................-..
F.hepatica (Fh) YG.CG.........U.C...G.A.U..............U..........-..
u iiii ui u uiu i iiiiii iiii i i i ii
60 70 80 90 100 110 120
Cc UAAGCAACAUUGGGCCUGGUUAGUACUUGGAUGGGUGACCGCUUGGGAACACCGGGUGUUGUAGGCUUU
Pr ......GUG.C.....CA.......................C......U..U......C.CUAC...-
Nh ......GUG.C.A...YA...U...................C......U..U......A..UA....-
Dj ......G.C..A.....C..........A...C....G..UC......U..GA....G....C.A...-
Dt ....U...UAAC...UGC..........AU...........A......U..GCA...CCA.U.A...-
Dl .....GUAGAAU...UAC...........U..U.....UA.GC......U..GUA....C....A...-
Pt .....GCGGCAA...UAC...........C..A......U.GC......U..GUA.....C.U.A...-
Mc ......CGG......UC..A.........C..........UG.......U....A......CU.A...-
Be ......CUG.C.A..G.............C..........UG.......U..UAC...C...U.A...U
Gg ..........C.A..G............UC..........UG.......U..UAC...C..C.CU...U
Pa ........G.C.A..ACU...........C..........UG.......U..A.U......C.CU...-
Ds ........G.......C.............C...........GC......U............CU....-
Ks .....U..G.......C..........C.............GC......U............CUC.A..-
Fh ........G.C.....C.............C...........GC......U............CGCCG..-
iiiiu ii iuiuiiiiii ii uiiii iiiiiiuii iii ii

Fig. 1. 5S rRNA sequences known from the platyhelminths. The sequence of *P. reticulata* (*) was published earlier; the majority consensus sequence for *D. japonica* is compiled from three sequences known from three populations (Ohama *et al.*, 1986; Hori *et al.*, 1988). Insertion in *P. torva* is proved by presence of a minor rRNA fraction which was partially sequenced and has in the same position an insertion 3 nucleotides long. Dots indicate the nucleotides identical to those in *C. convoluta*. i, invariant position; u, unique state.

severe test for its consistency, especially, in the root region.

To provide maximal homogeneity of the set of sequences in the analysis (see, e.g., Felsenstein, 1990), we used for rooting all the sequences known from the animals generally considered more closely related to the Platyhelminthes, that is *Brachionus plicatilis* (Rotatoria), *Echinorhynchus gadi* (Acanthocephala), *Emplectonema gracile*, and *Lineus geniculatus* (Nemertini). In addition, the majority consensus sequences for four more distant lower invertebrate groups were composed and used: cPorifera (3 species), cCoelenterata (6), cNematoda (3), and cAnnelida (3). The rooting sequences were rather diverse (with distances between them of 0.071–0.348 K_{nuc}), but did not seem to markedly increase the heterogeneity of the analyzed data sets — at least, the distances between the outgroup and platyhelminth sequences (0.090–0.597 K_{nuc}) varied within the same range as the distances between the platyhelminth sequences themselves (0.093–0.625 K_{nuc}).

Phylogenetic trees

Figure 2 shows that six of the eight used rooting sequences inferred maximum likelihood (ML) trees (Figs 2B,C) which correspond well to recent views on the phylogeny of the Turbellaria (Fig. 2A) and differ only in the position of the parasitic flatworms. The seventh tree was very similar, though *Echinorhynchus* 'branched off together with polyclads' instead of rooting the platyhelminth tree (Fig. 2B, arrowhead). The tree rooted with *Lineus* was very different (Fig. 2D), though this sequence in itself did not seem peculiar

Table 2. Comparison of several tree topologies applied to eight species sets rooted with different outgroup sequences

Tree topology	Rooting sequence							
	Brachionus		*Emplectonema*		cCoelenterata		*Echinorhynchus*	
	Ln *L*	*t*	Ln *L*	*t*	Ln *L*	*t*	Ln *L*	*t*
ML	−1038.459	–	−1044.60	–	−1053.06	–	−1041.01	–
C	−1045.945	1.06	−1051.56	0.96	−1060.18	1.00	−1057.90	1.68
D	−1047.508	1.49	−1052.96	1.17	−1061.75	1.23	−1059.54	1.91
B	−1047.020	1.37	−1053.20	1.49	−1061.71	1.49	−1059.82	1.99*
A1	−1059.044	1.66	−1065.21	1.61	−1073.40	1.64	−1071.49	2.07*
A2	−1064.366	1.94	−1068.82	1.76	−1076.89	1.75	−1074.84	2.09*
A3	−1067.434	2.13*	−1071.52	2.09*	−1080.14	2.13*	−1077.85	2.37*
Tree topology	Rooting sequence							
	cAnnelida		cPorifera		cNematoda		*Lineus*	
	Ln *L*	*t*	Ln *L*	*t*	Ln *L*	*t*	Ln *L*	*t*
ML	−1026.779	–	−1049.85	–	−1078.22	–	−1047.73	–
C	−1030.887	0.53	−1054.75	0.63	−1083.03	0.62	−1058.37	0.74
D	−1032.716	0.73	−1056.61	1.02	−1085.00	1.03	−1060.05	0.87
B	−1032.177	0.89	−1056.66	0.95	−1084.80	0.93	−1059.99	0.85
A1	−1042.349	1.21	−1068.74	1.41	−1097.17	1.42	−1071.06	1.24
A2	−1046.519	1.39	−1073.18	1.59	−1102.65	1.70	−1074.87	1.42
A3	−1050.101	1.82	−1077.08	2.05*	−1105.89	1.89*	−1078.69	1.70

Ln *L*., logarithm likelihood, *t* - Student's *t*-ratio for the difference in Ln *L* between given tree (see Fig. 3) and the ML tree for corresponding data set. * denotes that difference is statistically significant (t>1.96 for p<0.05).

(distances to the other rooting sequences 0.101–0.244 K_{nuc}, those to the platyhelminth species 0.142–0.475 K_{nuc}; see the previous paragraph). This confirms that our series of outgroup sequences was an effective test for the consistency of the dominant tree topology. Yet, two details of this topology are probably erroneus: *Dugesia* is definitely a monophyletic taxon, and the Cestoda is generally considered more closely related to the Monogenea, than to the Trematoda. This questions the reliability of other branching patterns and outlines the importance of statistical estimation of the tree topologies (see below).

On the position of the Neodermata

Parasitic flatworms could be grouped together in the same way in all ML trees, though two branches with zero length are to be taken in account here (Fig. 2). Maximum parsimony (MP) and Fitch trees nearly always support monophyly of the Neodermata. The phylogenetic relationships of the Neodermata remain controversial. The Neodermata is most often considered either the sister group for the Dalyellioida, rhabdocoels with doliiform pharynx (see, e.g., Ehlers, 1985; Brooks, 1989), or as an earlier branch of rhabdocoels which retained the less specialized rosulate pharynx (see Joffe *et al.*, 1987, Joffe & Chubrik, 1988; Kotikova & Joffe, 1988 for review and evidence from the morphology of the pharynx and nervous system). Two other hypotheses which could be tested with our data would consider the Neodermata the sister group either for the Neorhabdocoela + Seriata (c.f. Ruitort *et al.*, 1992b), or for the Seriata (note again that the hypotheses suggested by Rohde, (1990 and present volume) cannot be discussed here).

Phylip allows a statistical test of whether the difference between logarithm likelihoods or lengths of two given trees is significant. We estimated 6 topologies representing these four hypotheses (Fig. 3) in each of our 8 species sets. Statistical estimates (Student's t-

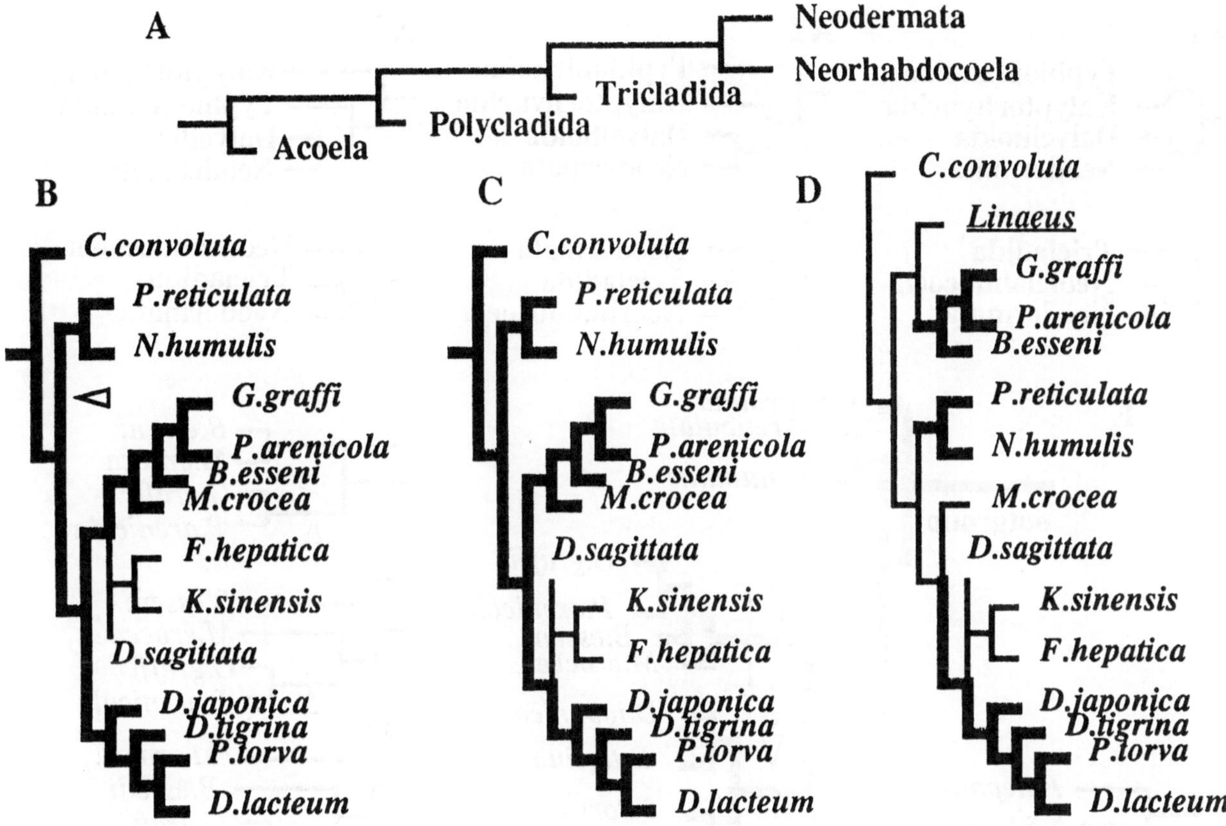

Fig. 2. A. Phylogenetic tree representing recent views on the evolution of the Turbellaria and one of widely accepted views on the position of the Neodermata. B-D. Topologies of maximum likelihood trees for eight sets of species with different outgroup sequences. B. *Brachionus, Emplectonema,* cCoelenterata and *Echinorhinchus*; the last occupies a position shown with arrowhead (and not in the root). C. cAnnelida, cPorifera, cNematoda. D. *Lineus* (the branches present in the trees on Figs B and C are shown by the wide lines).

ratio) for the trees based on the 'dalyellioid' hypothesis were 1.15–2 fold higher than for three other hypotheses, and its variant (Fig. 3,A3) was significantly worse than ML tree in 6 of 8 data sets (Table 2). Thus, the 'dalyellioid' hypothesis fits poorly with the available 5S rRNA data compared with other tested assumptions. None of the 3 other hypotheses may be preferred.

Compatability of recent hypotheses on the phylogeny of the Turbellaria with rRNA molecular data

Statistical estimates described in the previous paragraph also demonstrated that the tree topologies representing widely accepted recent views on the phylogeny of the Turbellaria, with Neodermata not branching with the Dalyellioida (Figs 2A, 3E, b,c,d), fit well with 5S rRNA data (Table 2).

The maximum parsimony (MP) method returned 1–10 shortest trees, depending of the species set. Estimates of various hypotheses about the relationships of Neodermata were the same as obtained from ML method. The trees corresponding to recent views on the platyhelminth phylogeny, provided that the Neodermata were not derived together with the Dalyellioida, were less than 5% longer than MP trees, and the difference was far from being statistically significant.

In addition, we united the data published by Ruitort *et al.*, 1992a and Baverstock *et al.*, 1991, corrected alignement and, thus, obtained two data sets including 16 and 13 platyhelminth species with 443 and 536 well aligned positions correspondingly. For the sake of comparison, they were rooted as in Baverstock *et al.* (1991). Studied both by ML and MP methods, these data also demonstrated no statistically significant disagreement with phylogenetic concepts corresponding to the tree shown on Fig. 2A.

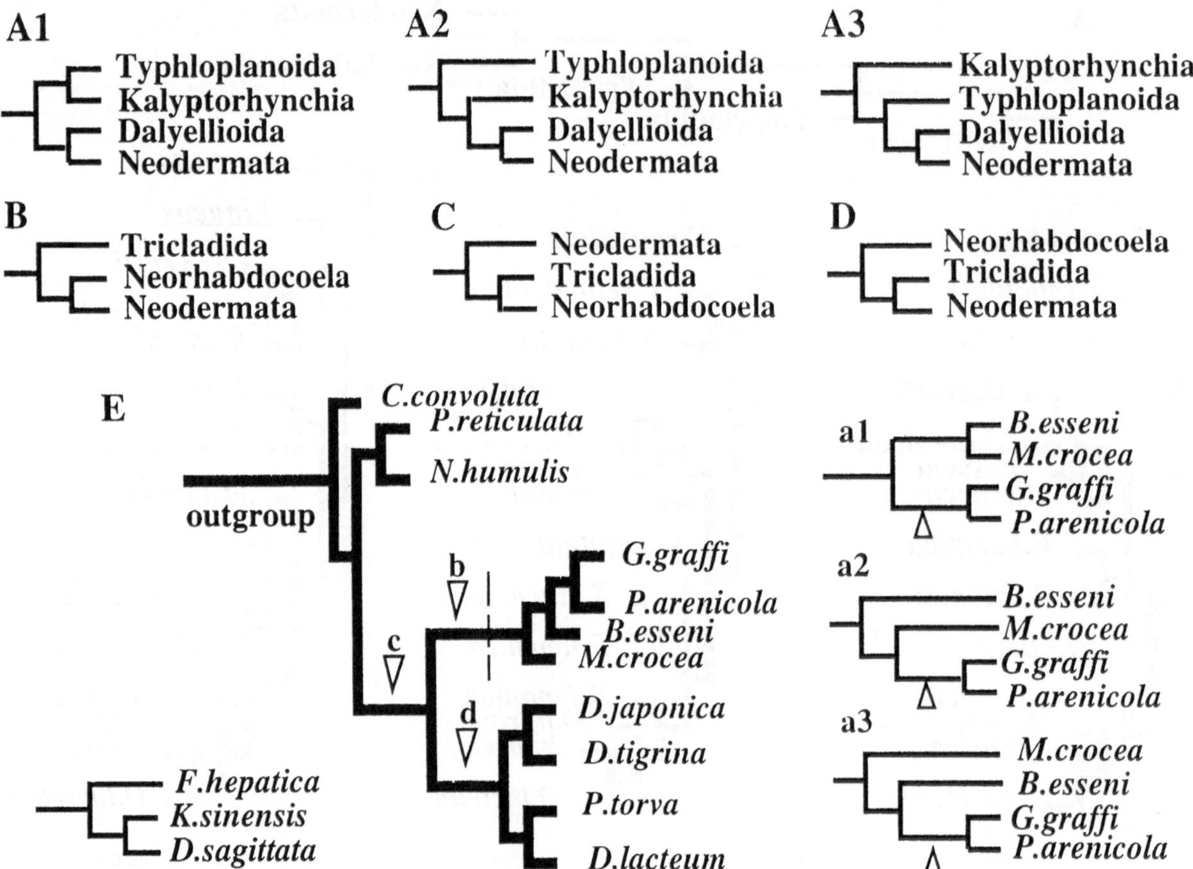

Fig. 3. A–D. Trees, representing four hypotheses about the relationships of the Neodermata with the Turbellaria, the 'dalyellioid' hypotheses is shown in 3 variants, A1–A3. E, topologies of the trees used for the statistical estimation of these hypotheses. To make trees corresponding to hypotheses shown in B,C,D the neodermate subtree (bottom, left) was grafted to the base turbellarian tree (bottom, center) at the positions marked with arrowheads b,c,d accordingly. To compose the trees corresponding to A1–A3, the neorhabdocoel subtree (on the right of the dashed line) was replaced by subtrees a1–a3 correspondingly, with the neodermate subtree grafted at the positions shown by arrowheads.

The rates of nucleotide base replacement and phylogenetic trees

The relative rate of change may be approximated from the lengths of the branches in the trees (ML, MP, or Fitch-Margoliash) or directly from a distance matrix. With any method, the rate of base replacement in 5S rRNA of planarians was estimated as high, and that of parasitic platyhelminths as very low. These results completely agree with the estimates obtained from 18S rRNA by Ruitort *et al.* (1992b). A high rate of evolution may cause inappropriate early divergence of a group in the tree or incorrect clustering of the 'quick clocks' together (see, e.g., Lake, 1991). Therefore, the uncertainty of the position of the Neodermata and/or a relatively too early position of the planarian branch in various trees constructed from 5S and 18S rRNA data (see Hendriks *et al.*, 1986; Hori & Osava, 1987; Baverstock *et al.*, 1991; Ruitort *et al.*, 1992a,b) may result, correspondingly, from too low and too high rates of base replacement. The influence of the heterogeneity in the rates of base replacement is important for it can drastically affect the topology of phylogenetic trees independently of the sequence lengths.

Acknowledgments

We are very grateful to Dr O. N. Yunchis (Research Institute for Lake and River Fishery, St-Petersburg) and to Dr T. A. Timofeeva (Zoological Institute, St-Petersburg) who collected for us two parasitic species. Dr E. N. Grusov from the same Institute let us use his computer, though it notably hindered his own work. We are also very grateful to our colleagues from Moscow State University: to Dr V. K. Bobrova for the help in experimental work, to Miss Katya Drusina for perfect and willing technical assistance, and to Prof. A. S. Antonov for constant attention to this work and critical remarks to the manuscript.

References

Baverstock, P. R., R. Fielke, A. M. Johnson, R. A. Bray & I. Beveridge, 1991. Conflicting phylogenetic hypothesis for the parasitic platyhelminths tested by partial sequencing of 18S ribosomal RNA. Int. J. Parasitol. 21: 329–339.

Brooks, D. R., 1989. The phylogeny of the Cercomeria (Platyhelminthes: Rhabdocoela) and general evolutionary principles. J. Parasitol. 75: 606–616.

Ehlers, U., 1985. Das phylogenetishe System der Plathelminthes. G. Fisher, Stuttgart - New York, 317 pp.

Felsenstein, J., 1990. PHYLIP 3.3. University of California Herbarium, Berkley, California.

Hendriks, L., E. Huysmans, A. Van den Berghe & R. De Wachter, 1986. Primary structure of the 5S ribosomal RNAs of 11 arthropods and applicability of 5S RNA to the study of metazoan evolution. J. Mol. Evol. 24: 103–109.

Hori, H., A. Muto, S. Osawa, M. Takai, K.-Y. Lue & M. Kawakatsu, 1988. Evolution of turbellaria as deduced from 5S ribosomal RNA sequences. Fortschritte Zool./Progress Zool. 36: 163–167.

Hori, H. & S. Osawa, 1987. Origins and evolution of organisms as deduced from 5S rRNA sequences. Mol. Biol. Evol. 4: 445–472.

Joffe, B. I., G. S. Slusarav & T. A. Timofeeva, 1987. Stroenie glotki mongenei i ih phylogeneticheskie sviasi s turbellariyami. [Pharynx structure in the monogeneans and their phylogenetic relationships with the turbellarians.] Parasitologia (St-Petersburg) 21: 472–481.

Joffe, B. I. & G. K. Chubrik, 1988. Stroenie glotki trematod i phylogeneticheskie sviasi mezhdu Trematoda i Turbellaria. [The structure of the pharynx in trematodes and phylogenetic relations between Trematoda and Turbellaria]. Parasitologia (St-Petersburg) 22: 297–303.

Kotikova, E. A. & B. I. Joffe, 1988. On the nervous system of the dalyellioid turbellarians. Fortschritte Zool./Progress Zool. 36: 191–194.

Lake, A. J., 1989. Origin of the eukaryotic nucleus determined by rate-invariant analyses of ribosomal RNA genes. In B. Fernholm, K. Bremer & H. Jornvall (eds), The hierarchy of life. Elsevier Science Publishers, Amsterdam: 87–101.

Ohama, T., T. Kumazaki, H. Hori, S. Osawa & M. Takai, 1983. Fresh water planarians and a marine planaria are relatively dissimilar in the 5S rRNA sequences. Nucleic Acids Res. 11: 433–436.

Rieger, R. M., S. Tyler, J. P. S. Smith & G. E. Rieger, 1990. Platyhelminthes: Turbellaria. In F. W. Harrison & B. J. Bogitsh (eds), Platyhelminthes and Rhynchocoela, vol. 3 in F. W. Harrison (ed.) Microscopic anatomy of invertebrates. Wiley-Liss, New York: 7–140.

Rohde, K., 1990. Phylogeny of the Platyhelminthes, with special reference to parasitic groups. Int. J. Parasitol. 20: 979–1007.

Ruitort, M., K. G. Field, J. M. Turbeville, R. A. Raff & J. Baguñà, 1992a. Enzyme electrophoresis, 18S rRNA sequences, and levels of phylogenetic resolution among several species of freshwater planarians (Platyhelminthes, Tricladida, Paludicola). Can. J. Zool. 70: 1425–1439.

Ruitort, M., K. G. Field, R. A. Raff & J. Baguñà, 1992b. 18S rRNA sequences and phylogeny of Platyhelminthes. Biochem. Syst. Ecol. 21: 71–77.

Specht, T., J. Wolters & V. A. Erdmann, 1990. Compilation of 5S rRNA and 5S rRNA gene sequences. Nucl. Acids Res. 18: 2215–2230.

Troitsky, A. V., Yu E. Melekhovets, G. M. Rakhimova, V. K. Bobrova, K. M. Valiejo-Roman & A. S. Antonov, 1991. Angiosperm origin and early stages of seed plant evolution deduced from rRNA sequence comparisons. J. Mol. Evol. 32: 253–261.

Wainwright, P., G. Hinkle, M. L. Sogin & S. K. Stickel, 1993. Monophyletic origins of the Metazoa: an evolutionary link with Fungi. Science 260: 340–342.

Hydrobiologia **305**: 45–47, 1995.
L.R.G. Cannon (ed.), Biology of Turbellaria and some Related Flatworms.

Brackish-water Plathelminthes from the Faroe Islands

P. Ax
II. Zoologisches Institut und Museum der Georg-August- Universität, Berliner Str. 28, 37073 Göttingen, Germany

Key words: Faroe Islands, Plathelminthes, brackish-water, distribution mechanisms

Abstract

Species-rich communities of brackish-water plathelminths of the European continental coast and of the north American Atlantic coast are known to be most similar. In contrast nothing has been recorded about a possible settlement of corresponding faunal elements between the continents on volcanic islands which were covered by glaciers in the Pleistocene. At the Faroe Islands 2 freshwater species, 11 marine-euryhaline species and 12 genuine brackish-water organisms were found in minute brackish-water biotopes with highly variable salinity. *Macrostomum curvituba* Luther, *Minona baltica* Karling & Kinnander and *Coronhelmis lutheri* Ax are the most abundant species. Several brackish-water plathelminths are able to tolerate freshwater conditions. Most probably they have evolved mechanisms which are comparable to those of limnetic organisms allowing long distance transport across the sea in an inactive state.

Introduction

Historically the Baltic is the centre of research on the biology of brackish-water animals. Due to the monumental work of A. Luther and T. G. Karling, the Gulf of Finland represents the brackish-water area of the world whose freeliving plathelminths are best studied. If one adds the comprehensive studies of C. den Hartog in estuaries of the Netherlands and the efforts at the German North Sea coast and in the Western Baltic by the Göttingen school then about 100 brackish-water flatworms are known from northern Europe (Luther, 1943–1963; Karling, 1963, 1974; den Hartog, 1977; Ax, 1956; Armonies, 1987).

Comparative investigations in Canada and Alaska have revealed the existence of 27 brackish-water species common to both continents (Ax & Armonies, 1987, 1990; Ax, 1991). Presumably they are relics of a very old fauna that existed 200 million years ago in northern coastal brackish-waters of the former supercontinent Pangaea (Sterrer, 1973). Subsequently because of the creation of the North Atlantic by means of plate tectonics, the formerly connected community was separated into two more or less identical sets of European and North American populations.

But what about the existence or absence of this community on islands between the continents — on volcanic islands that arose after the dissolution of Pangaea and that were covered by glaciers in the Pleistocene?

I adressed the problem at the Faroe Islands halfway between Scotland and Iceland. A lagoon and several bights and fjords (Table 1) with incoming freshwater and highly variable salinity were investigated in August 1992.

Results

Composition of species:

The 25 determined species could be classified in 3 ecological groups.

1. Freshwater animals — *Macrostomum distinguendum* Papi, 1951 and *Prorhynchus stagnalis* Schultze, 1851.
2. Widely distributed marine-euryhaline organisms —including the proseriates *Itaspiella helgolandica* (Meixner, 1938) and *Notocaryoplana arctica* Steinböck, 1935, the rhabdocoels *Promesostoma marmoratum* (Schultze, 1851), *Acrorhynchides robustus* (Karling, 1931) and *Provortex balticus* (Schultze, 1851).

Table 1. Occurrence of brackish-water plathelminths at the Faroe Islands. a = In the description of Luther (1943) from Finland. b = Species identification uncertain. c = see (Ax, 1993).

	Faroe Islands					Europe	North America
		Bights		Fjords			
	Saksŭn	Hvalvik	Hosvik	Kaldbaksfjord	Skalafjord		
Macrostomum curvituba Luther, 1947	+	+	+	+	+	+	+
Coronhelmis lutheri Ax, 1951	+	+	+	+	+	+	+
Minona baltica Karling & Kinnander, 1953	+	+	+	+	+	+	+
Paramonotus hamatus (Jensen, 1978)	+	+	+	+		+	
Prognathorhychus canaliculatus Karling, 1947	+	+	+	+		+	
Coelogynopora schulzii Meixner, 1938	+	+	+			+	+
Beklemischeviella contorta (Beklemischev, 1927)[a]	+			+	+	+	
Provortex pallidus Luther, 1948	+				+	+	
Stygoplanellina halophila Ax, 1954[b]	+					+	+
Vejdovskya mesostyla Ax, 1954	+					+	
Macrostomum balticum Luther, 1947		+				+	
Coelogynopora hangoensis Karling, 1953[c]		+				+	Greenland

1 = Saksŭn (Lagoon Pollur at Saksun: Island Streymoy). 2 Hvalvik (Bight of Hvalvik: Island Steymy at the sound between Streymoy and Eystŭroy. Medium sand in salt marsh area. August 8, 1992 (salinity 5‰)). 3 Hosvik (Bight of Hosvik: Island Streymoy at the sound between Streymoy and Eystŭroy. Medium and coarse sand. August 4 and 25, 1992 (Brackish water with low salinity and freshwater)). 4 Kaldbaksfjord (Island Streymoy). 5 Skalafjord (Island Eysturoy. small lagoon behind a sandy beach at the inner end of the fjord. August 13, 1992 (freshwater)).

3. True brackish-water plathelminths — very common species such as *Macrostomum curvituba* and *Coronhelmis lutheri* as well as the rare species *Macrostomum balticum* and *Coelogynopora hangoensis* (Table 1). These species all exist on the continental coasts of Europe and five species are known from North America.

Analysis of two biotopes

Saksun

The lagoon Pollur in the northwest of the Streymoy Island formerly a harbor, is a large freshwater basin with coarse sand and gravel influenced by seawater only occasionally via a small water course. Pure freshwater conditions existed at the inner end of the lagoon during the investigations on August 6 and 19, 1992.

With 10 different brackish-water plathelminths, this site has the highest number of species coexisting in one place in the Faroes. Exactly the same association of species exists in sandy beaches of the Gulf of Finland.

Kaldbaksfjord (Island Streymoy)

A beach at the inner end of the Kaldbaksfjord with coarse sand and gravel is influenced strongly by a river collecting water out of the mountains. Depending on the input of freshwater, of air pressure and the direction of wind, rapid changes of the salinity were observed. During 26 days of August, 1992, marine conditions existed on 6 days, freshwater on 3 days and brackish-water on 17 days — commonly with a salinity between 5 and 10‰.

Coronhelmis lutheri was the dominant species in the beach found in all samples from 5 m below the water edge upto 5 m above it. *Beklemischeviella contorta* on the other hand was restricted to the end of the river with freshwater or brackish-water of low salinity.

Species abundance

Macrostomum curvituba, *Coronhelmis lutheri* and *Minona baltica* were the most abundant species They were found regularly at the 5 main stations of the investigation and sometimes in high densities. On the other hand the five species *Provortex pallidus*, *Stygoplanellina halophila*, *Vejdovskya mesostyla*, *Macrostomum balticum* and *Coelogynopora hangoensis* were detected in one or two places only with low numbers of individuals. Identification of one single specimen of *V. mesostyla* confirms the existence of a population of this species in the Faroe Islands.

Conclusions

The brackish-water plathelminths of the Faroes may be part of an extremely old brackish-water fauna from the time of the supercontinent Pangaea. In principle this hypothesis remains uneffected by our findings at the Faroe Islands

Nevertheless we must look in addition at explanations for long distance transport over a short time. If suitable brackish-water biotopes for the present fauna at the Faroes did not emerge before the end of the Pleistocene colonisation must have occurred in no more than the last several thousand years.

According to longstanding field observations, individuals of brackish-water species cannot tolerate marine conditions continuously. An unprotected crossing of hundreds of kilometres in the sea is most improbable. On the other hand numerous brackish-water plathelminths are able to live in freshwater as in our example the ten species of the lagoon Pollur at Saksun. Comparable to this invasion into freshwater three of the species — *Macrostomum curvituba*, *Paramonotus hamatus*, *Coelogynopora schulzii* — have conquered the freshwater of the river Elbe in northern Germany together with other brackish-water plathelminths (Düren & Ax, 1993).

Like freshwater organisms, most probably genuine brackish-water plathelminths too have evolved suitable mechanisms for long distance passive transport across the sea in an inactive state. The capacity to encyst is known for many plathelminth species living in the harsh environment of salt marshes (Bilio, 1964; Armonies, 1986, 1987). Regarding the Faroe fauna encystment has been proved for *Macrostomum curvituba*, *Coelogynopora schulzii* and *Prognathorhynchus canaliculatus*. Experimental study is necessary to evaluate the significance of cysts for the distribution of brackish-water plathelminths.

Acknowledgment

Cordial thanks to Dr Arne Nørrevang for his hospitality and support of the work at the laboratory BIOFAR, Kaldbak, Faroe Islands.

References

Armonies, W., 1986. Free-living Plathelminthes in North Sea salt marshes: Adaptations to environmental instability. An experimental study. J. exp. mar. Biol. Ecol. 99: 181–197.

Armonies, W., 1987. Freilebende Plathelminthen in supralitoralen Salzwiesen der Nordsee: Ökologie einer borealen Brackwasser-Lebensgemeinschaft. Microfauna Marina 3: 81–156.

Ax, P., 1956. Das ökologische Verhalten der Tŭrbellarien in Brackwassergebieten. XIV. Int. Congr. of Zoology, Danish Science Press, LTD. Copenhagen: 462-464.

Ax, P., 1991. Northern circumpolar distribution of brackish-water plathelminths. Hydrobiologia 227 (Dev. Hydrobiol. 69): 365–368.

Ax, P., 1993. Die Brackwasserart *Coelogynopora hangoenis* (Proseriata, Plathelminthes) von Grönland und den Faröer. Microfauna Marina 8: 145–152.

Ax, P. & W. Armonies, 1987. Amphiatlantic identities in the composition of the boreal brackish water community of Plathelminthes. Microfauna Marina 3: 7–80.

Ax, P. & W. Armonies, 1990. Brackish water Plathelminthes from Alaska as evidence for the existence of a boreal brackish water community with circumpolar distribution. Microfauna Marina 6: 7–109.

Bilio, M., 1964. Die aquatische Bodenfauna von Salzwiesen der Nord- und Ostsee. I. Biotop und ökologische Faunenanalyse: Tŭrbellaria. Int. Revue ges. Hydrobiol. 49: 509–562.

Düren, R. & P. Ax, 1993. Thalassogene Plathelminthen aus Sandstränden von Elbe und Weser. Microfauna Marina 8: 267–280.

Hartog, C. den, 1977. Tŭrbellaria from intertidal flats and salt-marshes in the estuaries of the southwestern part of the Netherlands. Hydrobiologia 52: 29–32.

Karling, T. G., 1963. Die Turbellarien Ostfennoskandiens. V. Neorhabdocoela 3. Kalyptorhynchia. Fauna Fennica 17: 1–59.

Karling, T. G., 1974. Turbellarian fauna of the Baltic proper. Identification, ecology and biogeography. Fauna Fennica 27: 1–101.

Luther, A., 1943. Untersuchungen an rhabdocoelen Turbellarien. IV. Ueber einige Repräsentanten der Familie Proxenetidae. Acta zool. Fenn. 38: 1–95.

Luther, A., 1960. Die Turbellarien Ostfennoskandiens. I. Acoela, Catenulida, Macrostomida, Lecithoepitheliata, Prolecithophora und Proseriata. Fauna Fennica 7: 1–155.

Luther, A., 1962. Die Turbellarien Ostfennoskandiens. III. Neorhabdocoela 1. Dalyellioida, Typhloplanoida: Byrsophlebidae und Trigonostomidae. Fauna Fennica 12: 1–71.

Luther, A., 1963. Die Turbellarien Ostfennoskandiens. IV. Neorhabdocoela 2. Typhloplanoida: Typhloplanidae, Solenopharyngidae und Carcharodopharyngidae. Faune Fennica 16: 1–163.

Sterrer, W., 1973. Plate tectonics as a mechanism for dispersal and speciation in the interstitial sand fauna. Netherlands J. Sea Res. 7: 200–222.

Hydrobiologia **305**: 49–53, 1995.
L.R.G. Cannon (ed.), Biology of Turbellaria and some Related Flatworms.

Platyhelminths as paleogeographical indicators

Ronald Sluys
Expert-Center for Taxonomic Identification, Institute for Systematics and Population Biology, Zoological Museum, University of Amsterdam, P.O. Box 94766, 1090 GT Amsterdam, The Netherlands

Key words: Turbellaria, biogeography, vicariance, continental drift, trans-Pacific tracks, new record *Kontikia bulbosa*

Abstract

Turbellarians do not feature as examples in the present discussions on the theory and method of analytical biogeography. It is argued, however, that turbellarian distributional records form good examples of large-scale biogeographic patterns resulting from continental breakup. Some turbellarian taxa also indicate biogeographic links across the Pacific Ocean, which can be visualized readily by means of track construction. Amphi-pacific organismal distributions form the ingredients of trans-Pacific biogeographic tracks. Such tracks may be explained historically either as the result of dispersal or of vicariance. In the case of the flatworm examples, as well as many other organisms, dispersal explanations are the least satisfactory. However, under a vicariance paradigm the classical pre-drift reconstruction of Pangea cannot adequately explain trans-Pacific tracks. Therefore, alternative paleogeographic models may be invoked as explanatory hypotheses: the lost continent Pacifica, island integration, a new reconstruction of eastern Gondwanaland, an expanding earth. None of these alternative models is fully compatible with all geological and biogeographic data available at present. It is stressed that biogeographic data and theories should not be made subservient to geological theories. Biogeographical data on flatworms may indicate paleogeographical relations which are worthy of examination by geologists.

Introduction

Currently, the theory and practice of analytical biogeography form the subjects of an ongoing debate in the discipline of systematic biology (Myers & Giller, 1988; Brooks & McLennan, 1991; Humphries, 1992). The number of practical examples analysed in many of these theoretical papers revolves around a limited number of recurring cases, e.g. Rosen's poecilid fishes, Brundin's midges, *Nothofagus* (see Humphries & Parenti, 1986; Humphries, 1992). It is evident that present biogeographic theory owes much to detailed analyses of these taxa, particularly those with world-wide distributional patterns reflecting the sequence of continental breakup. Therefore, it is surprising that biogeographic patterns in such an old group as flatworms hardly ever have been used in methodological studies in analytical biogeography, either as clear examples or as interesting problematic cases. An exception is the parasitic helminths studied by Brooks (see Brooks & McLennan, 1991 and references therein), which formed the starting point for a method in historical biogeography and coevolutionary studies now known as Brooks Parsimony Analysis (Brooks, 1990). Turbellarians, however, do not feature in recent discussions on the theory and method of biogeography.

In this paper I want to draw attention to the fact (1) that turbellarians exhibit large-scale biogeographic patterns reflecting continental breakup, and (2) that some turbellarian distributional patterns require alternative vicariance explanations and suggest paleogeographic connections between areas which, according to conventional theory, never have been in proximity.

Method

At present there is no single biogeographic method all workers agree upon, but a consensus is emerging that a phylogenetic analysis of taxa and a subsequent cladistic biogeographic study form the basic ingredients for analysing and comparing biotic patterns in space and time (see Humphries, 1992 and references therein). The extent to which these two requirements are met in certain cases depends on the state of systematic knowledge of particular taxa and on the specific questions asked. It is the author's opinion that in less than ideal cases considerable insight still can be gained by applying a more 'generalized technique' (Humphries, 1992) than cladistic biogeography, *i.e.* track analysis. Individual tracks are drawn between sister taxa or concern an entire monophyletic group when relationships within that group have not yet been analysed. Coincident individual tracks make up a generalized track, assumed to link areas that once constituted a single ancestral biota. In one of the following sections I shall deploy track analysis in order to visualize some interesting and important biogeographic patterns in turbellarians.

Previous studies featuring planarians

In the past a number of workers have used planarians as examples of large-scale biotic relations that may have resulted from altered continental configurations.

Harrison (1928) used land planarians as one of his animal examples showing the former connections between South America, Australia, and New Zealand by way of Antarctica. According to Harrison, Wegener's hypothesis is capable of elucidating the biogeography of a large number of disparate taxa. Harrison also discussed biotic distributions that do not fit the Wegener hypothesis. In particular, he pointed to endemic distributions on Pacific islands, which he explained by postulating a 'Polynesian arc', running from Antarctica to Hawaii via Tonga and Samoa.

To the best of my knowledge, the first planarian systematist putting triclad biogeography into the perspective of continental drift was Marcus when he wrote that 'Wegener's theory or hypothesis of Continental Drift cannot be proved by the hitherto known distribution of the Terricola, but the recent maps of their distribution become more intelligible with this theory' (Marcus, 1953: 53). In similar vein, Ball & Fernando (1969) and Ball (1975) concluded some years later that the distributional patterns of southern hemisphere dugesids are best explained by the vicariance process of continental drift.

Apart from the monographic study of the marine triclads (Sluys, 1989) there have been no other studies analyzing planarian biogeography from a plate tectonic perspective.

Distributional patterns in selected turbellarian taxa

Three genera of land planarians

According to the taxonomic review of Ogren & Kawakatsu (1991) the terricolan genus *Kontikia* contains 22 species, of which a generalized distribution map is presented in Fig. 1 [it must be noted that the taxonomy of this genus has not yet stabilized – see Winsor (1991), Ogren *et al.* (1993)]. It is clear from Fig. 1 that the pattern is mainly Gondwanian, with only isolated records from Laurasia and that the track bypasses the African continent (the latter most likely due to a collecting artefact). Under the assumption that the animals are poor dispersers, such a disjunct distribution only can be explained adequately by assuming that the vicariant process of plate tectonics operated during the evolution of the genus *Kontikia.* As it happens, the assumption of poor dispersal capacity in triclads is the least supported for land planarians. All available evidence suggests that freshwater and marine triclads have poor dispersal capacities (Ball & Fernando, 1969; Sluys, 1989), but there is ample circumstantial evidence that the situation in land planarians can be different. For example, land planarians have been recorded from greenhouses and botanic gardens all over the world. To some extent, this is the case also in the genus *Kontikia. K. ventrolineata* (Dendy, 1892) has been found in botanic and private gardens in Victoria, Tasmania, and southeastern Queensland (Winsor, 1979); *K. orana* Froehlich, 1955 has been reported from urban gardens in Queensland (Winsor, 1986) while it was already known from man-modified areas in Brazil. Passive anthropochore dispersal is invoked to account for the occurrence of these two species in Australia (Winsor, 1979; 1986).

Other records have been explained also as the result of passive dispersal. Froehlich (1955; 1967) and De Beauchamp (1961) invoked dispersal to explain the distribution of *K. kenneli* (Von Graff, 1899) and *K. orana* in the new world and Jones (1981) speculated that the occurrence of *K. andersoni* Jones, 1981 in northern Ireland resulted from introduction of the animal in pots of soil. Specimens of *K. mexicana* (Hyman, 1939) in California were presumably introduced (Hyman, 1943). The occurrence of *K. bulbosa* Sluys, 1983 on Madeira has been attributed to introduction by means of banana rhizomes (Marcus & Marcus, 1959; Sluys, 1983). A new record for *K. bulbosa* concerns the Canary Islands, where it was collected

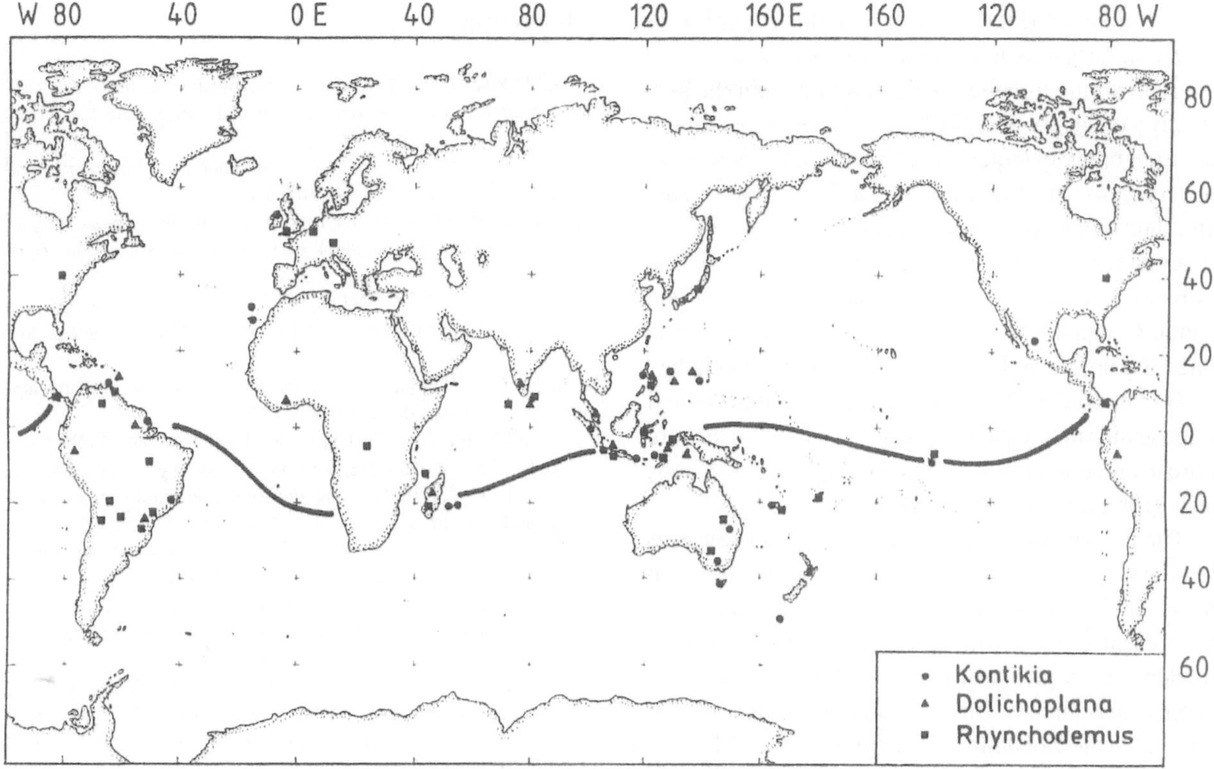

Fig. 1. Generalized distribution and track of the land planarian genera *Kontikia*, *Dolichoplana*, and *Rhynchodemus*; records from greenhouses omitted.

from Las Palmas (collection Natural History Museum, London, BMNH 1965.3.16.1, Las Palmas, August 1962).

The number of *ad hoc* hypotheses postulated for the distributional records of the genus *Kontikia* obscures the fact that the major track conforms to a well founded generalized track across former parts of Gondwanaland, *viz.* South America, Africa, India, Australia, and New Zealand. Within the Terricola this track is exemplified also by the genera *Dolichoplana* and *Rhynchodemus* (Fig. 1).

Trans-Pacific tracks

For both *Kontikia* and *Rhynchodemus* species have been recorded from localities well within the Pacific basin, suggesting biogeographic links across the Pacific. On the basis of a cladistic analysis of the taxa involved, such trans-Pacific tracks have been demonstrated also for the marine triclads (Sluys, 1989). These trans-Pacific tracks have been demonstrated for many other groups of organisms (see Croizat, 1958; Sluys, 1989; Matile, 1990) and I am convinced that they hold true not only for the turbellarians mentioned above but also for many other taxa. However, other studies suggesting trans-Pacific tracks in turbellarians appear to be scarce. At present, I know only of Tajika's (1991) study of the polyclad genus *Discoplana. D. pacificola* (Plehn, 1896) from the East Pacific is the sister species of *D. gigas* (Schmarda, 1859), which is broadly distributed in the Indo-West Pacific region; the sister group relationship between these two species evidently suggests a trans-Pacific track within the genus *Discoplana*.

Vicariance explanations and alternative paleogeographic models

Because of the biology of some turbellarians, workers may choose dispersal hypotheses for explaining particular distributional records. Such has been the case for the genus *Kontikia* (see above). In similar vein one could try to explain the trans-Pacific sister group relationship between the polyclads *D. pacificola* and *D. gigas* as a result of allopatric speciation after dispersal of larvae. However, the larvae of *Discocelis* are of the direct type, which generally is found only among inshore plankton (Prudhoe, 1985). In this paper I wish to de-emphasize such quantum-dispersal explanations and to examine turbellarian distribution from the perspective of vicariance. Under a vicariance paradigm, the usual breakup sequence of Pangea can explain adequately trans-Atlantic, trans-Indian Ocean, and trans-Antarctic biogeographic tracks in turbellarians and other organisms. However, trans-Pacific tracks pose a problem because according to conventional theory the Pacific Ocean always has been present since Pangean times, its precursor being the EoPacific or Panthalassa. This has induced several workers to search for and propose alternative paleogeographical models that are more in agreement with the biogeography of the Pacific basin (for a review, see Sluys, 1994).

The most controversial alternative model concerns the theory of an expanding earth, proposing that the earth has increased in size over the ages. In the present context, the fascinating consequence of Shields' (1979) reconstruction of the supercontinent Pangea on a smaller earth at Jurassic times is that it obliterates the EoPacific. As a consequence, Shields' hypothesis shows a perfect fit between trans-Pacific tracks and breakup of the supercontinent. But this fit may in part be due to the fact that Shields' reconstruction is based on geological, paleontological, as well as biogeographic information. Other continental assemblies on a smaller earth are different (Owen, 1983) and can explain only biogeographic tracks in the Indo-West Pacific.

This alternative model involving an expanding earth, as well as others (e.g. lost continent Pacifica, new reconstruction of eastern Gondwanaland, island integration; see Sluys, 1994) are still debated among geologists. Therefore, I suggest that in our explanations of the historical biogeography of the Platyhelminthes we do not feel constrained by any geological model, either conventional or alternative. We should let the biogeographic data speak for themselves; the still developing methods of the discipline of historical biogeography shall enable us to compare biogeographic generalizations with paleogeographic hypotheses and look for mutual consistency. In this way, biogeographers may point to possible paleogeographic connections not taken into consideration by geologists and thus stimulate particular lines of geological research. It is highly likely that an evolutionarily old group as flatworms still indicate in their present-day distributional record geographic situations of old.

References

Ball, I. R. & C. H. Fernando, 1969. Freshwater triclads (Platyhelminthes, Turbellaria) and continental drift. Nature 221: 1143–1144.

Ball, I. R., 1975. Nature and formulation of biogeographical hypotheses. Syst. Zool. 24: 407–430.

Brooks, D. R., 1990. Parsimony analysis in historical biogeography and coevolution: methodological and theoretical update. Syst. Zool. 39: 14–30.

Brooks, D. R. & D. A. McLennan, 1991. Phylogeny, ecology, and behavior. The University of Chicago Press, Chicago & London, XII + 434 pp.

Croizat, L., 1958. Panbiogeography. Published by the author, Caracas. vols. I, IIa, IIb. 1018, 771, 1731 pp.

De Beauchamp, P., 1961. Classe des Turbellariés. In: P. P. Grassé (ed.), Traité de Zoologie, Tome IV, Masson et Cie, Paris: 35–212.

Froehlich, C. G., 1955. Sôbre morfologia e taxonomia das Geoplanidae. Bolm. Fac. Fil. Ciênc. Letr., Zool. 19: 195–251.

Froehlich, C. G., 1967. A contribution to the zoogeography of neotropical land planarians. Acta zool. Lilloana 23: 153–162.

Harrison, L., 1928. The composition and origin of the Australian fauna with special reference to the Wegener hypothesis. Rep. Australas. Ass. Advmt. Sci. 18: 332–396.

Humphries, C. J., 1992. Cladistic biogeography. In Forey, P. L., C. J. Humphries, I. J. Kitching, R. W. Scotland, D. J. Siebert & D. M. Williams (eds), Cladistics, Clarendon Press, Oxford: 137–169.

Humphries, C. J. & L. R. Parenti, 1986. Cladistic biogeography. Oxford monographs on Biogeography no. 2, Clarendon Press, Oxford, XII + 98 pp.

Hyman, L. H., 1943. Endemic and exotic land planarians in the United States with a discussion of necessary changes of names in the Rhynchodemidae. Am. Mus. Novit. 1241: 1–21.

Jones, H. D., 1981. A new species of land planarian from Northern Ireland (Platyhelminthes: Turbellaria). J. Zool., Lond. 193: 71–79.

Marcus, E., 1953. Turbellaria Tricladida. Exploration du Parc National de l'Upemba, Fasc. 21: 1–62.

Marcus, E. & E. Marcus, 1959. Turbellaria from Madeira and the Azores. Bolm. Mus. munic. Funchal 12: 15–42.

Matile, L., 1990. Récherches sur la systématique et l'évolution des Keroplatidae (Diptera, Mycetophiloidea). Mém. Mus. nat. Hist. nat., sér. A, 148: 1–682.

Myers, A. A. & P. S. Giller, 1988. Analytical Biogeography – An integrated approach to the study of animal and plant distributions. Chapman and Hall, London & New York, XIII + 578 pp.

Ogren, R. E. & M. Kawakatsu, 1991. Index to the species of the family Geoplanidae (Turbellaria, Tricladida, Terricola). Part II: Caenoplaninae and Pelmatoplaninae. Bull. Fuji Women's Coll. 29: 25–102.

Ogren, R. E., M. Kawakatsu & E. M. Froehlich, 1993. Addendum II. Hallez's (1890–1893) classification system of land planarians. Addendum III. Winsor's (1991b) provisional classification of Australian and New Zealand caenoplanid land planarians. Bull. Fuji Women's Coll. 31: 61–86.

Owen, H. G., 1983. Atlas of continental displacement. Cambr. Earth Sci. Ser., Cambridge University Press, Cambridge. X + 159 pp.

Prudhoe, S., 1985. A monograph on Polyclad Turbellaria. British Museum (Natural History), London, 259 pp.

Shields, O., 1979. Evidence for initial opening of the Pacific Ocean in the Jurassic. Palaeogeogr., Palaeoclimat., Palaeoecol. 26: 181–220.

Sluys, R., 1983. A new species of land planarian from Madeira (Platyhelminthes, Turbellaria, Tricladida). J. Zool., Lond. 201: 433–443.

Sluys, R., 1989. A monograph of the marine triclads. A. A. Balkema, Rotterdam & Brookfield, XII + 463 pp.

Sluys, R., 1994. Explanations for biogeographic tracks across the Pacific Ocean: a challenge for paleogeography and historical biogeography. Progr. Phys. Geogr. 18: 42–58.

Tajika, K.-I., 1991. Polyclad turbellarians collected on the Osaka University expedition to Viti Levu, Fiji, in 1985, with remarks on distribution and phylogeny of the genus Discoplana. Hydrobiologia 227 (Dev. Hydrobiol. 69): 333–339.

Winsor, L., 1979. Land planarians (Tricladida: Terricola) of the Royal Botanic Gardens, Melbourne, Victoria. Victorian Nat. 96: 155–161.

Winsor, L., 1986. Land planarians (Turbellaria: Tricladida: Terricola) introduced into Australia-2. Kontikia orana Froehlich, 1955. Victorian Nat. 103: 9–11.

Winsor, L., 1991. A provisional classification of Australian terrestrial geoplanid flatworms (Tricladida: Terricola: Geoplanidae). Victorian Nat. 108: 42–49.

Hydrobiologia **305**: 55–61, 1995.
L.R.G. Cannon (ed.), Biology of Turbellaria and some Related Flatworms.

Taxonomy and geographical distribution of *Dugesia japonica* and *D. ryukyuensis* in the Far East

Masaharu Kawakatsu[1]*, Iwashiro Oki[2] & Sachiko Tamura[3]
[1]*Biological Laboratory, Fuji Women's College, Kita-16, Nishi-2, Kita-ku, Sapporo (Hokkaidô) 001, Japan (*Author for correspondence);*
[2]*Ôsaka Environmental Project Association, MFC 7F, Uchihon-machi 1-2-15, Chû'ô-ku, Ôsaka 537, Japan;*
[3]*Ôsaka Prefectural Institute of Public Health, Nakamichi-1-chôme 3-69, Higashinari-ku, Ôsaka 537, Japan*

Key words: Turbellaria, Tricladida, *Dugesia*, taxonomy, Japan, Far East

Abstract

Dugesia japonica Ichikawa et Kawakatsu, 1964, is a common and polymorphic species of freshwater planarian distributed widely in the Far East. In 1976 the geographic populations were separated into 2 subspecies (*D. j. japonica* and *D. j. ryukyuensis*). The taxonomy of this species is reconsidered once again from the morphological, anatomical, histological, and karyological viewpoints. From the result of these studies, *D. j. ryukyuensis* is elevated to the rank of species: *D. ryukyuensis* Kawakatsu, 1976. *D. japonica* ($n=8$, $2x=16$, $3x=24$) differs from *D. ryukyuensis* ($n=7$, $2x=14$, $3x=21$) in having an asymmetrical penis papilla without a well-developed valve surrounding its basal part, and a well-developed vagina (distribution: the Japanese Islands, Taiwan, the Korean Peninsula, China, and Primorskiy, Northeast Siberia, in Russia). *D. ryukyuensis* is characterized by an asymmetrical penis papilla with a well-developed valve surrounding its basal part, and a less-developed vagina (distribution: the Southwest Islands of Japan).

Introduction

Dugesia japonica Ichikawa et Kawakatsu, 1964, is a polymorphic species widely distributed in the Far East including the Japanese Islands. During the past 40 years, Kawakatsu has studied the life history, vertical and geographical distributions, and taxonomy of this species (Kawakatsu, 1965, 1967, 1974, 1991). The species was erroneously known in the literature as *Dugesia gonocephala* (Dugès, 1830) (*olim Planaria, Euplanaria*) until its taxonimic position was determined by Ichikawa & Kawakatsu (1964).

D. japonica was separated into 2 subspecies in a paper by Kawakatsu *et al.* (1976): *D. j. japonica* Ichikawa et Kawakatsu 1964 ($n=8$, $2x=16$, $3x=24$) and *D. j. ryukyuensis* Kawakatsu, 1976 ($n=7$, $2x=14$, $3x=21$). The latter was described from Okinawa Island, the Southwest Islands of Japan (*i.e.*, Nansei Shotô).

During the past 17 years, a series of cytotaxonomic papers on *D. japonica* have been published by our team (Kawakatsu *et al.*, 1979, 1980; Oki *et al.*, 1981; Tamura *et al.*, 1985, 1987, 1988, 1991; Yamayoshi *et al.*, 1980). Recently, we reviewed our previous taxonomic and karyological results and added new data on *D. japonica* (*D. j. japonica* and *D. j. ryukyuensis*) from the Southwest Islands of Japan. As a result, it is now concluded, there are sufficient morphological differences between these 2 subspecies to elevate *D. j. ryukyuensis* to the rank of a species. In the present paper, of which this is the first of twin publications, we will give differential diagnoses for these 2 *Dugesia* species of the Far East, together with an account of their geographical distributions. Karyological problems will be discussed in another paper (Tamura *et al.*, 1995).

Materials and methods

Previously prepared serial sections retained in Kawakatsu's laboratory were used. Newly prepared sections (7–8 micrometers) of the samples from the Southwest Islands of Japan were stained with Delafield's hematoxylin and erythrosin. The Specimen Lot Numbers given for each stock of samples were those registered in Kawakatsu's fixing notebook according to his permanent recording system. (No list of the collecting sites of animals is given here.)

Results

Dugesia japonica Ichikawa et Kawakatsu, 1964

Description of animals from the type locality was given in the paper by Ichikawa & Kawakatsu (1964, pp. 187–193, Figs 1 A–D, 2 A–B, 3, 4 A–E); redescription of the type series was given by Kawakatsu *et al.* (1976, pp. 84–90, Figs 1A–E, 2 A–C, 3).

Serial sections of animals from 47 stations (from 10 islands) of the Southwest Islands of Japan were examined; sexually mature specimens were found from 32 stations. Among these samples *D. japonica* was found in 7 stations: Yakushima Island (Specimen Lot No. 38, in Kawakatsu & Iwaki, 1967); Amami-Ôshima Island (Nos. 1868, 1869); Okinawa Island (No. 1292, in Kawakatsu & Tanaka, 1976; No. 1913); Iriomotejima Island (No. 1991, not fully matured); Yonaguni Island (No. 2005). Schematic figures of the copulatory apparatus of 3 specimens from 3 islands (Amami-Ôshima Island, Okinawa Island and Yonaguni Island) are shown in the Seventh ISBT preprint (Kawakatsu *et al.*, 1993, pp. 10–11, Fig. 12 E, F, R).

Judging from karyological data (Tamura *et al.*, 1988, 1991), *D. j. japonica* has been recorded from Tanegashima Island, Yakushima Island (2 stations). Amani-Ôshima Island (3 stations), north and central parts of Okinawa Island (4 stations), Ishigakijima Island (3 stations), Iriomotejima Island (a single station), and Yonaguni Island (2 stations). Sexual specimens for taxonomic examination were not obtained from Tanegashima Island. Non-sexual specimens (*Dugesia* sp.) from Uotsurishima Island, the Senkaku Islands, could not be classified taxonomically.

Additional descriptions of morphology, anatomy and histology of *D. japonica* with schematic figures of the copulatory apparatus from different localities in Japan and neighboring countries can be found in the following papers.

1. Japan (exclusive of the Southwest Islands): Kawakatsu & Iwaki (1967, pp. 182–184, Figs 3 A–B; Kagoshima in Kyûshû and Yakushima Island); Kawakatsu & Miyazaki (1972, p. 85, Fig. 1, pp. 91–117, pls. VII–XXXII; Nîmi, Okayama Pref. in Honshû); Kawakatsu *et al.* (1972, pp. 121–125, Figs 2 A–F, 3 A–F, 4 A–C; the Izu Peninsula in Honshû and the Izu Islands); Kawakatsu *et al.* (1980, pp. 5–7, Fig. 1, pl. 1 A–K; the Tsushima Islands).
2. The Southwest Islands of Japan: Kawakatsu & Tanaka (1976, pp. 74–77, Figs 1 B–C, 2 A–B; the Benoki River, Okinawa Island).
3. Taiwan: Ichikawa & Kawakatsu (1967, pp. 178–180, Fig. 2 A–D; Taipei and Ulai); Kawakatsu & Iwaki (1968, pp. 132–136, Figs 2 A–E, 3 A–C, 4 A–E; Taichung, etc.); Kawakatsu *et al.* (1979, pp. 65–71, Figs 2 A–I, 3 A–D, 4 A–N; Taipei, Hsinchu, etc.); Kawakatsu *et al.* (1986, pp. 76–83, Figs 1 A–D, 2 A–F, 3 A–U; Mt. Alishan, etc.); Kawakatsu *et al.* (1989, pp. 26–29, Figs 1 A–G, 2 A–I; Liukei, Chipên, etc.).
4. South Korea: Kawakatsu & Kim (1966, pp. 104–107, Figs 1 A–D, 2, 3 A–D; Seoul and the suburbs); Kawakatsu & Kim (1967, pp. 250–251, Fig. 1, pls. 1, Figs A–C, 2, Fig. A; the vicinities of Hwanseon-gul Cave and Kwaneum-gul Cave); Kawakatsu *et al.* (1967, pp. 188-189, Fig. 2 A–C; Quelpart Island); Kawakatsu *et al.* (1976, pp. 95–96; Seoul).
5. China: Ichikawa & Kawakatsu (1967, pp. 179, 182–185, Figs 2 J, 3 C, 4 H; Hangchow); Katô (1950, pp. 188–189, Fig. 1; Shansi-sheng); Kawakatsu & Wong (1975, pp. 265–270, Figs 2 A–G, 3 A–D; Hong Kong); Kawakatsu *et al.* (1976, pp. 102–109, Figs 13–14; Hangchow); Liu, (1989, pp. 39–41, Figs 1, 2 B; Harbin); Tu (1934, pp. 202–203, pl. III; Peking).
6. Northeast Siberia of Russia: Dyganova & Porfirjeva (1990, pp. 27–30, Figs 2–3); Porfirjeva & Timoshkin (1984, pp. 64–65, Figs 4–5; Bikin in Primorskiy, the Bolshaya Ussurka River and the Byra River).

Taxonomic remarks and geographical distribution: D. japonica differs from *D. ryukyuensis* in having a penis papilla without a well-developed valve surrounding its basal part, and a well-developed vagina. Sometimes, the vagina is surrrounded by a halo-like structure, which consist of mesenchymal tissue transversed by several coarse rows of longitudinal muscles and less-developed radial ones. In some populations from Taiwan and China, animals may have a less-developed valve surrounding the basal part of the penis papilla. The species has a chromosome number of $n = 8$, $2x = 16$ and $3x = 24$ (heteroploidy may occur).

D. japonica is distributed in the Japanese Islands, Taiwan, the Korean Peninsula, the south and northeastern areas of China, and the Primorskiy area of Northeast Siberia in Russia (Fig. 1).

Dugesia ryukyuensis Kawakatsu, 1976

The species description based on animals from the type locality was given in the paper by Kawakatsu *et al.* (1976, pp. 96–102, Figs 9 A–H, 10 A–B, 11 A–G, 12).

Based on examination of serial sections of animals from the Southwest Islands of Japan (see the section of *D. japonica*), *D. ryukyuensis* was found in the following 25 stations: Tanegashima Island (Specimen Lot Nos. 1779, 1780); Amani-Ôshima Island (No. 1403); Okino-erabujima Island (No. 928, in Kawakatsu & Tanaka, 1971); Ihiyajima Island (No. 1412); Okinawa Island (Nos. 285, 331, 390, 413, in Ichikawa & Kawakatsu, 1967; Nos. 1103, in Kawakatsu & Tanaka, 1971; Nos. 1387, 1388, 1389, 1491, 1624, 1989); Kumejima Island (Nos. 1409, 1764); Miyakojima Island (No. 1293, in Kawakatsu &Tanaka, 1976; Nos. 1407, 1765, 1766, 1842); Ishigakijima Island (No. 1104, in Kawakatsu & Tanaka, 1971; No. 1990). Schematic figures of the copulatory apparatus of 15 specimens from 7 islands (Tanegashima Island, Amani-Ôshima Island, Okinawa Island, Ihiyajima Island, Kumejima Island, Miyakojima Island, and Ishigakijima Island) are shown in the Seventh ISBT preprint (Kawakatsu *et al.*, 1993, pp. 10–11, Fig. 12 A–D, G–Q).

Non-sexual specimens examined from the following 13 stations are believed to be *D. ryukyuensis*: Okino-erabujima Island (No. 927, in Kawakatsu & Tanaka, 1971); Ihiyajima Island (Nos. 1410, 1411); Okinawa Island (No. 330, in Ichikawa & Kawakatsu, 1967; No. 1156, in Kawakatsu & Tanaka, 1976; Nos. 1371, 1390, 1391). Keramajima Island (No. 1407); Kumejima Island (No. 703, in Kawakatsu & Tanaka, 1971); Ishigakijima Island (No. 1294, in Kawakatsu & Tanaka, 1976; Nos. 1405, 1406). Judging from the karyological data (Tamura *et al.*, 1988, 1991), *D. ryukyuensis* was recorded from Tanegashima Island, central and southern parts of Okinawa Island (3 stations), Miyakojima Island, and Ishigakijima Island.

Additional descriptions of morphology, anatomy and histology of *D. ryukyuensis* with schematic figures of the copulatory apparatus from different localities in the Southwest Islands of Japan, other than the type locality, can be found in the following papers.

(i) Okino-ertabujima Island: Kawakatsu & Tanaka (1971, pp. 48–49, Fig. 2 A, pls. I, Figs A–B, II, Fig. A). / (ii) Okinawa Island: Ichikawa & Kawakatsu (1967, pp. 179–184, Figs 2 E–I, 3 A–B, 4 A– C); Kawakatsu & Tanaka (1971, p. 50, pls. I, Figs C–E, II, Figs B– C). / (iii) Kumejima Island: Kawakatsu & Tanaka (1971, p. 50, pl. I, Figs F–G). / iv) Miyakojima Island: Kawakatsu & Tanaka (1976, pp. 74–77, Figs 1 D, 2 B). / v) Ishigakijima Island: Kawakatsu & Tanaka (1971, pp. 48–50, Fig. 2 B, pls. I, Figs H–I, II, Figs D–E).

Taxonomic remarks and geographical distribution: *D. ryukyuensis* differs from *D. japonica* in having a penis papilla with a well-developed valve surrounding its basal part, and a less developed vagina. Usually, the body of living sexually mature specimens is smaller and more slender than that of *D. japonica*. The species has a chromosome number of $n = 7$, $2x = 14$ and $3x = 24$ (heteroploidy may occur).

D. ryukyuensis is distributed only in the Southwest Island of Japan; the species is not recorded from Yakushima Island, Iriomotejima Island and Yonaguni Island (Fig. 1). Previous records of this species from the alpine region of Taiwan (Kawakatsu *et al.*, 1979) and Hangchow in Southeastern China (Kawakatsu *et al.*, 1976) are incorrect (see the section 'Corrective remarks to the previous paper' in Kawakatsu *et al.*, 1986, p. 80).

Discussion

The present paper suggests that diversity exists in what was once believed to be one phenotype of *D. japonica*. This diversity was originally explained as due to the presence of a geographical subspecies identified as *D. j. ryukyuensis*. However, recent study and reevaluation, as in the present paper, has convinced the authors that the differences indicate a separate species rather than only a subspecies. As was already discussed in the previous paper (Tamura *et al.*, 1991), we can explain the current geographical distribution of these 2

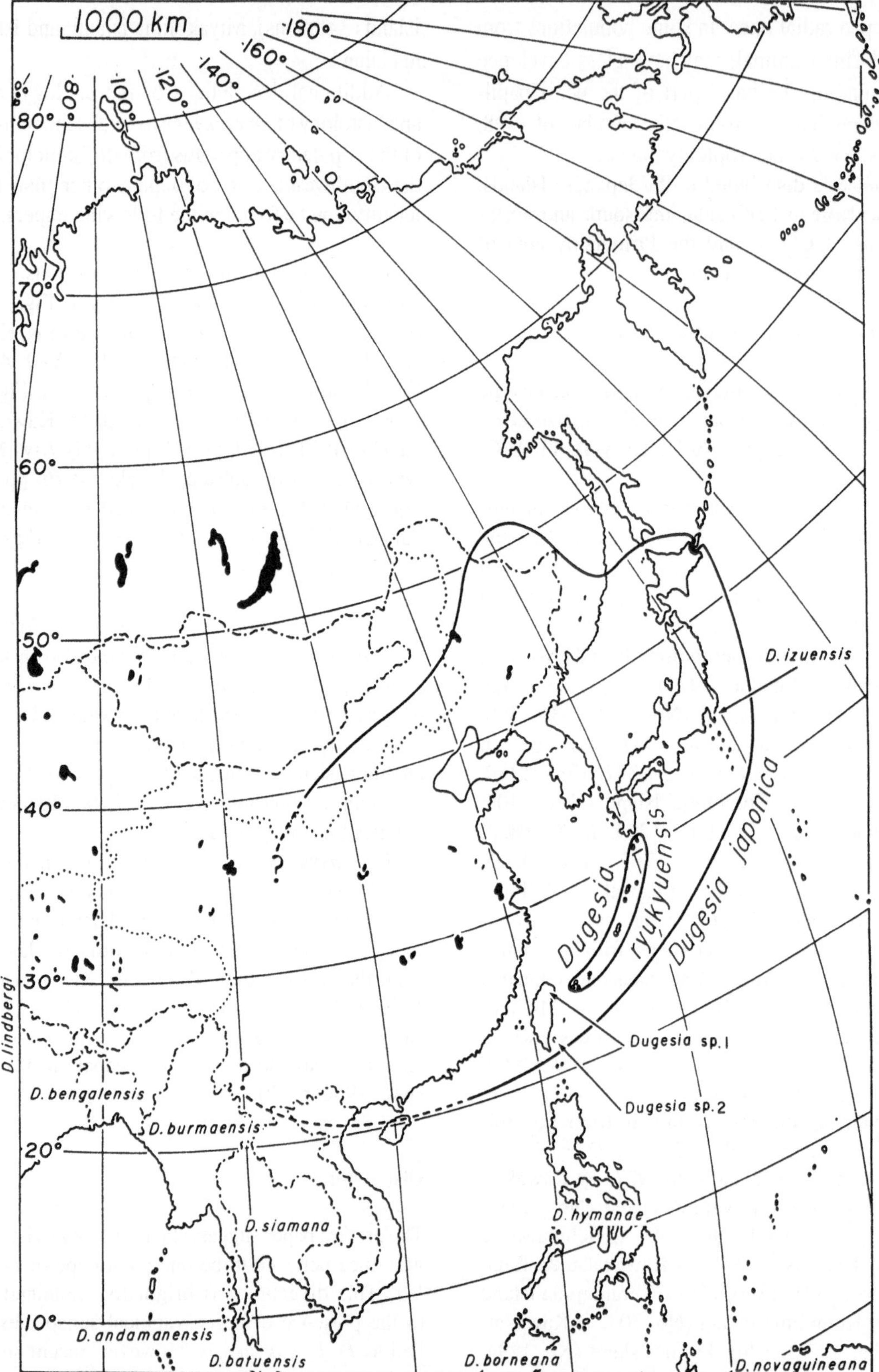

Fig. 1. Map of the Far East, showing the geographical distribution of *Dugesia japonica* and *Dugesia ryukyuensis* (heavy solid lines). Distributions of several Asiatic *Dugesia* species are also indicated.

species in the Southwest Islands of Japan as the result of two separate invasions by 2 proto-*Dugesia* species, *D. ryukyuensis* in the Miocene and *D. japonica* after the early Quaternary (see also 8 palaeogeographical maps of the Far East printed in the Sixth ISBT preprint by Kawakatsu *et al.*, 1990, pp. 13–16, Figs 22–28).

Among the known 22 *Dugesia* species from Asia, the chromosome numbers of the following 7 species are known in addition to *D. japonica* and *D. ryukyuensis* reported in the present paper (see Kawakatsu *et al.*, 1989, p. 11, Fig. 9); they are: *D. tamilensis* Kawakatsu, 1980, from South India (2x = 16; Kawakatsu *et al.*, 1980) and *D. bengalensis* Kawakatsu, 1983, from West Bengal in East India (2x = 16; Kawakatsu, 1983); *D. siamana* Kawakatsu, 1980, from Thailand (2x = 16, 3x = 24; Kawakatsu *et al.*, 1980); *D. batuensis* Ball, 1970, from Malaysia (n = 7, 2x = 14; Kawakatsu *et al.*, 1989); *D. indonesiana* Kawakatsu, 1973, from Indonesia (2n = 16, 3n = 24; Benazzi & Gourbault, 1975); *D. lindbergi* de Beauchamp, 1959, from North India, Afghanistan and Pakistan (probably n = 8, 2n = 16, 3n = 24; Handa & Vashisht, 1977; see Kawakatsu *et al.*, 1980, p. 266, footnote 5); *D. austroasiatica* Kawakatsu, 1985, from Japan (an exotic species, 2x = 16; Kawakatsu *et al.*, 1986). Of these 9 species mentioned above, the karyotype of *D. lindbergi* is not yet fully clarified.

The correlation between the anatomical character of the penis papilla and the karyotypes of *Dugesia* species from Asia is not yet solved. However, the following findings may have phylogenetic importance. Animals of *D. japonica* from some populations have a moderately developed valve at the basal part of the dorsal lip of the asymmetrical penis papilla. This character is also found in some specimens of *D. lindbergi*, *D. indonesiana* and *D. austroasiatica* (see Kawakatsu, 1973a, b; Kawakatsu & Ôgawara, 1974; Kawakatsu *et al.*, 1985). These 4 *Dugesia* species have a haploid number of n = 8 and a diploid number of 2x = 16. Except for *D. austroasiatica*, karyotypes of *D. japonica*, *D. lindbergi* and *D. indonesiana* show striking resemblances; no subtelocentric chromosomes are observed in these species (Benazzi & Gourbault, 1975; Kawakatsu *et al.*, 1980, 1986). No difference of the karyotype is observed in animals of *D. japonica* with or without valves surrounding the basal part of the penis papilla.

Animals of *D. tamilensis*, *D. bengalensis* and *D. siamana* have an asymmetrical penis papilla without basal valves (Kawakatsu *et al.*, 1980, 1983). They have chromosome numbers of n = 8 and 2x = 16; their karyotypes consist of 7 pairs of meta- or submetacentric chromosomes and 1 pair of subtelocentric ones. Animals of *D. austroasiatica* without a valve at the basal part of the penis papilla have a similar karyotype (Kawakatsu *et al.*, 1986).

Animals of *D. batuensis* and *D. ryukyuensis* have an asymmetrical penis papilla with a well-developed valve surrounding its basal part; the same anatomical character is also conspicuous in animals of *D. hymanae* (Šivickis, 1928) from the Philippines (Kawakatsu, 1972). As was already pointed out in a previous paper (Kawakatsu *et al.*, 1989), the chromosome numbers (n = 7, 2x = 14) and karyotypes of *D. batuensis* and *D. ryukyuensis* show a close resemblance. However, the 5th chromosome in *D. batuensis* is submetacentric whereas that in *D. ryukyuensis* is subtelocentric (Kawakatsu *et al.*, 1989, p. 9, Fig. 6). The chromosomes of *D. hymanae* are not known.

Acknowledgments

The cytotaxonomy of *Dugesia* species from the Far East was a matter of primary concern for both of Kawakatsu's and Dr Sluys' teams. Our frank discussion on the subject at Amsterdam (August 17, 1987: Kawakatsu, Oki, Drs Sluys, de Vries & Ball) was useful for the progress of the present study. For this reason, the authors are indebted to Prof. Dr Ronald Sluys and Dr Elizabeth de Vries. They are also grateful to Prof. Dr Robert E. Ogren (Wilkes University, Pennsylvania, USA) for his reviewing an early draft of the present paper.

References

Benazzi, M. & N. Gourbault, 1975. Cytotaxonomical study of *Dugesia indonesiana* Kawakatsu (Tricladida Paludicola). Acc. Naz. Lincei, ser. VIII, 58: 237–243.

Dyganova, R. Y. & N. A. Porfirjeva, 1990. Planarii Aziatskoĭ Chasti SSSR. Morfologiia, Sistematika, Rasprostranenie. Izdatel'stvo Kazanskogo Univ., 1990: 1–52. (In Russian)

Handa, S. M. & K. Vashisht, 1977. Abstract in Sect. Zool., Ent. and Fish., 64th Ind. Sci. Congress, Budhneshwar (Orissa), India. Typewritten copy.

Ichikawa, M. & M. Kawakatsu, 1964. A new freshwater planarian, *Dugesia japonica*, commonly but erroneously known as *Dugesia gonocephala* (Dugès). Annot. Zool. Japon. 37: 185–194.

Ichikawa, A. & M. Kawakatsu, 1967. Report on freshwater planaria from the East China Sea area. Nature & Life in Southeast Asia, 5: 175–188.

Katô, K., 1950. On some turbellarians from Sanshi, North China. Zool. Mag., Tôkyô, 59: 188–190. (In Japanese)

Kawakatsu, M., 1965. On the ecology and distribution of freshwater planarians in the Japanese Islands, with special reference to their vertical distribution. Hydrobiologia 26: 349–408.

Kawakatsu, M., 1967. On the ecology and distribution of freshwater planarians in the Japanese Islands, with special reference to their vertical distribution (Revised Edition). Bull. Fuji Women's College, no. 5: 117–177.

Kawakatsu, M., 1972. The freshwater planaria from the Philippines. Annot. Zool. Japon. 45: 234–244.

Kawakatsu, M., 1973a. Report on freshwater planaria from Indonesia (Sumatra and Java). Contr. Biol. Lab. Kyoto Univ., 24: 87–103 + pls. 1–5.

Kawakatsu, M., 1973b. Report on freshwater planarians from Pakistan. Bull. Fuji Women's College, no. 11, ser. II: 79–95.

Kawakatsu, M., 1974. Further studies on the vertical distribution of freshwater planarians in the Japanese Islands. In: N. W. Riser & M. P. Morse (eds), The Hyman Memorial Volume — Biology of the Turbellaria. McGraw-Hill Book Co., New York: 291–338.

Kawakatsu, M., 1991. History of the study of Turbellaria in Japan. Hydrobiologia, 227 (Dev. Hydrobiol. 69): 389–398.

Kawakatsu, M., I. Horikoshi & H. Akama, 1972. Report on freshwater planarians from the Izu Peninsula and the Izu Islands in Japan. Zool. Mag., Tôkyô, 81: 119–126. (In Japanese, with English summary)

Kawakatsu, M. & S. Iwaki, 1967. Report on freshwater planarian from the Satsunan Islands and Kagoshima (Kyûshû) in South Japan. Bull. Fuji Women's College, no. 5: 179–185.

Kawakatsu, M. & S. Iwaki, 1968. Report on freshwater planaria from Taiwan (Formosa). Bull. Fuji Women's College, no. 6: 129–137.

Kawakatsu, M., S. Iwaki & W.-J. Kim, 1967. Report on freshwater planaria from Quelpart (Cheju) Island, Korea. Zool. Mag., Tôkyô, 76: 187–189. (In Japanese, with English summary.)

Kawakatsu, M. & W.-J. Kim, 1966. Morphological studies on the freshwater planarian, *Dugesia japonica* Ichikawa et Kawakatsu, from Korea. Zool. Mag., Tôkyô, 75: 103–107. (In Japanese, with English summary)

Kawakatsu, M. & W.-J. Kim, 1967. Results of the speleological survey in South Korea 1966. VI. Freshwater planarians from limestone caves in South Korea. Bull. Natn. Sci. Mus. Tokyo, 10: 247–258 + pl. 1–3.

Kawakatsu, M. K.-Y. Lue, M. Takai, H. Hori, A. Muto & S. Osawa, 1986. A new series of studies on the freshwater and land planarians from Taiwan. IV. *Dugesia japonica japonica* Ichikawa et Kawakatsu from the alpine region. Bull. Fuji Women's College, no. 24, ser. II: 75–85.

Kawakatsu, M., R. W. Mitchell, I. Oki, S. Tamura & S. Yussof, 1989. Taxonomic and karyological studies of *Dugesia batuensis* Ball, 1970. (Turbellaria, Tricladida, Paludicola), from the Batu Caves, Malaysia. J. Speleol. Soc. Japan, 14: 1–14 + pl. 1.

Kawakatsu, M. & T. Miyazaki, 1972. Effect of different fixatives on a common Japanese freshwater planarian, *Dugesia japonica* Ichikawa et Kawakatsu. Bull. Fuji Women's College, no. 10, ser. II: 81–117 (+ pls. VII–XXXII). (Both in English and Japanese)

Kawakatsu, M. & G. Ôgawara, 1974. Additional report on freshwater planarians from North Borneo, Malaya, Sri Lanka, and South Africa. Bull. Fuji Women's College, no. 12, ser. II: 69–86.

Kawakatsu, M., I. Oki, S. Tamura, R. E. Ogren, T. Yamada & H. Murayama, 1990. Preprint of papers given at the Sixth International Symposium on the Biology of Turbellaria, Hirosaki, Japan. Occ. Publ., Biol. Lab. Fuji Women's College, no. 22: 1–16.

Kawakatsu, M., I. Oki, S. Tamura & H. Sugino, 1976. Studies on the morphology, karyology and taxonomy of the Japanese freshwater planarian *Dugesia japonica* Ichikawa et Kawakatsu, with a description of a new subspecies, *Dugesia japonica ryukyuensis* subspec. nov. Bull. Fuji Women's College, no. 14, ser. II: 81–126.

Kawakatsu, M., I. Oki, S. Tamura, T. Takai, O. A. Timoshkin & N. A. Porfirjeva, 1993. Preprint of papers given at the Seventh International Symposium on the Biology of Turbellaria, Åbo, Finland. Occ. Publ., Biol. Lab. Fuji Women's College, no. 25: 1–20.

Kawakatsu, M., I. Oki, S. Tamura & T. Yamayoshi, 1985. Reexamination of freshwater planarians found in tanks of tropical fishes in Japan, with a description of a new species, *Dugesia austroasiatica* sp. nov. Turbellaria; Tricladida; Paludicola). Bull. Biogeogr. Soc. Japan 40: 1–19.

Kawakatsu, M., I. Oki, S. Tamura, T. Yamayoshi & K. A. Aditya, 1983. A new freshwater planarian from West Bengal, India. Bull. Biogeogr. Soc. Japan 38: 1–12.

Kawakatsu, M., I. Oki, S. Tamura, T. Yamayoshi, K.-Y. Lue & M. Hagiya, 1979. Additional report on freshwater planarians from Taiwan. Bull. Fuji Women's College, no. 17, ser. II: 59–91.

Kawakatsu, M., I. Oki, S. Tamura, T. Yamayoshi & N. Takahashi, 1980. Morphological, karyological and taxonomic studies of *Dugesia japonica* Ichikawa et Kawakatsu from the Tsushima Islands. Proc. Jap. Soc. Syst. Zool., no. 19: 1–10 + pls. 1–2.

Kawakatsu, M., M. Takai, H. Hori, A. Muto, S. Osawa & K.-Y. Lue, 1989. A new series of studies, etc. VI. *Dugesia japonica japonica* Ichikawa et Kawakatsu, 1964, from two additional localities and *Dugesia* sp. from Lanhsü Island. Bull. Fuji Women's College, no. 27, ser. II: 25–33.

Kawakatsu, M., M. Takai, I. Oki, S. Tamura & M. Aoyagi, 1986. A note on an introduced species of freshwater planarian, *Dugesia austroasiatica* Kawakatsu, 1985, collected from culture ponds of Tirapia mossambica in Saga City, Kyûshû, Japan (Turbellaria, Tricladida, Paludicola). Bull. Fuji Women's College, no. 24, ser. II: 87–94.

Kawakatsu, M., S. Tamura, T. Yamayoshi & I. Oki, 1980. The freshwater planarians from Thailand and South India. Annot. Zool. Japon. 53: 254–268.

Kawakatsu, M. & I. Tanaka, 1971. Additional report on freshwater planaria from the Southwest Islands of Japan. Biol. Mag. Okinawa, no. 8: 46–52 + pls. 1–2.

Kawakatsu, M. & I. Tanaka, 1976. Additional report on freshwater planaria from the Southwest Islands of Japan, II. Zool. Mag., Tôkyô, 85: 73–77.

Kawakatsu, M. & M. H. Wong, 1975. The freshwater planaria from Hong Kong. Annot. Zool. Japon. 48: 262–273.

Liu, D.-Z., 1989. Freshwater planarians (Tricladida) of China. Chinese J. Zool. 24, no. 38–43, 31. (In Chinese)

Oki, I., S. Tamura, T. Yamayoshi & M. Kawakatsu, 1981. Karyological and taxonomic studies of *Dugesia japonica* Ichikawa & Kawakatsu in the Far East. Hydrobiologia 84: 53–68.

Porfirjeva, N. A. & O. A. Timoshkin, 1984. K faune planariĭ (Tricladida, Paludicola) Sikhotė-Alinia. In: Akad. Nauk SSSR, Dal'nevostochnyi Nauchnyi Tsentr Biologo-pochnvennyĭ Inst., Biologiia Presnykh vod Dal'nego Vostoka. Vladivostok: 62–68. (In Russian)

Tamura, S., I. Oki & M. Kawakatsu, 1988. Karyological and taxonomic studies of *Dugesia japonica* from the Southwest Islands of Japan. Fortsch. Zool. 36: 123–127.

Tamura, S., I. Oki & M. Kawakatsu, 1991. Karyological and taxonomic studies of *Dugesia japonica* from the Southwest Islands of Japan-II. Hydrobiologia 227 (Dev. Hydrobiol. 69): 157–162.

Tamura, S., I. Oki & M.Kawakatsu, 1995. A review of chromosomal variation in *Dugesia japonica* and *D. ryukyuensis* in the Far East. Hydrobiologia 305 (Dev. Hydrobiol. 108): 79–84.

Tamura, S., I. Oki, M. Kawakatsu, K.-Y. Lue, M. Takai, H. Hori, A. Muto & S. Osawa, 1985. A new series of studies, etc. II. Chromosomes of *Dugesia japonica japonica* Ichikawa et Kawakatsu, 1964, from Mt. Alishan and the Kenting National Park. Bull. Fuji Women's College, no. 23, ser. II: 127–132.

Tamura, S., I. Oki, M. Kawakatsu, K.-Y. Lue, M. Takai, H. Hori, A. Muto & S. Osawa, 1987. A new series of studies, etc. V. Chromosomes of *Dugesia japonica japonica* Ichikawa et Kawakatsu, 1964, from four Additional localities and *Dugesia* sp. from Lanhsü Island. Bull. Fuji Women's College, no. 25, ser. II: 55–65.

Tu, T.-J., 1934. Notes on some turbellarians from the Tsing Hua Campus. Sci. Rep. Nat. Tsing Hua Univ., B, 1: 191–205 + pls. I–III.

Yamayoshi, T., S. Tamura, I. Oki & M. Kawakatsu, 1980. A survey of the physico-chemical analysis of river waters in relation to the chromosomal variation of freshwater planarians in Ôsaka Prefecture: The Mino'o River and the Kinyûji River. Proc. Ôsaka Prefect. Inst. Public Health, Ed. Public Health, no. 18: 147–158. (In Japanese)

Hydrobiologia **305**: 63–70, 1995.
L.R.G. Cannon (ed.), Biology of Turbellaria and some Related Flatworms.

Geographical distribution of *Phagocata vivida* in the Far East

Masaharu Kawakatsu[1]*, Oleg A. Timoshkin[2], Nina A. Porfirjeva[3] & Masayuki Takai[4]
[1]*Biological Laboratory, Fuji Women's College, Kita-16, Nishi-2, Kita-ku, Sapporo (Hokkaidô) 001, Japan (*Author for correspondence)*
[2]*Limnological Institute, Russian Academy of Sciences, P.O. Box 4199, Ulanbatorskaya 3, Irkutsk 664033, Russia*
[3]*Department of Invertebrate Zoology, Kazan State University, Lenina 18, Kazan 12008, Russia*
[4]*Biological Laboratory, Saga Medical School, Nabeshima 5-1-1, Saga 849, Japan*

Key words: Turbellaria, Tricladida, *Phagocata*, taxonomy, distribution, Japan, Far East

Abstract

Phagocata vivida (Ijima et Kaburaki, 1916) is common in cold-water habitats in mountainous and hilly regions in Japan; in Northern Japan it occurs in lowland areas. Comparative studies of the material from South Korea and Primorskiy in Northeast Siberia, Russia, show that *Ph. vivida* is distributed widely in these geographic areas. *Phagocata miyadii* Okugawa, 1939, reported from North Korea and Northeastern China is a synonym of *Ph. vivida*. Geographical distribution of this species in the Japanese Islands now becomes very clear. Judging by its geographical and vertical distributions, the species probably is a preglacial faunal element that entered Japan by the northern route to Old Honshû Island along the coast of the Old Sea of Japan.

Introduction

Phagocata vivida (Ijima et Kaburaki, 1916) is a common freshwater planarian in Japan mainly inhabiting cold-water streams and springs, both in mountainous and hilly areas (Kawakatsu, 1965a, 1967, 1974; see also 1969, pp. 54–55, Fig. 4). The species was first described by Ijima & Kaburaki (1916) under the name of *Planaria vivida*; their material was obtained from central and northern areas of Honshû (*i.e.*, Kantô and Tôhoku Regions), Japan. The species was also redescribed in detail by Kaburaki (1922a). Kenk (1930) transferred the species into the genus *Fonticola* Komárek, 1926; then, Okugawa (1939, 1940) assigned the species to *Phagocata* Leidy, 1847. Some Russian taxonomists classified *Ph. vivida* under the genus *Penecurva* Livanov & Zabusova, 1940 (see also Zabusova-Zudanova, 1970). The genera *Fonticola* and *Penecurva* (the latter was employed as a subgenus in a paper by Porfirjeva & Timoshkin, 1980) are now considered as synonyms of *Phagocata* (see Kenk, 1974, p. 40). A modern taxonomic redescription of *Ph. vivida*, based upon many living and preserved specimens collected from the vicinity of Hirosaki in Aomori Prefecture, the northernmost area of Honshû, Japan, was published by Kawakatsu *et al.*, (1982); the principal literature for this species was listed (pp. 31–32).

In the summer of 1992, Kawakatsu borrowed many sets of sections of freshwater planarians from Russia (loc. Kamchatka, Sakhalin and Northeast Siberia) from Porfirjeva and Timoshkin; slides of *Ph. vivida* from Russian localities in Primorskiy were included (the material used for their 1984 paper; see also Dyganova & Porfirjeva, 1990). From the results of comparative anatomical and histological studies of serial sections of *Ph. vivida* now retained in Kawakatsu's laboratory, it is now possible to discuss the taxonomic position of this widely distributed species and related species in the Far East.

In neighboring countries of Japan, several *Phagocata* species have been reported from Korea (Kawakatsu, 1967, 1970; Kawakatsu & Kang, 1969; Kawakatsu & Kim, 1967; Kawakatsu & Liu, 1987; Kim, 1967, 1968; Tu & Po, 1959), Northeastern China (Arndt, 1922; Kawakatsu & Liu, 1987; Liu, 1989, 1990; Okugawa, 1939, 1940) and Primorskyi (or Primorskiy) and

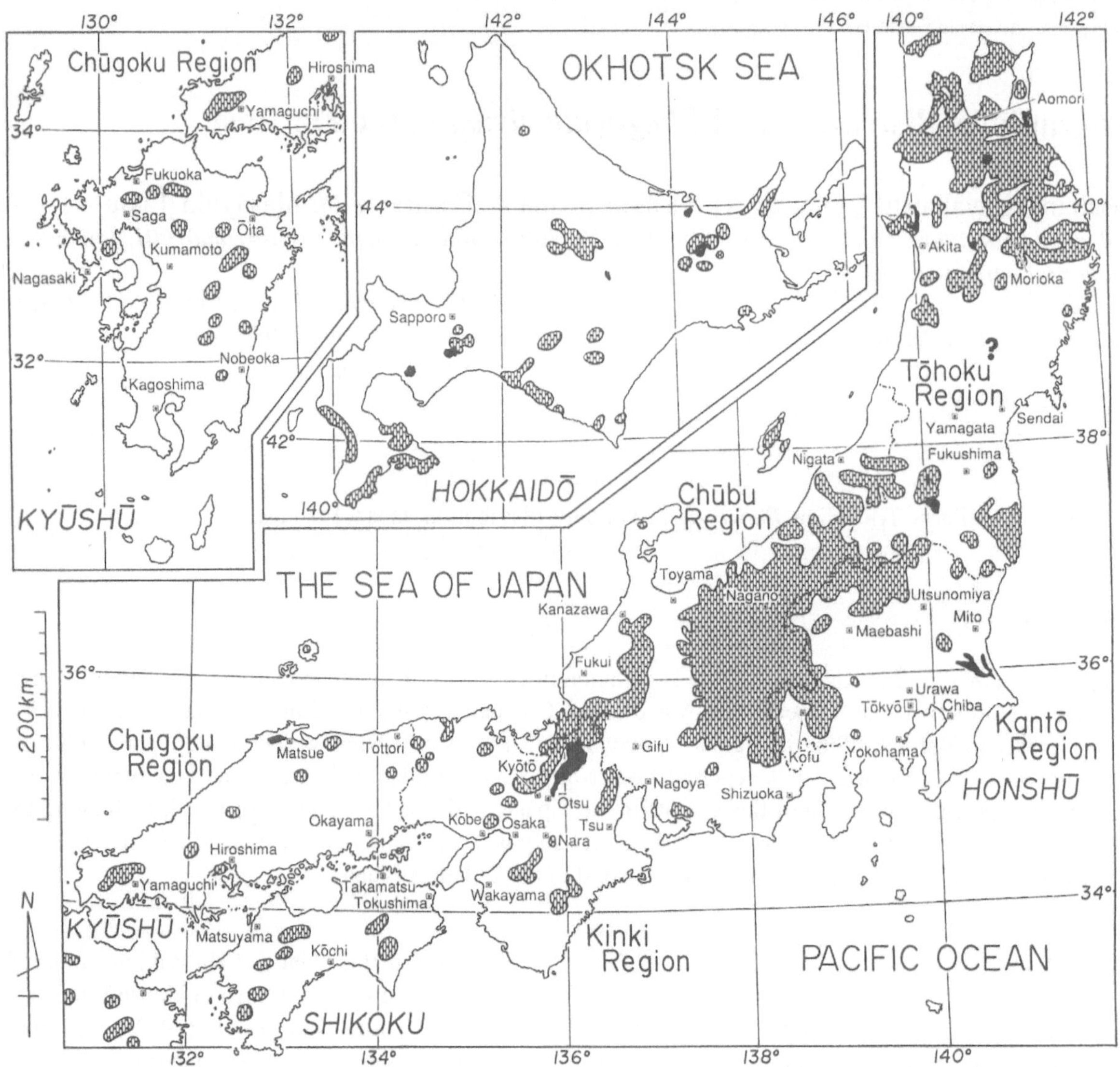

Fig. 1. Map of Japan (Hokkaidô, Honshû, Shikoku, and Kyûshû), showing geographical distribution of *Phagocata vivida* (brick area). For main mountains and volcanic ranges, see Kawakatsu *et al.* (1993, Fig. 4).

in the northeastern area of Siberia, Russia (Dyganova & Porfirjeva, 1990; Porfirjeva & Timoshkin, 1984; Zabusov, 1903; Zabusova-Zudanova, 1962).

Among these records of *Phagocata* species in the Far East, Kawakatsu (1970) identified *Phagocata* sp. (species of Korea) as conspecific with *Ph. vivida* based upon serial sections of animals from several localities in South Korea, including Ullung Do Island in the Sea of Japan. More recently, the occurrence of *Ph. vivida* in Primorskiy (Dyganova & Porfirjeva, 1990; Porfirjeva & Timoshkin, 1984) and Northeastern China (Liu, 1989, 1990) was also reported. *Ph vivida* has a pair of well-developed, pointed auricles on the head. Its external appearance is very similar to that of the following *Phagocata* species reported from the Far East: *Phagocata sibirica* (Zabusov, 1903) (*olim Planaria*) from Northeast Siberia; *Phagocata uenoi* Okugawa, 1939, and *Phagocata miyadii* Okugawa, 1939, from Northeastern China. *Phagocata coarctata* (Arndt, 1920) (*olim Planaria*) from the vicinity of Vladivostok in Primorskiy and *Phagocata pellucida* (Ijima et Kaburaki, 1916) (*olim Planaria*) from the vicinity of Yuzhno-Sakhalinsk (old name, Vladimirofka), Sakhalin.

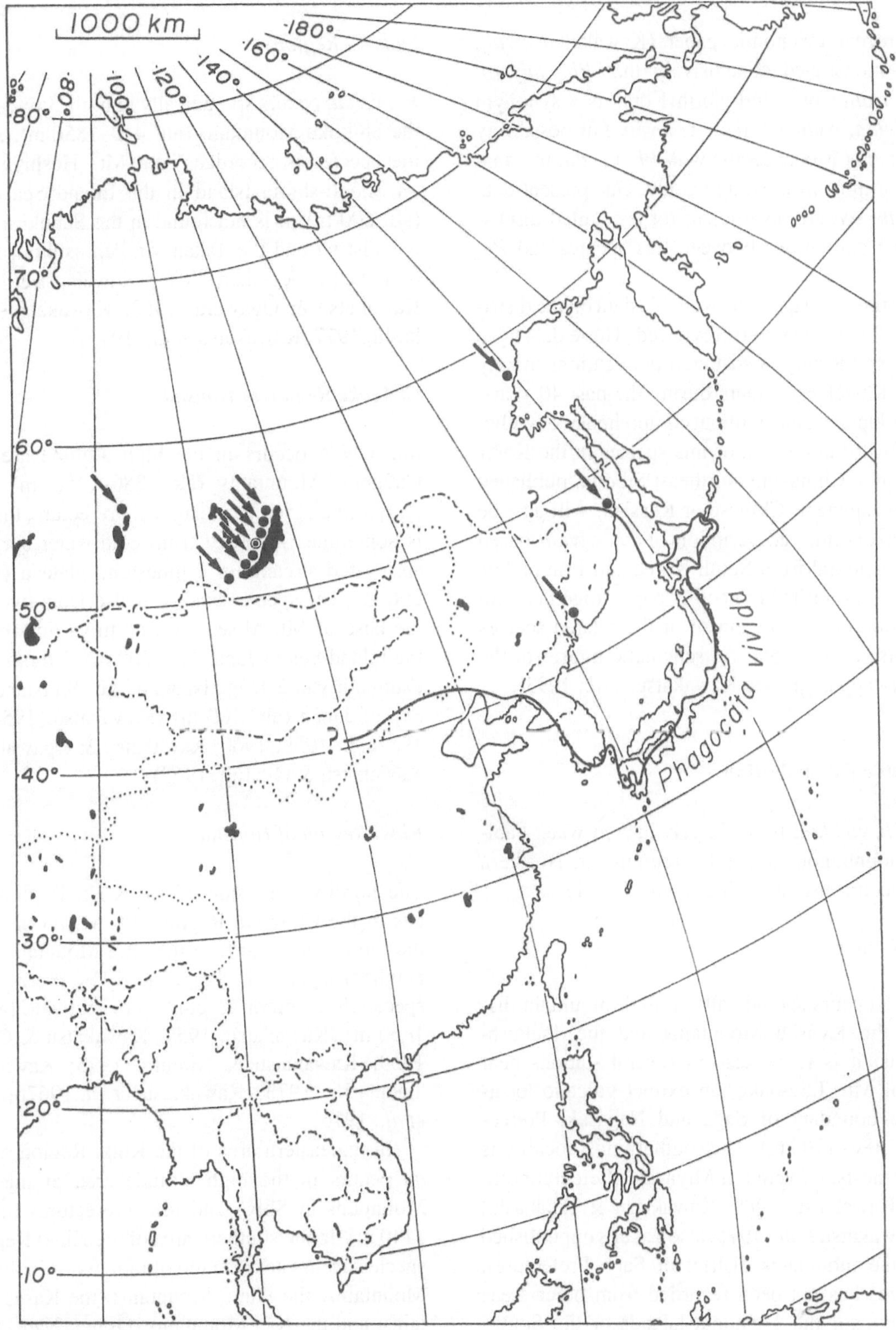

Fig. 2. Map of the Far East, showing the distributional limits of *Phagocata vivida*. The known localities of *Phagocata sibirica* are also shown (solid circles with arrows; double circles, the type locality).

Our recent taxonomic paper (Kawakatsu *et al.*, 1994) demonstrated conclusively that *Ph. miyadii* reported from China and North Korea is a synonym of *Ph. vivida*. Although there is a very fair possibility that *Ph. uenoi* is conspecific with *Ph. vivida*, the final judgement must await future study. Our present data on *Ph. sibirica* are insufficient for a detailed discussion about similarities between this species and *Ph. vivida*.

In the present paper, the available data on the distribution of *Ph. vivida* will be reviewed. These data have appeared so far only in scattered publications mainly from the Kawakatsu's team during the past 40 years, largely in Japanese and without comprehensive synthesis. The distribution data of this species in the Korean Peninsula, China and Northeast Siberia, published mainly in Japanese, Chinese or Russian, will also be given. Taxonomic redescription of *Ph. vivida* based upon the material from South Korea and Primorskiy, Russia, was published in another paper, together with a new taxonomic arrangement of the related species (Kawakatsu *et al.*, 1994; for schematic figures of the copulatory apparatus, see Kawakatsu *et al.*, 1993).

Geographical distribution

Japan: *Ph. vivida* is found largely in cold-water habitats in mountainous and hilly regions; in Northern Japan it occurs also in lowland areas (Fig. 1).

Kyûshû Region

Ph. vivida occurs sporadically in high mountain districts of the Kyûshû Mountains and the Tsukushi Mountains; it is found also in several stations near the top of Mt. Tara-dake, an extinct volcano located at the boundary of Saga and Nagasaki Prefectures (alt. 480–1700 m). The southernmost locality is Mt. Wani-no-tsuka-yama in Miyazaki Prefecture (alt. 550 m) (Kawakatsu, 1989; Kawakatsu & Takahashi, 1977; Kawakatsu *et al.*, 1975; also Takai's unpublished data on the mountains district of Saga Prefecture). The species has not been recorded from other areas of Kyûshû, nor the Tsushima Islands or the Southwest Islands of Japan (Kawakatsu, 1976; Kawakatsu & Tarui, 1955, 1957; Okugawa & Kawakatsu, 1956b).

Shikoku Region

Ph. vivida occurs sporadically in high altitude areas of the Shikoku Mountains (alt. 440–1850 m). Although the species is recorded from Mt. Hoshiga-jô-yama on Shôdo-shima Island in the Inland Sea of Japan (alt. 600 m), it is not found in the Sanuki mountainous district and the Takanawa Peninsula in Northern Shikoku (alt. less than 990 m) (Kawakatsu & Itô, 1963; Kawakatsu & Ôgawara, 1968; Kawakatsu & Takahashi, 1977; Kawakatsu *et al.*, 1975).

Chûgoku Region of Honshû

Ph. vivida occurs in the high altitude area of the Chûgoku Mountains (alt. 380–1450 m). In the Akiyoshi district in Yamaguchi Prefecture, the species is sometimes recorded from cold-water springs and spring-fed streams in a limestone plateau (alt. over 200 m). The species is known also from two islands: the base of Mt. Misen, Aki-no-miya-jima Island, in the Inland Sea of Japan (alt. 200 m); a foothill of Mt. Daimanji-yama, Dôgo Island of the Oki Islands, in the Sea of Japan (alt. 100 m) (Kawakatsu, 1954, 1955, 1957a, b, 1959, 1960; Kawakatsu & Ôgawara, 1969; Kawakatsu & Okafuji, 1965).

Kinki Region of Honshû

The northwestern side of the Kinki Region continues to the Chûgoku Region; *Ph. vivida* occurs near the top of mountains. In the Hira Mountains and the neighboring areas in Shiga and Kyôto Prefectures, the species is common in cold-water habitats (alt. 120–1460 m) (Kawakatsu, 1959; Kawakatsu & Ôgawara, 1969; Kawakatsu & Nîmura, 1975; Kawakatsu & Takahashi, 1973; Kawakatsu *et al.*,1967b; Nakano *et al.*, 1976).

In the eastern area of the Kinki Region, *Ph. vivida* occurs in the high altitude area of the Suzuka Mountains in Shiga and Mie Prefectures (alt. 610–1140 m). In the southern area of the Kinki Region, the species occurs near the top of mountains in the Kongô Mountains, the Izumi Mountains, the Kasagi Mountains, and the Ki'i Mountains (Ôsaka, Nara, Mie and Wakayama Prefectures; alt. 600–1580 m) (Kanasaki, 1969; Kawakatsu & Takahashi, 1973; Kawakatsu *et al.*, 1967a, 1968). The southern limit of distribution of this species is in Mt. Gyôjakaeri-dake of the Ki'i Mountains in the Yoshino-Kumano National Park (alt. 480–950 m) (Kawakatsu & Takakhashi, 1973). *Ph.*

vivida is not recorded from Awaji-shima Island in the Bay of Ôsaka (Kawakatsu & Ôgawara, 1968).

Chûbu Region of Honshû

The Chûbu Region includes 3 districts: the Hokuriku district situated north on the Sea of Japan, the central, high mountainous district (so-called Japan Alps) and the Tôkai district south on the Pacific Ocean.

The Ryôhaku Mountains located in the northwestern part of the Hokuriku district is a continuation of the Hira Mountains and the Suzuka Mountains of the Kinki Region. *Ph. vivida* is common in these mountains areas (alt. 100–2200 m). In the Hokuriku district, *Ph. vivida* occurs sporadically in cold-water habitats in the Noto Peninsula, the Higashi-kubiki hilly district, the Uonuma hilly district, and other low mountainous districts in Toyama and Nîgata Prefectures (alt. 100–1520 m). *Ph. vivida* is also common on the western side of the Echigo Mountains in the northwestern part of the Hokuriku district (alt. 70– 1500 m; including Mr Murayama's unpublished data). The species is also distributed in mountains of Sado Island in the Sea of Japan (alt. 5-840 m) (Honma *et al.*, 1972; Kawakatsu, 1954; Kawakatsu & Murayama, 1975; Kawakatsu *et al.*, 1971, 1978, 1979).

The central district of the Chuûbu Region is characterized by many high mountains belonging to the Hida Mountains, the Kiso Mountains and the Akaishi Mountains (there are several peaks over 3000 m in altitude); the Hakusan Volcanic Zone, the Norikura Volcanic Zone and the Fuji Volcanic Zone are there. *Ph. vivida* is the commonest species in these mountain areas (alt. 200–2700 m) (K. Aoki, 1973; Y. Aoki, 1957; Hara, 1968, 1969, 1970; Hirose, 1954; Horasawa & Imafuku, 1935; Ichikawa & Kawakatsu, 1967; Kasai, 1940; Kawakatsu, 1959, 1965a, 1967; Kawakatsu & Iwaki, 1967; Kawakatsu & Nîmura, 1977; Kawakatsu *et al.*, 1971, 1974, 1975, 1985; Murayama, 1977, 1993; Nîmura (=Ni-imura), 1973, 1980, 1984; Nîmura & Murayama, 1972; Ochiai & Kurata, 1959; Okugawa & Kawakatsu, 1956a; Okugawa *et al.*, 1955; Sonehara, 1957; Uéno, 1929, 1931, 1932, 1934, 1935a, b). According to Mr Murayama's unpublished data on the distribution of freshwater planarians in the Mts Shirouma-Tateyama-Kurobe area, the northern part of the Hida Mountains, *Ph. vivida* occurs in many stations (alt. 320–2280 m).

In the Tôkai district, *Ph. vivida* is very rare; it is now only recorded from several localities in the Mino-Mikawa Heights (Mt. Dando-san and Mt. Hongû-san, near Okazaki in Aichi Prefecture; alt. 200–960 m) (Kawakatsu *et al.*, 1978). *Ph. vivida* is not found in the other lowland areas in Aichi and Shizuoka Prefectures (Kawakatsu *et al.*, 1985); it is not distributed in the Izu Peninsula and in the Izu Islands (Kawakatsu & Horikoshi, 1972; Kawakatsu *et al.*, 1972, 1985).

Kantô Region of Honshû

Ph. vivida is common in the middle and high mountain areas; a few populations of this species are found in Mt. Tanzawa-yama (Kanagawa Prefecture), Mt. Bukô-yama (Saitama Prefecture), Mt. Haruna-san (Gunma Prefecture), Mt. Tsukuba-san, and Mt. Yamizu-san (Ibaraki Prefecture) (alt. 400–2400 m) (Akama, 1969; Chinone, 1961; Horikoshi, 1975; Kaburaki, 1922, 1922, 1936; Katayama, 1972, 1978a, b; Kawai, 1954; Kawakatsu & Horikoshi, 1969; Kawakatsu *et al.*, 1977, 1978; Kugaya, 1963; Nishimura & Horikoshi, 1972; Yamada, 1980, 1984).

Tôhoku Region of Honshû

In the southern part of the Tôhoku Region (especially in Fukushima Prefecture), *Ph. vivida* occurs in many habitats in mountainous areas. The species is also distributed widely in mountain areas in Akita and Iwate Prefectures. In the Aomori Prefecture (the northernmost part of Honshû), the species is very common in cold-water habitats both in the lowland and high altitude areas (alt. 0–1200 m) (Furu'uchi, 1968; Furu'uchi *et al.*, 1970; Furu'uchi & Mutô, 1968, 1970; Kamiguchi, 1967; Kawakatsu, 1959, 1961a, b, 1978, 1985; Kawakatsu *et al.*, 1967, 1968, 1969, 1970a, b, 1976, 1982; Naoki, 1987, 1988, 1989, 1990, 1991: Terayama *et al.*, 1968; Teshirogi, 1974; Teshirogi & Ishida, 1980; Teshirogi *et al.*, 1978a, b, 1980, 1981).

In the Tôhoku Region, the distribution data for freshwater planarians are scanty in Yamagata and Miyagi Prefectures; we have a few data on *Ph. vivida* in this geographical area. *Ph. vivida* is not distributed on Awashima Island in the Sea of Japan (Murayama, 1970).

Hokkaidô Region

The distribution range of *Ph. vivida* in the Hokkaidô Region is distinct. In the southern part of Hokkaidô (*i.e.*, the Oshima Peninsula), the species is only found in the lowland area and the foot of mountains. In the southeastern part (the vicinity of Sapporo, the Shikotsu-Tôya National Park and the Erimo district), the species occurs sporadically both in lowland and mountainous areas. In the northeastern part (*i.e.*, Kushiro, the Akan National Park and the Shiretoko National Park), localities of *Ph. vivida* are scanty and limited. In the central part of Hokkaidô (the Daisetsuzan National Park), *Ph. vivida* is rather common in the alpine area (only in the Ishikari River system) (alt. 0–1770 m) (Ichikawa, 1953; Kawakatsu, 1959, 1965; Kawakatsu & Shimamura, 1974; Kawakatsu & Yamada, 1965, 1966, 1967, 1967, 1968; Kawakatsu *et al.*, 1969; Shimamura *et al.*, 1968; Yamada, 1955; Yamada & Kawakatsu, 1965, 1966a; Yamada & Tanji, 1964; Yanagita, 1969). The northernmost locality of this species is a cold-water spring at On'nenai in northern Hokkidô (Kawakatsu, 1974, pp. 330–332, Fig. 21, tables 10A and B; Yamada, 1966).

Ph. vivida is not found in the following areas in Hokkaidô: the Oshima-Hiyama district (Kawakatsu *et al.*, 1968), the Shakotan Peninsula and the Niseko mountainous district (Kawakatsu *et al.*, 1969; Yamada & Kawakatsu, 1966b), the Tôya district in the Shikotsu-Tôya National Park (Yamada & Kawakatsu, 1966c); the Yûbari Mountains (Yamada & Kawakatsu, 1967), the Uryû-numa district (Kawakatsu *et al.*, 1969; Yûmen, 1968), the southwestern part of the Daisetsuzan National Park (Yamada & Tanji, 1962, 1963; Yanagita, 1974, 1975), and Mt. Wenshiri-dake and Mt. Piyashiri-dake in the Teshio and Kitami districts (Yamada, 1961, 1962, 1964, 1965b, 1966). *Ph. vivida* is also not distributed on islands off Hokkaidô: Okushiri Island, Teure Island, Yangeshiri Island, Rishiri Island, and Rebun Island (Kawakatsu, 1958a, b, Kawakatsu & Tarui, 1959; Kawakatsu & Yamada, 1966).

Other countries in the Far East

Ph. vivida (including the data of *Ph. miyadii*, a synonym of *Ph. vivida* is recorded from South Korea, North Korea, Northeastern China, and Primorskiy in Russia (Fig. 2).

South and North Korea

In South Korea, *Ph. vivida* is recorded from cold-water stations of the Sobaek and the Taebaek Mountains (alt. 600– 1500 m) (Kawakatsu, 1970; Kawakatsu & Kang, 1969; Kawakatsu & Liu, 1987; Kim, 1967, 1968). The species is also recorded from a cave-stream of Simboggul Cave, Kumdae-ri, located in the western part of South Korea (alt. 450 m; Kawakatsu & Kim, 1967) and Ullung-do Island on the Sea of Japan (Kawakatsu, 1970; Kawakatsu & Kang, 1969; Kim, 1968). *Ph. vivida* is not distributed in lowland areas in the vicinty of Seoul (Kawakatsu & Kim, 1966; Kim, 1964) and Cheju-do Island on the East China Sea (Kawakatsu *et al.*, 1967).

In North Korea, the species was recorded from 2 localities under the name of '*Ph. miyadii*' (Yangdeg in Pyeongan-nam-do and Fouhsi, near Pyonggang, in Gangweon-bug-do (Kawakatsu & Liu, 1987; Tu & Po, 1959).

China

The main mountains of northern China are the Chang-Kuang-tsi-Ling Mountains, the Hisiao-Hsing-an-Ling Mountains (the Lesser Range) and the Ta-Hsing-an-Ling Mountains (the Greater Khingan Range). *Ph. vivida* is distributed widely in mountainous areas: the species is also recorded from Neimêngku (Inner Mongolia) (Kawakatsu & Liu, 1987; Liu, 1989, 1990).

Primorskiy Region of Russia

The Sikhoté-Alin Mountains are located in the Primorskiy district (altitude of the highest peak is 2077 m). *Ph. vivida* is recorded from the following localities: Rudnaja, Gorbuscha, Frolovka, Sihote, Edinka, Kedrovaja, Bira, Popova & Russkij Island (Dyganova & Porfirjeva, 1990; Porfirjeva & Timoshkin, 1984).

Ph. vivida is not recorded from the Chukotskii Peninsula in northeastern Siberia (Porfirjeva & Timoshkin, 1980), Sakhalin (Zabusova-Zudanova, 1960), nor the Kuril Islands (or Kurilskiye Ostrava) (Miyadi, 1937).

Distribution range and zoogeography

The boundaries of the geographical distribution of *Ph. vivida* can now be described. Its range is bounded on the south, in the Japanese Islands, by a line running from Mt. Tara-dake, over Mt. Wani-no-tsukasa-yama in the Kyûshû Mountains, the Shikoku Mountains, the Kongô-Izumi-Ki'i Mountains, the Suzuka Mountains (Kinki Region), the Mino-Mikawa Heights, the Akaishi Mountains (Chûbu Region), Mt. Tanzawa-yama, Mt. Bukô-yama, Mt. Haruna-san, Mt. Tsukuba-san (Kantô Region), to the Abukuma-Kitagami Mountains (Tôhoku Region). To the north, in Hokkaidô, the range is bounded by a line from the seashore region of the Oshima Peninsula, over the Lake Shikotsu-ko area, Mts. Daisetsu-zan, On'nenai, to the northern base of the Shiretoko Peninsula. The boundary line mentioned here is shown in Fig. 2 (see also Kawakatsu & Takahashi, 1973, p. 118, Fig. 2, 1977, p. 93, fig. 1; Kawakatsu & Yamada, 1967, p. 55, Fig. 2). The presumed distributional limits in other areas of the Far East is added in Fig. 2.

Ph. vivida is a stenothermal species among stream-dwelling planarians in the Far East. The vertical distribution of this species in the Japanese Island is described and discussed by Kawakatsu (1965, 1967, 1974). Judging from its geographical distribution in the Far East, the species probably entered the Old Honshû Island by northern routes along the coast of the Old Sea of Japan (*i.e.*, the third transmigrating route of freshwater planarians in the Far East proposed by Kawakatsu, 1974, p. 301; see also Kawakatsu, 1965, 1967; Kawakatsu *et al.*, 1991).

Acknowledgments

The authors wish to thank Prof. Dr Robert E. Ogren (Wilkes University, Pennsylvania, USA) for his reviewing of an early draft of the present paper.

References

Because of limitations in space, it is impossible to fully list the publications referred to above. As a rule the articles excluded from the following list may be found in Kawakatsu (1968–1993; Bull. Fuji Women's College, nos. 6–31, ser. II). However, several Japanese articles including important distribution maps and taxonomy of freshwater planarians in the Far East are given here.

Arndt, W., 1922. Untersuchungen an Bachtricladen. Ein Beitrag zur Kenntnis der Paludicolen Korsikas, Rumäniens und Sibiriens. Z. wiss. Zool. 120: 98–146 + pl. IV.

Dyganova, R. Y. & N. A. Porfirjeva, 1990. Planarii Aziatskoĭ Chasti SSSR. Morfologiîa, Sistematika, Rasprostranenie. Izdatel'stvo Kazanskogo Univ. 1990: 1–52. (In Russian)

Ichikawa, A. & M. Kawakatsu, 1967. Records of two planarian species of the family Kenkiidae from Japanese subterranean waters. Arch. Hydrobiol. 63: 512–519.

Ijima, I. & T. Kaburaki, 1916. Preliminary descriptions of some Japanese triclads. Annot. Zool. Japon. 9: 153–171.

Kaburaki, T., 1922. On some Japanese freshwater triclads; with a note on the parallelism in their distribution in Europe and Japan. J. Coll. Sci. Imp. Univ. Tokyo, 44: 1–71 + pl. I.

Kawakatsu, M., 1960. Notes on the freshwater planarians found in the subterranean waters of the Akiyoshi district. Jap. J. Zool. 12: 609–620.

Kawakatsu, M., 1965. On the ecology and distribution of freshwater planarians in the Japanese Islands, with special reference to their vertical distribution. Hydrobiologia 26: 349–408.

Kawakatsu, M., 1967. On the ecology and distribution of freshwater planarians in the Japanese Islands, with special reference to their vertical distribution (Revised Edition). Bull. Fuji Women's College, no. 5: 117–177.

Kawakatsu, M., 1968–1993. A list of publications on Japanese turbellarians, etc. Bull. Fuji Women's College, no. 6–31, ser. II.

Kawakatsu, M., 1969. An illustrated list of Japanese freshwater planarians in color. Bull. Fuji Women's College, no. 7, ser. II: 45–91 (+ pls. VII-VIII).

Kawakatsu, M., 1970. Methods of the ecological survey of freshwater planarians. Korean J. Limnol. 3, nos. 1–2: 11–14. (In Korean, with English summary)

Kawakatsu, M., 1974. Further studies on the vertical distribution of freshwater planarians in the Japanese Islands. In N. W. Riser & M. P. Morse (eds), The Hyman Memorial Volume — Biology of the Turbellaria. McGraw-Hill Book Co., New York: 291–338.

Kawakatsu, M. & S. Iwaki, 1967. Studies on the morphology, taxonomy and ecology of freshwater planarian, Phagocata kawakatsui Okugawa, with remarks on distribution. Jap. J. Ecol. 17: 214–224.

Kawakatsu, M., S. Iwaki & W.-J. Kim, 1967. Report of freshwater planaria from Quelpart (Cheju) Island, Korea. Zool. Mag., Tôkyô, 76: 187–189. (In Japanese, with English summary)

Kawakatsu, M. & S.-W. Kang, 1969. Annotated bibliography of the Korean Turbellarians. Korean J. Limnol. 2, nos. 3–4: 42–49. (In English, Korean and Japanese)

Kawakatsu, M. & W.-J. Kim, 1966. Morphological studies on the freshwater planarian, Dugesia japonica Ichikawa et Kawakatsu, from Korea. Zool. Mag., Tôkyô, 75: 103–107. (In Japanese, with English summary)

Kawakatsu, M. & W.-J. Kim, 1967. Results of the speleological survey in South Korea 1966. VI. Freshwater planarians from limestone caves in South Korea. Bull. Natn. Sci. Mus. Tokyo, 10: 247–258 + pls. 1–3.

Kawakatsu, M. & D.-Z. Liu, 1987. History of study of Trbellaria in China. Part. 3: Supplementary notes on the turbellariology in the People's Republic of China. Bull. Fuji Women's College, no. 25, ser. II: 39–54.

Kawakatsu, M., I. Oki, S. Tamura, M. Takai, O. A. Timoshkin & N. A. Porfirjeva, 1993. Prepringt of papers given at the Seventh International Symposium on the Turbellaria, Åbo, Finland. Occ. Publ., Biol. Lab. Fuji Women's College, no. 25: 1–20.

Kawakatsu, M. & N. Takahashi, 1973. Report on the ecological survey of freshwater planarians in the mountainous district near Ôsaka City, the Kongô Mountains, and the Ki'i Mountains, Honshû. Zool. Mag., Tôkyô, 82: 114–120. (In Japanese, with English summary)

Kawakatsu, M. & N. Takahashi, 1977. Report on the ecological survey of freshwater planarians in the southern part of Shikoku and the southeastern part of Kyûshû, with an additional note on the southern limit of distribution of Phagocata vivida in Southwest Japan. Bull. Fuji Women's College, no. 15, ser. II: 91–96. (In Japanese, with English summary)

Kawakatsu, M., N. Takahashi, G. Okafuji & H. Yoshida, 1975. Report on the ecological survey of freshwater planarians in the Mt. Onigajô district in Shikoku and Kyûshû, with a note on the Southern limit of distribution of Phagocata vivida in Southwest

Japan. Bull. Fuji Women's College, no. 13, ser. II: 79–91 + pls. III–VII. (In Japanese, with English summary)

Kawakatsu, M., W. Teshirogi & S. Ishida, 1982. Morphology of Phagocata vivida (Ijima et Kaburaki, 1916) with supernumerary eyes collected from Mt. Iwaki-san in Aomori Prefecture, Tôhoku Region, Honshû, Japan. Sci. Rep. Hirosaki Univ. 29: 30–46.

Kawakatsu, M., W. Teshirogi & T. Tokui, 1976. Record of a freshwater planarian, Polycelis sapporo (Ijima et Kaburaki, 1916), from the bottom of Lake Towada-ko in North Japan, with a note on the southern limit of distribution of this species. Physiol. Ecol. Japan, 17 (Dedicated to Professors Syuiti Mori and Masaaki Morishita in Commemoration of their retirements from Kyoto University): 477–483.

Kawakatsu, M., O. A. Timoshkin, N. A. Porfirjeva & M. Takai, 1994. Taxonomic notes on Phagocata vivida (Ijima et Kaburaki, 1916) from South Korea and Primorskiy, Russia (Turbellaria: Tricladida: Paludicola). Bull. Biogeogr. Soc. Japan, 49, 1.

Kawakatsu, M. & T. Yamada, 1967. Report on the ecological survey of freshwater planarians in the southern part of Shikotsu-Dôya National Park (Lake Shikotsu-ko district), Hokkaidô, with a note on the distribution of Phagocata vivida in North Japan. Zool. Mag., Tôkyô, 76: 50–56. (In Japanese, with English summary)

Kawakatsu, M., T. Yamada, H. Murayama & Y. Naoki, 1991. Geographical distribution of Polycelis (Seidlia) auriculata in Japan. Hydrobiologia 227 (Dev. Hydrobiol. 69): 355–363.

Kenk, R., 1930. Beiträge zum System der Probursalier (Tricladida Paludicola). Zool. Anz. 89: 145–162, 289–302.

Kenk, R., 1974. Index of the genera and species of the freshwater triclads (Turbellaria) of the world. Smith. Contr. Zool., no. 183: i–ii, 1–90.

Kim, W.-J., 1964. Notes on freshwater planaria from Korea (with an appendix written by M. Kawakatsu). Collect. & Breed., Tôkyô, 26: 261–263. (In Japanese)

Kim, W.-J., 1967. Notes on freshwater planarians from the Sobaek Mountains, Korea (with an appendix written by M. Kawakatsu). Collect & Breed., Tôkyô, 29: 323–325. (In Japanese)

Kim, W.-J., 1968. Notes on freshwater planarians from the Taebaek Mountains, Korea (with an appendix written by M. Kawakatsu). Collect. & Breed., Tôkyô, 30: 59–62. (In Japanese)

Liu, D.-Z., 1989. Freshwater planarians (Tricladida) of China. Chinese J. Zool. 24, no. 6: 38–43, 31 (In Chinese)

Liu, D.-Z., 1990. On the discovery of the family Dendrocoelidae and Phagocata vivida from China. Acta Zootax. Sinica, 15: 124–127. (In Chinese, with English summary)

Miyadi, D., 1937. Limnological survey of the North Kuril Islands. Arch. Hydrobiol. 31: 433–483.

Okugawa, K. I., 1939. Probursalia (Tricladida-Paludicola) of Manchoukuo. Annot. Zool. Japon. 18: 155–164 + pl. 7.

Okugawa, K. I., 1940. Probursalia of Manchoukuo. In: Reports of the Limnological Survey of Kwantung and Manchoukuo. Kyôto: 437–444. (In Japanese)

Porfirjeva, N. A. & O. A. Timoshkin, 1980. Planariĭ (Tricladida, Paludicola) Chukotskogo poluostrova i ikh zoogeograficheskie svi͡azi. In: Akad. Nauk SSSR, Dal'nevostochnyi Nauchnyi Tsentr Biologo-pochvennyĭ Inst., Fauna Presnykh vod Dal'nego Vostoka. Vladivostok: 37–43. (In Russian)

Porfirjeva, N. A. & O. A. Timoshkin, 1984. K faune planariĭ (Tricladida, Paludicola) Sikhotė-Alinia. In: Akad. Nauk SSSR, Dal'nevostochnyi Nauchnyi Tsentr Biologo-pochvennyĭ Inst., Biologiia Presnykh vod Dal'nego Vostoka. Vladivostok: 62– 68. (In Russian)

Teshirogi, W., S.-I. Sasaki & M. Kawakatsu, 1981. Freshwater planarians from Lake Usoriyama-ko and its lake-side area, the Shimokita Peninsula, in North Japan. Sci. Rep. Hirosaki Univ. 28: 84–96.

Tu, T.-J. & H.-K. Po, 1959. Distribution of Dugesia (=Euplanaria) gonocephala (Dugès) in China and North Korea. Chinese J. Zool. 9: 416–419. (In Chinese)

Zabusova-Zudanova, Z. I., 1960. Planarii dal'nego Vostoka. Trudy Obshch. Estestvoisp. Kazanskom Gosud. Univ. 63: 112–138. (In Russian)

Zabusova-Zudanova, Z. I., 1962. Russelenie presnovodnykh planariĭ v Krasnoiarskom Krae. Trudy Obshch. Estestvoisp. Kazanskom Gosud. Univ. 65: 87–101. (In Russian)

Zabusova-Zudanova, Z. I., 1970. Rasselnie planariĭ po sovetskomu soiuzu. In: Voprosy Ėvoli͡ut͡sionnoi Morfologii i Biotsenologii. Izdatel'stvo Kazanskogo Univ. 1970: 164–172. (In Russian)

Hydrobiologia **305**: 71–77, 1995.
L.R.G. Cannon (ed.), *Biology of Turbellaria and some Related Flatworms.*

Chromosomes of *Temnocephala minor*, an ectosymbiotic turbellarian on Australian crayfish found in Kagoshima Prefecture, with karyological notes on exotic turbellarians found in Japan

Iwashiro Oki[1], Sachiko Tamura[2], Masayuki Takai[3] & Masaharu Kawakatsu[4*]
[1]*Ôsaka Environmental Project Association, MFC 7F, Uchihon-machi 1-2-15, Chû'ô-ku, Ôsaka 540, Japan*
[2]*Ôsaka Prefectural Institute of Public Health, Nakamichi-1-chôme 3-69, Higashinari-ku, Ôsaka 537, Japan*
[3]*Biological Laboratory, Saga Medical School, Nabeshima 5-1-1, Saga 849, Japan*
[4]*Biological Laboratory, Fuji Women's College, Kita-16, Nishi-2, Kita-ku, Sapporo (Hokkaidô) 001, Japan*
*(*Author for correspondence)*

Key words: Turbellaria, Temnocephalida, Tricladida, *Dugesia*, *Bipalium*, *Platydemus*, Japan

Abstract

Records of exotic turbellarian species found in Japan are reviewed from taxonomic and karyological viewpoints. *Temnocephala minor* Haswell, 1888, an ectocommensal on a freshwater crayfish of Australia, was found from culture ponds of *Cherax tenuimanus* (introduced from W. Australia) in Kagoshima Prefecture. *T. minor* had the chromosome number of 2x = 18 (2sm + 2m + 2m + 2sm + 2m + 2m + 2m + 2sm + 2m). The following 3 species of exotic freshwater triclads were recorded from tanks and ponds used for tropical fish culture: *Dugesia austroasiatica* Kawakatsu, 1985 (2x = 16), *Dugesia tigrina* (Girard, 1850) (2x = 16) and *Rhodax*? sp. (3x = 24; 3x = 24 & 3x + 1LB + 1SB = 25 + 1SB). The following 3 species of exotic terrestrial triclads were recorded: *Bipalium nobile* Kawakatsu et Makino, 1982 (2x = 10), *Bipalium kewense* Moseley, 1878 (2x = 18), and *Platydemus manokwari* de Beauchamp, 1962 (*n* = 6, 2x = 12). An extensive occurrence of *P. manokwari* in the Southwest Islands of Japan may be due to an unexpected introduction of the animal in very recent years.

Introduction

After World War II, in Japan, several species of supposedly exotic freshwater and terrestrial triclad turbellarians have been reported by our team (see Table 1). Species identification was made by examination of serial sections of sexual specimens; chromosomal analysis of sexual and/or asexual specimens was employed in some cases. In 1985, we studied chromosomes of a dugesiid species collected from culture ponds containing freshwater crayfish from Western Australia, *Cherax tenuimanus* (called marron in Australia), at the Ibusuki Branch of the Kagoshima Prefectural Fisheries Experimental Station (KPFES), Kyûshû, Japan. These non-sexual specimens were identified as *Dugesia tigrina* (Girard, 1850) based upon the pigmented pharynx and the karyotype (Tamura *et al.*, 1985). At that time many live specimens of a temnocephalid species, ectocommensal on marron, were collected. Although chromosomes of that species were observed by Oki and Tamura, the data have not been published. Judging from the comparative observation of whole-mounted specimens of the temnocephalid species in question, and other specimens on marrons of Western Australia received from Dr N. Morrissy, it is diagnosed as *Temnocephala minor* Haswell, 1888.

In our paper, we first describe the karyological data of this species, together with some remarks about the known karyotypes of this animal group reported from Uruguay. Secondarily, we review available data on exotic turbellarians because they have appeared in scattered publications.

Table 1. Records of exotic Turbellarians in Japan

Species	Chromosome	Localities and References
Temnocephalida		
Temnocephalidae		
Temnocephala minor	2x=18	Ibusuki, Kagoshima (Tamura *et al.*, 1985; Kawakatsu, Oki, Tamura, Takai *et al.*, 1993; present paper)
Tricladida		
Dugesiidae		
*Dugesia austroasiatica**	2x=16	Sapporo (Kawakatsu & Hirai, 1968); Musashino, near Tôkyô (Kawakatsu *et al.*, 1985); Chiba (Kawakatsu, Oki, Tamura, Takai *et al.*, 1993; present paper); Nîgata (Hirao *et al.*, 1970); Saga (Kawakatsu, Takai *et al.*, 1986; Takai *et al.*, 1986)
Dugesia tigrina	2x=16	Yokohama (Kawakatsu *et al.*, 1985); Nagoya (Kawakatsu *et al.*, 1985); Nagasaki (Kawakatsu, Tamura *et al.*, 1993); Kagoshima (Tamura *et al.*, 1985)
Dimarcusidae		
Rhodas sp.	3x=24	Sapporo (unpublished data); Nagoya (Kawakatsu *et al.*, 1985)
Bipaliidae		
*Bipalium nobile***	2x=10	Akita (Kawakatsu, Oki, Tamura & Takai *et al.*, 1993); Tôkyô (Kawakatsu & Aoki, 1968, 1969; Kawakatsu, Makino & Shirasuma, 1982); Toyonaka, near Ôsaka (Kawakatsu *et al.*, 1990; Oki, Tamura, Ogren & Kawakatsu, 1991); Kôchi (Kawakatsu, Oki, Tamura & Takai *et al.*, 1993)
Bipalium kewense	2x=18	Tôkyô (Seo *et al.*, 1988); Chichijima Island (Kawakatsu *et al.*, 1990; Oki, Tamura, Ogren & Kawakatsu *et al.*, 1991); Okinawa (Kawakatsu, 1983; Kawakatsu, Oki, Tamura, Itô *et al.*, 1993)
Rhynchodemidae		
Rhynchodeminae		
Platydemus manokwari	2x=12	Okinawa Islands (Kawakatsu, Oki, Tamura, Itô *et al.*, 1993; Oki *et al.*, 1991)

* In early publications, Kawakatsu misidentified the species as *D. tigrina*.
** In early publications, Kawakatsu misidentified the species as *D. kewense*.

Materials and methods

T. minor was collected by Mr M. Ninagawa on May 8, 1985; loc.: culture ponds of the Ibusuki Branch of the KPFES. Specimen whole mounts for taxonomic study (Kawakatsu's Specimen Lot Nos. 1782–1785) were prepared with or without staining. Others were used for chromosomal observation using the following method: (1) the animals were cut transversally near the basal level of tentacles and were kept in a covered petri dish filled with tap water for 2 days (*ca* 18 °C); (2) the body including the regenerated tissues was treated by solution of $10^{-6}\sim10^{-7}$ M colchicine (*ca* 30 min.); (3) the tissues were soaked with 45% acetic acid prior to the staining with aceto-orcein. Detailed techniques for the squashes were the same as for freshwater planarians (Oki *et al.*, 1980).

Numerous, preserved specimens of *T. minor* were received from Dr N. Morrissy; loc.: culture ponds of the West Australian Marine Research Laboratories,

Pemberton, Western Australia (collected and fixed with absolute ethanol on June 4, 1985; Kawakatsu's Specimen Lot No. 1781).

Observations and discussion

Table 1 gives a synopsis of the exotic turbellarians found in Japan together with their chromosome numbers.

1. *Temnocephala minor* (Fig. 1 A and B): According to the collector's (Mr M. Ninagawa) information, about 20 specimens of marron were introduced into the Ibusuki Branch of the KPFES from Western Australia, in June, 1984. Occurrence of *T. minor* was first observed in the beginning of April, 1985. After the disinfection of culture ponds and marron by a 30 ppm formalin solution, the *T. minor* populations decreased sharply.

Alive *T. minor* is slender, approximately 5 mm long by 1 mm wide with 5 tentacles. *T. minor* can be separated easily from the host body, and may reattach to the body of the Japanese freshwater crab, *Geothelphusa dehaani*, according to the collector (in litt.). In natural habitats of *T. minor*, hosts are reported as *Cherax destructor* (called 'yabbie' in Australia), *C. dispar*, and *C. punctatus* (see Cannon & Jennings, 1987; Jennings, 1971; Mills, 1983; Schaefer, 1971; Williams, 1980).

Five specimens were used for chromosomal observation (Fig. 1 C). Although division figures were very few, 18 mitotic figures were observed. The chromosome number of the diploid cells was $2x = 18$, with a karyotype of 2sm + 2m + 2m + 2sm + 2m + 2m- + 2m + 2sm + 2m, shown in descending order of size (Fig. 2). Meiosis was not observed in our material.

The known chromosome numbers and karyotypes in the temnocephalid group are very limited. The data of 4 Uruguayan species were reported by González *et al.* (1987) and Ponce de León (1988, 1990) as follows: *Temnocephala iheringi* Haswell, 1893 (host: a freshwater snail, *Pomacea canaliculata*) ($2n = 18$); *Temnocephala haswelli* Ponce de León, 1989 (host: *Pomacea canaliculata*) ($2n = 20$); *Temnocephala axenos* Monticelli, 1899 (hosts: crabs *Aegla* spp. and crayfishes *Paratax* spp.) ($n = 6$, $2n = 12$); *Temnocephala brevicornis* Monticelli, 1899 (hosts: crabs *Pseudothelphusa* spp. and turtles *Hydromedusa* spp.) ($n = 9$, $2n = 18$).

Approximately 30 species of *Temnocephala* Blanchard, 1849, are now reported: they are known from Australia (including Tasmania), New Zealand, Indonesia (including the Aru Kepulauan Islands), Malaya, and South America (Baer, 1961; Jennings, 1971). Additional chromosomal data are necessary to support discussion of karyological evolution in this animal group.

2. *Dugesia austroasiatica*: Originally mistakenly considered *Dugesia tigrina*, it has a karyotype of 2m + 2m + 2sm + 2sm + 2st + 2sm + 2m + 2m (Kawakatsu, Takai *et al.*, 1986). It was presumably introduced into Japan from Southeast Asia, although its original habitat is unknown. It is now widespread in Japan (Table 1). However, the species is only recorded from tanks and ponds used for tropical fish culture. *D. austroasiatica* is morphologically like *Dugesia* sp. from Taiwan with which it shares a similar karyotype (Kawakatsu *et al.*, 1979). Other dugesiid species from Southeast Asia with a similar karyotype are *Dugesia siamana* Kawakatsu, 1980, *Dugesia bengalensis* Kawakatsu, 1983, and *Dugesia tamilensis* Kawakatsu, 1980 (Kawakatsu *et al.*, 1980, 1983).

3. *Dugesia tigrina*: This North America species was found in tanks and ponds used for tropical fish culture; no sexual specimens were collected in Japan (Table 1). The pharynx pigmentation is conspicuous even in non-sexual specimens. The chromosome number of diploid cells was $2x = 16$, with a karyotype of 2m + 2m + 2m + 2m + 2m + 2sm + 2m + 2m. In many European countries, *D. tigrina* is already a naturalized species; it is also distributed in Brazil and Uruguay in South America (Kawakatsu *et al.*, 1981, 1982, 1990, 1992); Kawakatsu, Hauser & Friedrich, 1983, 1986). Recently, a naturalized population of *D. tigrina* at Nagasaki, Kyûshû, Japan, was recorded (Kawakatsu, Tamura *et al.*, 1993). The introduction route of this species into Japan is not known.

4. *Rhodax*? sp.: This undescribed species is tentatively placed under the genus *Rhodax* Marcus, 1946, in the family Dimarcusidae Mitchell et Kawakatsu, 1972 (Kawakatsu & Mitchell, 1983; Sluys, 1990). This species was found in tanks for tropical fish culture at Nagoya, Central Japan (Kawakatsu *et al.*, 1985). Externally, this species shows a striking resemblance to *Rhodax evelinae* Marcus, 1946, from São Paulo, Brazil. A single, non-sexual specimen examined karyologically, had triploid cells: $3x = 24$. The other 2 non-sexual specimens had two different types of cells intermingled in

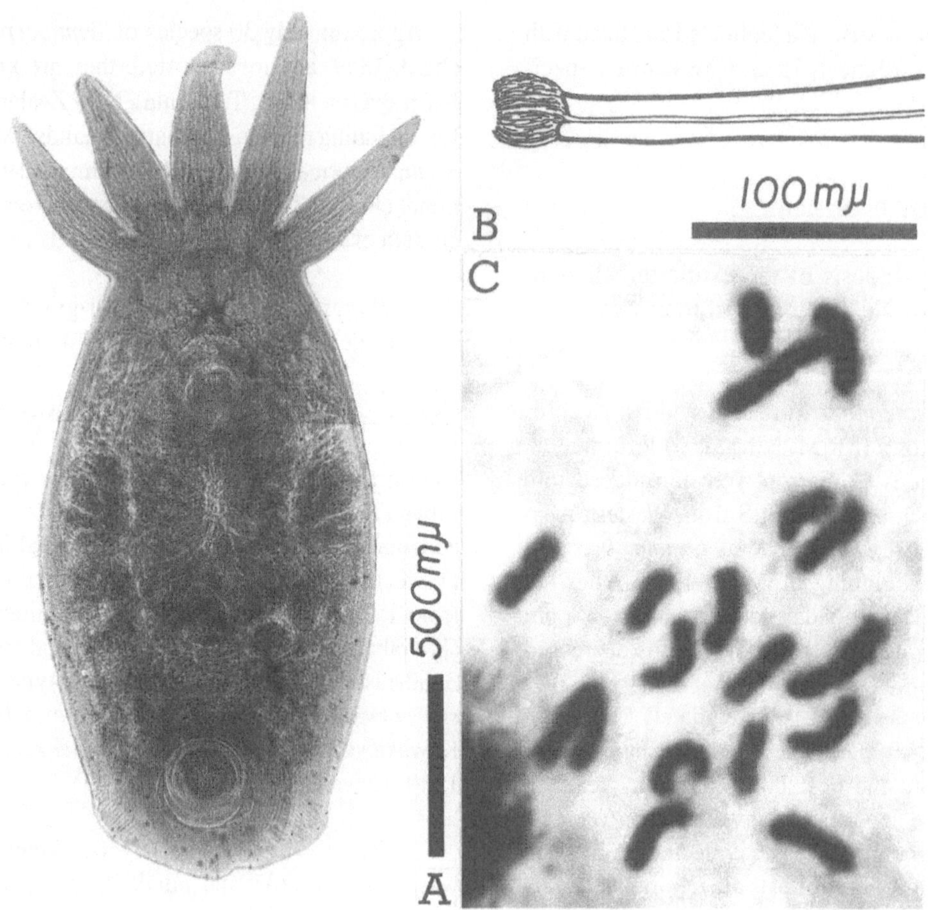

Fig. 1. Photomicrographs (A, C) and a sketch (B) of *Temnocephala minor*. (A) View of a preserved specimen (Specimen Lot No. 1781). (B) Penis stylet. (C) Chromosomes (2x = 18).

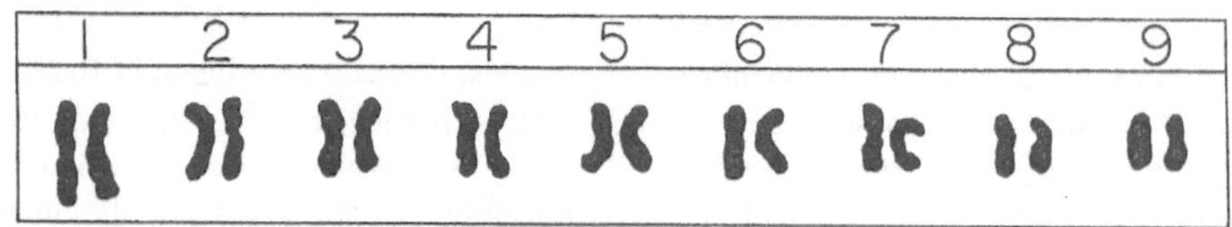

Fig. 2. Idiograms of *Temnocephala minor* (2x = 18).

one body: triploid cells and triploidic aneuploid cells (3x = 24 & 3x + 1LB + 1SB = 25 + 1SB).

Recently, the same species was found in tanks at Sapporo, Northern Japan (our unpublished data).

5. *Bipalium nobile*: In recent years, this species has been found in several localities other than the type locality in Tôkyô (Table 1). Although *B. nobile* is a noticeable animal because of its giant size (300–800 mm or more in length and 5–6.5 mm in width), no record on this large land planarian in Japan is known before the 1960's. Consequently we consider *B. nobile* to be a recently naturalized animal in Japan.

The chromosome number of the diploid cells was 2x = 10. Two karyotypes were found: 2m + 2sm + 2sm + sm & st + 2sm in the Yokohama population and 2m + 2sm + sm & st + 2st + m & sm in the Toyonaka population (Kawakatsu *et al.*, 1990; Oki *et al.*, 1988). According to Seo *et al.* (1988), their animals of *B. nobile* from the Hino population showed a karyotype of 2sm + 2m + 2sm + 2st + 2sm. For other literature on this species, see Ogren & Kawakatsu

(1987, 1988a) and Ogren *et al.* (1992).

6. *Biplium kewense*: This is a cosmopolitan species recorded from various areas of the world; the latest data for wildwide distribution of this species are shown in a paper by Ogren *et al.* (1992). Taxonomic redescriptions of this species from several areas of the world were given by Winsor (1983), Ogren (1984) and Kawakatsu (1985). Although the reliable records for occurrence of this species in Japan are scanty, the localities shown in Table 1 seem to be correct. *B. kewense* and *B. nobile* show a striking resemblance in general appearance, but the anatomy and histology of their copulatory structures are different (Kawakatsu, 1985; Kawakatsu, Makino & Shirasawa, 1982).

In animals from the Chichijima population of *B. kewense* (?), the chromosome number of the diploid cells was 2x = 18, with a karyotype of 2m + 2m + 2m + 2sm + 2st + 2sm + 2sm + 2sm + 2sm. Data of Seo *et al.* (1988) for specimens from Tôkyô can be summarized as 2x = 18 (2m + 2m + 2m + 2sm + 2st + 2sm + 2st + 2st + 2sm) (Kawakatsu *et al.*, 1990; Oki *et al.*, 1991). Additional discussion on the karyology of this wide spread species was given in the previous papaer (Oki, Tamura, Ogren & Kawakatsu, 1991). For other literature, see Ogren & Kawakatsu (1987) and Ogren *et al.* (1992).

7. *Platydemus manokwari*: In the autumn of 1990, numerous specimens of a rhynchodemid land planarian (with one pair of eyes) was found in the southern part of Okinawa Island. Kawakatsu, who received both live and preserved specimens, identified the animal as *P. manokwari*. This species is known as a predator of giant African snails; its original habitat may be considered as Irian Jaya (formerly Netherlands New Guinea), Indonesia. This land planarian has recently been introduced to several Pacific Islands by unknown human activities (Kawakatsu, Ogren & Muniappan, 1992).

After the autumn of 1990, 386 stations in the Southwest Islands of Japan were searched for land planarians, of which 132 stations in the following islands were inhabited by numerous specimens of *P. manokwari*: Okinawa Island, Zamamijima Island, Kumejima Island, Miyakojima Island, and Irabujima Island (Kawakatsu, Oki, Tamura, Itô, *et al.*, 1993). Repeated sampling (from 1961 onwards) indicated that this species was not present in samples sent to Kawakatsu from residents of the Okinawa Islands and colleagues who visited this area. *P. manokwari* is a noticeable species in the field (30–50 mm or more in length and 4–8 mm in width). We believe that the extensive occurrence of this species in the Southwest Islands of Japan mentioned above may be due to the unintentional introduction of the animal in very recent years (Kawakatsu, Oki, Tamura, Itô *et al.*, 1993).

Animals from Bugsuk Island, the Philippines, had a diploid chromosome number of 2x = 12 (n = 6), with a karyotype of 2sm + 2sm + 2st + 2sm + 2m + 2m (Kawakatsu *et al.*, 1987; Oki *et al.*, 1988). The result of the chromosomal study of specimens from the Okinawa population is coincident with the karyotype of the Bugsuk Island specimens (Kawakatsu, Tamura *et al.*, 1993). For other literature, see Ogren & Kawakatsu (1988b) and Ogren *et al.* (1992).

Acknowledgments

We wish to thank the following for their assistance in making collections of turbellarian samples used for this paper, as well as for supplying information, maps, photographs, copies of papers, etc.: Dr N. M. Morrissy (*Temnocephala*: Western Australian Marine Research Laboratories, Department of Fisheries and Wildlife, Perth, Western Australia, Australia); Prof. Dr Hiroshi Suzuki (*Temnocephala*: Yokohama National University, Yokohama); the late Mr Mitsuo Ninagawa (*Temnocephala*: Kagoshima); Mr Akihiko Zenyôji (*D. austroasiatica*: Chiba); staff members of the Naha Plant Protection Station (*P. manokwari*: Naha, Okinawa). We are also grateful to Prof. Dr Robert E. Ogren (Wilkes University, Wilkes-Barre, Pennsylvania, USA) for reviewing early draft of the present paper.

References

Baer, J. G., 1961. Classe des Temnocéphales. In P.- P. Grassé (ed.), Traité de Zoologie, IV. Masson et Cie Éditeurs, Paris: 213–241.

Cannon, L. R. G. & J. B. Jennings, 1987. Occurrence and nutritional relationships of four ectosymbiotes of the freshwater crayfishes Cherax dispar Riek and *Cherax punctatus* Clark (Crustacea: Decapoda) in Queensland. Aust. J. mar. Freshwat. Res. 38: 419–427.

González, L. E., R. Ponce de León & E. S. de Vaio, 1987. Chromosome differences between two species of Temnocephala (Platyhelminthes). Cytobis 49: 85–88.

Hirao, Y., M. Kawakatsu & W. Teshirogi, 1970. Records of an exotic freshwater planarian species, *Dugesia tigrina* (Girard), found in tanks of tropical fishes in Japan. J. Jap. Ass. Zool. Gardens & Aquariums, 11: 25–27. (In Japanese, with English summary)

Jennings, J. B., 1971. Parasitism and commensalism in the Turbellaria. In: Advances in Parasitology, 9. Academic Press, London: 1–32.

Kawakatsu, M., 1983. Notes on land planarians. The Heredity (Iden), Tôkyô, 37, no. 2: 52–61. (In Japanese)

Kawakatsu, M., 1985. A note on the morphology of *Bipalium kewense* Moseley, 1978, and *Bipalium adventitium* Hyman, 1943 (Turbrllaria, Tricladida, Terricola). Bull. Fuji Women's College, no. 23, ser. II: 85–100.

Kawakatsu, M. & J.-I. Aoki, 1968. Notes on *Bipalium kewense* Moseley collected from the garden of the Imperial Palace in Tôkyô. The Heredity (Iden), Tôkyô, 22, no. 10: 45–47. (In Japanese)

Kawakatsu, M. & J.-I. Aoki, 1969. Additional notes on *Bipalium kewense* Moseley collected from the garden of the Imperial Palace in Tôkyô. Collect. & Breed., Tôkyô, 31: 374– 377. (In Japanese)

Kawakatsu, M., J. Hauser & S. M. G. Friedrich, 1983. Morphological, karyological and taxonomic studies of freshwater planarians from South Brazil. V. *Dugesia tigrina* (Girard, 1850) from Municipio Botucatu, Estado de São Paulo, and *Dugesia schubarti* (Marcus, 1946) from the vicinity of São Paulo (Turbellaria, Tricladida, Paludicola). Bull. Fuji Women's College, no. 21, ser. II: 147–163.

Kawakatsu, M., J. Hauser & S. M. G. Friedrich, 1986. Morphological, etc. VIII. Four *Dugesia* species (*D. tigrina*, *D. schubarti*, *D. anderlani*, and *D. arndti*) collected from several localities in Estado de Rio Grande do Sul (Turbellaria, Tricladida, Paludicola). Bull. Fuji Women's College, no. 24, ser. II: 41–62.

Kawakatsu, M., J. Hauser, S. M. G. Friedrich & O. de Souza Lima, 1982. Morphological, etc. III. *Dugesia tigrina* (Girard, 1850) and *Dugesia schubarti* (Marcus, 1946) from the vicinities of São Carlos, Estado de São Paulo (Turbellaria, Tricladida, Paludicola). Bull. Fuji Women's College, no. 20, ser. II: 73–90.

Kawakatsu, M., J. Hauser & R. Ponce de León, 1992. Freshwater planarians from Uruguay and Rio Grande do Sul, Brazil: *Dugesia ururiograndeana* sp. nov. and *Dugesia tigrina* (Girard, 1850) (Turbellaria, Tricladida, Paludicola). Bull. Biogeog. Soc. Japan, 47: 33–50.

Kawakatsu, M. & T. Hirai, 1968. On an exotic freshwater planarian species, *Dugesia tigrina* (Girard) found in tanks of tropical fishes in Sapporo City. The Heredity (Iden), Tôkyô, 22, no. 7: 31–32. (In Japanese)

Kawakatsu, M., N. Makino & Y. Shirasawa, 1982. *Bipalium nobile* sp. nov. (Turbellaria, Tricladida, Terricola), a new land planarian from Tokyo. Annot. Zool. Japon, 55: 236–262.

Kawakatsu, M. & R. W. Mitchell, 1983. Record of a cave-adapted planarian (Turbellaria, Tricladida, Maricola) from Guatemala. Annot. Zool. Japon, 56: 291–298.

Kawakatsu, M., R. E. Ogren & R. Muniappan, 1992. Redescription of *Platydemus manokwari* de Beauchamp, 1962 (Turbellaria, Tricladida, Terricola), from Guam and the Philippines. Proc. Jap. Soc. Syst. Zool., 47: 11–25.

Kawakatsu, M., I. Oki, S. Tamura, H. Itô, Y. Nagai, K. Ogura, S. Shimabukuro, F. Ichinohe, H. Katsumata & M. Kaneda, 1993. An extensive occurrence of a land planarian, *Platydemus manokwari* de Beauchamp, 1962, in the Ryûkyû Islands, Japan (Turbellaria, Tricladida, Terricola). Biol. Inl. Wat., Nara, 8: 5–14.

Kawakatsu, M., I. Oki, S. Tamura, R. E. Ogren, T. Yamada & H. Murayama, 1990. Preprint of papers given at the Sixth International Symposium on the Biology of Turbellaria, Hirosaki, Japan. Occ. Publ., Biol. Lab. Fuji Women's College, 22: 1–16.

Kawakatsu, M., I. Oki, S. Tamura, K. Sekiguchi & R. E. Ogren, 1987. Preprint of papers given at the Fifth International Symposium on the Biology of Turbellaria, Göttingen, Bundesrepublik Duetschland. Occ. Publ., Biol. Lab. Fuji Women's College, 18: 1-24.

Kawakatsu, M., I. Oki, S. Tamura, M. Takai, O. A. Timoshkin & N. A. Porfirjeva, 1993. Preprint of papers given at the Seventh International Symposium on the Biology of Turbellaria, Åbo, Finland. Occ. Publ., Biol. Lab. Fuji Women's College, 25: 1–20.

Kawakatsu, M., I. Oki, S. Tamura & T. Yamayoshi, 1985. Reexamination of freshwater planarians found in tanks of tropical fishes in Japan, with a description of a new species, *Dugesia austroasiatica* sp. nov. (Turbellaria, Tricladida, Paludicola). Bull. Biogeogr. Soc. Japan, 40: 1–19.

Kawakatsu, M., I. Oki, S. Tamura, T. Yamayoshi & A. K. Aditya, 1983. A new freshwater planarian from West Bengal, India. Bull. Biogeogr. Soc. Japan, 38: 3–10.

Kawakatsu, M., I. Oki, S. Tamura, T. Yamayoshi, J. Hauser & S. M. G. Friedrich, 1981. Morphological, etc. II. *Dugesia tigrina* (Girard, 1850) (Turbellaria, Tricladida, Paludicola). Bull. Fuji Women's College, no. 19, ser. II: 113–136.

Kawakatsu, M., I. Oki, S. Tamura, T. Yamyoshi, K.-Y. Lue & M. Hagiya, 1979. Additional report on freshwater planarians from Taiwan. Bull. Fuji Women's College, no. 17, ser. II: 59–91.

Kawakatsu, M. & R. Ponce de León, 1990. The occurrence of *Dugesia tigrina* (Girard, 1850) (Turnellaria, Tricladida, Paludicola) in Uruguay. Proc. Jap. Soc. Syst. Zool. 41: 5–14.

Kawakatsu, M., M. Takai, I. Oki, S. Tamura & M. Aoyagi, 1986. A note on an introduced species of freshwater planarian, *Dugesia austroasiatica* Kawakatsu, 1985, collected from culture ponds of *Tirapia mossambica* in Saga City, Kyûshû, Japan (Turbellaria, Tricladida, Paludicola). Bull. Fuji Women's College, no. 24, ser. II: 87–94.

Kawakatsu, M., S. Tamura, M. Takai, K. Yamamoto, R. Ueno & I. Oki, 1993. The first record of occurrence of a naturalized population of *Dugesia tigrina* (Girard, 1850) at Nagasaki, Kyûshû, Japan (Turbellaria, Tricladida, Paludicola). Bull. Biogeogr. Soc. Japan, 48: 5–11.

Kawakatsu, M., S. Tamura, T. Yamayoshi & I. Oki, 1980. The freshwater planarians from Thailand and South India. Annot. Zool. Japon, 53: 254–268.

Mills, B. J., 1983. A review of diseases of freshwater crayfish, with particular reference to the yabbie, *Cherax destructor*. Fish. Res. Pap., Dept. Fish. S. Aust., no. 9: 1–18.

Ogren, R. E., 1984. Exotic land planarians of the genus *Bipalium* (Platyhelminthes: Turbellaria) from Pennsylvania and the Academy of Natural Sciences, Philadelphia. Proc. Pa. Acad. Sci. 58: 193–201.

Ogren, R. E. & M. Kawakatsu, 1987. Index to the species of the genus *Bipalium* (Turbellaria, Tricladida, Terricola). Bull. Fuji Women's College, no. 25, ser. II: 79–119.

Ogren, R. E. & M. Kawakatsu, 1988a. Index to the species of the genus *Bipalium* (Turbellaria, Tricladida, Terricola): Additions and corrections. Occ. Publ., Biol. Lab. Fuji Women's College, 19: 1–16.

Ogren, R. E. & M. Kawakatsu, 1988b. Index to the species of the family *Rhynchodemidae* (Turbellaria, Tricladida, Terricola). Part I: *Rhynchodeminae*. Bull. Fuji Women's College, no. 26, ser. II: 39–91.

Ogren, R. E., M. Kawakatsu & E. M. Froehlich, 1992. Additions and corrections of the previous land planarian indices of the world (Turbellaria, Tricladida, Terricola). Bull. Fuji Women's College, no. 30, ser. II: 59–103.

Oki, I., S. Tamura, H. Itô, Y. Nagai, K. Ogura, S. Shimabukuro, F. Ichinohe, H. Katsumata, M. Kaneda & M. Kawakatsu, 1991. An extensive occurrence of a land planarian, *Platydemus manokwari*, in the Okinawa Islands, Japan. Zool. Sci., Tôkyô, 8: 1200.

Oki, I., S. Tamura, R. E. Ogren & M. Kawakatsu, 1988. Karyological and taxonomic studies of three species of the genus *Bipalium* from Japan and the United States and *Platydemus manokwari* from the Philippines. Fortsch. Zool. 36: 139–143.

Oki, I., S. Tamura, R. E. Ogren & M. Kawakatsu, 1991. Karyology of four land planarian species of the genus *Bipalium* from Japan. Hydrobiologia 22: 163–167.

Oki, I., S. Tamura, T. Yamayoshi & M. Kawakatsu, 1980. Preprint of a paper given at the Third International Symposium —The Biology of the Turbellaria in Honour of Prof. Tor G. Karling, Hasselt (Diepenbeek), Belgium. Occ. Publ., Biol. Lab. Fuji Women's College, no. 2: 1–23.

Ponce de León, R., 1988. Karyotypes of two species of *Temnocephala* (Platyhelminthes). Fortsch. Zool. 36: 145–149.

Ponce de León, R., 1990. Male meiosis in Neotropical *Temnocephala*. The Sixth Internat. Symp. on the Biol. of the Turbellaria, Programme/Abstracts. Hirosaki Univ., Hirosaki: 71.

Schaefer, C. W., 1971. Observations on temnocephalid hosts and distributions. Z. Zool. Syst. Evolutionsforsch. 9: 139–143.

Seo, N., N. Makino & Y. Shirasawa, 1988. The karyotypes of the genus *Bipalium* (Platyhelminthes: Terricola) — Three species in the fission type and two species in the non-fission type —. Bull. Tokyo Med. College, no. 14: 13–41. (In Japanese, with English summary)

Sluys, R., 1990. A monograph of the Dimarcusidae (Platyhelminthes, Seriata, Tricladida). Zool. Scripta, 19: 13–29.

Takai, M., M. Aoyagi, I. Oki, S. Tamura & M. Kawakatsu, 1986. Karyological and taxonomic studies of the *Dugesia* species in Southeast Asia. X. *Dugesia austroasiatica* from culture ponds fishes in Saga City. Zool. Sci., Tôkyô, 3: 1108.

Tamura, S., I. Oki, M. Kawakatsu, M. Ninagawa, M. Matsusato & H. Suzuki, 1985. A note on an introduced species of freshwater planarian, *Dugesia tigrina* (Girard, 1850), found in culture ponds of Australian crayfish in Kagoshima Prefecture, Japan. Bull. Fuji Women's College, no. 23, ser. II: 133–137.

Williams, W. D., 1980. Australian Freshwater Life. Savage & Co., Brisbane: xi, 321 pp.

Winsor, L., 1983. A revision of the cosmopolitan land planarian *Bipalium kewense* Moseley, 1978 (Turbellaria, Tricladida, Terricola). Zool. J. Linn. Soc., Lond. 79: 61–100.

Hydrobiologia **305**: 79–84, 1995.
L.R.G. Cannon (ed.), Biology of Turbellaria and some Related Flatworms.

A review of chromosomal variation in *Dugesia japonica* and *D. ryukyuensis* in the Far East

Sachiko Tamura[1], Iwashiro Oki[2] & Masaharu Kawakatsu[3]*
[1]*Ôsaka Prefectural Institute of Public Health, Nakamichi-1-chôme 3-69, Higashinari-ku Ôsaka 537, Japan*
[2]*Ôsaka Environmental Project Association, MFC 7F, Uchihon-machi 1-2-15, Chu'ô-ku, Ôsaka 540, Japan*
[3]*Biological Laboratory, Fuji Women's College, Kita-16, Nishi-2, Kita-ku, Sapporo (Hokkaidô) 001, Japan*
*(*Author for correspondence)*

Key words: Turbellaria, Tricladida, *Dugesia*, karyology, Japan

Abstract

The chromosome numbers of *Dugesia japonica* Ichikawa et Kawakatsu, 1964, are $n = 8$, $2x = 16$ and $3x = 24$; those of *Dugesia ryukyuensis* Kawakatsu, 1976, are $n = 7$, $2x = 14$ and $3x = 21$. The karyotypes of both species include diploid, triploid and mixoploid; aneuploidic and mixoaneuploidic karyotypes may occur. In 785 specimens studied of *D. japonica*, the occurrence rates of specimens having each karyotype are substantially the same (29–37%). Diploid sexual specimens represented nearly 10% of the total and virtually no triploid or mixoploid sexual specimens were found. The diploid karyotype can be inherited by both sexual and asexual reproduction; the triploid and mixoploid karyotypes will be inherited only by asexual reproduction. In 51 specimens studied of *D. ryukyuensis*, the different karyotypes are: diploid (*ca* 39%), triploid (*ca* 57%) and mixoploid (*ca* 4%). Diploid sexual specimens represented nearly 25% of the total; sexual specimens with triploidic karyotypes made up nearly 27%. The diploid, triploid and mixoploid karyotypes were also found in juveniles hatched from cocoons. The diploid karyotype is inherited by both sexual and asexual reproductions; the other karyotypes may be inherited by parthenogenesis or self-fertilization (including pseudogamy) and asexual reproduction.

Introduction

The chromosome numbers and karyotypes of *Dugesia japonica* Ichikawa et Kawakatsu, 1964, and *Dugesia ryukyuensis* Kawakatsu, 1976, have been studied by our team since 1975. The former, a polymorphic species distributed widely in the Far East including the Japanese Islands, has chromosome numbers of $n = 8$, $2x = 16$ and $3x = 24$. The latter, for a time considered a subspecies of *D. japonica*, is found only in the Southwest Islands of Japan (Kawakatsu *et al.*, 1976, 1994); it has chromosome numbers of $n = 7$, $2x = 14$ and $3x = 21$. The karyotypes of both species show a high degree of variation (Tamura *et al.*, 1991 and references therein).

We review the karyology of *D. japonica* and *D. ryukyuensis* and give new karyological data of juveniles of the latter observed in laboratory cultures.

Material and methods

For examination of chromosome numbers of planarians, the squash method (Oki *et al.*, 1980) was employed. For the form of chromosomes the scale-method (Oki *et al.*, 1991) was used.

Results

Dugesia japonica

The karyotypes of 785 specimens collected from 136 habitats (Tamura *et al.*, 1991) in the Far East were examined; the results are shown in Table 1. Various karyotypes detected are grouped in 3 classes as follows: (1) diploid (including diploidic aneuploid and mixoaneuploid), (2) triploid (including triploidic aneuploid and mixoaneuploid) and (3) mixoploid (includ-

Table 1. Karyotypes of *Dugesia japonica* classified into 3 classes with groups based upon the data of 785 specimens collected from 136 habitats.[1]

Karyotypes	No. of stations where each karyotype was detected					No. of specimens examined			
	Total	Items	Sex.	Asex.	Sex. & asex.	Total	Items	Sex.	Asex.
Diploid	75					293			
2x		68	29	56	17		275	69	206
2x*		4	2	4	2		14	4	10
2x**		3	1	2	–		4	1	3
Triploid	83					242			
3x		7	0	7	–		24	0	24
3x*		36	0	36	–		97	0	97
3x**		40	1	40	1		121	1	120
Mixoploid	55					250			
2x & 3x		46	1	46	1		226	8	218
2x & 3x & 4x		1	0	1	–		3	0	3
2x & 4x		3	1	2	0		9	1	8
2x** & 3x**		3	0	3	–		10	0	10
3x* & 4x*		2	0	2	–		2	0	2
Total		213	35	199	21		785	84	701

[1] If two different karyotypes were detected in animals from a single habitat, the number of stations surveyed is counted 2.
*: aneuploid; **: mixoaneuploid

Table 2. Karyotypes of *Dugesia ryukyuensis* classified into 3 classes with groups based upon the data of 51 specimens collected from 11 habitats.[1]

Karyotypes	No. of stations where each karyotype was detected					No. of specimens examined			
	Total	Items	Sex.	Asex.	Sex. & asex.	Total	Items	Sex.	Asex.
Diploid	8					20			
2x		5	4	3	2		17	11	6
2x*		1	0	1	–		1	0†	1†
2x**		2	2	0	–		2	2	0
Triploid	11					29			
3x**		8	5	5	2		21	7	14 =
3x***		3	3	0	–		8	7	1
Mixoploid	1					2			
2x** & 3x**		1	1	0	–		2	2	0
Total		20	15	9	4		51	29	22

[1] For the method of calculating the number of stations in this table, see Table 1.
† A single asexual specimen colllected became mature under laboratory culture conditions.
*: heteromorphy; **: aneuploid; ***mixoaneuploid

ing mixoploidic mixoaneuploid). Each of the above-mentioned 3 classes of karyotypes is subdivided into 3 to 5 fundamental groups. The total number of stations where each karyotype were detected is also shown. Furthermore, the number of animals classified according to different karyotypes is listed (for the number of cells studied, see Kawakatsu *et al.*, 1993).

As will be seen from Table 1, the total number of stations where the karyotypes were clarified was 213. The number of stations of the karyotypes in the diploid, triploid and mixoploid classes are 75 (*ca* 35 %), 83 (*ca* 39%), and 55 (*ca* 26%), respectively. In the diploid class, 91% of stations had orthoploidic diploid populations. The other karyotypes in this class were negligible (only 4–5% each). In the triploid class, in contrast, only 7 (*ca* 8%) were orthoploidic triploid, but the triploidic aneuploid karyotypes and the triploidic mixoaneuploid karyotypes formed nearly 92%. Various triploidic mixoaneuploid karyotypes were found. Finally, in the mixoploid class, nearly 84% were orthoploidic mixoploid karyotype (2x & 3x); other karyotypes were negligible (2–5%).

In animals having the mixoploid karyotype (2x & 3x), the occurrence rate of diploidic cells in one body was counted. The ratio between them varied according to each animal even in the same habitat (4–97%). Planarians having the diploidic & triploidic & tetraploidic karyotype (2x & 3x & 4x) and diploidic & tetraploidic karyotype (2x & 4x) showed very few tetraploidic cells.

Photomicrographs of chromosomes and idiograms (and further references to them) of *D. japonica* from Japan, Taiwan and Korea will be found in the following papers: Kawakatsu *et al.* (1987, 1990, 1993); Oki *et al.* (1981); Tamura *et al.* (1991).

Dugesia ryukyuensis

Karyotypes of 51 specimens collected from 11 habitats in the Southwest Islands of Japan were examined (Tanegashima Island, Okinawa Island, Miyakojima Island, and Ishigakijima Island); the results are shown in Table 2. The geographical distribution of the karyotypes of *D. ryukyuensis* is shown elsewhere (Tamura *et al.*, 1991).

As will be seen from Table 2, the total number of stations where the karyotypes were clarified was 20. The numbers of stations with karyotypes in the diploid, triploid and mixoploid classes were 8 (*ca* 40%), 11 (*ca* 55%) and 1 (*ca* 5%), respectively. In the diploid class, a little over 60% were orthoploidic diploid (2x). In the triploid class, 73% had triploidic aneuploid karyotypes (including 3x with a few number of small B-chromosomes) and 27% had triploidic mixoaneuploid karyotypes. In the mixoploid class, only the diploidic & triploidic aneuploid karyotype is known.

Photomicrographs of chromosomes and idiograms (and further references to them) of *D. ryukyuensis* from the Southwest Islands of Japan will be found in the following papers: Kawakatsu *et al.* (1987, 1990, 1993); Oki *et al.* (1981); Tamura *et al.* (1988, 1991).

Table 3. Karyotypes of *Dugesia ryukyuensis* classified into 3 classes with groups based upon the data of 52 juveniles observed in laboratory cultures

Karyotypes	No. of stations Total	No. of stations Items	No. of specimens examined Total	No. of specimens examined Items
Diploid	5		19	
2x		4		12
2x*		1		7
Triploid	3		32	
3x*		2		19
3x**		1		13
Mixoploid	1		1	
2x* & 3x*		1		1
Total		9		52

[1] Animals laying cocoons were collected from 3 populations on Okinawa Island and a single population on Tanegashima Island. For the method of calculating the number of stations in this table, see Table 1.
*: aneuploid; **: mixoaneuploid

Juveniles of D. ryukyuensis

The following observations of the chromosomal analysis have been made using animals from 3 populations in Okinawa Island: Kinchô, Ginowan-1 and Ginowan-2 (Kawakatsu *et al.*, 1993).

(1) The karyotypes of 19 juveniles hatched from each of 5 cocoons (Kinchô and Ginowan-1 stocks) showed 4 combinations: (i) 4 specimens diploidic aneuploid only (1 cocoon), (ii) 6 specimens triploidic aneuploid only (2 cocoons), (iii) 5 specimens orthoploidic diploid and triploidic aneuploid (1 cocoon), (iv) 4 specimens orthoploidic diploid, diploidic aneuploid and triploidic aneuploid (1 cocoon).

(2) The karyotype of 6 juveniles hatched from 2 cocoons laid by orthoploidic diploid specimens (Ginowan-2 stock) was only orthoploidic diploid.

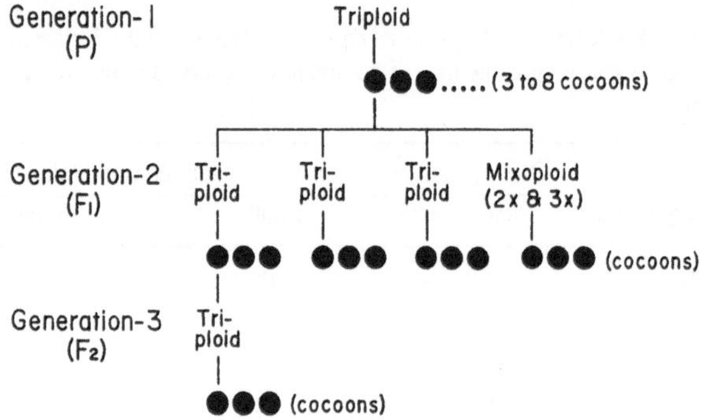

Fig. 1. Karyotypes of 3 generations of *Dugesia ryukyuensis* from the Kinchô population, Okinawa, cultured under isolated conditions.

(3) One triploidic aneuploid specimen (P) from the Kinchô stock cultured under isolated conditions reached maturity and laid several cocoons. Karyotypes of 4 juveniles (F_1) from a single cocoon were: triploidic aneuploid (3), diploidic and triploidic mixoploid (1). One triploidic aneuploid juvenile cultured under isolated conditions reached maturity and laid cocoons. A single juvenile (F_2) from one of these cocoons had a triploidic aneuploid karyotype. The result is shown in Fig. 1.

Table 3 shows all karyological data of 52 juveniles from the Okinawa populations (Kinchô, Ginowan-1 and Ginowan-2 stocks), and the Tanegashima population (Minamitane stock from the southern part of the Island) which are classified into 3 classes, diploid, triploid and mixoploid. The diploid karyotype (2x) is common in juveniles from every stock. Juveniles in the Kinchô and Ginowan-1 stocks had the triploidic aneuploid karyotype, the triploidic mixoaneuploid karyotype and the diploidic & triploidic aneuploid karyotype.

Discussion

Karyotypes of D. japonica and D. ryukyuensis

Among the *Dugesia* species of the world, the following haploid numbers are known: $n = 4$, $n = 7$, $n = 8$, and $n = 9$. The haploid number of $n = 4$ is known from Europe, North and South America, including the Caribbean area (Benazzi & Benazzi-Lentati, 1976, 1992; Kawakatsu *et al.*, 1984). The haploid number of $n = 7$ is known from Asia (2 species, of which one is *D. ryukyuensis* discussed in the present paper) and Europe (Pala *et al.*, 1981; see also Kawakatsu *et al.*, 1989). Most species have the haploid number of $n = 8$; they occur in Asia (karyotypes of 6 species are now known, of which one, *D. japonica*, is discussed in the present paper), Europe, North and South America (Benazzi & Benazzi-Lentati, 1976, 1992; Kawakatsu *et al.*, 1984). The haploid number of $n = 9$ is known only from the Caribbean area (Guadeloupe, West Indies) and South America (Brazil) Gourbault, 1980; Kawakatsu *et al.*, 1984).

As was reviewed in the present paper, a high degree of karyological variations is found in *D. japonica* and *D. ryukyuensis*. Karyological variation is also reported in several other dugesiid species as follows: *D. polychroa* (Schmidt, 1861) from Italy (2x = 8~6x = 24; Benazzi & Benazzi-Lentati, 1976, 1992); *D. schubarti* (Marcus, 1946) from Brazil (2x = 8 & 3x = 12, 2x = 8 & 3x = 12 & 4x = 16, 2x = 8 & 4x = 16, 4x = 16; Kawakatsu *et al.*, 1984); *D. siamana* Kawakatsu, 1980, from Thailand (2x = 16 & 3x = 24; Kawakatsu *et al.*, 1980); *D. tigrina* (Girard, 1850) from European countries (2x = 16 ~ 3x = 24; Dahm, 1958; Benazzi & Benazzi-Lentati, 1976, 1992); *D. benazzii* Lepori, 1951, from the Tyrrhenian Islands, Italy (2x = 16 ~6x = 48; Benazzi & Benazzi-Lentati, 1976, 1992); *D. dorotocephala* (Woodworth, 1897) from U.S.A. (2x = 16 & 3x = 24; Benazzi & Benazzi-Lentati, 1976; Benazzi & Puccinelli, 1982); *D. anderlani* Kawakatsu et Hauser, 1983, from Brazil (2x = 18 & 3x = 27; Kawakatsu *et al.*, 1983).

It is important to observe from the above species list, that significant variation of karyotypes is found in those *Dugesia* species having haploid numbers of $n = 4$, $n = 7$, $n = 8$, and $n = 9$ from various ares of the world. Their karyological variations cited above in most species are the triploid, tetraploid, hexaploid, and orthoploidic mixoploid karyotypes; the polysomic and aneuploid karyotypes are known in only a few dugesiid species (Benazzi & Benazzi-Lentati, 1976). It is clear that the karyological variations of *D. japonica* and *D. ryukyuensis* are remarkably similar to those of other species.

Do sexual specimens of *D. japonica* and *D. ryukyuensis* occur regularly in the 3 classes of the karyotypes? The data calculated from Tables 1 and 2 are shown in Fig. 2. It is clear that the diploid (*ca* 37%), triploid (*ca* 31%) and mixoploid (*ca* 32%) karyotypes in *D. japonica* are evenly distributed. Among the total specimens, nearly 10% are sexual diploid. However, sexual specimens are hardly found in the triploid class (less than 1%). In *D. ryukyuensis*, different karyotypes are diploid (*ca* 39%), triploid (*ca* 57%) and mixoploid

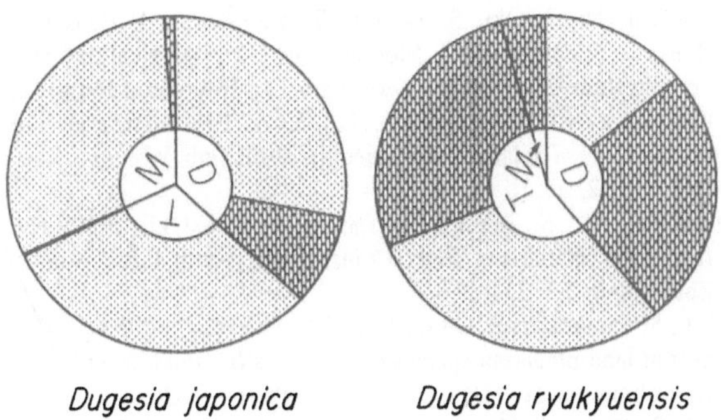

Fig. 2. Occurrence of diploidy (D), triploidy (T) and mixoploidy (M) in *Dugesia japonica* and *Dugesia ryukyuensis*. Brick areas indicate percentage of sexual specimens; dotted areas, that of asexual specimens.

(*ca* 4%) (Fig. 2). Sexual specimens are 25% diploid and 27% triploid (including the aneuploidic karyotypes and the mixoaneuploidic ones). No asexual specimen with the mixoploid karyotype was found.

Although data for *D. ryukyuensis* are less satisfactory than those of *D. japonica*, it appears that triploidy is more common in *D. ryukyuensis* than *D. japonica*; triploid sexual specimens are also more common in *D. ryukyuensis* than in *D. japonica*. The percentage of specimens with mixoploid, however, is greater in *D. japonica* than in *D. ryukyuensis*, whereas diploidy is substantially the same in both species (Fig. 2).

The relationship between karyotype and propagation

In *D. japonica*, diploid karyotypes will be inherited from generation to generation by both sexual (cocoon-laying) and asexual (fission or fragmentation in some cases) reproduction. Since sexual specimens are very rare in animals having the triploid and mixoploid karyotypes, these karyotypes will be inherited only by asexual reproduction. The following observations may support this view.

Bessho *et al.* (1992) examined mitochondrial DNA heterogeneity in a population of *D. japonica* known to have 2 karyotypes: 2x and 2x & 3x (Mie-Seihôji stock, in Tamura *et al.*, 1979). They recognized 2 types of planarians: (1) type A (having almost identical sequences), and (2) collectively called type D (showing some heterogeneity even in a single individual). Bessho *et al.* (1992) suggested that type A may represent sexual specimens (2x) and type D mixoploid planarians (2x & 3x). This supports our view about the propagation of animals having the mixoploid karyotype by asexual reproduction.

In *D. ryukyuensis*, no essential differences were observed between the karyotypes of animals collected from field localities and those of juveniles obtained in laboratory cultures (see Tables 2 and 3). It was observed that many specimens from various habitats of *D. ryukyuensis* with triploid karyotypes have sexual organs. Moreover, as will be seen in Fig. 1, we have exact laboratory data that one mixoploidic and triploidic planarians hatched from one cocoon that was laid by a triploidic animal culture isolated from other planarians.

The following supposition may explain the abovementioned findings: (1) the diploid karyotype will be inherited by sexual and asexual reproduction; (2) the other karyotypes (including triploid and mixoploid) will be inherited by parthenogenesis, self-fertilization or pseudogamy, and asexual reproduction.

Additional discussion on the karyology of *D. japonica* and *D. ryukyuensis* from a taxonomic viewpoint will be found in the paper by Kawakatsu *et al.* (1995).

Acknowledgments

The authors are grateful to Prof. Dr Robert E. Ogren, (Wilkes University, Pennsylvania, USA) for his reviewing of an early draft of the present paper.

References

Benazzi, M. & G. Benazzi-Lentati, 1976. In: B. John (ed.), Animal Cytogenetics, 1, Platyhelminthes. Gebrüder Borntraeger, Berlin: vii + 182 pp.

Benazzi, M. & G. Benazzi-Lentati, 1992. Pseudogamy (gynogenesis) in planarians: annotations some forty years on. Parts I & II. In: R. Dallai (ed.), Sex Origin and Evolution, Selected Symposia and Monographs U. Z. I. 6: 87–102. Mucchi, Modena.

Benazzi, M. & I. Puccinelli, 1982. Nuovo contributo alla cariologia ed alla biologia riproduttiva del Nord America. Atti Soc. Tosc. Sci. Nat., Mem., ser. B, 88: 203–209.

Bessho, Y., T. Ohama & S. Osawa, 1992. Planarian mitochondria I. Heterogeneity of cytochrome c oxidase subunit I gene sequences in the freshwater planarian, *Dugesia japonica*. J. Molec. Evol. 34: 324–330.

Dahm, A. G., 1958. Taxonomy and Ecology of Five Species Groups in the Family Planariidae (Turbellaria, Tricladida, Paludicola). Nya Litografen, Malmö: 241 pp. (pls. I–V, appendices I–V).

Gourbault, N., 1980. Morphologie et cytotaxonomie d'une planaire nouvelle de Guadeloupe (Turbellarié, Triclade). Bull. Mus. Natn. Hist. Nat. Paris, 4e ser. 2, A, no. 3: 751–757.

Ichikawa, A. & M. Kawakatsu, 1964. A new freshwater planarian, *Dugesia japonica*, commonly but erroneously known as *Dugesia gonocephala* (Dugès). Annot. Zool. Japon. 37: 185–194.

Kawakatsu, M., J. Hauser, S. M. G. Friedrich, I. Oki, S. Tamura & T. Yamayoshi, 1983. Morphological, karyological and taxonomic studies of freshwater planarians from South Brazil. IV. *Dugesia anderlani* sp. nov. (Turbellaria, Tricladida, Paludicola), a new species from São Leopoldo in Estado de Rio Grande do Sul. Annot. Zool. Japon. 56: 196–208.

Kawakatsu, M., R. W. Mitchell, I. Oki, S. Tamura & S. Yussof, 1989. Taxonomic and karyological studies of *Dugesia batsensis* Ball, 1970 (Turbellaria, Tricladida, Paludicola), from the Batu Caves, Malaysia. J. Speleol. Soc. Japan, 14: 1–14 + pl. 1.

Kawakatsu, M., I. Oki & S. Tamura, 1995. Taxonomy and geographical distribution of *Dugesia japonica* and *D. ryukyuensis* in the Far East. Hydrobiologia 305 (Dev. Hydrobiol. 108): 55–61.

Kawakatsu; M., I. Oki, S. Tamura, R. E. Ogren, T. Yamada & H. Murayama, 1990. Preprint of papers given at the Sixth International Symposium on the Biology of Turbellaria, Hirosaki, Japan. Occ. Publ., Biol. Lab. Fuji Women's College, no. 22: 1–16.

Kawakatsu, M., I. Oki, S. Tamura, K. Sekiguchi & R. E. Ogren, 1987. Preprint of papers given at the Fifth International Symposium on the Biology of Turbellaria, Göttingen, Bundesrepublik Deutschland. Occ. Publ., Biol. Lab. Fuji Women's College, no. 18: 1–24.

Kawakatsu, M., I. Oki, S. Tamura & H. Sugino, 1976. Studies on the morphology, karyology and taxonomy of the Japanese freshwater planarian *Dugesia japonica* Ichikawa et Kawakatsu, with a description of a new subspecies, *Dugesia japonica ryukyuensis* subspec. nov. Bull. Fuji Women's College, no. 14, ser. II: 81–126.

Kawakatsu, M., I. Oki, S. Tamura, M. Takai, O. A. Timoshkin & N. A. Porfirjeva, 1993. Preprint of papers given at the Seventh International Symposium on the Biology of Turbellaria, Åbo, Finland. Occ. Publ., Biol. Lab. Fuji Women's College, no. 25: 1–20.

Kawakatsu, M., I. Oki, S. Tamura, T. Yamayoshi, J. Hauser & S. M. G. Friedrich, 1984. Morphological, karyological and taxonomic studies, etc. VI. *Dugesia schubarti* (Marcus, 1946) from the vicinity of São Leopoldo, Estado de Rio Grande do Sul (Turbellaria, Tricladida, Paludicola). Bull. Fuji Women's College, no. 22, ser. II: 45–62.

Kawakatsu, M., S. Tamura, T. Yamayoshi & I. Oki, 1980. The freshwater planarians from Thailand and South India. Annot. Zool. Japon. 53: 254–268.

Oki, I., S. Tamura, R. E. Ogren & M. Kawakatsu, 1991. Karyology of four land-planarian species of the genus *Bipalium* from Japan. Hydrobiologia 227 (Dev. Hydrobiol. 69): 163–167.

Oki, I., S. Tamura, T. Yamayoshi & M. Kawakatsu, 1980. Preprint of a paper given at the Third International Symposium — The Biology of the Turbellaria in Honour of Prof. Tor G. Karling, Hasselt (Diepenbeek), Belgium. Occ. Publ., Biol. Lab. Fuji Women's College, no. 2: 1–23.

Oki, I., S. Tamura, T. Yamayoshi & M. Kawakatsu, 1981. Karyological and taxonomic studies of *Dugesia japonica* Ichikawa et Kawakatsu in the Far East. Hydrobiologia 84: 53–68.

Pala, M., S. Casu & R. A. Vacca, 1981. *Dugesia hepta*, nuova specie di planaria di acqua dolce di Sardegna appartenente alla superspecie *Dugesia gonocephala* (Dugès) (Turbellaria, Tricladida). Bull. Soc. Sarda Sci. Nat. 20 (for 1980): 97–107.

Tamura, S., I. Oki & M. Kawakatsu, 1988. Karyological and taxonomic studies of *Dugesia japonica* from the Southwest Islands of Japan. Fortsch. Zool. 36: 123–127.

Tamura, S., I. Oki & M. Kawakatsu, 1991. Karyological and taxonomic studies of *Dugesia japonica* from the Southwest Islands of Japan-II. Hydrobiologia 227 (Dev. Hydrobiol. 69): 157–162.

Tamura, S., T. Yamayoshi, I. Oki & M. Kawakatsu, 1979. Karyological and taxonomic studies of *Dugesia japonica* Ichikawa et Kawakatsu. II. Chromosomes of *Dugesia japonica japonica* collected from eighteen localities in Japan. Proc. Jap. Soc. Syst. Zool., no. 17: 1–14 + pls. 1–2. (In Japanese, with English summary)

Hydrobiologia 305: 85–90, 1995.
L.R.G. Cannon (ed.), Biology of Turbellaria and some Related Flatworms.

Chromosome bands in freshwater triclads

R. Canovai[1], G. Stocchino[2], I. Privitera[1], A. Alberti[2], M. Pala[2] & L. Galleni[1]
[1]*Dipartimento di Coltivazione e Difesa delle Specie Legnose, sezione: Entomologia Agraria, Università di Pisa, Italy*
[2]*Istituto di Zoologia, Università di Sassari, Italy*

Key words: Platyhelminthes, triclads, chromosome banding, speciation

Abstract

Four species of Triclads belonging to three families, *Dugesia polychroa* and *Dugesia mediterranea* (Dugesiidae), *Planaria torva* (Planariidae) and *Dendrocoelum lacteum* (Dendrocoelidae) were studied for chromosome banding: C-banding for all the species, ASG banding for *Dugesia polychroa* and G-banding for *Dugesia polychroa* and *Dugesia mediterranea.* C-banding results for *Dugesia lugubris-polychroa* group suggest a high level of heterochromatin evolution in the group. Small bands strictly limited to the centromeric region were seen in *Dugesia lugubris* and this pattern is similar to the patterns of some species belonging to other families, notably *Planaria torva* and *Dendrocoelum lacteum.* This could be considered a primitive character. In contrast numerous heterochromatin bands which are useful to characterize the different karyotypes were seen in *Dugesia polychroa* and *Dugesia mediterranea.* Our data suggest that heterochromatin is important in freshwater triclad speciation.

Introduction

Many of the problems related to karyological evolution in fresh water triclads were studied by Benazzi and his co-workers (papers reviewed in Benazzi 1982, Benazzi and Benazzi-Lentati 1976) through karyometric analysis. Nevertheless fine characterization of turbellarian chromosomes, heterochromatin evolution and correct identification of chromosome sets in aneuploid strains are still problematical.

Recently new techniques for the fine characterization of turbellarian chromosomes have been developed by Benazzi *et al.* (1981), Galleni *et al.* (1986, 1989, 1991), Sluys & De Jong (1984) and Teshirogi & Ishida (1981). In particular C-banding was applied to *Dugesia gonocephala s. l.* by Sluys & De Jong (1984), *Dugesia polychroa* (O. Schmidt) by Galleni *et al.* (1986), *Dugesia lugubris* (O. Schmidt) by Benazzi *et al.* (1981); Galleni *et al.* (1989), *Dugesia nova* Benazzi by Benazzi *et al.* (1981), *Polycelis nigra* (Müller) by Galleni *et al.*, (1991) and *Polycelis auricolata* Ijima & Kaburaki by Teshirogi & Ishida (1981), but often results have not been definitive or have been related to the different technique used. For this reason these results cannot be compared easily.

The *Dugesia lugubris-polychroa* group is composed of several biotypes (for a review see Benazzi (1982); Benazzi & Benazzi-Lentati (1976). The first four biotypes, indicated by the letters A, B, C, D, belonging to the species *D. polychroa* are characterized by different levels of ploidy and different mechanisms of reproduction. Other biotypes are good taxonomic species: *D. lugubris* (E biotype), *D. nova* (F biotype) derived from the E biotype by a Robertsonian translocation (Benazzi, 1982), *D. mediterranea* Benazzi, Baguñà, Ballester, Puccinelli & Del Papa (G biotype) (Benazzi *et al.*, 1975). Biotype H, not classified at the moment, was described by Ball (1979) from specimens collected on the island of Corfù.

C-banding was attempted by Benazzi *et al.* (1981) on *D. lugubris* and *D. nova.* They observed no bands in *D. lugubris* and large pericentromeric and telomeric bands in *D. nova.* Determination of amounts of DNA in these species (Benazzi *et al.*, 1981) confirmed the Robertsonian translocation proposed by Benazzi & Puccinelli (1973) on the basis of karyometric data.

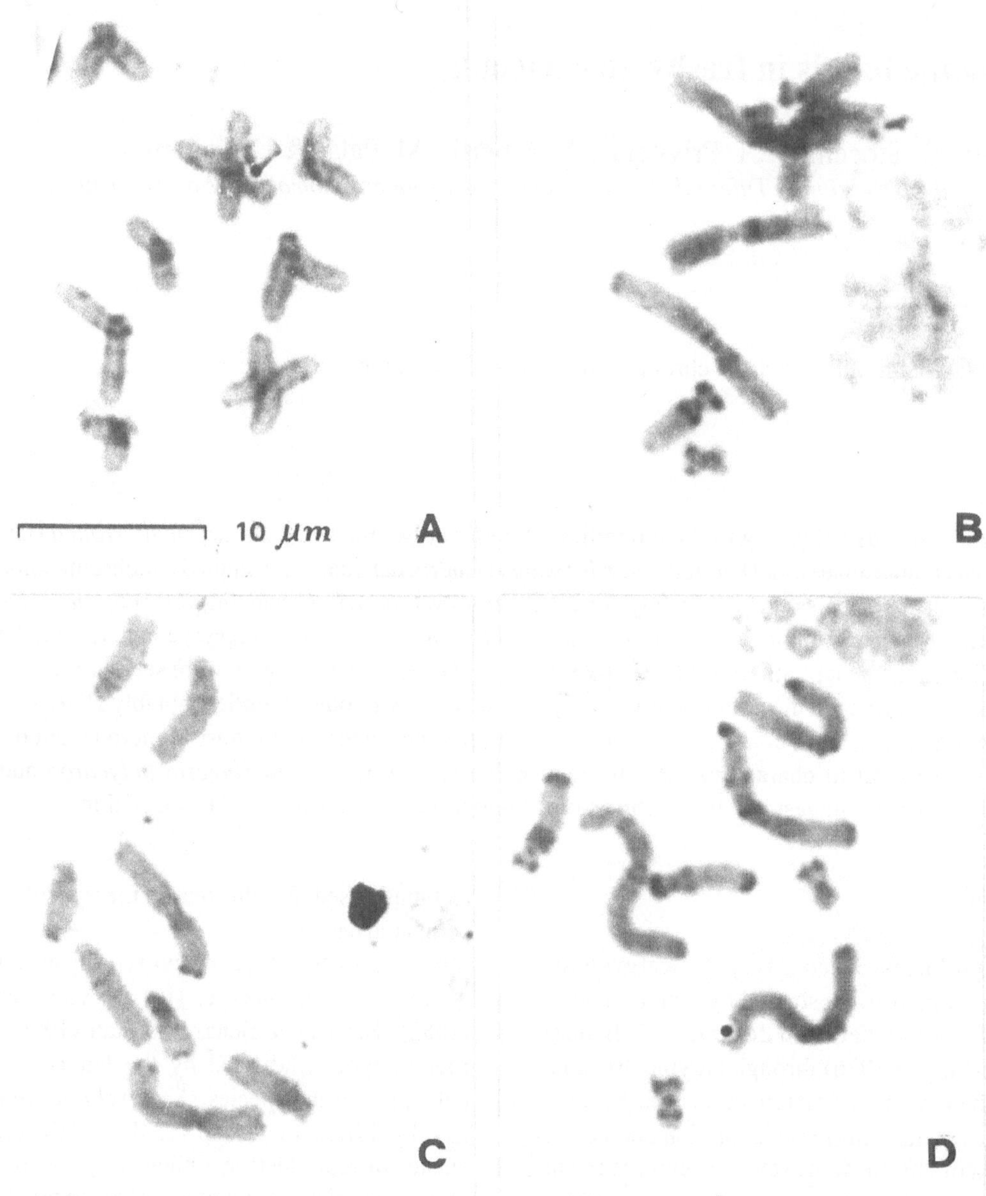

Fig. 1. C-banding of planarian metaphasic plates. Technique I –A, *Dugesia polychroa*, B, *D. mediterranea*; by technique 2 – C, *D. polychroa*, D, *D. mediterranea*.

Pellicciari *et al.* (1986) showed, by microdensitometric analysis of the Feulgen staining DNA, that *D. lugubris* has much more DNA than *D. polychroa*, in spite of small differences in karyometric parameters between the species.

C-banding obtained by Galleni *et al.* (1986) from *D. polychroa* and *D. lugubris* by Galleni *et al.* (1989) showed small centromeric bands for all the chromosomes of *D. lugubris* and bands involving not only pericentromeric regions, but also telomeric and paracentromeric regions in *D. polychroa*. The bands obtained from *D. polychroa* were considerably larger than those obtained from *D. lugubris* (even allowing for artefacts).

The studies cited above show bands obtained by different and not yet standardized methods vary and

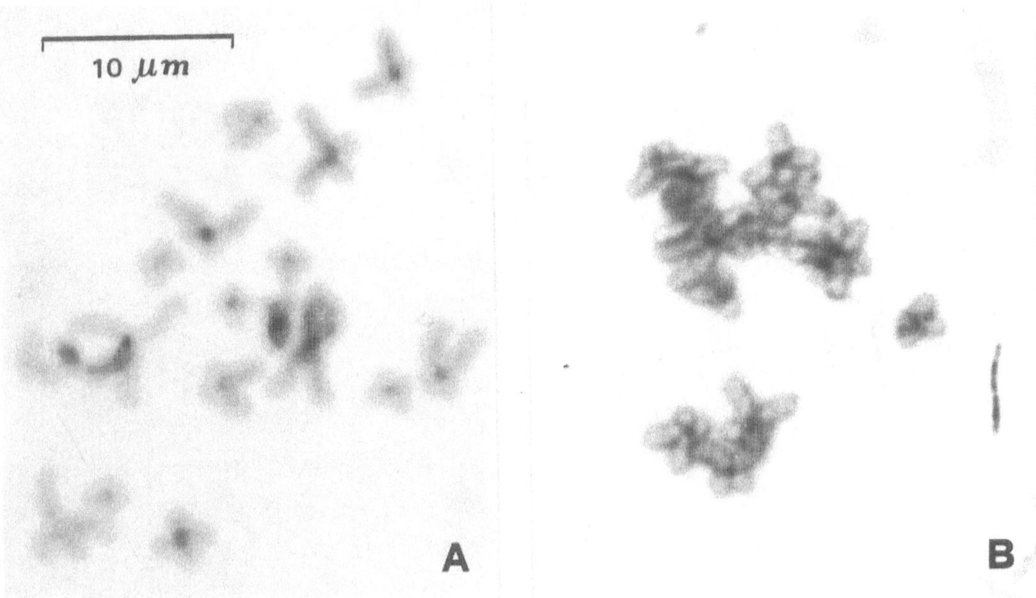

Fig. 2. C-banding of planarian metaphasic plates by technique 1 –A, *Planaria torva*; B, *Dendrocoelum lacteum*

therefore a comparison between bands obtained by different workers is not possible. The absence of C-bands in *D. lugubris* reported by Benazzi *et al.* (1981) and the confirmation of their presence by Galleni *et al.* (1989) provides proof of these difficulties. Looking for heterochromatin by using different banding techniques gave us the opportunity to reach both a good level of reliability and some preliminary conclusions about heterochromatin evolution.

Materials and methods

Species examined: *Dugesia polychroa*: biotype A and *Dugesia mediterranea*: sexual strain (Dugesiidae), *Dendrocoelum lacteum* (Müller) (Dendrocoelidae), *Planaria torva* (Müller) (Planariidae).

All the specimens studied belong to strains reared in the laboratories of D. C. D. S. L. sez. Entomologia Agraria, University of Pisa and of the Istituto di Zoologia, University of Sassari. The chromosome plates were obtained with two different techniques, the first one standardized by Galleni *et al.* (1986) and the second one by Ribas *et al.* (1988). A third technique, standardized by the laboratory of Pisa, was also employed in order to obtain plates of better quality. According to this third method the end of every specimen was cut and left in fresh water for three days in order to obtain good regenerative blastemas. On the third day the regenerating piece was transferred to colchicine (GIBCO) diluted in fresh water, 0.15% final solution, and kept for 18–24 h at room temperature. The preparation was then transferred for 20 min to a 1% solution of sodium citrate and fixed in Carnoy (3:1 methanol:acetic acid) for 30 min. Chromosome plates were made by dispersing the cells in a few drops of 60% acetic acid; preparations were air dryed and aged for at least ten days.

C-Banding technique n. 1

The preparations were treated with 0.2N HCl. They were then transferred to a saturated solution of barium hydroxide at 50 °C for 90–120 sec., rinsed in acid water and incubated in 2 × SSC (8.9 g of sodium citrate and 17.5 g of sodium chloride/l) at 60 °C for 25 min. The preparations were then rinsed with distilled water and stained with 5% Giemsa solution, dissolved in Sorensen's buffer. Preparations were stained for 20 min.

C-Banding technique n. 2

Preparations were incubated in 1N HCl for 3 min at room temperature and rinsed in distilled water, then treated with a saturated solution of barium hydroxide at 50 °C for 8 min and rinsed with tap water. Preparations were treated again in 1N HCl for several seconds and

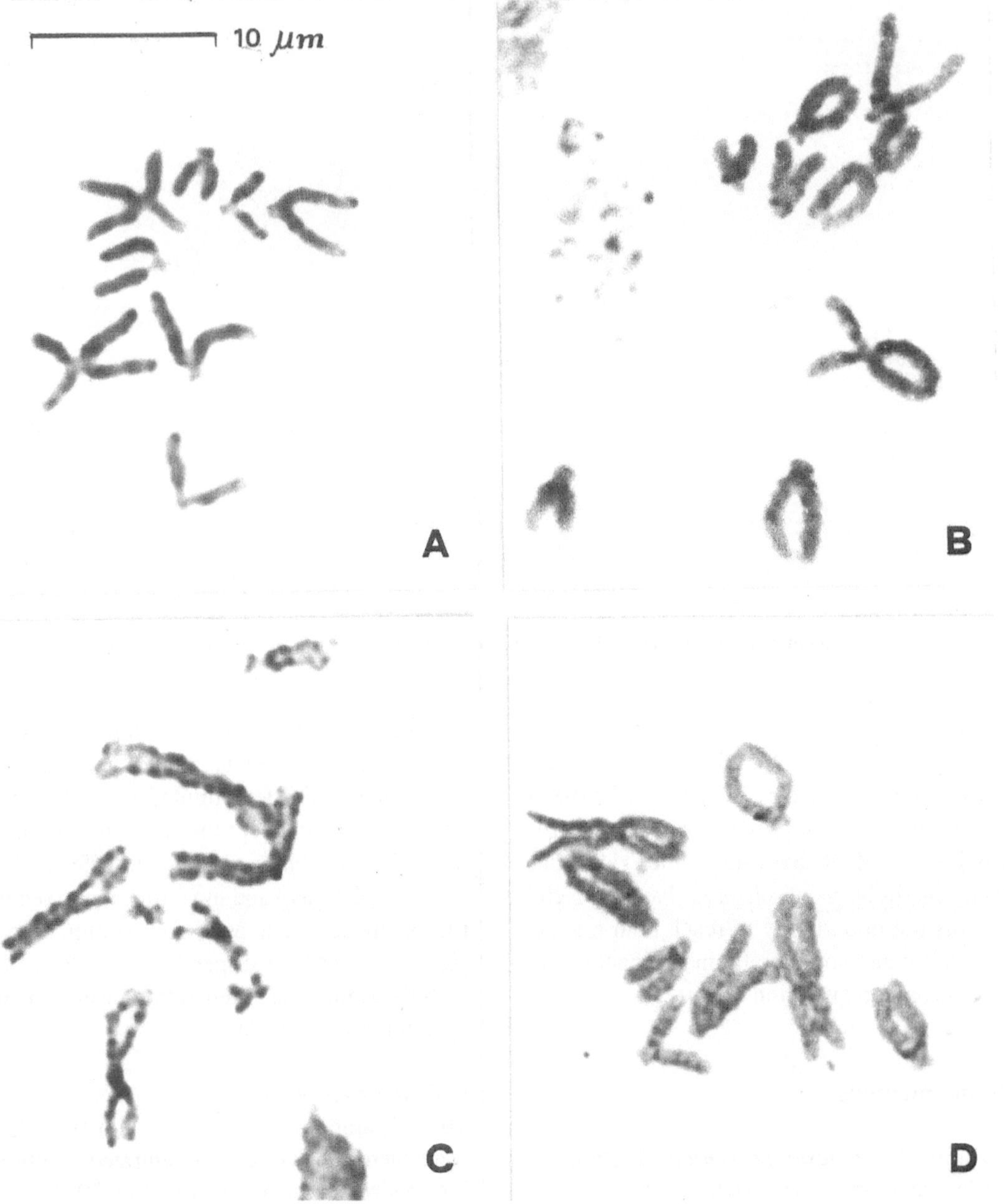

Fig. 3. ASG or G banding of planarian metaphasic plates – A, ASG with 2 × SSC: *D. polychroa*, B, ASG with 3 × SSC: *D. polychroa*; G-bands – C, *Dugesia mediterranea*; D, *Dugesia polychroa*.

then rinsed with distilled water. They were incubated in 2 × SSC solution at 65 °C for 2 h and then rinsed with distilled water. Preparations were stained with 5% Giemsa solution for 90 min.

ASG-Banding

Preparations were either incubated in 2 × SSC (see above) at 60 °C for 54–59 min, then rinsed in distilled water and stained in a Giemsa solution for 20 min. On some occasions preparations were incubated in a saline solution (3 × SSC = 13.35 g of sodium citrate and 26.25 g of sodium chloride/l) at 65 °C for 22–26 minutes.

Table 1. C-banding on the chromosomes of the planarians *Dugesia polychroa* and *D. mediterranea.*

Chromosome I		
	Large centromeric band.	Small centromeric band.
	Often telomeric bands on both telomeres.	A pericentromerlc band on both sides of the centromere. A band near the middle of one of the arms. A telomeric band on one telomere.
Chromosome II		
	Centromeric band.	Small centromeric band.
	Sometimes faintly stained telomeric bands.	A pericentromeric band on both
	A paracentromeric band on the long arm.	sides of the centromere. Sometimes a band in the middle of one of the arms. Telomeric band on both telomeres.
Chromosome III		
	Centromeric band.	Small centromeric band.
	Sometimes telomeric bands.	Telomeric band on both arms.
	Paracentromeric and paratelomeric bands on the long arm.	Distinct pericentromeric band.
Chromosome IV		
	Large centromeric and telomeric band.	Distinct centromeric and
	A band in the middle of the long arm.	telomeric band.

G-Banding

Preparations were treated in 1N HCl for 5 minutes and then rinsed in distilled water. They were treated in Trypsin (Difco) solution (0.25 g in 100 ml of isotonic saline solution) for 15 sec and rinsed twice with 0.02 M phosphate buffer. Finally, they were stained in 10% Giemsa solution for 5 min and rinsed in 0, 02 M phosphate buffer.

Results

C-Banding

The results of C-banding in *Dugesia polychroa* and *D. mediterranea* are summarized in Table 1 and Fig. 1. With *D. lacteum* and *P. torva* only small bands limited to the centromeric regions were seen (Fig. 2).

ASG-banding

The technique was limited to *D. polychroa* (Fig. 3A, B). With 3 × SSC C-like bands (centromeric, pericentromeric and paracentromeric) were obtained, while with 2 × SSC scattered and often ill-defined bands appeared.

G-banding

This technique was limited to *D. mediterranea* (Fig. 3C) and *D. polychroa* (Fig. 3D). In the latter C-like bands were obtained, while in the former, scattered ill-defined bands often appeared.

Discussion

The C-banding techniques used in this study show bands in species from different families. For this reason we assume that the techniques are now reliable.

ASG-banding and G-banding sometimes show C-like bands. Localization of these bands is similar to that obtained with C-banding techniques I and II. This is a further demonstration that heterochromatin regions are not a technical artefact. In contrast the G-like bands obtained with these techniques are still difficult to interpret. It is very difficult to obtain G bands in invertebrates. According to Holmquist (1990) this is due to the absence of compartmentalization of the invertebrate genome. In contrast, this compartmentalization is present in vertebrates and shows clear G bands.

With ASG techniques scattered bands were obtained also in Dipteran mitotic chromosomes (Canovai *et al.*, in prep.), but also in this group conclusions are far from being unequivocal. In our opinion the problem of G-banding in invertebrates is still an open question.

Some preliminary conclusions can be made on heterochromatin bands in freshwater triclads. C-bands were found in all the freshwater triclad species examined. In many cases they are strictly limited to the centromeric area as in *D. lugubris* (see Galleni *et al.*, 1989), *Dendrocoelum lacteum* and *Planaria torva*. Sometimes the bands are also present along the arms of some chromosomes or in some telomeric regions, as in *P. nigra* (see Galleni *et al.*, 1991) and these data are in good agreement with those obtained by Teshirogi & Ishida (1981) on *P. auricolata*.

The *Dugesia gonocephala* group is still problematical. Bands were obtained by Sluys & De Jong (1984) in *D. gonocephala* as well as in *D. sicula* by Pala *et al.* (unpublished data). Moreover in this group banding is very difficult and comparisons of the results obtained by various workers are not possible.

Various banding patterns are present in telomeric and centromeric regions as well as in regions along the arms of *D. polychroa* and *D. mediterranea*. These species show a high level of heterochromatin evolution. In the *Dugesia lugubris-polychroa* group *D. lugubris* has heterochromatin bands strictly limited to the centromeric regions. This banding pattern is similar to other species from different families and could be considered as a primitive character. Further genomic evolution in this group of species is suggested by the differentiated patterns of bands in *D. polychroa* and *D. mediterranea*. Our data suggest that heterochromatin evolution is important in species differentiation in freshwater triclads. A task for the future will be to ascertain the DNA composition of heterochromatin bands and the role of this heterochromatin differentiation in the process of speciation.

Acknowledgments

This work was supported by grants from Italian MURST (60% obtained from the Universities of Pisa and Sassari).

References

Ball, I. R., 1979. The karyotypes of two Dugesia species from Corfù, Greece (Platyhelminthes, Turbellaria). Bijd. Dierkunde 48: 187–190.

Benazzi, M., 1982. Speciation events evidenced in Turbellaria. In C. Barigozzi (ed.) Mechanisms of speciation. Alan R. Liss New York: 301–304.

Benazzi, M. & G. Benazzi-Lentati, 1976. Animal cytogenetics. I. Platyhelminthes. Gebr. Borntraeger, Berlin & Stuttgart, 182 pp.

Benazzi, M. & I. Puccinelli, 1973. A Robertsonian translocation in the freshwater Triclad Dugesia lugubris: karyometric analysis and evolutionary inferences. Chromosoma, 40: 193–198.

Benazzi, M., J. Baguñà, R. Ballester, I. Puccinelli & R. Del Papa, 1975. Further contributions to the taxonomy of the Dugesia lugubris-polychroa group, with description of Dugesia mediterranea n. sp.. Boll. Zool. 42: 81–89.

Benazzi M., D. Formenti, M. G. Manfredi Romanini, C. Pellicciari & C. A. Redi, 1981. Feulgen DNA content and C-banding of Robertsonian transformed karyotypes in Dugesia lugubris. Caryologia 34: 129–139.

Galleni, L., R. Canovai, A. Esposito & R. Stanyon, 1989. Characterization of turbellarian chromosomes. II. C-banding in Dugesia lugubris (Tricladida: Paludicola). Trans. am. Microsc. Soc., 108: 304–308.

Galleni, L., R. Canovai, M. Gualandi, C. Pellicciari, S. Garagna & R. Stanyon, 1986. Fine characterization of turbellarian chromosomes. I. Giemsa and Quinacrine banding in Dugesia polychroa (O. Schmidt). Genetica 71: 47–50.

Galleni, L., R. Canovai & R. Stanyon, 1991. Characterization of turbellarian chromosomes. III. C-banding in Polycelis nigra (Tricladida-Paludicola). The nucleus 34: 119–122.

Holmquist, G. P., 1990. Contents of G and R bands defy contemporany paradigms. In T. Sharma (eds), Trends in Chromosome Research, Springer-Verlag, Narosa Publishing House, New Delhi: 39–52.

Pellicciari, C., S. Garagna, D. Formenti, C. A. Redi, M. G. Manfredi Romanini & M. Benazzi, 1986. Feulgen-DNA amounts and karyotype lengths of three planarian species of the genus Dugesia. Experientia 42: 75–77.

Ribas M., M. Pala, R. A. Vacca, M. Riutort & J. Baguñà, 1988. Taxonomic status of the western Mediterranean asexual populations of the Dugesia (D.) gonocephala group. Morphological, karyological and biochemical data. In Ax/Ehlers/Sopott-Ehlers (eds), Free living and Symbiotic Platyhelminthes, Gustav Fischer Verlag, Stuttgart, New York 36: 129–137.

Sluys, R. & H. De Jong, 1984. Chromosome morphological studies of Dugesia gonocephala s. l. (Platyhelminthes, Tricladida). Caryologia, 37 n. 1–2: 9–20.

Teshirogi, W. & S. Ishida, 1981. Studies on the speciation of japanese freshwater planarian Polycelis auricolata based on the analysis of its karyotypes and constitutive proteins. Hydrobiologia 84: 69–77.

Hydrobiologia **305**: 91–99, 1995.
L.R.G. Cannon (ed.), *Biology of Turbellaria and some Related Flatworms*.

Where Turbellaria? Concerning knowledge and ignorance of marine turbellarian ecology

P. J. S. Boaden
The Queen's University of Belfast, School of Biology and Biochemistry, Marine Biology Station, Portaferry, Co. Down, BT22 1PF, N. Ireland, U.K.

Key words: Turbellaria, meiobenthos, ecology

Abstract

Present knowledge of marine free-living turbellarians is reviewed in the light of a standard ecological text. Little is known of their ecological adaptations or fitness but they are clearly successful animals. Diversity within communities can be considerable, possibly relating to the small size of individuals and temporal and spatial resource gradients. Little is known about genetic diversity, environmental tolerances and demography. Local distribution patterns are better known but rarely proven with appropriate statistics. New data from work in Northern Ireland on behaviour and habitat choice of *Procerodes littoralis* and on species richness, population aggregation and microdistribution of meiofaunal turbellarians is presented.

Nothing appears known about intra-specific competition but more about inter-specific relationships including responses to biogenic structures. Dietary information is sparse and mostly observational with little known about feeding efficiency. Studies at Sylt have revealed mainly univoltine or polyvoltine annual life cycles partly related to habitat preference and possibly giving some evidence of r and K selection. Factors controlling local abundance are largely uninvestigated and the reasons for the apparent rarity of many species not understood.

Lists and counts of turbellarians within particular habitats provide some knowledge of community structure but very few numerical analyses, e.g. of diversity and similarity, exist. Contribution to energy flux and to structuring communities has not been determined in spite of the turbellarians' assumed major role as meiofaunal predators. Nevertheless marine turbellarians appear particularly fitted to studies such as habitat heterogeneity, community stability and biogeography and are surely suitable subjects for field and laboratory experiments. There is no lack of direction for studies to proceed. Impulsion is required.

Introduction

This paper attempts to assess the current state of knowledge concerning the ecology of free-living marine turbellarians. In particular, to help determine where present areas of information and ignorance lie, it compares available literature with a standard modern ecological text-book (Begon *et al.*, 1986). Martens & Schockaert (1986) briefly assessed the importance of meiobenthic turbellarians but there is no recent general review of their ecology. This paper is intended to fill this gap and follows the book's format by discussing factors affecting individual organisms, then species populations and interactions and concluding with a section on communities, including structure, stability and distribution.

Autecology

Evolutionary factors

It is clear from their distribution and abundance that many turbellarian species are successful animals. It can be assumed that any abundant species has become adapted to the environment in which it is common. Abundance can be phenomenal. Dörjes (1968) sampling a 5.5 cm^{-2} area found up to 1500 *Haplogonaria syltensis* Dörjes (equivalent to 2 7 × 10^6 m^{-2}

and Pawlak (1969) and Murina (1981) have reported densities of over 115 000 m^{-2} for *Paramalostomum fusculum* Ax and *Pseudomonocelis ophoicephala* (O. Schmidt) respectively. Boaden (1981) hypothesised evolutionary adaptation from plesiomorphic anoxic or micro-oxic to apomorphic oxic conditions and (Boaden, 1989) that the platyhelminths' plesiohabitat was fine and/or detritus rich sand. Previously Riedl (1963) suggested adoption of habitats starting from marine mud and sand through to others such as rock and phytal.

Little is known about convergent or parallel evolution within turbellarians. It appears that some, such as production of chordoid tissue (Ax, 1966) or proboscis evolution (Karling, 1961), has resulted from habitat/niche factors. Common ancestry may predispose adaptations such as spicule production (Rieger & Sterrer, 1975). Presumably species are abadapted (*i.e.* have inherited adaptations) to many aspects of their environment; this would be the case for example in Galapagos 'sister-species' (Ax, 1977).

Little is known about genetic diversity, phenotypic variation, polymorphism and the production of ecotypes and sibling species in marine flatworms. Exceptions are work by Reuter (1961) on genetic adaptation of *Gyratrix hermaphroditus* Ehrenburg populations from different salinities, Curini-Galetti & Martens (1990) on the Monocelididae and Curini-Galetti (1994) on *Pseudomonocelis*. Ax (1959) noted an eyeless form of *Monocelis lineata* (Müller) from coastal groundwater.

Abiotic factors

Information is limited and comes mainly from (1) observation and/or measurement of environmental parameters (often in relation to faunistic studies) and (2) laboratory-, or occasionally field-, experiments on behaviour.

1. The simplest and most widespread information comes from habitat characterisation in original or supplementary species descriptions. Examples, too numerous to cite, may be found in papers by Ax, den Hartog, Luther, Karling, Marcus, Westblad and others. These often contain information on zoogeography and factors such as salinity and substratum. Ax (1959) and Karling (1974) exemplify this faunistic approach. Bilio (1964, 1967) discussed salt marsh turbellarian distribution in relation to salinity.

Many 'Ax school' papers relate distribution of species to sediment type, depth in sediment, height on beach and temporal factors such as temperature. Examples are Hoxhold (1974) on Kalyptorhynchia and Sopott (1973) on Proseriata. The latter group were affected by temperature, salinity, pore water content and sulphide layer depth; both seasonal and tidal vertical migration occurred. Ax & Ax (1970) related vertical and horizontal migration of brackish beach turbellarians to ground-water-table level fluctuation. Rieger & Ott (1971) found active and passive horizontal and vertical migrations of sand beach flatworms were associated with tides and resultant pore water movement. Boaden & Platt (1971) and Boaden (1968, 1977b) described vertical migration in a sand beach, citing tidal or mechanical disturbance (*Cicerina*, *Coelogynopora*) and light intensity (*Haplogonaria*) as possible factors. Other examples of migration can be added from the classical work on *Convoluta roscoffensis* Graff (Bohn, 1903; Gamble & Keeble, 1904) and observations such as from Ax (1951). Armonies (1990) found that emergence from the sediment was particularly strong at night. Thus, many sandy-shore flatworms have migration patterns at least partly mediated by abiotic factors.

2. In laboratory work Boaden (1968) showed that vibration induced downward migration of *Paramolostomum*. Bilio (1964) recorded positive geotaxis in salt-marsh *Coelogynopora schulzii* Meixner. In choice-experiments this species did not select particular sediment grain-sizes (Jansson, 1967). However it is likely that other sediment dwelling species will show such selection – as indicated by graded-sand colonisation (Boaden, 1962). Jansson, (1968) found salinity fluctuation tolerance and a relationship between the upper limits of temperature tolerance and environmental temperature in beach turbellarian populations. The marine triclad *Procerodes littoralis* (Ström) was the subject of some classical eco-physiological work on salinity tolerance (Pantin, 1931a, b; Weil & Pantin, 1931) but little further work has been undertaken on this common and widely occurring species (den Hartog, 1968).

Substratum selection by *P. littoralis* from an intertidal stream-bed in N. Ireland has been shown in recent work (Boaden, unpublished). In experiments the species entered coarse gravel in preference to other sediments ranging from large stones to sand. Specimens exhibited positive rheotaxis and geotaxis in fresh-water but negative rheotaxis and geotaxis in sea-water; dark-adapted worms were negatively phototaxic but light-adapted were not and dark-adapted worms moved faster (but did not turn more) than light-adapted specimens. Such responses presumably helped keep worms in their preferred habitat Boaden (1977b) showed how responses to light, vibration, gravity, water current and oxygen gradient helped maintain the distribution pattern of *Monocelis lineata* in swash-zone sediment.

Habitat oxygen content is important in helping to determine species occurrence. Most marine free-living turbellarians are restricted to oxic habitats, as indicated by the positive relationship between high oxygen, redox potential and turbellarian abundance found by McLachlan (1978). Many phytal and sediment-dwelling flatworms die rapidly after collection if the sample becomes anaerobic. However some turbellarian species show tolerance of, or are specifically associated with, anaerobic or micro-oxic conditions even though they may belong to a mainly 'aerobic' group. Examples are the kalyptorhynchids *Neoschizorhynchus parvorostro* Ax & Heller and *Pseudoschizorhynchoides ruber* Schilke (Boaden, 1977a, 1981). The families Retronectidae (Catenulida) and Solenofilomorphidae (Acoela) are characteristic of such conditions (Sterrer & Rieger, 1974; Crezée, 1975 respectively) and must have specific eco-physiological adaptations such as efficient sulphide detoxification (Powell *et al.*, 1980). Some turbellarians occur within anoxic habitats inside stromatolitoid nodules (Westphalen, 1993).

It appears that understanding of some of the abiotic factors helping to define the ecological niche of free-living flatworms, particularly from sand beaches, is just beginning.

Biotic factors

Very little is known about anthropogenic factors and a modicum about other biotic ecological factors. There have been few studies of meiofauna in relation to pollution (for details and inherent difficulties see Moore & Bett, 1989). Turbellaria have rarely been considered in these. Data, where it exists, may be suspect due to sampling and extraction techniques. Thus, for example, Bleakley & Boaden (1974) had results indicating severe mortality of rhabdocoeles and seriates with population decline to less than 20% within 2–3 weeks of beach application of oil-spill remover. However they felt these insufficiently reliable to publish (personal communication). Johnston (unpublished) working at Doctor's Bay, a N. Ireland beach with some sewage contamination (see Boaden, 1978), found a decline in sand-dwelling turbellarian species richness with increasing nearness to a sewer outfall. No turbellarians were found at the outfall itself. At distances of 55, 110, 165 and 220 m seaward down the beach, the number of species actually found or expected (*i.e.* E(Sn′) from species richness curves) in large samples yielding about 250 individual turbellarians, were 7, 16, 17 and 23 respectively in summer and 9, 15, 16 and 19 in winter. It is not known whether the impoverishment was mediated through salinity, oxygenation or other factors. Reise (1983) found that sewage depressed turbellarian abundance and diversity through increased green algal (mainly *Enteromorpha*) growth.

Population ecology

Very little is known of marine free-living turbellarians' interactions either with their biotic (e.g. food) or abiotic (e.g. space) resource requirements, or of demographic processes such as birth and death rates.

Ball & Reynoldson (1981) found an average of 5 (s.d. 2.5) young per cocoon for *Procerodes littoralis*. In laboratory observations of *P. littoralis* in N. Ireland (Boaden, unpublished) fed-specimens produced no cocoons. However 6 specimens, which survived from an original batch of 50 which had been accidentally starved for 5 weeks, all produced cocoons. These took 4–5 weeks to hatch; 1 worm produced 3 cocoons in this period, 1 produced two and the others 1 each. The average young per cocoon was 4.4. In 1993 cocoon production in the shore population started in early February. Within 3 weeks a population of 726 worms (at the start of sampling) in a 1.5 m wide, 18 m long, shore transect had produced 114 capsules. It is not known how long this production rate lasted. As far as the author is aware the only year-long records of reproduction in marine turbellarians under controlled laboratory conditions are for the triclad *Cercyra hastata* Schmidt and the proseriate *Pseudomonocelis ophiocephala* Schmidt from Sevastopol (Murina, 1981). Population mean breeding rate (*sensu* Edmonson, 1960) for the triclad ranged from 1.4 in winter to 4.2 in summer, totalling 10.8 over the year. Maximum longevity was 7 months during which 7 generations were produced. Corresponding values for the proseriate were 0.8 (winter)–4.8 (autumn), totalling 12.2, 11 months and 5 generations.

Several papers on natural populations at Sylt give monthly percentage occurrence of juveniles, adolescent, adult and senescent individuals of particular species (Ehlers, 1973; Faubel, 1976; Hoxhold, 1974; Sopott, 1973). Earlier references to breeding include

Ax (1969), den Hartog (1964) and Schmidt (1969). The few available records of development times (egg-adult) range from 29 days to 3 months (Giesa, 1966; Ax, 1969). As yet no cohort life tables or survivor curves are available.

Distribution patterns within local populations have rarely been published with appropriate statistics but there are reports both of contagious and normal distributions for Turbellaria when considered as a single taxon (Gray & Rieger, 1971; Vitiello, 1968). Among meiobenthologists the extreme patchiness of many species-populations is well appreciated (Heip, 1975) – although this is hardly the right term since it causes considerable quantitative sampling difficulties (see later). Faubel (1976) gave crowding and patchiness data for 11 sand flat acoel species. Some showed marked aggregation, others were more evenly distributed or had a mixed pattern. Dittmann & Reise (1985) determined log variance-to-mean ratios for mud flat species. These showed only a little aggregation in small scale samples (from within 0.3 m^{-2}) but a considerable amount when considering the whole mud flat area. Reise (1987) calculated patchiness indices for 31 turbellarian taxa from an *Arenicola* sand flat. Most indices were very high and all departed from randomness, however the sampling pattern was specifically associated with microsites and not with patchiness within these.

Work at Doctor's Bay (see above) included multiple sampling within particular sediment areas. Thus there are sets of 30–50 core samples taken from tidal pool sand, from areas of exposed drained sediment from *Arenicola* feeding depressions and from *Arenicola* faecal mounds (each set collected during one low-tide period). These samples revealed considerable clumping within populations. 22 data-sets (e.g. the numbers of *Monocelopsis otoplanoides* Ax in 40 pool sand cores or *Promesostoma marmoratum* Schultze in 40 feeding depression cores) for turbellarian species were tested for fit to the negative binomial, Pólya-Aeppli and Newman type-A distributions. All showed good fit to the negative binomial, 8 best fitted this and 7 best fitted each of the other distributions. The method of Iwao & Kuno (1971) applied to combined data from all the samples for each turbellarian taxon considered showed that mean crowding increased significantly as mean density increased. This was the case for *M. otoplanoides**, *P. marmoratum**, *Proxenetes minimus* den Hartog, *Retronectes* sp., *Vejdovskya* sp.*, *Zonorhynchus seminascatus* Karling and an unidentified species (all at $p<0.001$) and in *Mecynostomum* sp. and *Pseudograffilia arenicola* Meixner (at $p<0.01$ and $<0.0.5$ respectively). This relationship could also be demonstrated within particular single-station data sets for those species asterisked* and for *Macrostomum pusillium* Ax and *Philactinoposthia* sp.. Other unpublished work on a *Paramalostomum dubium* (Beauchamp) population from Ballylumford Beach, N. Ireland showed that this also fitted the negative binomial distribution. Analysis, by Iyer's (1949) joins between clumps method, of this species' aggregation pattern within an apparently homogenous beach area indicated irregular patches of above mean density ranging in area from below 7 to at least 240 cm^{-2}. Thus, aggregation is a common feature of marine sediment flatworm populations. It is almost certainly biologically mediated and has been observed in laboratory conditions (Boaden, 1977b). Little is known of larger scale dispersion or migration except within individual beaches (see earlier). It is assumed to occur. Indications of this are that, although most marine flatworms have no planktonic larvae, there is little morphological intra-specific variation in at least some fairly widely separated populations (e.g. *Macrostomum hystricinum* Beklemischev from Europe and N. America – see Rieger, 1977); benthic turbellarians can colonize sediment traps suspended in water (Hagerman in Gerlach, 1977); there are amphi-atlantic species and, for some brackish water species at least, transport whilst encysted is possible (Ax & Armonies, 1987).

Interactions

The author knows of no published evidence for intra-specific competition in marine free-living flatworms although it seems obvious that competition for food is likely to affect survival, growth and reproduction. Some preliminary work (personal observation) has suggested that the numbers of an orange dalyelloid living and feeding on orange colonies of the tunicate *Botryllus* on *Fucus serratus* were limited by the colony perimeter, underneath which they sheltered, rather than by colony area. Thus spatial competition was indicated. Congeneric occurences within sand cores (see earlier) suggest speciation resulting from previous intra-specific competition.

Dörjes (1968) presented evidence for seasonal inter-specific competition by oligochaetes and archiannelids with the acoel *Haplogonaria syltensis* resulting in restriction of its spatial habitat. Faubel (1977) gave evidence for the competitive displacement of the acoel *Actinoposthia biaculeata* Faubel and the macrostomid

Myozona purpurea Faubel by other acoels. However there appears to be no other evidence for competitive heirarchies in marine turbellarians and no experimental determination of the importance of competition.

Most information on the reaction of marine turbellarians with other taxa falls into two categories.

1. the effect of biogenic structures. Reise (1981, 1984) showed that temporary algal mats depressed sand flat turbellarian abundance and diversity but that the presence of bivalves (*Macoma*) and sediment irrigating polychaetes (e.g. *Arenicola, Nereis*) enhanced the sub-surface occurrence of flatworms and other meiofauna. This enhancement will have been due in part to physical and chemical gradients and their related microbiota occurring in the burrows, tubes and their walls (Meyers *et al.*, 1987) although other factors may have been involved. Johnston (unpublished) found significantly more *Monocelopsis otoplanoides* in sand from *Arenicola* head shafts than from tail shaft or beach surface sands; in laboratory choice experiments the flatworm preferentially entered head- rather than tail-shaft sand; the attractive factor was destroyed by heat but restored by conditioning the sand by contact with *Arenicola*.

2. trophic relationships. Much information is incidental to taxonomic and faunistic papers too numerous to cite, for example by Ax, Lang, den Hartog, Karling, Luther, Marcus, Meixner and Westblad. Bilio (1967) specifically considered food relationships of salt-marsh flatworms. Most turbellarians are predators in the broad sense that they consume organisms (or parts of them) which are alive when first attacked. Many turbellarians are true predators, killing many prey; others are grazers on animal tissue, removing only a part of the prey individual or colony which consequently usually survives; yet others are parasites consuming part of, or resources from, only one or a few individuals; the latter two categories grade into each other (e.g. orange dalyelloid above). A few flatworms may feed principally on bacteria and micro-algae. Known prey items include diatoms, ciliates, hydroids nematodes, other turbellarians, small crustaceans, annelids and tunicates (see Martens & Schockaert, 1986; Jennings, 1957). Available trophic information mainly relates to sedimentary and salt-marsh turbellarians, but they are also important consumers in other habitats such as the phytal (Jansson, 1966). There is an extensive literature on the importance of predation in the structure and control of communities (see Begon *et al.*, 1986) but the author has not found any contribution of marine turbellarian ecology to this or *vice versa*. Obvious areas for research arise, for example in relation to predator/food patch distribution, diet specificity, diet switching and community equilibria.

Predation by meiofauna (including turbellarians) on juvenile and larval macrofauna (Watzin, 1986) may have been critical in favouring the evolution of macrofaunal pelago-benthic life cycles (Warwick, 1989).

Some free-living turbellarians are detritivores or scavengers (e.g. Bush, 1966), thus functioning in part as decomposers, and can be expected to have donor-controlled dynamics leading to stable community structure (Begon *et al., l.c.*). Perhaps this is the case for *Otoplana*- and surf-zone communities. Some laboratory evidence of bacteriovory has been presented by Meyer-Reil & Faubel (1980).

Mutualism with autotrophic protistans is known.The classical example is that of *Convoluta roscoffensis* (see Holligan & Gooday, 1975), others are the pelagic acoel *Haplodiscus* (see Dörjes, 1970) and sublittoral dalyelloid *Pogaina* spp (Ax, 1970). *Paracatenula* species have no mouth but apparently derive nutrition from prokaryotic symbionts occurring in entoderm cells (Ott *et al.*, 1982). At least two interstitial turbellarians can utilize dissolved amino-acids taken up via the body wall (Tempel & Westheide, 1980).

Turbellaria are eaten by some macrofaunal crustaceans and bivalves (Feller, 1984), by predatory polychaetes and apparently by deposit feeders. There is little information on this (see Martens & Schockaert, 1986) or on parasites or disease although sporozoans, mesozoans and ciliates have been observed in various turbellarians, for example in proseriates (Sopott, 1973).

Life cycles

Ax (1977) summarised work from Sylt which has provided most of our current knowledge of marine free-living turbellarian life cycles. A univoltine life cycle, *i.e.* one limited or extended period of reproduction (usually spring/summer) during the year, was found in 45 of the 74 species listed. However (principally in Acoela and Dalyellioida) there were bivoltine and polyvoltine cycles tending to have reproductive maxima in spring and autumn. Life cycle type was partly correlated with habitat – all species living at depth in the beach slope were univoltine but other types, as well as this, occurred in surface-sediment species. Much reproduction was associated with warmer temperatures as evidenced by field work (for example Faubel, 1976) and by laboratory study (Apelt, 1969). Dormancy of the egg capsule stage was found in 17 of the species, 10 being Kalyptorhynchia, and appeared to be correlated with geographical distribution.

In contrast with the resulting summer and autumn maximum abundances at Sylt, the maximum abundances recorded at Ballymaconell (see later) were in October and December, implying strong recruitment in autumn/winter. It appears, from the little known, that tropical beach turbellarian reproduction is not seasonal (Ax & Schmidt, 1973). The rapid and extreme abundance achieved by many acoel and macrostomid species together with their relatively wide habitat range (Faubel, *l.c.*; Dörjes, 1968; Dittmann & Reise, 1985) suggests r-selection. Conversely, K-selection probably occurs in more habitat-specialised, less abundant species such as many kalyptorhynchids (Hoxhold, 1974).

Abundance

Although a number of papers on general meiofaunal abundance have given figures for 'Total Turbellaria' much of this data is highly suspect. Unfortunately this is probably also true of specialised work citing more specific population counts. The problem arises from a number of sources including lack of appropriate sampling methods, sampling regimes and sample processing. As an example of the latter's effect, Coull *et al.* (1977) found an average of about 9 turbellarians in live-sorted sub-samples of deep sea meiofauna but only 1.5 in preserved material ($p<0.05$). Martens (1984) showed the marked effect of different extraction methods on determination of turbellarian abundance.

In work at Doctor's Bay core-size efficiency was checked since it is known that quadrat or core size affects numerical results (Elliot, 1971). The latter's methods for determining a) quadrat size imprecision and b) the number of samples necessary to obtain a reliable mean with a standard error of less than 20% were applied. For total turbellarians within specific micro-habitats (semi-drained-, pool-, *Arenicola* feeding funnel- and tail mound- and randomly collected-sediment) using a 2.9 cm internal diameter core the number of samples required varied from 4 to 19 (cf. the gnathostomulids 67–256!). For individual turbellarian species for particular sites and seasons it ranged from 6–166 samples.

It is therefore not surprising that quantitative work is somewhat problematical and that little is known from 'generalist papers' about turbellarian populations or their stability. Modern marine turbellarian ecology, such as evidenced in Dittmann & Reise (1985) and Reise (1987), requires a specialist approach so far rarely found. The latter paper, apparently the only one to have considered long term population stability, showed that species with larger population size and/or niche breadth were the most persistent. There is still little knowledge as to why some species appear rare and others common.

Communities

Turbellarian diversity within particular communities or habitats can be great, presumably reflecting environmental 'grain' (*i.e.* spatial and temporal heterogeneity and scale), and can match or exceed that of other meiofaunal taxa (Martens & Schockaert, 1986). Dittmann & Reise (1985) gave comparative figures of 83 and 49 species for Isle of Sylt sand and mud flats respectively; Maguire & Boaden (unpublished), working at a single station on a semi-exposed sand beach (Ballymaconell, N. Ireland – see Maguire, 1977) over a 15 month period, were able to distinguish 50 species of which 32 were Kalyptorhynchia. Turbellarians' perception of environmental gradients will relate to their body size and helps account for the difficulty of understanding details of their local distribution, for example congeneric occurrence of *Neoschizorhynchus* spp. within small cores (Boaden, 1977a). Apart from spatial gradients, diversity can also be influenced by temporal factors and by the presence of other organisms as detailed in several papers on sand flat ecology by Reise (1981, 1983, 1984, 1987).

Unfortunately it remains a rare occurrence to have platyhelminth assemblages investigated in detail over any length of time. Thus, little is known of their true diversity or species constancy within benthic communities although the latter appears to be recognised in classic works such as Remane (1933), Meixner (1938) and Reidl (1953, 1963). The species richness of the Turbellaria rivals that of the Nematoda in some areas (Martens & Schockaert, 1986) and diversity such as expressed by Shannon-Wiener's H' can be considerable (<3.35 Dittman & Reise, 1985; often >2 Reise, 1988).

As already stated, little is known about the role of free-living flatworms in helping to structure communities. The same may be said of their contribution to energetics. Their contribution to community biomass and production has been 'guesstimated' in several generalist papers (for example Rudnick *et al.*, 1985) but real dry weight, carbon content, growth and respiration values are largely lacking. Faubel (1982) measured dry weight values for various size classes and Widbom (1984) calculated a conversion factor of 78% from dry to ash-free weight. Boaden & El Hag (1984) determined respiration rates for *Monocelopsis otoplanoides* (mean dimensions were 0.88 mm long, 0.13 mm wide) ranging from about 50–300 nl O_2 individual h^{-1}; Beadle (1934) gives values for *Procerodes* at different salinities. We are a still a long way from being able to assess the turbellarians' real contribution to community metabolism.

Other issues

Begon *et al.* (1986) concluded their text with further discussion of topics, such as habitat disturbance and temporal heterogeneity, island biogeography and community stability, to which our present knowledge of marine free-living flatworm ecology can contribute little more than already mentioned.

Conclusion

It is no longer sufficient to proceed with turbellarian ecology on the Oscar Schmidt dictum 'Man braucht, wie es scheint, wo man will nur zuzugreifen und ist der Ausbeute sicher' as Gamble (1893) did 100 years ago although, with the probable exception of Sylt, it is still true of faunistics. As shown, marine turbellarians are an important component of the marine environment. Understanding and quantification of their ecology needs the application of proper theory, proper field and laboratory procedure and proper personell. If the impetus can be found for this, ecological knowledge of marine flatworms will be able to advance as rapidly as other aspects, such as systematics, have done recently. 'One needs, so it appears, only to sample where one will', and to apply some modern skill, 'and success is assured'.

Acknowledgments

I am indebted to a number of former students for inclusion of some previously unpublished information and particularly to Dr Nick Johnston for sampling statistics relating to Doctor's Bay.

References

Apelt, G., 1969. Fortpflanzungsbiologie, Entwicklungszyklen und vergleichende Frühentwicklung acoeler Turbellarien. Mar. Biol. 4: 267–325.

Armonies, W., 1990. Short-term changes of meiofaunal abundance in intertidal sediments. Helgol. Meeresunters. 44: 375–386.

Ax, P., 1951. Die Turbellarien des Eulitorals der Kieler Bucht. Zool. Jb. Syst. 80: 277–238.

Ax, P., 1959. Zur Systematik, Ökologie und Tiergeographie der Turbellarienfauna in den ponto-kapischen Brackwassermeeren. Zool. Jb. Syst. 89: 43–184.

Ax, P., 1966. Die Bedeutung der interstitiellen Sandfauna für allgemeine Probleme der Systematik, Ökologie und Biologie. Veröff. Inst. Meeresforsch. Bremerhaven 2: 15–66.

Ax, P., 1969. Populationsdynamik, Lebenszyklen und Fortpflanzungsbiologie der Mikrofauna des Meeressandes. Verh. deutsch. zool. Ges. Innsbruck 1968: 66–113.

Ax, P., 1970. Neue *Pogaina*-Arten (Turbellaria, Dalyelloida) mit Zooxanthellen aus dem Mesopsammal der Nordsee- und Mittelmeerküste. Mar. Biol. 5: 337–340.

Ax, P., 1977. Life cycles of interstitial Turbellaria from the eulittoral of the North Sea. Acta zool. fenn. 154: 11–20.

Ax, P. & W. Armonies, 1987. Amphiatlantic identities in the composition of the boreal brackish water community of Plathelminthes. A comparison between the Canadian and European Atlantic coast. Mikrofauna Marina 3: 7–80.

Ax, P. & R. Ax, 1970. Das Verteilungsprinzip des subterranen Psammon am Übergang Meer-Süsswasser. Mikrofauna Meeresboden 1: 1–51.

Ax, P. & P. Schmidt, 1973. Interstitielle Fauna von Galapagos I. Einführung. Mikrofauna Meeresboden 20: 1–38.

Ball, I. R. & T. Reynoldson, 1981. British Planarians. Synopses of the British fauna, 19. Cambridge University Press, Cambridge, 141 pp.

Beadle, L., 1934. Osmotic regulation in *Gunda ulvae*. J. exp. Biol. 11: 382–396.

Begon, M., J. L. Harper & C. R. Townsend, 1986. Ecology: individuals, populations and communities. Blackwell Scientific, Oxford, 876 pp.

Bilio, M., 1964. Die aquatische Bodenfauna von Salzwiesen der Nord- und Ostsee. I. Biotop und ökologische Faunenanalyse: Turbellaria. Int. Revue ges. Hydrobiol. 52: 487–533.

Bilio, M., 1967. Nahrungsbeziehungen der Turbellarien in Küstensalzwiesen. Helgoländer wiss. Meeresunter 15: 602–621.

Bleakley, R. J. & P. J. S. Boaden, 1974. Effects of an oil spill remover on beach meiofauna. Ann. Inst. oceanogr. 50: 51–58.

Boaden, P. S., 1962. Colonization of graded sand by an interstitial fauna. Cah. Biol. mar. 3: 245–248.

Boaden, P. J. S., 1968. Water movement – a dominant factor in interstitial ecology. Sarsia 34: 125–136.

Boaden, P. J. S., 1977a. Thiobiotic facts and fancies. (Aspects of the distribution and evolution of anaerobic meiofauna). Mikrofauna Meeresboden 61: 45–63.

Boaden, P. J. S., 1977b. The behaviour of *Monocelis lineata* (Müller) Turbellaria Proseriata) in a false otoplanid-zone. Acta zool. fenn. 154: 37–46.

Boaden, P. J. S., 1978. Biological studies of the north-east coast. In W. K. Downey & G. Ni Uid (eds), Coastal Pollution Assessment. Development of Estuaries, Coastal Regions and Environmental Quality. National Board for Science and Technology, Dublin: 93–99.

Boaden, P. J. S., 1981. Oxygen availability, redox and the distribution of some Turbellaria Schizorhynchidae and other forms. Hydrobiologia 84: 103–112.

Boaden, P. J. S., 1989. Meiofauna and the origins of the Metazoa. Zool. J. linn. Soc. 96: 217–227.

Boaden, P. J. S. & A.-G. El Hag, 1984. Meiobenthos and the oxygen budget of an intertidal sand beach. Hydrobiologia 118 (1) (Dev. Hydrobiol. 26): 39–47.

Boaden, P. J. S. & H. M. Platt, 1971. Daily migration patterns in an intertidal meiobenthic community. Thalassia jugoslav. 7: 1–12.

Bohn, G., 1903. Sur les mouvements oscillatoires des *Convoluta roscoffensis*. C. r. hebd. Séanc. Acad. Sci., Paris 137: 576–578.

Bush, L. F., 1966. Distribution of sand fauna in beaches at Miami, Florida. Bull. mar. Sci. 1: 58–75.

Crezée, M., 1975. Monograph of the Solenophilomorphidae (Turbellaria: Acoela). Int. Revue ges. Hydrobiol. 60: 769–845.

Coull, B. C., R. L. Ellison, J. W. Fleeger, R. P. Higgins, W. D. Hope, W. D. Hummon, R. M. Rieger, W. E. Sterrer, H. Thiel & J. H. Tietjen, 1977. Quantitative estimates of the meiofauna from the deep sea off North Carolina, USA Mar. Biol. 39: 233–240.

Curini-Galetti, M. C., 1993. *Pseudomonocelis ophiocephala* (Schmidt, 1861) (Platyhelminthes, Proseriata) is a complex of four sibling species. Abstract. 31, VIIth International Symposium on the Biology of the Turbellaria, Åbo/Turku, 17–22 June 1993.

Curini-Galetti, M. C. & P. M. Martens, 1990. Karyological and ecological evolution of the Monocelididae (Platyhelminthes, Proseriata). Mar. Ecol. (Pubbl. Stn. Zool. Napoli I) 11: 225–261.

Dittmann, S. & K. Reise, 1985. Assemblage of free-living Plathelminthes on an intertidal mud flat in the North Sea. Microfauna Marina 2: 95–115.

Dörjes, J., 1968. Zur Ökologie der Acoela Turbellaria) in der Deutschen Bucht. Helgoländer wiss. Meeresunters. 18: 78–115.

Dörjes, J., 1970. *Haplodiscus bocki* spec. nov., eine neue pelagische Turbellarie der Ordnung Acoela von der Molukken-See mit einer Diskussion der Gattung. Ark. Zool. 23: 255–266.

Edmondson, W. T., 1960. Reproductive rates of rotifers in natural conditions. Mem. Ist. ital. Idrobiol. 12: 21–77.

Ehlers, U., 1973. Zur Populationsstruktur interstitieller Typhloplanoida und Dalyelloida (Turbellaria, Neorhabdocoela). Mikrofauna Meeresboden 19: 1–105.

Elliot, J. M., 1971. Some methods for the statistical analyses of samples of benthic invertebrates. Freshwat. biol. Ass. Sci. Publ. 25, 144 pp.

Faubel, A., 1976. Populationsdynamik und Lebenszyklen interstitieller Acoela und Macrostomida (Turbellaria). Mikrofauna Meeresbodens 56: 1–107.

Faubel, A., 1977. The distribution of Acoela and Macrostomida (Turbellaria) in the littoral of the North Frisian Islands, Sylt, Romo, Jordsand, and Amrum (North Sea). Senkenbergiana maritima 9: 59–74.

Faubel, A., 1982. Determination of individual meiofauna dry weight values in relation to definite size classes. Cah. Biol. mar. 23: 339–345.

Feller, R. J., 1984. Serological tracers of meiofaunal food webs. Hydrobiologia 118 (1) (Dev. Hydrobiol. 26): 119–125.

Gamble, F. N., 1893. Contributions to a knowledge of british marine Turbellaria. Quart. J. microsc. Sci. 34: 433–528.

Gamble, F. W. & F. Keeble, 1904. The bionomics of *Convoluta roscoffensis*, with special reference to its green cells. Quart. J. microsc. Sci. 47: 363–427.

Gerlach, S. A., 1977. Means of meiofauna dispersal. Mikrofauna Meeresboden 61: 89–103.

Giesa, S., 1966. Die Embryonalentwicklung von *Monocelis fusca* Oersted (Turbellaria Proseriata). Z. Morph. Ökol. Tiere 57: 137–230.

Gray, J. S. & R. M. Rieger, 1971. A quantitative study of the meiofauna of an exposed sandy beach at Robin Hood's Bay, Yorkshire. J. mar. biol. Ass. U.K. 51: 1–19.

Hartog, C. den, 1984. Proseriate flatworms from the Deltaic area of the rivers Rhine, Meuse and Scheldt. I and II. Proc. Kon. Ned. Akad. Wetensch. C 67: 10–34.

Hartog, C. den, 1968. Marine triclads from the Plymouth area. J. mar. biol. Assoc. U.K. 45: 209–223.

Heip, C., 1975. On the significance of aggregation in some benthic marine invertebrates. In H. Barnes (ed.), Proceedings of the 9th European Marine Biology Symposium. Aberdeen University Press, Aberdeen: 527–538.

Holligan, P. M. & G. W. Gooday, 1975. Symbiosis in *Convoluta roscoffensis*. Symp. Soc. exp. Biol. 29: 205–227.

Hoxhold, S., 1974. Zur Populationsstruktur und Abundanzdynamik interstitieller Kalyptorhynchia (Turbellaria, Neorhabdocoela). Mikrofauna Meeresboden 41: 1–134.

Iwao, S. & E. Kuno, 1971. An approach to the analysis of aggregation pattern in biological populations. In G. P. Patil, E. C. Pielou & W. E. Waters (eds), Statistical ecology, I. Spatial patterns and statistical distributions. Pennsylvania State University Press, University Park & Lond: 461–504.

Iyer, K. P. V., 1949. The first and second moments of some probability distributions arising from points on a lattice and their application. Biometrika 36: 135–141.

Jansson, A.-M., 1966. Diatoms and microfauna – producers and consumers in the *Cladophara* belt. Veröff. Inst. Meeresforsch. Bremerhaven 2: 281–288.

Jansson, B.-O., 1967. The significance of grain size and pore water content for the interstitial fauna of sandy beaches. Oikos 18: 311–322.

Jansson, B.-O., 1968. Quantitative and experimental studies of the interstitial fauna in four Swedish sandy beaches. Ophelia 5: 1–71.

Jennings, J. B., 1957. Studies on feeding, digestion and food storage in free-living flatworms (Platyhelminthes: Turbellaria). Biol. Bull. 112: 63–80.

Karling, T. G., 1961. Zur Morpholoie, Entstehungsweise und Funktion des Spaltrüssels der Turbellaria Schizorhynchia. Ark. Zool. 13: 253–286.

Karling, T. G., 1974. Turbellarian fauna of the Baltic Proper. Identification, ecology and biogeography. Fauna fenn. 27: 1–101.

Maguire, C., 1977. Meiofaunal community structure and vertical distribution a comparison of some Co. Down beaches. In B. Keegan, P. O'Ceidigh & P. J. Boaden (eds), Biology of Benthic Organisms. Pergammon Press, Oxford: 425–431.

Martens, P. M., 1984. Comparison of three different extraction methods for Turbellaria. Mar. Ecol. Prog. Ser. 14: 229–234.

Martens, P. M. & E. R. Schockaert, 1986. The importance of turbellarians in the marine meiobenthos: a review. Hydrobiologia 132 (Dev. Hydrobiol. 32): 295–303.

McLachlan, A., 1978. A quantitative analysis of the meiofauna and chemistry of the redox potential discontinuity zone in a sheltered sandy beach. Estuar. coast. mar. Sci. 7: 275–290.

Meixner, J., 1938. Turbellaria (Strudelwürmer). I. Tierwelt N.-u. Ostsee 4b: 1–146.

Meyer-Reil, L.-A. & A. Faubel, 1980. Uptake of organic matter by meiofauna organisms and interrelationships with bacteria. Mar. Ecol. Prog. Ser. 3: 251–256.

Meyers, M., H. Fossing & E. N. Powell, 1987. Microdistribution of interstitial meiofauna, oxygen and sulfide gradients, and the tubes of macrofauna. Mar. Ecol. Prog. Ser. 35: 223–241.

Moore, C. G.. & B. J. Bett, 1989. The use of meiofauna in marine pollution impact assessment. Zool. J. linn. Soc. 96: 263–280.

Murina, G. V., 1981. Notes on the biology of some psammophile Turbellaria of the Black Sea. Hydrobiologia 84: 129–130.

Ott, J., G. Rieger, R. Rieger & F. Enderes, 1982. New mouthless interstitial worms from the sulfide system: symbiosis with prokaryotes. Proc. Stat. Zool. Nap. Ital.: Mar. Ecol. 3: 313–333.

Pantin, C. F. A., 1931a. The adaptability of *Gunda ulvae* to salinity. I. The environment. J. exp. Biol. 8: 63–72.

Pantin, C. F. A., 1931b. The adaptability of *Gunda ulvae* to salinity III The electrolyte exchange J. exp. Biol. 8: 82–94.

Pawlak, R., 1969. Zur Systematik und Ökologie (Lebenszyklen, Populationsdynamik) der Turbellarien-Gattung *Paromalostomum*. Helgoländer wiss. Meeresunters. 19: 417–453.

Powell, E. N., M. A. Crenshaw & R. M. Rieger, 1980. Adaptations to sulfide in sulfide system meiofauna. End products of sulfide detoxification in three turbellarians and a gastrotrich. Mar. Ecol. Prog. Ser 2: 169–177.

Reise, K., 1981. High abundance of small zoobenthos around biogenic structures in tidal sediments of the Wadden Sea. Helgoländer Meeresunters 34: 413–425.

Reise, K., 1983. Sewage, green algal mats anchored by lugworms, and the effects on Turbellaria and small Polychaeta. Helgoländer Meeresunters 36: 151–162.

Reise, K., 1984. Free-living Plathelminthes (Turbellaria) of a marine sand flat: an ecological study. Microfauna Marina 1: 1–62.

Reise, K., 1987. Spatial niches and long-term performance in meiobenthic Plathelminthes of an intertidal lugworm flat. Mar. Ecol. Prog. Ser. 38: 1–11.

Reise, K., 1988. Plathelminth diversity in littoral sediments around the island of Sylt in the North Sea. In P. Ax, U. Ehlers & B. Sopott-Ehlers (eds), Free-living and symbiotic Plathelminthes. Fortschr. Zool. Prog. Zool. 36: Gustav Fischer, Stuttgart/New York: 469–480.

Remane, A., 1933. Verteilung und Organisation der benthonischen Mikrofauna der Kieler Bucht. Wiss. Meeresunters., Abt. Kiel 21: 161–221.

Reuter, M., 1961. Untersuchungen über Rassenbildung bei *Gyratrix hermaphroditus* (Turbellaria Neorhabdocoela). Acta zool. fenn. 100: 1–32.

Riedl, R., 1953. Quantitativ ökologische Methoden mariner Turbellarienforschung. Öst. zool. Z. 4: 108–145.

Riedl, R., 1963. Probleme und Methoden der Erforschung des litoralen Benthos. Zool. Anz. Suppl. 26: 505–567.

Rieger, R. M., 1977. The relationship of character variability and morphological complexity in copulatory structures of Turbellaria-Macrostomida and -Haplopharyngida. Mikrofauna Meeresboden 61: 197–216.

Rieger, F. & J. Ott, 1971. Gezeitenbedingte Wanderungen von Turbellarien und Nematoden eines nordadriatischen Sandstrandes. In Troisième Symposium Européen de Biologie Marine, Vie Milieu, suppl. 22: 425–457.

Rieger, R. M. & W. Sterrer, 1975. New spicular skeletons in Turbellaria, and the occurrence of spicules in marine meiofauna. Z. zool. Syst. Evol.- forsch. 13: 207–248.

Rudnick, D. T., R. Elmgren, J. B. Frithsen, 1985. Meiofaunal prominence and benthic seasonality in a coastal marine ecosystem. Oecologia 67: 157–168.

Schmidt, P., 169. Die quantitative Verteilung und Populationsdynamik des Mesopsammons am Gezeitensandstrand der Nordseeinsel Sylt. II. Quantitative Verteilung und Populationsdynamik einzelner Arten. Int. Rev. ges. Hydrobiol. 54: 95–174.

Sopott, B., 1973. Jahreszeitliche Verteilung und Lebenszyklen des Proseriata (Turbellaria) eines Sandstrand der Nordseeinsel Sylt. Mikrofauna Meeresboden 15: 1–106.

Sterrer, W. & R. Rieger, 1974. Retronectidae – a new cosmopolitan marine family of Catenulida (Turbellaria). In N. W. Riser & M. P. Morse (eds), Biology of the Turbellaria. McGraw-Hill, New York, 63–92.

Tempel, D. & W. Westheide, 1980. Uptake and incorporation of dissolved amino acids by interstitial Turbellaria and Polychaeta and their dependence on temperature and salinity. Mar. Ecol. Prog. Ser. 41–50.

Warwick, R. M., 1989. The role of meiofauna in the marine ecosystem: evolutionary considerations. Zool. J. linn. Soc. 96: 229–241.

Watzin, M. C., 1986. Larval settlement into marine soft-sediment systems: interactions with the meiofauna. J. exp. mar. Biol. Ecol. 98: 65–113.

Weill, E. & C. F. A. Pantin, 1931. The adaptability of *Gunda ulvae* to salinity. II. The water exchange. J. exp. Biol. 8 73–81.

Westphalen, D., 1993. Stomatolitoid microbial nodules from Bermuda – a special microhabitat for meiofauna. Mar. Biol. 127: 145–157.

Widbom, B., 1984. Determination of average individual dry weights and ash-free dry weights in different sieve fractions of marine meiofauna. Mar. Biol. 84: 101–108.

Vitiello, P., 1968. Variations de la densite du microbenthos sur une aire restrainte. Recueil Trav. Stat. mar. Endoume 43: 261–270.

Hydrobiologia **305**: 101–104, 1995.
L.R.G. Cannon (ed.), Biology of Turbellaria and some Related Flatworms.

Stylochus tauricus, a predator of the barnacle *Balanus improvisus* in the Black Sea

G.-V. Murina, V. Grintsov & A. Solonchenko
Institute of Biology of the Southern Seas, Academy of Sciences of Ukraine, 2 Nakhimov Str., 335011, Sevastopol, Crimea, Ukraine

Key words: Stylochus, predator, *Balanus*, larvae

Abstract

The flatworm *Stylochus tauricus* Jacubova has been found associated with the barnacle *Balanus improvisus* Darwin, on which it feeds. The predation rate (the number of barnacles eaten by one polyclad in a month) ranges between 5–10. Inside the empty shells of *B. improvisus* some egg-plates of *S. tauricus* were observed. Pelagic Götte's larvae aged 2–3 days possess 4 lobes while those aged 7–8 days have 5 lobes. Flatworms can prey on the young of another species *Balanus eburneus* Gould, whereas predation on the mussels *Mytilus galloprovincialis* Lam. is rare. There is a direct correlation between predator abundance and prey ingested.

Introduction

Out of 100 species of *Stylochus*, predation is only known for 14. They are known to feed on bivalve molluscs and barnacles (Galleni *et al.*, 1980). Four species are monophagous, preying on barnacles alone: *S. alexandrinus* Steinbock, *S. neapolitanus* (Delle Chiaje), *S. tripartitus* Hyman, *S. zanzibaricus* Laidlaw (Galleni *et al.*, 1980).

Rzepishevskji (1979) provided a detailed description of how *Balanus improvisus* Darwin is ingested by a polyclad, and he identified this species as *Stylochus pilidium* Lang (Sevastopol Bay, Black Sea). However, size, abundance and position of the cerebral eyes, as well as other morphological characteristics suggest that this is another species. According to Tokynova (University of Kazan, Tatarstan, pers. ob.), the Black Sea *Stylochus* which predates on barnacles, is *Stylochus tauricus* Jacubova (Jacubova, 1909).

Materials and methods

S. tauricus was collected in two regions of the Black Sea: Laspy Bay, the South Crimea and Inkerman, Sevastopol Bay. To study the ecology and the abundance of polyclads from April 1991 to November 1992, we exposed flat round plastic collectors (75^x 35^x 9 mm) at different depths. Submergence depth ranged from 0.2–12 m from the water surface. The exposure continued for 1 to 8 months. To study biological aspects of these stylochids, flatworms were maintained in glass containers with sea water at 17–18°C from November 1992 to April 1993.

Results and discussion

Morphological aspects of *S. tauricus* were described in detail by Jacubova (1909). This species has an elongated-oval shape when moving and is almost round when at rest. Length varies from 2.5 to 16 mm, width from 1.0 to 6 mm (when moving). The dorsal surface coloration varies from light yellow to grey, while the ventral surface is lighter in color. Two long transparent nuchal tentacles with about 30 black eyes inside each are present (Fig. 1). Cerebral eyes (1 to 4 pairs) are located in a form of a rectangle. Marginal eyes occur along the entire margin, being more numerous in its anterior third. On the dorsal surface of larger individuals, white oval sacs containing spermatozoa have occasionally been observed.

After ingestion of a soft barnacle body, stylochids generally deposit an egg-plate inside the empty shell,

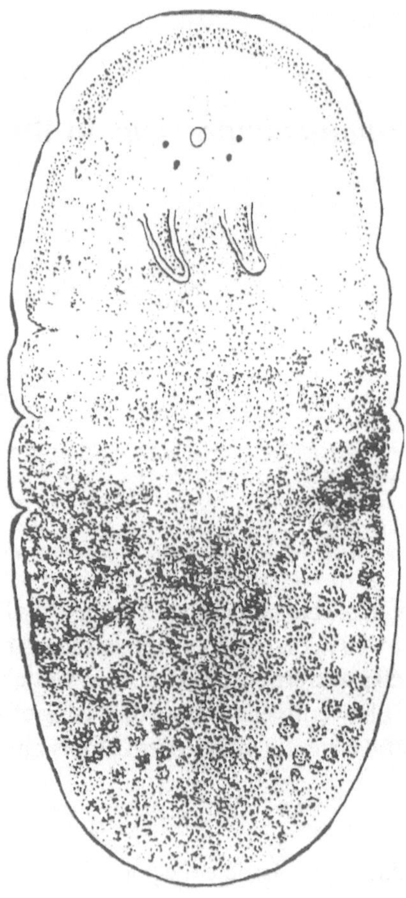

Fig. 1. Dorsal view of adult *Stylochus tauricus*.

in some cases on the wall of a glass container. The smallest mature individual was about 6.5 mm in length. The egg-plates are usually oval-shaped, the number of eggs about 3000 and these are white, translucent, round and 100–110 μm in diameter.

In our February experiments incubation continued for about 7–9 days with hatching of pelagic Götte's larvae at 17– 18°C. Larvae aged 2–3 days possess 4 lobes and while developing a 5-th lobe appeared (Fig. 2). When alive, larvae are about 90–100μm in length. During our study they were fed a mixture of algae (*Phaeodactylum tricornutum* Bohlin and *Monochrysis* sp.). In a tank larvae swim for about a month. After settling on a glass container they died within a week, although the shells of *Mytilus galloprovincialis* Lam. and those of both live and dead barnacles were provided as substratum.

Adult *S. tauricus* acclimatised in aquaria over 8 months and this provided the possibility to study the characteristics of their feeding and behaviour in detail.

The mode of penetration of the predator into the prey shell was described in detail by Hurley (1976). Attacking *Stylochus tripartitus* crawl into live *Balanus pacificus* Pilsbury, insert their pharynx between the opercular plates of the barnacle, digest the latter's muscles, and then penetrate the shell.

Stylochus ingests the fleshy parts of the prey, leaving the cirri, then the worm ejects cirri outside. According to Hurley (1976) and Rzhepishevskji (1979), the time needed to consume one barnacle is from 1 to 3 hours. Dead *Balanus* eaten by stylochids are recognized by the presence of open unmoving opercular plates.

To determine the maximum amount of food ingested by 10 *S. tauricus* (8–12 mm in length), each was placed in a separate glass container and fed once every two days for a month. The highest predation rate was 10 barnacles per month by a stylochid 12 mm in length. The average amount of food ingested by one flatworm (10–12 mm in length), feeding on barnacles 10–12 mm in diameter, was 0.42 mg of dry weight per day at 17–18°C.

As well, experiments on nutritional preference were carried out. *S. tauricus* was offered larger-size barnacles of another species *B. eburneus* Gould, as well as small, young mussels, *M. galloprovicialis*, 1–4 mm in length. Stylochids preferred small, young individuals of *B. eburneus*. Out of 40 specimens of *Mytilus*, only one individual (4 mm in length) was ingested.

As a rule, each dead barnacle housed only one invader. But when the number of barnacles provided was less than the number of polyclads, about 2–3 stylochids could be observed inside one empty shell. Rzhepishevskji (1979) also found about 30 flatworms of different size in one large *Balanus*.

Barnacles are not only the sole source of prey for *S. tauricus*, but they are also a home and a cradle for their young. Care for the offspring is shown by the constant water aeration within the empty shell created by the movement of cilia covering the body surface of the flatworm. *S. tauricus* does not abandon the egg-plates deposited on the bottom of a container for the week prior to hatching even though live barnacles have been provided as food. The egg-plate appears to be protected by the adult worm, otherwise it would be rapidly exposed to attack by infusoria, *Diophrys scutum* Dujardin. Due to their own apparent toxicity the adult stylochids do not have predators and are not eaten by other organisms.

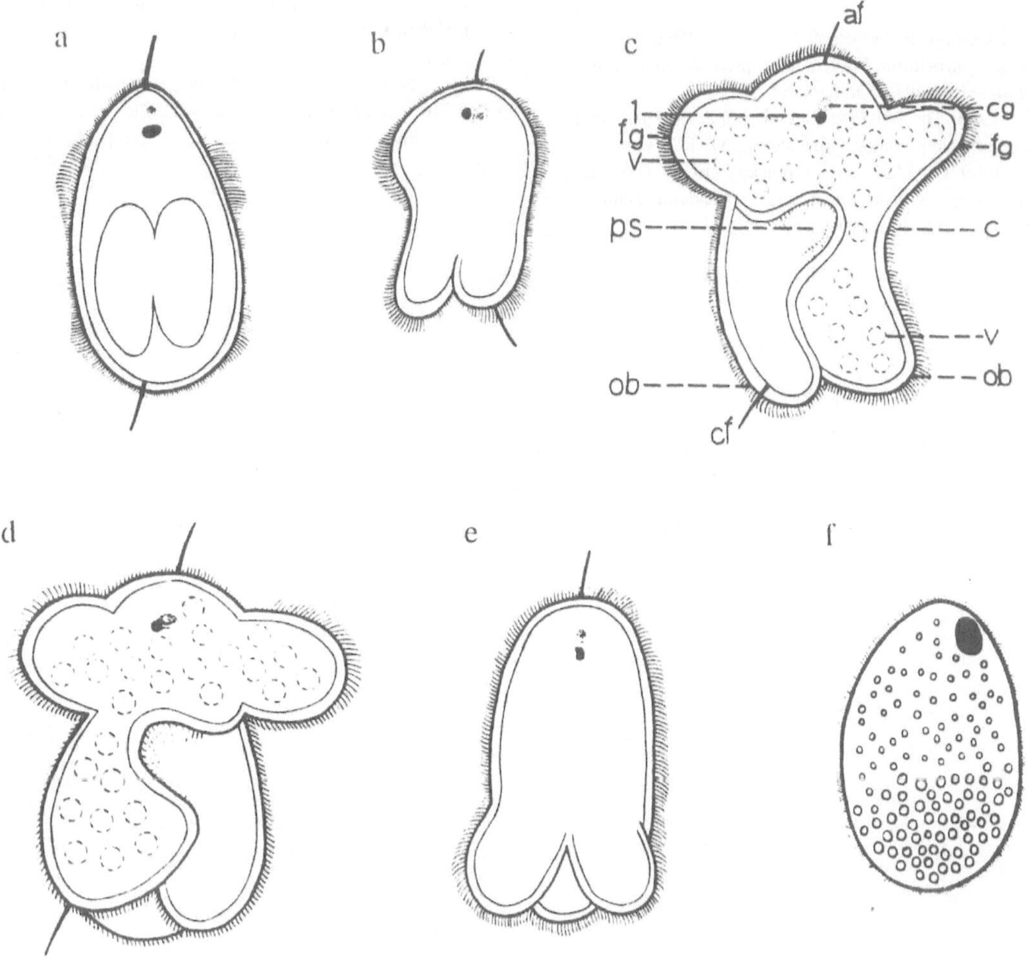

Fig. 2. Larvae of *S. tauricus* in different stages of development: a. Larva newly hatched from eggs: b. Bi-lobed larva several hours old; c. Four-lobed Götte's larva 2–3 days old; d. Five-lobed larva 7–8 days old; e. Larva prior to the settling on the bottom; f. Young flatworm aged over one month and crawling on the bottom; *af* - apical tuft; *cg* - cerebral ganglion; *fg* - frontal lobes, *v* - yolk cells; *ob*- oral lobes; *cf* - caudal tuft; *ps* - primary stomach; *1* - eyes-spot; *cf* - cilia.

Comparison of *S. tauricus* density on depth of collector has been made.For the autumn months, a maximum of 0.26 individuals cm^{-2} were found at a depth of 12 m. The proportion of dead barnacles constituted 64%. In the upper water column (0–2 m), stylochids were rare or non-existent. Direct correlation between the number of flatworms and the submergence depth of collectors is explained by the increase of prey barnacles with depth. The coefficient of correlation between the exposure depth and *B. improvisus* density was 0.88 (α=0.05).

Balanus improvisus is a common, widespread species inhabiting the shelf of the Black Sea. According to Tarasov & Zevina (1954), it is characterised by the annual settling of larvae, rapid growth, early onset of reproduction, euryhalinity, eurythermity and tolerance to pollution. The density of these barnacles reached 11 000 ind. m^{-2}. In our experiments, maximum density of barnacles on collectors exposed from August 20 to November 15 was 141 000 ind. m^{-2}. The longest exposure (8 months) in nature showed that the predator never consumed all the prey. Despite heavy losses experienced by the barnacles, there was a constant balance between the number of predators and prey since barnacle beds were continuously replenished from larval settlement and new growth.

References

Galleni, L., P. Tongiorgi, E. Ferrero & U. Salgetti, 1980. *Stylochus mediterraneus* (Turbellaria: Polycladida), predator on the mussel *Mytilus galloprovincialsi*. Mar. Biol. 55: 317–458.

Hurley, A. C., 1976. The polyclad flatworm *Stylochus tripartitus* Hyman as a barnacle predator. Crustaceana, 31: 110–111.

Jacubova, L. I, 1909. Polyclada Sevastopolskoy bukhty. (Polyclada of Sevastopol Bay). Sapiski Imperatorskoi Academii Nauk, St. -Petersburg, 24: 1–32. (in Russian)

Rzhepishevskji, I. K., 1979. Vyedanie balanusov turbellariey Stylochus pelidium. (Barnacles ingestion by turbellaria *Stylochus pillidium*). Biol. mory, Naukov Kumka, Kiev 48: 23–27. (in Russian).

Tarasov, N. I. & G. B. Zevina, 1957. Cirripedia Thoracacia morey SSSR. (Cirripedia Thoracacia of USSR Seas). Izd. Akad. Nauk SSR. Fauna SSSR. Crustacea. 6: 267 pp. [In Russian]

Hydrobiologia **305**: 105–111, 1995.
L.R.G. Cannon (ed.), Biology of Turbellaria and some Related Flatworms.

Predation behaviour of land planarians

Robert E. Ogren
Wilkes University, Wilkes-Barre, PA 18704, USA

Key words: land planarians, predation, capture methods, adhesive mucus, population control and depletion

Abstract

Predatory behaviour of land planarians is seldom observed or reported. Aspects reported are (1) finding prey; (2) attack behaviour; (3) capture using adhesive mucus, pharyngeal action, poisonous secretions, physical embrace; (4) feeding by extension of pharynx, releasing copious digestive fluid. The species *Bipalium kewense, B. adventitium* and *B. pennsylvanicum* attack earthworms, immobilizing them by physical holding, digesting by pharyngeal secretions and then ingesting the treated tissue. Group attacks on giant African land snails involving chemotactic tracking, occur in *Platydemus manokwari* and *Endeavouria septemlineata.* Specialized capture methods are used by some species; *Rhynchodemus sylvaticus* uses an expanded cephalic hood to capture small insects and in Africa, termites are captured by the elongated anterior of *Microplana termitophaga* as planarians wait within the colony air shaft openings to ensnare the workers in sticky mucus. The result of extensive predation by land planarians may seriously reduce the prey, e.g., providing effective population control of giant land snails by introduced *Platydemus manokwari*, or causing serious depletion of desirable earthworm populations by the exotic *Artioposthia triangulata* in North Ireland.

Introduction

Relatively few published observations exist describing the predatory behaviour of land planarians. One reason is that this behaviour is seldom observed, collectors being more interested in planarians as preserved taxonomic specimens than as subjects for behavioral studies. Moreover, land planarians perish so easily that preservation becomes a first priority. The earliest account of land planarian predatory behaviour is found in ancient Oriental literature written by Chinese and/or Japanese. These accounts from the pre-scientific age in China and Japan (*ca* 860), found in various manuscripts, woodcut prints and so-called 'Materia Medica', are reviewed by Kawakatsu (1969), Kawakatsu & Lue (1984) and Lue & Kawakatsu (1986). An early summary of habitats and feeding was provided by Moseley (1875), then by von Graff ((1899) and under the heading of ecology and nutrition (von Graff, 1917: 3347). A summary for Australian land planarians is given by Steel (1901). A brief summary by Barker (1989) reviews selected literature of land planarian predation and describes attack on terrestrial slugs in New Zealand. Predatory behaviour includes: A. accidental finding of prey, B. oriented predatory activities with use of tactile and chemical signals, C. capture specializations, e.g., anterior adhesive structures, toxic secretions, physical embrace, D. pharyngeal activities. The present review brings published accounts together. For a full account of the taxonomy (and synonymies) of species reported here see Ogren *et al.*, (1992) and references therein.

Accidental finding of prey

Planarians moving across surfaces with side to side swinging of the anterior end may stop when the food source is accidentally encountered. A classic example of predator behaviour shown by *Bipalium adventitium* Hyman is described by Dindal (1970), and for *Bipalium pennsylvanicum* Ogren by Ogren & Sheldon (1991). Once encountered accidentally the prey are captured and fed upon. Lehnert (1891) briefly described how *Bipalium kewense* Moseley encountered and fed on living earthworms. The Chinese observer 'Yn-Yang Tsa-Tsu' about 860 PE writing of the presumed bipaliid land planarian (Doko), describes how it chases and captures earthworms (Kawakatsu & Lue, 1984; Lue & Kawakatsu, 1986). In the case of *Microplana terrestris* (Müller), prey are found by chance contact through random movements, followed by quick capture (Jennings, 1959). *Rhynchodemus sylvaticus* employs random discovery for capture of small springtails (Wallner, 1937) and small flies placed nearby in its environment (Arndt, 1938; Schremmer, 1955). In nature *B. adventitium, B. kewense, B. pennsylvanicum* individuals are found on the soil beneath stones or logs, presumably quiescent between meals with live earthworms nearby (Chandler, 1976; Neck, 1987; Ogren, 1986; Ogren & Sheldon, 1991). If contact with food is made, and the planarian ready to feed, arousal and predatory behaviour will occur (Neck, 1987). Pfitzner (1958) using *Dolichoplana feildeni* von Graff demonstrated, and illustrated with photographs, the capture of earthworms (*Lumbricus sp.*). Planarians placed in a petri dish with one earthworm, made physical contact, moving over and around the earthworm. After 25 min several, but not all the planarians were coiled around the prey, pressed against its body with extended pharynges. In 45 min the prey was motionless and the planarians were busy feeding. E.M. Froehlich (in Litt., letter of 4 July, 1987) mentions that Rodriques (1972) observed *Geoplana burmeisteri* Schulz & Müller under laboratory conditions capture and feed on garden snails.

An especially interesting and successful opportunistic predator on earthworms is the geoplanid *Artioposthia triangulata* (Dendy). When ready to feed it captures, by encirclement, an earthworm encountered in its environment, extends the pharynx releasing digestive fluids and feeds, often consuming the entire worm (Willis & Edwards, 1977; Blackshaw, 1990, 1991, 1992; Blackshaw & Stewart, 1992: 212). There is no evidence that *A. triangulata* aggressively tracks the earthworm prey (Blackshaw & Stewart, 1992: 212).

Oriented predatory activities

Tactile and chemical sensations in planarians are important in orienting toward prey, recognizing food and capturing prey (Hyman, 1951: 200). The anterior end is important for this behaviour. Ogren (1957) showed that *Rhynchodemus sylvaticus* (Leidy) was attracted to biological fluids such as beef liver juice. The worms moving randomly in a dish, stopped and began feeding on the first liver juice available rather than searching for a larger sample, which was near by. Prey may be tracked by chemical signals, e.g., from mucus trails of snails, or tissue exudates from injured prey as examplified by the planarians *Platydemus manokwari* de Beauchamp and *Endeavouria septemlineata* (Hyman). Once the food signal is sensed, the planarian moves toward the source. A similar feeding orientation, possibly mediated by surface diffusion, was observed for *Bipalium pennsylvanicum* which stopped to feed on mashed and torn slugs (Ogren & Sheldon, 1991). Sensitivity of various regions of the body to nutritive material was tested by Ogren (1957) for *R. sylvaticus* and revealed that all parts of body (pre-eye, post-eye, pre-pharynx, post- pharynx) stimulated the planarian to move its body near the stimulus and extend the pharynx.

Certain Geoplanidae show very aggressive oriented behaviour. *E. septemlineata* follow and capture the giant African land snail. Mead (1963: 306) described how the anterior is moved in a searching manner apparently sensing the snail's slime trail. One or several planarians overtake the snail, crawl upon the body and attack it. It is known that the rhynchodemid *P. manokwari* also effectively preys on the large African land snail and is used to control its populations (Muniappan, 1987, 1990; Kawakatsu *et al.*, 1992; 1993). It is reported that *P. manokwari* uses chemical and tactile senses to locate prey (Kaneda, *et al.*, 1990). Other such effective predations on snails and earthworms are given for geoplanid species by Froehlich (1955b).

Prey preference was demonstrated by Sheldon in Ogren & Sheldon (1991) who showed *B. pennsylvanicum* preferred live earthworms over slugs. In Africa the land planarian, *Microplana termitophage* Jones *et al.*, 1990, uses worker termites as prey but avoids soldier termites. Prey preferences are suggested by the following: *R. sylvaticus* feeding on earthworms (Lehnert, 1891), feeding on small snails, slugs, arthropods and beef liver (Ogren, 1955), *Microplana scharffi* (von Graff) feeding on earthworms (Garnett, 1928), and *Caenoplana coerulea vaga* (Hyman) feeding on isopods (Olewine, 1972). With *Artioposthia triangulata* earthworms of several species are accepted as food. There is no evidence the planarians prefer one earthworm species (Blackshaw & Stewart, 1992: 213), but availability of an earthworm species is important. Isopods are common prey for *Parakontikia ventrolineata* (Dendy) (olim *Geoplana*) according to Barker (1989: 76). However, individual *P. ventrolineata* also attack the slug *Deroceras panormitanum* (Lessona & Pollonera). Feeding follows massive attack by as many as 7 planarians.

Coexistance and orientation of predator and prey in the same habitat is important to capture, making more certain the discovery of preferred food. Land planarians are part of the soil ecosystem (Dendy, 1890b; Ogren, 1955; Ogren & Sheldon, 1991; Ball & Sluys, 1990), belonging to the soil 'heterotrophic microcommunity' of Dindal (1990: 8). Land planarians also have specialized physiological adaptations that assure their success in precarious terrestrial environments (Little, 1983).

Habitat relationships between predator and prey are described by Sheppe (1970: 214; Jones *et al.*; Darlington & Newson, 1990; Jones, 1993) for *M. termitophaga* which places its body purposively within earth openings to capture worker termites. The red geoplanids *Geobia subterranea* (Schultz & Müller) live within numerous burrow passages in the soil habitat of the earthworm *Lumbricus corethrurus* which are attacked. While the planarian body restrains the prey, its extruded pharynx sucks earthworm blood for food (Shultz & Müller, 1857; Schultz, 1857). An interesting ecological study of vertical distribution of 5 species of earthworm prey available to *Artioposthia triangulata* has revealed that surface active species were more frequently used as food (Blackshaw & Stewart, 1992: 214).

Capture techniques

Techniques may include anterior specialization, entrapment in sticky mucus, activity of the pharynx and physical immobilization by holding, finally subduing the live prey. Anterior features such as head musculature, mucoid glands, chemical and tactile receptors are of fundamental importance for capturing prey. The capture of snails by geoplanid species is described by Froehlich (1955b; 267). The planarian adhers to the snail's shell by its anterior end, eventually surrounding the prey, then extending the pharynx beneath the shell to begin feeding. If the head of *Bipalium kewense, B. adventitium* or *B. pennsylvanicum*, successfully contacts the earthworm prey, the planarian usually adhers to the epithelium, then moves onto the earthworm and glides its head along the body of its prey (Barnwell *et al.*, 1965; Dindal, 1970; Neck, 1987; Ogren & Sheldon, 1991). In examining the role of the head in behaviour of *B. kewense*, Barnwell (1966) revealed that the head is important for detection of prey, but not for capture and ingestion. Head ablation experiments were performed followed by regeneration and feeding experiments. The planarians did not attack when the head region was incompletely regenerated. Capture and ingestion of prey, however, did not require an intact head but was possible by action of the intact pharynx.

In his biological studies of land planarians, Froehlich (1955b) observed *Rhynchodemus* sp. using its expansive cephalic hood to capture prey, e.g., small springtails (Collembola). I can confirm capture of Collembola with the cephalic hood by *R. sylvaticus* from my unpublished field notes (1956) as also observed by Wallner (1937). E.M. Froehlich (in litt.) mentions that the geoplanid *Pasipha pasipha* (Marcus) captures isopods with its cephalic musculo-glandular organ. In a taxonomic paper, Froehlich (1955a) stated that the capture of prey may be the function for the anterior glandulo-muscular organs found in *Choeradoplana* von Graff (e.g., *C. marthae* C.G. Froehlich) and *Issoca* Froehlich (e.g., *I. piranga* C.G. Froehlich). Winsor (1991) described *Pimea monticola* Winsor from New Caldonia which possesses an anterior ventral pad of glandular-muscular organs with possible adhesive functions involved in capture of prey and also in holding to a substrate.

The head is frequently used as grasping organ whose adhesive ventral surface can strongly attach to the prey or substrate or both, and in *Bipalium* species the broad semilunar head encircled by the sensory tract serves as a device for detection and adhesive attachment. In the case of *B. pennsylvanicum* the planarian uses this organ to capture earthworm prey (Ogren & Sheldon, 1991) by first gliding over, then firmly attaching to the body surface with the underside of its head. This was also observed by Neck (1987) for *B. kewense*. According to Dindal (1970), once head contact is made by *B. adventitium* on the earthworm, the planarian crawls onto the prey and extends its pharynx while the body encloses the prey. This planarian behaviour is described in a 9th century Chinese manuscript translated by Lue & Kawakatsu (1986: 318). Sheppe (1970: 214) and Jones, *et al.* (1990) described how in Africa the land planarian *M. termitophaga*, situated at the rim of colony air shafts, actively captures worker termites by extending its head, touching the prey, quickly withdrawing the extended anterior, then crawling over the termite body which becomes entrapped in slime. This was graphically illustrated by Jones (1993). Pfitzner (1958) observed how *Dolichoplana feildeni* captures and feeds on earthworms (*Lumbricus* sp.). Planarians made physical contact by their anterior then moved over and around the earthworm for capture. The head of *Artioposthia triangulata* is the first region to contact and adhere to earthworm prey and is followed by contact by the mouth and coiling of the planarian around the earthworm (Blackshaw & Stewart, 1992: 212).

Entrapment in sticky mucus is important. Observations by Steel (1901) for Australian land planarians indicate they are effective carnivores, capturing annelids, molluscs and arthropods by crawling over the body which then becomes entangled in mucus. Capture of beetles and crustaceans using mucus entrapment and physical force by *Caenoplana spenceri* (Dendy) is related briefly by Spencer (1891). Feeding began when the pharynx penetrated the arthropod's joints (Spencer, 1891: 86). Mucus entrapment has been observed for other geoplanids (Brittlebank, 1888; Dendy, 1890a, b). In Rhynchodemidae, *R. sylvaticus* entraps fruit flies in adhesive mucus (Arndt, 1938) and similarly with other insect prey (Schremmer, 1955). According to Jennings (1959) mucus entrapment does not aid initial prey capture by *M. terrestris* because mucus dries quickly in the habitat. Nevertheless, mucus provides a role later in preventing escape.

Planarian pharyngeal activities and released digestive fluids are of fundamental importance to successful capture and immobilization of prey. For *R. sylvaticus* the importance of the pharynx to predation and feeding is confirmed by Schremmer (1955) and Arndt (1938). Descriptions by Jennings (1959) of predation by *M. terrestris* demonstrate the importance of the cylindrical pharynx being thrust through the body wall quickly without apparent aid of solvent juices. Ogren (1957), for *R. sylvaticus*, revealed that an extension response of the pharynx occurs to nutrients found by the planarian.

The experiments of Barnwell (1966) using *B. kewense* demonstrated that capture and ingestion required the intact pharynx rather than the head. Dindal (1970) reported it was the extended and secreting pharynx of *B. adventitium* contacting the earthworm that provoked violent prey reaction, rather than initial contact by the planarian. Moreover, the released digestive secretions quickly liquefied portions of the earthworm surface. Therefore, the pharynx was of primary importance to successful capture. In *B. adventitium, B. kewense*, and *B. pennsylvanicum* the pharynx is of the collared, or plicate type (Dindal, 1970) which spreads out over the surface forming a strong attachment. The earthworm prey reacts very quickly to escape the contact of the pharynx (Dindal, 1970; Ogren & Sheldon, 1991) but once strong contact is made the planarian's hold is seldom broken.

Physical holding is common, e.g., the capture of snails and slugs by geoplanid species. Froehlich (1955b: 267) described how planaria hold and surround prey and *Bipalium* sp. is known to enclose snail prey for feeding (Miyoshi, 1955). Jennings (1959) described how the prey is held down, by the arched body of *M. terrestris* which attaches to the substrate and by mucus secretion. Physical holding of small prey by *R. sylvaticus* also was shown by Schremmer (1955). Capture of lumbricid earthworms by *B. adventitium, B. Kewense* and *B. pennsylvanicum*, as recorded by Dindal (1970), Neck (1987) and Ogren & Sheldon (1991), and by *Artioposthia triangulata*, as reported by Blackshaw & Stewart (1992: 213), results in the planarian effectively attaching, holding and enclosing the prey.

Immobilization may involve paralysis as well as physical holding for Johri (1952), studying *B. kewense*, observed that following physical contact, the earthworm prey reacted violently to the planarian but was soon paralyzed. The abundant sticky mucus is presumably an important factor in immobilization. In other cases release of digestive fluids, which contain proteolytic enzymes (Jennings, 1959, 1962) and other biochemicals possibly toxic to the prey, is important to immobilization and successful capture. This has been observed in examples above and studied in depth for *M. terrestris* by Jennings (1959, 1962). According to Dindal (1970) *B. adventitium* feeding on earthworms, immobilizes prey while releasing copious digestive fluids. However, immobilization may not always occur. Restudy of predation by *B. pennsylvanicum*, prepared on video tape by Dr Joe Sheldon, revealed that, while physical holding is important to successful capture and restraint, the earthworm may continue to struggle and move while the pharynx is applied and feeding takes place.

Discussion

The above citations provide good evidence that land planarians attack and capture prey much larger than themselves employing different techniques such as physical force, adhesive mucus, pharyngeal action, and very effective digestive secretions poured over the surface of the live prey by the protrusile pharynx. This is followed by active feeding. Land planarians can apparently find food by waiting in ambush for prey to make physical contact, by chance encounter during random locomotion, or by oriented behaviour tracking the prey be sensory cues. Perhaps they can use each of these strategies at various times to acquire food, however, orientated searching may be a more restricted behaviour.

A question related to food preference still not completely answered is: do land planarians prey on each other showing cannibalistic behaviour? A brief summary on cannibalism is provided by von Graff (1917: 3348). Cannibalism in fresh water planarians was observed by Hull (1947) who believed crowding was partly responsible. Although it has not been commonly observed, land planarians can be cannibalistic (Froehlich, 1955b: 268) which may occur as a result of limited food supply.

Other topics are described elsewhere such as feeding behaviour of other Turbellaria (Hyman, 1951: 200–202; Jennings, 1963, 1968), possible toxic effects of planarian secretions on prey and/or on vertebrates by accidental contact or ingestion (Arndt, 1925; Arndt & Manteufel, 1925; Hyman, 1951: 203; Lue & Kawakatsu, 1986: 381 Chinese lit), or possible neurotoxic effects (Blackshaw & Stewart, 1992: 213). Considerations raised by Barnwell *et al.* (1965) are the frequency of predation and the need for food in relation to environmental temperatures. Experimentation on attack rate by *Artioposthia triangulata* by Blackshaw (1991: 690), revealed that planarians attacked earthworms placed in their experimental environment with consumption decreasing over the second and third feeding periods. An average of 1.4 earthworms was consumed per week under experimental conditions. The larger planarians attacked more frequently, thus consumed more prey (Blackshaw, 1991: 692). Toxicity of mucus to prey remains unconfirmed and may not be an important factor (Barnwell *et al.*, 1965, Hyman, 1951: 202) for *B. kewense*. Studies on feeding and digestion, including digestive enzymes such as a collagenase (Landsperger *et al.*, 1981) and external vs internal digestion, have been reviewed and compared for various flatworms by Jennings (1959, 1962, 1963, 1968).

The effects of regular, effective predation on the prey population can have different ecological consequences. Mead (1963, 1979) and Muniappan (1987, 1990) reported planarians have been used to reduce and control giant snail populations. On the other hand, serious depletion of the earthworm populations in Northern Ireland has resulted from the predation by planarians accidentally introduced from New Zealand (Blackshaw, 1990, 1991, 1992; Blackshaw & Stewart, 1992).

Acknowledgments

Assistance of the following in obtaining pertinant literatuur is greatly appreciated: Eugene Sheddon Farley Library, Wilkes University; Library, Academy of Natural Sciences, Philadelphia. Special gratitude goes to Dr J. B. Jennings, retired, University of Leeds, Leeds, England for helpful suggestions concerning the topic; and Prof. Dr Masaharu Kawakatsu, Fuji Women's College, Sapporo, Japan, who reviewed the manuscript.

References

Arndt, W., 1925. Über die Gifte der Plattwürmer. Verh. dt. zool. Ges. 30: 135–145.

Arndt, W., 1938. Über den Beutefang von Rhynchodemus. Blätt. Aquarien u. Terrarienkunde 49: 12–13.

Arndt, W. & P. Mantenfel, 1925. Die Turbellarien als Träger von Giften. Z. Morph. Okol. Tiere 3: 344–357.

Ball, I. R. & R. Sluys, 1990. Turbellaria: Tricladida: Terricola. In D. L. Dindal (ed.), Soil Biology Guide: 137–153. John Wiley & Sons, N.Y.

Barker, G. M., 1989. Flatworm predation of terrestrial molluscs in New Zealand and a brief review of previous records. N. Z. Entomol. 12: 75–79.

Barnwell, G. M., 1966. The role of encephalization in the feeding behavior of a land planarian, Biplium kewense. J. Biol. Psych. 8: 41–47.

Barnwell, G. M., L. J. Peacock & R. E. Taylor, 1965. Feeding behavior of a land planarian, Bipalium kewense. Unpublished manuscript presented to the Southern Society for Philosophy and Psychology, Atlanta, Ga, April, 15–17, 1965 (8 pages photocopy available).

Blackshaw, G. P., 1990. Studies on Artioposthia triangulata (Dendy) (Tricladida: Terricola), a predator of earthworms. Ann. appl. Biol. 116: 169–176.

Blackshaw, R. P., 1991. Mortality of the earthworm Eisenia fetida (Savigny) presented to the terrestrial planarian Artioposthia triangulata (Dendy) (Tricladida: Terricola). Ann. appl. Biol. 118: 689–694.

Blackshaw, R. P., 1992. The effect of starvation on size and survival of the terrestrial planarian Artioposthia triangulata (Dendy) (Tricladida: Terricola). Ann. appl. Biol. 120: 573–578.

Blackshaw, R. P. & V. I. Stewart, 1992. Artioposthia triangulata (Dendy, 1894), a predatory terrestrial planarian and its potential impact on lumbricid earthworms. Agr. Zool. Rev. 5: 201–219.

Brittlebank, C. C., 1888. Food of planarians.Vitorian Nat. 5: 48.

Chandler, C. M., 1976. Field observations on a population of the land planarian, Bipalium kewense (Turbellaria, Tricladida) in middle Tennessee. J. Tenn. Acad. Sci. 51: 73–75.

Dendy, A., 1890a. The anatomy of an Australian land planarian. Trans. R. Soc. Vict. 1 (for 1889): 50–95 +pls. 7–10.

Dendy, A., 1890b. Zoological notes on a trip to Walhalla. Victorian Nat. 6: 128–136.

Dindal, D. L., 1970. Feeding behavior of a terrestrial turbellarian Bipalium adventitium. Am. Midl. Nat. 83: 635–637.

Dindal, D. L. 1990. Soil Biology Guide, John Wiley & Sons, N.Y., i-xviii + 1349 pp.

Froehlich, C. G., 1955a. Sôbre morfologia e taxonomia das Geoplanidae. Bol. Fac. Fil. Ciênc. Letr. Univ. São Paulo, Zoologia (19): 195–251 + plls. 1–14.

Froehlich, C. G., 1955b. On the Biology of land planarians. Bol. Fac. Fil. Ciênc. Letr. Univ. São Paulo, Zoologia (20): 263–271 + pl. 1.

Garnett, H., 1928. Rhynchodemus britannicus Percival. North West. Nat. Arbroath 3: 186–187.

Hull, F. M., 1947. Observations on cannibalism in planarians. Trans. Am. Microsc. Soc. 66: 96–98.

Hyman, L. H., 1951. The Invertebrates, II. Platyhelminthes and Rhynchocoela. McGraw-Hill, N.Y., 550 pp.

Jennings, J. B., 1959. Observations on the nutrition of the land planarian Orthodemus terrestris (O. F. Müller). Biol. Bull. 117: 119–124.

Jennings, J. B., 1962. Further studies on feeding and digestion in the triclad turbellaria. Biol. Bull. 124: 571–581.

Jennings, J. B., 1963. Some aspects of nutrition in the Turbellaria, Trematoda and Rhynchocoela. In Dougherty, E. C. *et al.* (eds), The Lower Metazoa. Univ. Calif. Press, Berkeley: 345–353.

Jennings, J. B., 1968. Nutrition and Digestion, Ch. 2., In Florkin, M. & B. T. Scheer (eds), Chemical Zoology, II. Porifera, Coelenterata and Platyhelminthes. Academic Press, Lond., 303–326.

Johri, L. N., 1952. A report on a turbellaria Placocephalus kewense, from Delhi State and its feeding behaviour on the live earthworm, Pheretima posthuma. Sci. Cult. Calcultta 18: 291.

Jones, H. D., 1993. The feeding behaviour of Microplana termitophaga in Zimbabwe. VIIth International Symposium on the Biology of the Turbellaria, Abst. 55, Åbo/Turku, Finland, 17–22 June, 1993.

Jones, H. D., J. P. E. C. Darlington & R. M. Newson, 1990. A new species of land planarian preying on termites in Kenya (Platyhelminthes: Turbellaria: Tricladida: Terricola). J. Zool., Lond. 220: 249–256.

Kaneda, M., K. I. Kitagawa & F. Ichinohe, 1990. laboratory rearing method and biology of Platydemus manokwari de Beauchamp (Tricladida: Terricola: Rhynchodemidae). Appl. Ent. Zool. 25: 524–528.

Kawakatsu, M., 1969. A list of publications on Japanese turbellarians (1968). Including titles of publications on foreign turbellarians written by the Japanes authors. Bull. Fuji Women's Coll, (7), ser. 2: 23–42.

Kawakatsu, M.& K. Y. Lue, 1984. History of the study of turbellarians in China. Part 2. Age of studies by Japanese and Chinese turbellariologists. Bull. Fuji Women's Coll., (22) ser. 2: 105–117.

Kawakatsu, M., S. Tamura & K.-Y. Lue 1984.Preprint of papers given at the fourth International Symposium on the Turbellaria. Occ. Publ., Biological Laboratory of Fuji Women's College, Sapporo (Hokkaido), Japan. (12): 20 pp.

Kawakatusu, M., I. Oki, S. Tamura, H. Itô, Y. Nagai, K. Ogura, S. Shimabukuro, F. Ichinohe, H. Katsumata and M. Kaneda, 1993. An extensive occurrence of a land planarian, Platydemus manokwari de Beauchamp, 1962, in the Ryûkyû Islands, Japan (Turbellaria, Tricladida, Terricola). Biol. Inl. Wat. No. 8: 5–14.

Kawakatsu, M., R. E. Ogren & R. Munippan, 1992. Redescription of Platydemus manokwaris de Beauchamp, 1962 (Turbellaria: Tricladida: Terricola), from Guam and the Philippines. Proc. Jap. Soc. Syst. Zool. (47): 11–25.

Landsperger, W. J., E. H. Peters & M. D. Dresden, 1981. Properties of a collagenolytic enzyme from Bipalium kewense. Biochim. Biophys. Acta, 661: 213–220.

Lehnert, G. H., 1891. Beobachtungen an Landplanarien. Arch. Naturgesch. 57: 306–350.

Little, C., 1983. The Colonisation of Land. Origins and adaptations of terrestrial animals. Cambridge University Press, New York, 290 pp.

Lue, K.-Y. & M. Kawakatsu, 1986. History of the study of turbellarians in China. Part. 1. Ages of Materia Medica and of early expeditions by westerns. Hydrobiologia 132 (Dev. Hydrobiol. 32): 317–322.

Mead, A. R., 1963. A flatworm predator of the giant African snail Achatina fulica in Hawaii. Malacologia 1: 305–311.

Mead, A. R., 1979. Economic Malacology with Particular Reference to Achatina fulica. Turbellarians, pp. 70–74. In: Fretter, V. & J. Peaks, (eds), Pulmonates, 2B, Academic Press, N.Y., pp. i-ix + 1–150.

Miyoshi, Y., 1955. Observations on the food habit of Bipalium. Collect. & Breed., Tokyo 17: 377. (In Japanese).

Moseley, H. N., 1875. On the anatomy and histology of the land-planarians of Ceylon, with some account of their habits, and a description of two new species, and with notes on the anatomy of some European acquatic species. Phil. Trans. r. Soc., Lond. 4: 105–171 + pls. 10–15.

Muniappan, R., 1987. Biological control of the giant African snail, Achatina fulica Bowdich, in the Maldives. FAO Plant Bull. 35: 127–133.

Muniappan, R., 1990. Use of the planarian Platydemus manokwari, and other natural enemies to control the giant African snail. In FECT Book Series (40): The Use of Natural Enemies to Control Agricultural Pests: 179–183.

Neck, R. W., 1987. A predatory terrestrial flatworm, Bipalium kewense, in Texas: Feral populations and laboratory observations. Texas J. Sci. 39: 267–271.

Ogren, R. E., 1955. Ecological observations on the occurrence of Rhynchodemus, a terrestrial turbellarian. Trans. am. Microsc. Soc. 74: 54–60.

Ogren, R. E., 1957. Physiological observations on movement and behavior of the land planarian Rhynchodemus sylvaticus (Leidy). Proc. Pa. Acad. Sci. 30: 218–225.

Ogren, R. E., 1986. The human factor in the spread of an exotic land planarian in Pennsylvania. Proc Pa. Acad. Sci. 117–118.

Ogren, R. E., M. Kawakatsu & E. M. Froehlich, 1992. Additions and corrections of the previous land planarian indices of the world (Turbellaria, Tricladida, Terricola). Bull. Fuji Women's College, (30), Ser. 2: 59–103.

Ogren, R. E. & J. K. Sheldon, 1991. Ecological observations on the land planarian Bipalium pennsylvanicum Ogren, with references to phenology, reproduction, growth rate and food niche. J. Pa. Acad. Sci. 65: 3–9.

Olewine, D. A., 1972. Further observations in Georgia on the land planarians, Bipalium kewense and Geoplana vaga (Turbellaria: Tricladida: Terricola). Assoc. Southeast. Biol. Bull. 19: 88.

Pfitzner, I., 1958. Die Bedingungen der Fortbewegung bei den deutschen Landplanarien. Zool. Beitr. N.F. 3: 235–311.

Rodriques, R. M., 1972. Ciclo Biologicco de Geoplana burmeisteri Max Schultze, 1857 (Turbellaria, Tricladida, Terricola). Dissertaçao apresentada ao Deparatmento de Zoologia do Instituto de Biociências para obtencçao do titulo de mestre em Zoologia.

Schremmer, F., 1955. Freilandfund der Landplanarie Rhynchodemus bilineatus Metsch. Verhandl. Zool. Bot. Ges. in Wein 94: 45–58.

Schultze, M., 1857. Contributions to the knowledge of terrestrial planariae, from communications from Dr. Fritz Müller of Brazil and personal investigations. Ann. Mag. nat. Hist. ser. 2, 20: 1–13.

Schultze, M. & F. Müller, 1857. Beiträge zur Kenntniss der Landplanarien nach Mittheilungen des Dr. Fritz Müller in Brasilien und nach eigenen Untersuchungen von Dr Max Schultze. Abhandl. Naturf. Ges. zu Halle 4: 19–38.

Sheppe, W., 1970. Invertebrate predation on termites of the African savanna. Insectes Soc., Paris,17: 205–218.

Spencer, W. B., 1891. Notes on some Victorian land planarians. Proc. r. Soc. Vict. N.S. 3 (for 1890): 84–93 + pls. 11–12.

Steel, T., 1901. Australian land planarians: Descriptions of new species and notes of collecting and preserving. no. 2. Proc. Linn. Soc. N.S. Wales 25: 563–580 + pl. 34.

von Graff, L., 1899. Monographie der Turbellarien. II. Tricladida Terricola (Landplanarien). Leipzig: Wilhelm Engelmann, 575 pp.

von Graff, L., 1912–1917. Turbellaria. In Bronn, H.G. (ed.), Klassen und Ordnungen des Tier-Reichs, 4, Wermer: Vermes, Abt. 1c (Turbellaria), Abt. 2, 1–37 + 2601–3369 pp. C.F. Winter'sche Verlagshandlung, Leipzig.

Wallner, W., 1937. Rhynchodemus terrestris, eine Landplanarie. Blätt. Aquar. Terarienkunde 48: 224–227.

Willis, R. J. & A. R. Edwards, 1977. The occurrence of the land planarian Artioposthia triangulata (Dendy) in Northern Ireland. Ir. Nat. J. 19: 112–116.

Winsor, L., 1991. A new genus and species of terrestrial flatworm from the central highlands of New Caledonia (Tricladida Terricola). In Chazeau, J. & S. Tillier (eds), Zoologia Neocaledonica, II, Mém. Mus. natn. Hist. nat., (A), 149: 19–30. Paris.

Hydrobiologia 305: 113–117, 1995.
L.R.G. Cannon (ed.), Biology of Turbellaria and some Related Flatworms.

The influence of body size on immediate reproductive success in *Dugesia gonocephala* (Tricladida, Paludicola)

C. Vreys[1] & N. Michiels[2]
[1]*Zoology Research Group, Department S.B.G., Limburgs Universitair Centrum, B-3590, Diepenbeek, Belgium*
[2]*Max-Planck-Institut für Verhaltensphysiologie, D-82319 Seewiesen (Starnberg), Germany*

Key words: reproductive success, body size, reproductive trade-off, *Dugesia gonocephala*

Abstract

The individual reproductive output of the stream-dwelling flatworm *Dugesia gonocephala* was investigated. Various measures of reproductive success were related to body size. (I) For the first 30 days in the laboratory small individuals produced no cocoons, individuals of intermediate size produced unfertilized cocoons and large individuals usually produced fertilized cocoons. (II) In individuals that produced a cocoon, no correlation was found between the number of cocoons produced in one month and body size. (III) Large individuals, however, produced larger cocoons. This was not due to the fact that unfertilized cocoons were smaller. (IV) Large cocoons tended to contain more young. (V) The average size of young hatching from large cocoons was larger. (VI) Large individuals produced their first cocoon soon after their arrival in the laboratory and seemed to have a higher chance of producing a fertilized first cocoon. (VII) A trade-off existed between producing many small versus few large young

Introduction

Numerous studies on the population dynamics of freshwater flatworms revealed that intra-specific competition for food is the most limiting factor in the life-cycle of planarians (Boddington & Mettrick, 1971, 1974, 1977; Callow & Woollhead, 1977; Herrmann, 1979, 1984, 1985; Adams, 1980). This will have an effect on how an individual divides its resources between survival and reproduction: the more food an individual can obtain, the more energy it can invest in reproductive activities. Since planarians vary considerably in size, we expected this feature to represent the relative competitiveness to obtain food and consequently to reproduce.

We investigated the production of cocoons and young of specimens of *Dugesia gonocephala*, a large (max: 25 mm) simultaneous hermaphroditic flatworm. Working under laboratory conditions gave us the opportunity to monitor the reproductive output of each worm individually. The main questions are: (1) do large individuals have more resources available for cocoon production and what aspects of reproductive output are affected most by an individual's size? (2) Is there a trade-off between the number of young in a cocoon and their size?

Materials and methods

On 15 March (N = 36) and 28 October 1992 (N = 97), 133 specimens of *Dugesia gonocephala* were collected in the River Noorbeek in the north-eastern part of Belgium (50°07′ N, 5°09′ E). On the date of sampling their length was measured to the nearest 1 mm while gliding normally. Worms were kept individually in small plastic vials (0.3 litre), placed in 2 large containers (35 litre). The latter were filled with continuously filtered and aerated tap water (without chlorine). In every vial 2 holes (diameter = 34 mm) covered by a mesh (0.5 × 0.5 mm) provided a constant water flow and allowed chemical signalling between the worms. The container was kept in a semi-dark room at a natural dark/light cycle. Temperature ranged between 15 ° and 18 °C. Worms were fed one punctured fresh *Chironomus* larva two or three times a week.

The following variables were recorded: number and volume of the cocoons deposited per individual, number of unhatched cocoons and number and average size of hatchlings per cocoon. Except for weekends, individual cocoon production was monitored daily. Two hundred and seventy one cocoons were collected in total (24 from 15/03/1992 to 17/04/1992 and 247 from 28/10/1992 to 20/05/1993). The diameter of 251 of these cocoons (the other 20 were broken) was measured to the nearest 0.01 mm using a Wild measuring eyepiece mounted on a Wild M8 stereomicroscope (enlargement: 50 ×). The formula 'Volume = $1.33.\pi.(\text{diameter}/2)^3$' was used to estimate cocoon volume (mm^3). Cocoons were kept in isolation until hatching 3 to 6 weeks later. Cocoons that did not hatch after 8 weeks were ruptured to check their content. None of them contained any young and therefore we considered them unfertilized. Young were counted and measured within two days. From 15/03/1992 to 17/04/1992 these young were killed with Steinmann's fluid (1 part saturated $HgCl_2$ solution in 5% NaCl, 1 part concentrated HNO_3, 1 part H_2O) and measured in the same way as the cocoons. From 28/10/1992 to 20/05/1993 young were filmed with a camcorder (Sony: SSC-M370CE) and a picture of each worm was taken using a Video Graphic Printer (Sony: UP-860/860CE). These pictures were used to measure the young. For both periods total and average size of the young (volume in mm^3) was estimated using the formula: 'Volume = $0.042.\pi.\text{Length}.\text{Width}^2$' (Heller & Hauser, personal communication). Although measuring techniques differed, there was no significant difference in the mean volume of young per cocoon between the periods 15/03/1992–17/04/1992 and 28/10/1992–20/05/1993 ($t = 1.46$; $df = 23.23$, $P = 0.16$).

All analyses are based upon individuals that survived in the laboratory for at least one month. These individuals were considered healthy. We wanted to investigate the instantaneous effect of size (in the field) on subsequent cocoon production. Therefore we only used cocoons produced during the first 30 days after sampling in analyses where body size is involved. In that way we tried to exclude laboratory effects (e.g. growth). Statistically distinct outliers were eliminated from the analyses.

Results

What size class of individuals reproduces?

Out of 133 individuals, 128 (96.7%) survived in the laboratory for at least one month. During the first thirty days of isolation these individuals produced 56 cocoons in total. On the basis of this cocoon production we could subdivide the worms into three size classes: small individuals producing no cocoons, individuals of intermediate size always producing unfertilized cocoons and large individuals usually producing fertilized cocoons (Table 1 and Fig. 1). An analysis of variance indicates that the difference between the size-classes is highly significant (Table 1).

Body size in relation to the production of cocoons and young

Production of cocoons

Within the group of individuals that produced cocoons, large individuals did not produce more cocoons than small ones ($r = -0.05$, $N = 30$, $P = 0.77$). Within one month a maximum of 4 cocoons was produced per individual ($N = 30$, $\bar{x} = 1.83 \pm 0.87$). The interval between two successive cocoons was on average 8.15 days ($N = 20$, $SD = 3.07$ days, min. = 3 days, max. = 14 days).

The mean cocoon volume per individual was calculated ($N = 30$, $\bar{x} = 6.53 \pm 1.90$ mm^3). There was a significant correlation between cocoon volume and the initial length of the individuals on the sampling date: large individuals produced larger cocoons (Fig. 2A, r,0.42, $N = 30$, $P = 0.022$). This was not due to the fact that unfertilized cocoons were smaller (volume fertilized versus unfertilized cocoons: $t = -0.19$, $df = 49$, $P = 0.85$).

Production of young

One hundred and twenty two out of 251 cocoons hatched (49%). Nine hundred and eighty four young hatched in total, an average of 8.07 ± 4.54 per cocoon (max = 19). Although not significant, there was a tendency for large cocoons to contain more young (Fig. 2B, $r = 0.18$, $N = 122$, $P = 0.053$).

The mean volume of the emerging young was calculated for 39 cocoons. Two hundred and seventy six young were measured in total. Their size varied between 0.6 and 6.9 mm^3 ($\bar{x} = 2.4 \pm 1.3$ mm^3). There was a significant correlation between cocoon volume and mean size of the young per cocoon: young hatching from large cocoons were on average larger (Fig. 2C, $r = 0.38$, $N = 39$, $P = 0.016$).

Table 1. Size classes of worms collected in March and October 1992, producing no cocoons, unfertilized cocoons and fertilized cocoons during their first month in the laboratory. Only individuals that survived for at least one month are taken into account.

	Number of individuals	Mean size (mm)	Std. dev. (mm)	Min. size (mm)	Max. size (mm)
Individuals producing					
- no cocoons	97	13.9	3.6	5	23
- always unfertilized cocoons	7	16.6	4.9	7	23
- ≥ 1 fertilized cocoon	24	19.0	2.9	14	24
F = 19.96; df = 2, 125; P = 0					

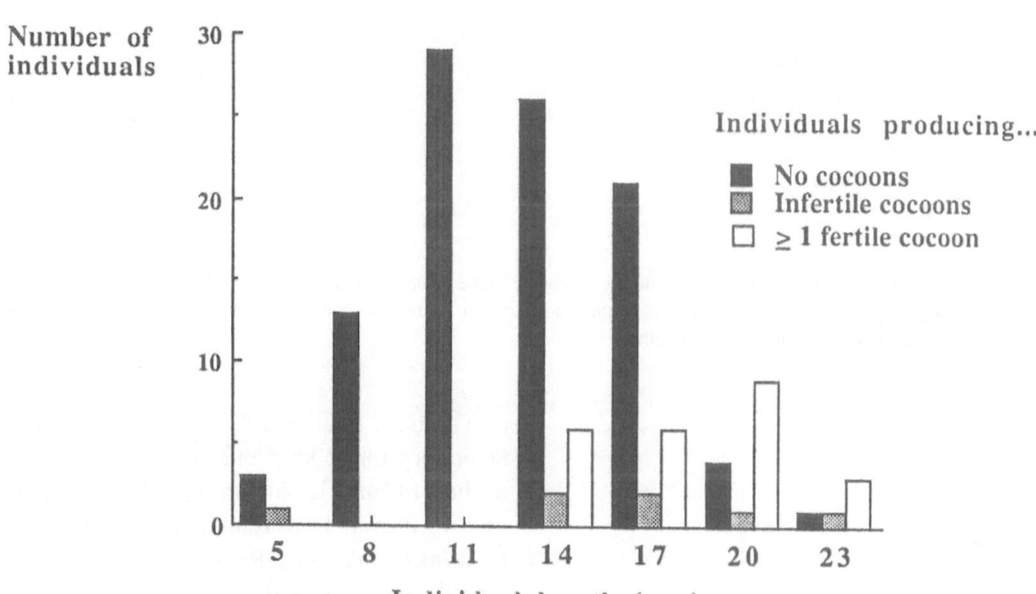

Fig. 1. Size distribution of *D. gonocephala* individuals kept in isolation, subdivided according to their cocoon production.

When is the first cocoon in the laboratory produced?

One individual produced its first cocoon 4 days after its arrival in the laboratory, whereas another individual waited for 90 days ($\bar{x}=33.48\pm20.63$ days). The correlation between individual length and the time before producing the first cocoon in the laboratory was highly significant: small individuals waited longer before producing their first cocoon (Fig. 2D, $r=-0.60$, $N=64$, $P=0.000$). Although statistically not significant, the chance to produce a fertilized first cocoon tended to be higher for larger individuals (length of fertilized versus unfertilized: $t=1.92$, $df=62$, $P=0.06$).

Trade-off: many small young versus few large young

When a cocoon contains many young, these young are on average smaller (Fig. 3A, $r=-0.44$, $N=38$, $P=0.006$). This indicates that a trade-off exists between producing many small versus few large young. An important consequence of producing many small young, is the fact that the hatching date is negatively correlated with the number of young in the cocoon: the more young a cocoon contains, the sooner it hatches (Fig. 3B, $r=-0.41$, $N=115$, $P=0.000$).

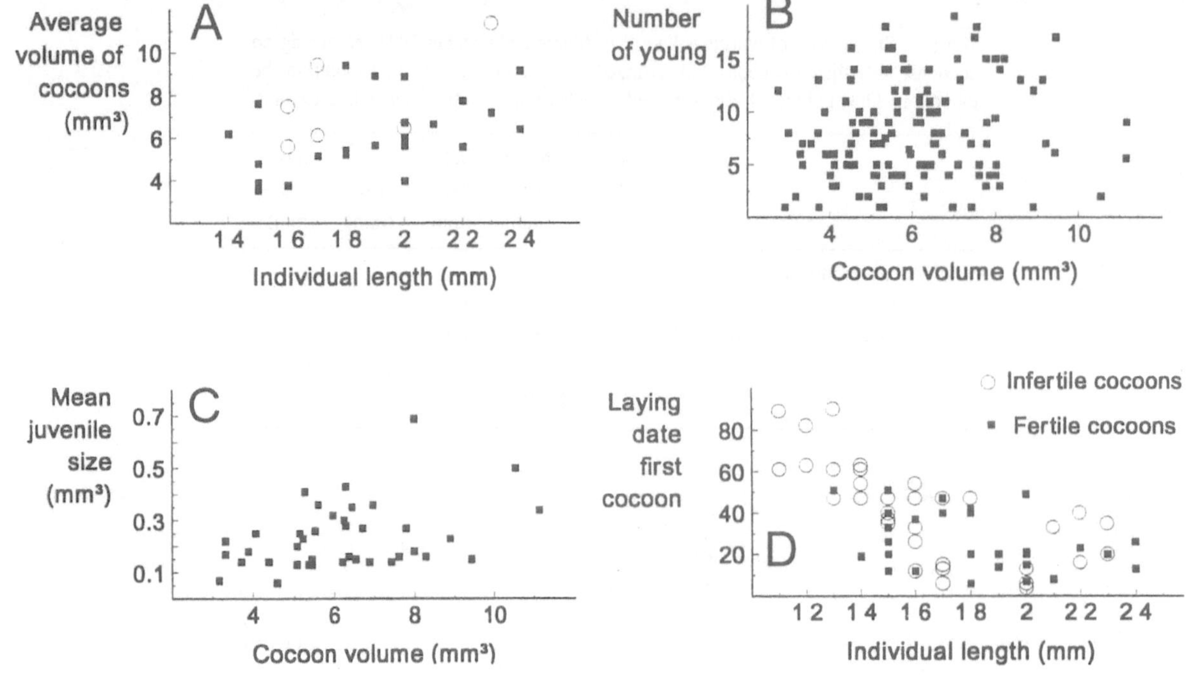

Fig. 2. A. Relationship between an individual's length and the average volume of the cocoons it produces. B. Number of young per cocoon as a function of cocoon volume. C. Average young size as a function of cocoon volume. D. Number of days in the laboratory before the first cocoon was produced as a function of an individual's length.

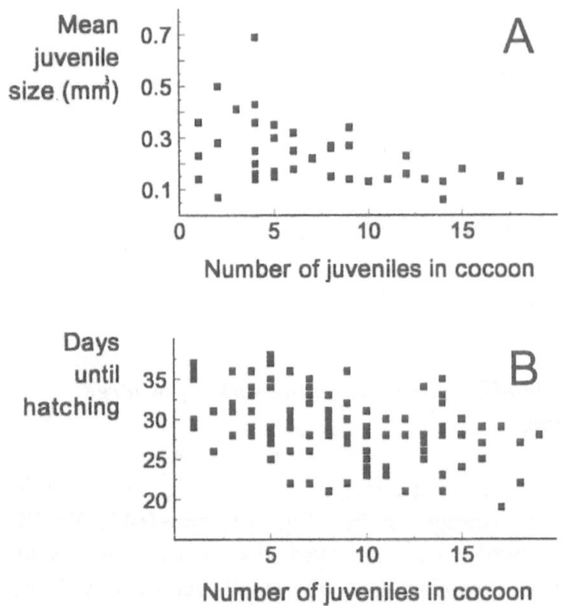

Fig. 3. A. Size of the young versus number in the same cocoon. B. Number of days until hatching as a function of the number of young in a cocoon.

Discussion

In this study we have demonstrated that in *Dugesia gonocephala* immediate reproductive success increases with body size. Of 128 individuals surviving in the laboratory for at least one month, 56% were smaller than 14 mm. In this group only one cocoon (volume = 5.06 mm^3) was found. Since planarians are able to shrink (Kenk, 1922; Reynoldson, 1968), small, non-reproducing individuals could either be growing young with undeveloped sexual organs or shrunken adults of which the sexual organs had regressed.

On the other hand, 54% of the individuals in the size-class of 14 to 24 mm ($N = 56$) deposited 271 cocoons in total. Apparently, worms first have to grow to a minimum length of 14 mm before they can invest in reproductive activities. However, cocoons deposited by the individuals of intermediate size were usually unfertilized. Since all cocoons were kept under the same conditions, it is unlikely that infertility was caused by laboratory conditions. It is more reasonable to believe that individuals producing unfertilized cocoons did not receive spermatozoa from copulation in the field, despite the fact that they reached maturity. It is important to point out that according to our own observations, self-fertilization does not occur in *D. gonocephala*. To determine the presence or absence of donor sperm, we made histological sections of two individuals: the first one produced 6 fertilized cocoons, the second one 5 unfertilized cocoons. In both occasions the last cocoon was deposited 1 day before fixation. The oviducts of the first individual contained many spermatozoa near the ovaries. In the oviducts of the second individual no spermatozoa were found. However, the question remains: why are individuals producing unfertilized cocoons?

Twenty five percent of the individuals that produced fertilized cocoons were about 20 mm long. In the group of fertile individuals, the results indicate that large individuals can spend more energy in the production of cocoons and young and therefore have a higher reproductive success than smaller conspecifics: large individuals start laying cocoons immediately after they arrive in the laboratory. These cocoons in turn are bigger and contain more and larger hatchlings. These results agree with the finding of Taylor and Reynoldson (1962) that body-length and cocoon volume are positively correlated and with the result of Skaer (1971) that a relation exists between the volume of a cocoon and the number of hatchlings it contains. Since there is a trade-off between many small young versus a few large ones, it appears that individuals can vary the number and quality of the young they produce. Whether being large at birth is an advantage in further life is still under investigation.

Summary

1. During the first month in the laboratory, 133 specimens of *D. gonocephala* were followed individually on their cocoon and juvenile production. The relation between body size and reproductive success was observed.
2. It is shown that large individuals can invest more energy in the production of cocoons and young: the cocoons are produced sooner, they are larger and contain more and larger young. Furthermore, the smallest individuals produce no cocoons at all, whereas the chance to produce unfertilized cocoons seems to be the highest for individuals of intermediate size.
3. A trade-off exists between producing many small versus few large young.

Acknowledgments

We wish to thank Natascha Steffanie for help with feeding the worms and collecting cocoons and young. C.V. was supported by a Research Assistantship of the Belgian Institute for Encouraging Scientific Research in Industry and Agriculture.

References

Adams, J., 1980. The role of competition in the population dynamics of a freshwater flatworm, *Bdellocephala punctata* (Turbellaria, Tricladida). J. anim. Ecol. 49: 565–579.

Boddington, M. J. & D. F. Mettrick, 1971. Seasonal changes in the chemical composition and food reserves of the freshwater triclad *Dugesia tigrina* (Platyhelminthes: Turbellaria). J. Fish Res. Bd Can. 28: 7–14.

Boddington, M. J. & D. F. Mettrick, 1974. The distribution, abundance, feeding habitats and population biology of the immigrant triclad *Dugesia polychroa* (Platyhelminthes: Turbellaria) in Toronto Harbour, Canada. J. anim. Ecol. 43: 681–699.

Boddington, M. J. & D. F. Mettrick, 1977. A laboratory study of the population dynamics and productivity of *Dugesia polychroa* (Turbellaria, Tricladida). Ecology 58: 109–118.

Callow, P. & S. Woollhead, 1977. The relationship between ration, reproductive effort and age-specific mortality in the evolution of life-history strategies. Some observations on freshwater triclads. J. anim. Ecol. 46: 765–781.

Herrmann, J., 1979. Population dynamics of *Dendrocoelum lacteum* (O. F. Müller) (Tricladida, Turbellaria) in a South Swedish lake. Arch. Hydrobiol. 85: 482–510.

Herrmann, J., 1984. Prey choice and its impact on the reproductive output in *Dendrocoelum lacteum* (Turbellaria, Tricladida) in a South Swedish lake. Arch. Hydrobiol. 100: 355–370.

Herrmann, J., 1985. Reproductive phenology in *Dendrocoelum lacteum* (Turbellaria) in South Sweden. Freshwat. Biol. 15: 127–132.

Kenk, R., 1922. Die normale und regenerative Entwicklung des Copulationsapparates paludicoles Tricladen. Zool. Anz. Bd. 54.

Reynoldson, T. B., 1968. Shrinkage Thresholds in freshwater triclads. Ecology 49: 583–586.

Skaer, R., 1971. Planarians. In G. Reverberi (ed.), Experimental Embryology of Marine and Fresh-water Invertebrates 104–124.

Taylot, M. & T. B. Reynoldson, 1962. The population biology of lake-dwelling *Polycelis* species with special reference to *P. nigra* (Müller) (Turbellaria, Tricladida). J. anim. Ecol. 31: 273–291.

Hydrobiologia **305**: 119, 1995.
L.R.G. Cannon (ed.), Biology of Turbellaria and some Related Flatworms.

Reproductive behaviour in *Dugesia gonocephala* (Tricladida, Paludicola)

C. Vreys[1] & N. Michiels[2]
[1]*Zoology Research Group, Department S.B.G., Limburgs Universitair Centrum, B-3590 Diepenbeek, Belgium*
[2]*Max-Planck-Institut für Verhaltensphysiologie, D-82319 Seewiesen (Starnberg), Germany*

The mating behaviour of the stream-dwelling flatworm *Dugesia gonocephala* was investigated. Specimens were collected during 3 sampling periods in March and October 1992, and in March 1993. A first series of individuals was placed together in groups of 2–5 individuals after an isolation period of at least 25 days. Their behaviour was filmed over at least 4 hours. A second series of individuals was subdivided into 6 groups, each of 2–15 individuals, without an isolation period. They were observed many times a day. Every copulation was recorded and whenever possible, videorecorded as well. In those pairs where a mating was observed, a clear and consistent asymmetry existed in the behaviour of both individuals before, during and after a copulation or attempt at copulation (N = 15). In the pre-copula phase, one individual (A) glides around normally, while the other (B) makes active searching movements with the head. Next, the 'active' searcher (B) comes to a sudden halt and the 'passive' individual (A) glides on top. When (B) does not respond by flattening its body, the 'climber' loses interest and leaves. When (B) however does flatten its body, (A) stays and flattens its body too. Copulation only follows when (B) raises the distal part of its body, thus exposing its gonopore. When so, both individuals start moving their tails in order to make genital contact. When a pair separates before a proper copulation (defined as genital contact for more than 10 min), it is termed an unsuccessful copulation attempt. Such attempts last 0.35 to 4.13 min. ($\bar{x}$ = 1.62 min, N = 27). Copulation duration varies from 10 to 410 min ($\bar{x}$ = 142.90 min, N = 10). A copulation always ends when individual (A) starts gliding away from individual (B). Both animals can go through the pre-copula phase and run through the same behavioural sequences for up to 30 times until a successful copulation occurs. Within a sequence, each role (A) and (B) is fixed, but individuals can switch roles from one sequence to the next. When a sequence ends in a successful copulation, the individuals lose interest in each other, at least temporarily.

Hydrobiologia **305**: 121–126, 1995.
L.R.G. Cannon (ed.), *Biology of Turbellaria and some Related Flatworms.*

The anatomy of *Microplana termitophaga* from Zimbabwe and Kenya: confirmation of the subfamily and genus

Hugh D. Jones[1], Meg S. Cumming[2] & John H. Kennaugh[1]
[1]*School of Biological Sciences, University of Manchester, Manchester, M13 9PL, UK*
[2]*19, Walmer Drive, Highlands, Harare, Zimbabwe*

Key words: Turbellaria, Terricola, musculature, sexual maturity, copulatory apparatus

Abstract

Microplana termitophaga has been observed feeding on termites in Harare, Zimbabwe. In small specimens the pharynx is about two-thirds of the way along the body, but in large specimens the pharynx is only one quarter of the way along the body. Details of the subepidermal musculature of specimens from Kenya and Zimbabwe are described which confirm the previously uncertain assignment of the species to the subfamily Microplaninae. Sexually mature specimens are found towards the end of the wet season in Zimbabwe. The anatomy of the reproductive system is described. The gonopore is about two-thirds of the way along the body in mature specimens, but sexual maturity seems not to be related simply to size. There is a genito-intestinal duct which confirms that the species is of the genus *Microplana*.

Introduction

Jones *et al.* (1990) described *Microplana termitophaga* from immature specimens found in Kenya. Triclad taxonomy at the generic level is partly dependent upon the anatomy of the copulatory apparatus and the generic assignment of this species was tentative, the species being placed in the most likely genus within the probable subfamily. The family Rhynchodemidae Graff, 1896, consists of two subfamilies: Rhynchodeminae Hyman, 1943 (part definition – strong subcutaneous longitudinal muscle aggregated into distinct bundles) and Microplaninae Pantin, 1953 (part definition – weak subcutaneous longitudinal muscle not aggregated into bundles). There was some doubt about the subfamily because, in the Kenyan specimens, the subcutaneous longitudinal muscle was indistinguishable. The specimens were tentatively placed within the Microplaninae. Subsequent to publication, further preserved specimens were received from Kenya that had been preserved in glutaraldehyde. One of these specimens has been examined using transmission electron microscopy which has allowed details of the subcutaneous musculature to be resolved.

Meanwhile, one of us, M. S. Cumming, has been monitoring invertebrate predators of termite mounds (*Odontotermes transvaalensis* Sjöstedt) in Harare, Zimbabwe over several years. Amongst the predators are terrestrial planarians with similar feeding habits to the Kenyan specimens and these worms are also *Microplana termitophaga*. Moreover, specimens of worms collected over the wet season (December–April) showed that later on in the season some of the specimens were sexually mature. Thus it is now possible to describe the copulatory apparatus.

This paper reports anatomical details additional to those in Jones *et al.* (1990) and discusses their taxonomic significance.

Materials and methods

The specimen from Kenya was collected from the type locality (Jones *et al.*, 1990) during May 1990. It had been fixed in 4% glutaraldehyde and preserved in 80% alcohol. It was subjected to routine EM preparation and examined and photographed with a Philips 201 TEM.

40 specimens from Zimbabwe were collected from termite mounds in or near the residential garden of M. S. Cumming in Harare between November 1990 and April 1991. The altitude is about 1600 m. They were preserved without prior narcotization in 70% alcohol with 5% glycerol and sent to H. D. Jones for further examination. All were examined by clearing in creosote and drawn to scale using the tracing attachment of a binocular microscope. In most specimens the position of the pharynx and copulatory apparatus (where the latter was present) could only be deduced after clearing when they were visible by translucency. The apertures of the mouth and gonopore were indistinguishable in most specimens even after clearing. Body length and, in specimens where they could be determined, the distances from the anterior end to the centre of the pharynx and the copulatory apparatus were measured. (Due to coiling of some specimens the position of the pharynx could not be determined.) These are expressed as the ratio of their distance from the anterior end to the body length (Ph-L ratio and G-L ratio respectively). One immature and three apparently mature specimens were embedded in wax, sectioned in various planes and stained in heamatoxylin and eosin. One specimen (the most mature sectioned, 13.4 mm long, collected 12.03.1991) is deposited in the Natural History Museum, London (accession number 1993.10.25.1). Others are retained by H. D. Jones.

Results

Pharyngeal position

In Zimbabwean specimens, the position of the pharynx along the body varies with the size of the specimen (Table 1, Fig. 1). In large specimens the pharynx is only one-fifth of the way along the body (Ph-L ratio 0.2). However, in the smallest specimens, the pharynx is much nearer to the posterior (Ph-L ratio 0.65).

Gut

In the sectioned mature specimens the two posterior rami of the gut fuse immediately in front of and behind the copulatory apparatus to form a very narrow (about 0.1 mm) dorsal anastamosis. Otherwise the gut is of the typical triclad plan.

Musculature

A layer of subepidermal circular muscle lies immediately beneath the basement membrane. Inside the

Table 1. Specimens from Zimbabwe. Collection date, body length (mm), distance from the anterior to the centre of the pharynx (A–Ph) where determinable, and from anterior to the centre of the copulatory apparatus (A–G) where present. Variously incomplete specimens indicated with +; specimens with sperm-filled male ducts visible by translucency are indicated with *; specimens sectioned indicated with (S).

Date	Body length	A-Ph	A-G
25.11.90:	8.9	3.7	–
	15.8	4.5	–
	20.2	7.8	–
	20.2	6.3	–
	20.7	5.3	–
	21.4	6.1	–
12.01.91:	4.3	–	–
	5.7	–	–
	15.7	3.5	–
	18.1	6.3	–
	22.5	3.8	–(S)
04.02.91:	12.3+	–	–
	14.9	3.4	–
	16.6	7.8	–
	18.7	4.4	–
	21.5	5.4	–
	29.2	7.6	–
	32.9	6.6	–
17.02.91:	2.8	1.8	–
	3.6	2.4	–
	13.2	4.2	9.2
	14.0	3.4	–
	14.5	4.5	–
	15.9	3.8	–
	21.7	–	–
	22.6	–	–
12.03.91:	4.5	2.4	–
	6.9	2.5	–
	13.4	3.6	8.9 (S)
	17.7	4.7	–
	24.1+	–	–
23.03.91:	10.3	6.6	–
	21.5	5.2	–
	22.2	8.4	16.5 (S)
	24.6	6.1	17.4
01.04.91:	13.6	–	–
	17.9+	7.4	16.5*
	20.8	–	–
	22.3	4.8	13.0*(S)
28.04.91:	14.1	5.1	9.1

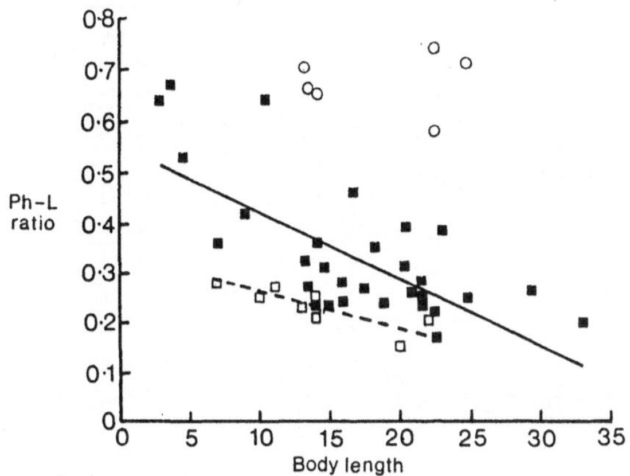

Fig. 1. Morphometric ratio of *Microplana termitophaga.* The Ph-L ratio plotted against preserved body length (mm) of worms from Zimbabwe (solid squares, solid line. $y = 0.55 - 0.0134x$, $r = 0.707$ ($p < <0.01$)) and of the type specimens from Kenya (open squares, dashed line. $y = 0.33 - 0.0072x$, $r = 0.84$ ($p < 0.01$)). Kenyan data from Table 1 of Jones *et al.*, (1990). The G-L ratios (open circles) are of the six mature specimens from Zimbabwe.

circular muscle are discrete longitudinal muscle fibres. These are visible in histological sections of specimens from Zimbabwe and in the TEM sections of the Kenyan specimen (Figs 2A, B).

Sexual maturity

In 7 of the 40 specimens from Zimbabwe some development of the copulatory apparatus was evident after clearing. In 2 of these, sperm stored in the male ducts was clearly visible after clearing. There is no correlation between the length of mature worms and the G-L ratio (Fig. 1). The G-L ratio of the 6 intact (one was damaged posteriorly) mature specimens is 0.67 (±0.05 s.d.) so that the copulatory apparatus is typically two-thirds of the way along the body. Though this is a small sample, more sexually mature individuals were found on later dates, towards the end of the wet season (Table 1).

Female reproductive system

There is a single pair of ovaries near the anterior end, and the two ovovitelline ducts run posteriorly adjacent to the ventral nerve cord on each side (Figs 2C, 3A). Vitelline glands are ventral to the gut diverticula and presumably discharge into the ovovitelline ducts. Posterior to the copulatory apparatus and gonopore, the ovovitelline ducts converge and unite (Fig. 3B). The common female duct continues anteriorly for a short distance to enter the common atrium but is joined by a short duct which connects postero-dorsally to the posterior anastamosis of the gut rami. This is a genito-intestinal duct (Figs 2D, 3B). There is no bursa present in the genito-intestinal canal.

Male reproductive system

The numerous testes are ventral and several may be seen in a single transverse section (Figs 2C, 3A). Testes occur post-pharyngeally to just in front of the copulatory apparatus. It is not possible to certainly distinguish the vas deferens except near the copulatory apparatus where each expands to form a seminal vesicle on each side and, in some specimens, may be full of ripe sperm. Here they can be seen to run dorsal to each ventral nerve cord, just lateral to the oviduct. The vas deferens on each side converge at a small papilla which discharges into a bulbous ejaculatory duct. This has a muscular wall and contains foliaceous (in section) processes projecting into the cavity (Figs 2D, 3B). The penis discharges though an apparently blunt papilla into the common atrium.

Discussion

Immature specimens from Zimbabwe and Kenya are indistinguishable in feeding behaviour, external features and internal anatomy as seen in histological transverse sections. Thus the Zimbabwe specimens are also *Microplana termitophaga.*

Jones *et al.* (1990) calculated the mean Ph-L ratio for the available specimens from Kenya as 0.23. However, it is clear from the larger sample and size range of specimens from Zimbabwe that the Ph-L ratio changes from small to large animals (Fig. 1). Co-plotting of results (Fig. 1) shows that the Ph-L ratio of the type specimens from Kenya also alters with size but with a slightly different relationship. The significance of this difference between specimens from the two localities is not clear. It is unlikely to be the result of difference in fixation, for fixation was carried out by flooding live worms with 70% alcohol in both cases. Even if fixation conditions were not identical in the two samples, the Ph-L ratio would be expected to be similar since it is a ratio and not an absolute distance. The change in Ph-L ratio is the result of disproportionate growth behind the pharynx as animals increase in size so that in small specimens the pharynx is nearer the rear, but nearer the front in larger specimens. The anterior end also grows but to a lesser extent. This change with growth suggests the use of the mean Ph-L ratio particularly over

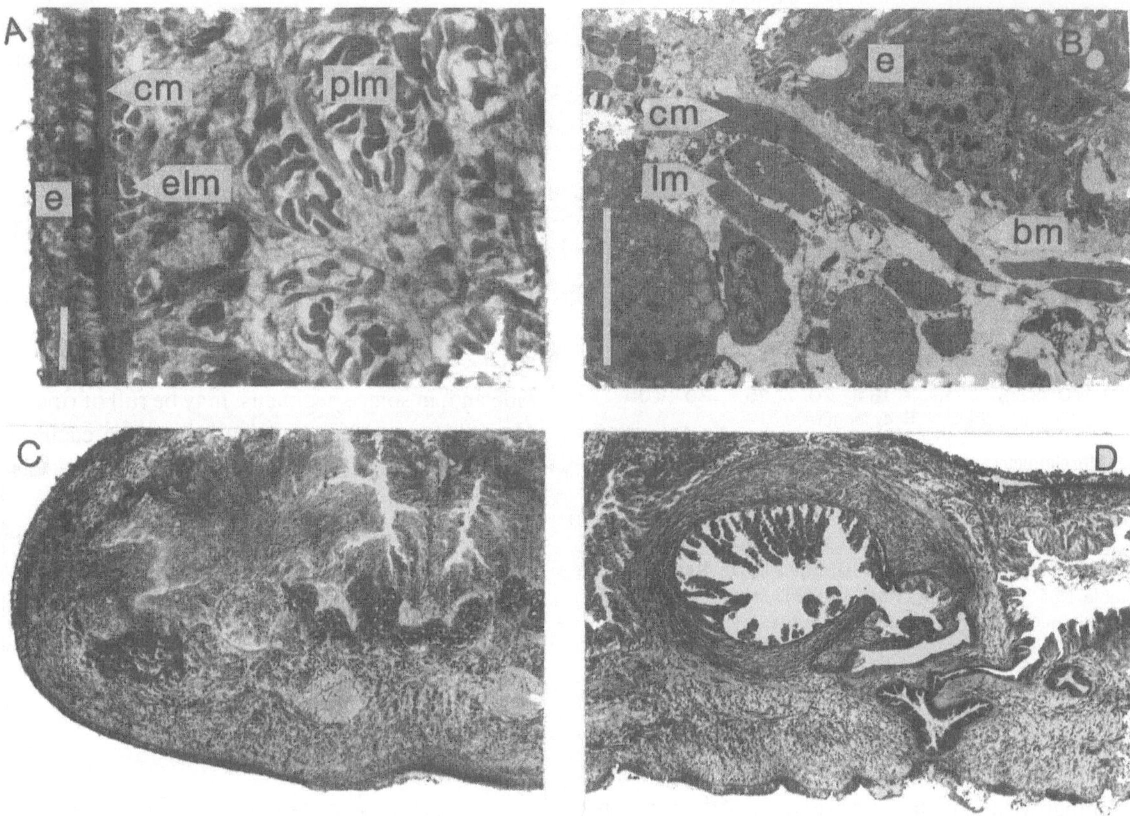

Fig. 2. A – micrograph of a portion of a transverse section of a Zimbabwe specimen showing the sub-epidermal circular and longitudinal muscle fibres. Scale bar 20 μm. B –transmission electron micrograph of a portion of a transverse section of a specimen from Kenya. Scale bar 5 μm. C – micrograph of a transverse section of a mature specimen from Zimbabwe (collection date 13.3.91). For scale see Fig. 3A. D – micrograph of a longitudinal section of the copulatory apparatus in the mid-line of the same specimen. For scale see Fig. 3B. The entire length of the genito-intestinal canal can be seen, with the joining oviducts just to the posterior. Key: e = epidermis; bm = basement membrane; cm = circular muscle; elm = sub-epidermal longitudinal muscle; lm = longitudinal muscle fibres; plm = parenchymal longitudinal muscle.

a considerable size range of specimens may be of little value.

The presence of a thin layer of discrete sub-epidermal longitudinal muscle fibres, not aggregated into bundles, confirms that the species is of the sub-family Microplaninae, rather than the Rhynchodeminae where the sub-epidermal longitudinal muscles are aggregated into bundles. The presence of a genito-intestinal canal in the Zimbabwe specimens confirms that the species is of the genus *Microplana* Vejdovsky, 1890.

Nothing is known of the reproduction of *M. termitophaga*. The occurrence of mature specimens towards the end of the wet season (March–April) suggests that period as the breeding season. From the admittedly small sample number, it appears that size alone does not determine maturity since the two largest specimens were immature, nor is there any correlation between size and maturity (Table 1, Fig. 1). No egg capsules belonging to *M. termitophaga* have been found. It is possible that the worms may also reproduce by fission and regeneration.

Whether the worms survive the 8 or 9 month dry season as worms, eggs or both is not known, though large worms are seen very early in the wet season which suggests that worms do survive the dry season. This does not exclude the possibility that egg capsules may also survive the dry season. The occurrence of very small worms over a particular period might indicate hatching of egg capsules, but it would require systematic recording of worm sizes by date to positively determine this. The collections documented here were simply initial dated collections and not intended to be representative of the size range of specimens on each occasion. A further complication is that rainfall in Zimbabwe is erratic even during the

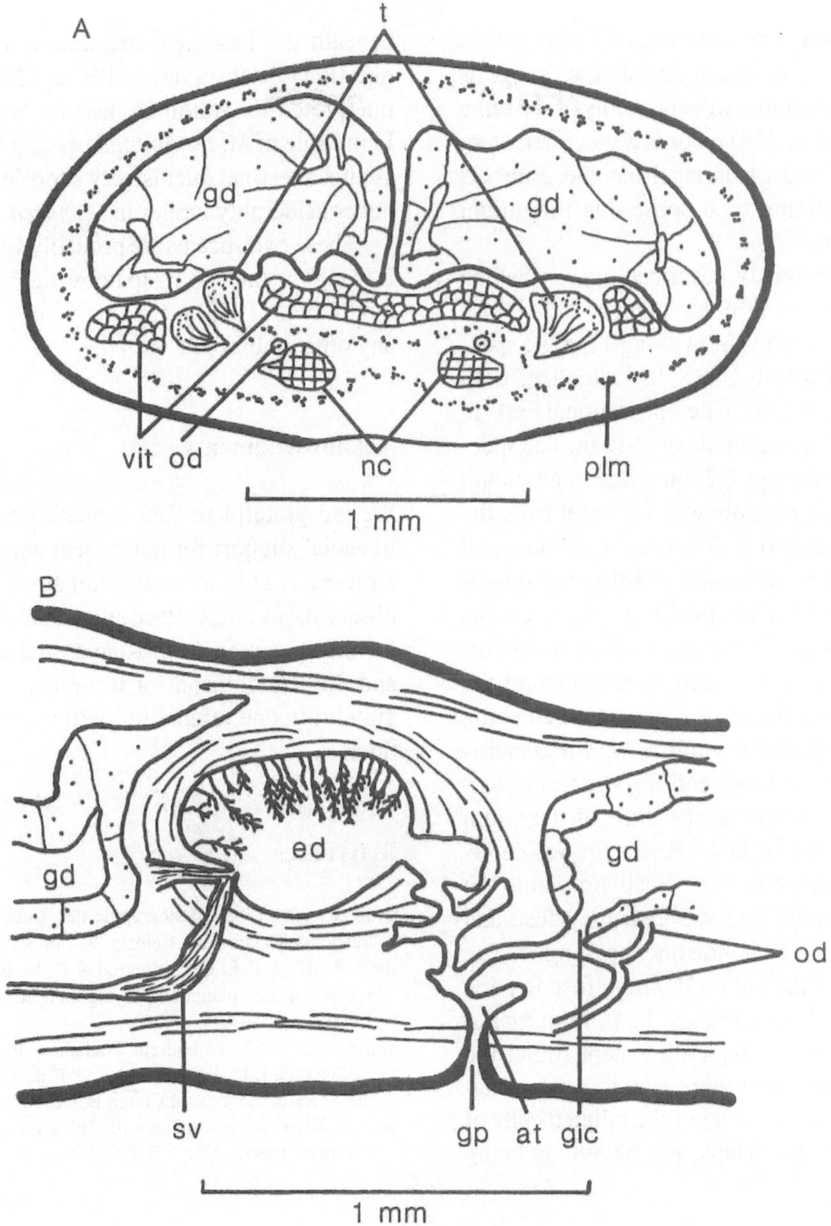

Fig. 3. A – interpretation of Fig. 2C of a mature specimen from Zimbabwe. B – diagrammatic reconstruction of the copulatory apparatus of the same specimen based largely on Fig. 2D. Key: at = common atrium; ed = ejaculatory duct; gd = gut diverticula; gic = genito-intestinal canal; gp = gonopore; nc = nerve cords; od = ovovitelline duct; plm = parenchymal longitudinal muscle; sv = seminal vesicle; t = testis; vit = vitelline gland cells.

wet season. There are thus discontinuous and unpredictable periods of feeding and growth for the worms and reproduction may vary considerably from year to year. The worms may occasionally have to survive over two years due a drought year. In 1992 a severe drought affected southern Africa and relatively few worms were seen in Harare over the normally active period. In 1993 there was a more normal wet season and plenty of worms were seen.

Sheppe (1970) described the feeding behaviour of similar worms feeding on termites in Zambia. The worms were thought to be geoplanids but were not properly identified and preserved specimens are lost (Jones *et al.*, 1990). Geoplanids in any case do not naturally occur in Africa. The occurrence of *Microplana termitophaga* from neighbouring Zimbabwe suggests that the worms from Zambia were probably of the same species. Graff (1899, p. 254) records a specimen of an unidentified black land planarian from the Zambezi region and it is tempting to suppose that this might have been *M. termitophaga.*

There is one previously described species which is similar to *Microplana termitophaga. M. harea* was described by Marcus (1953) from a single mature specimen found in what is now Zaire. It was collected on 18th March 1948 at Mukana, Upemba National Park, in 'forét marécageuse' at an altitude of 1810 m. The specimen was 15 mm long and 1.2 mm wide. The mouth was 7.5 mm and the gonopore was 11.5 mm from the anterior end (Ph-L ratio 0.5, G-L ratio 0.77). General colouring, musculature and testes position are similar in the two species. The main differences between this specimen and *M. termitophaga* are: 1. The central dorsal dark stripe of *M. harea* starts further behind the eyes than in *M. termitophaga.* 2. The ventral creeping sole occupies about half the width in *M. harea* and is narrower towards the anterior end. In *M. termitophaga* the creeping sole is about one third of the width along the whole length. 3. In *M. harea,* the vas deferentia enter and discharge into the ejaculatory bulb via separate papillae. In *M. termitophaga* they fuse and discharge singly through an indistinct opening. 4. The two posterior rami of the gut of *M. harea* fuse in parts to form several broad anastomoses. In *M. termitophaga* the posterior rami are separate except round the copulatory apparatus where there is a narrow anastamosis. 5. In *M. harea* the nuclei of the ciliated cells of the sole are stated to be insunk and figured as being beneath the basement membrane. (This has to be a questionable observation. Perhaps Marcus (1953) misinterpreted longitudinal muscle fibres in this region.) In the sole of *M. termitophaga* nuclei are distal. 6. The genito-intestinal duct is very short in *M. termitophaga* but considerably longer in *M. harea.*

These two species are probably very closely related. It would be interesting to know the habits of *M. harea,* but Marcus (1953) makes no mention of the feeding or any other habit of *M. harea.*

Acknowledgments

We are grateful to The Lindeth Charitable Trust for financial support for equipment and travel which has allowed H.D.J. to visit Zimbabwe to make further observations and collections; to Dr Robin M. Newson for the specimen from Kenya; and to Jackie Hallows and Sue Hutchinson for technical assistance. We are grateful to one referee for particularly helpful suggestions.

References

Graff, L. von, 1899. Monographie der Turbellarien. II. Tricladida Terricola. Engelmann, Leipzig, 574 pp + atlas.

Jones, H. D., J. P. E. C. Darlington & R. M. Newson, 1990. A new species of land planarian preying on termites in Kenya. J. Zool., Lond. 220: 249–256.

Marcus, E., 1953. Turbellaria Tricladida. In Exploration du Parc National de L'Upemba, Mission G. F. de Witte. Fasc. 21. Institut des Parcs Nationaux du Congo Belge, Brussels.

Sheppe, W., 1970. Invertebrate predation on termites of the African savanna. Insectes Soc. 17: 205–218.

Hydrobiologia **305**: 127–133, 1995.
L.R.G. Cannon (ed.), *Biology of Turbellaria and some Related Flatworms.*

Comparison of the nervous system of the rhabdocoel *Mesostoma ehrenbergii* with that of the polyclad *Notoplana acticola*

Harold Koopowitz, Mark Elvin & Tony Bae
Ecology and Evolutionary Biology, University of California, Irvine, CA 92717, USA

Key words: comparative neuroanatomy, neuronal evolution, neuron morphology, primitive nervous systems, Turbellaria

Abstract

The neuromuscular organization of the free-living marine polyclad *Notoplana acticola* (Boone) can be compared with that of the fresh water rhabdocoel *Mesostoma ehrenbergii* (Focke). These examples act as outgroups for each other, belonging to different clades of the Turbellaria and similarities between the two can be taken as synapomorphies to indicate features that might be primitive and general features of flatworm nervous systems. The fluorescent dye Lucifer Yellow was used to fill cells in the brains of the two species and the neuronal anatomies were photographed or drawn. Many different cell types have been found in both species. The predominant cell types in *Mesostoma* were heteropolar bipolar and heteropolar multipolar cells with isopolar bipolar and isopolar multipolar cells in much lower numbers; only a few unipolar cells were found. In *Notoplana*, many of the cells penetrated were also heteropolar bi- or multipolar cells. There were also neurons and architectural features that were morphologically very similar to each other, in the two animals, such as the BRA cells and commissural tracts. These shared features suggest that many of the 'peculiarities' of polyclad nervous systems may be general flatworm features.

Introduction

There now appéars to be some agreement that the Platyhelminthes emerged close to the base of the metazoan tree and that it was probably one of the earliest phyla to evolve (Barnes *et al.*, 1988, Hori *et al.*, 1988). However, the exact relationships between the Platyhelminthes and other phyla are still debatable. It is often assumed that the flatworms represent the most primitive extant animals to possess both discrete muscles and a central nervous system. If this is true one could argue that generalizations about their neuromuscular arrangements would give insight into the earliest evolution of neuromuscular systems. On the other hand, as the Turbellaria probably have had a very long evolutionary history, they may have had time to diverge from the primitive condition and thus give a very false image of that early evolutionary phase.

Investigations of the nervous system of the polyclad *Notoplana acticola* revealed an abundance of heteropolar neurons in the rind of the brain that resembled chordate neurons (Koopowitz, 1986; Koopowitz,

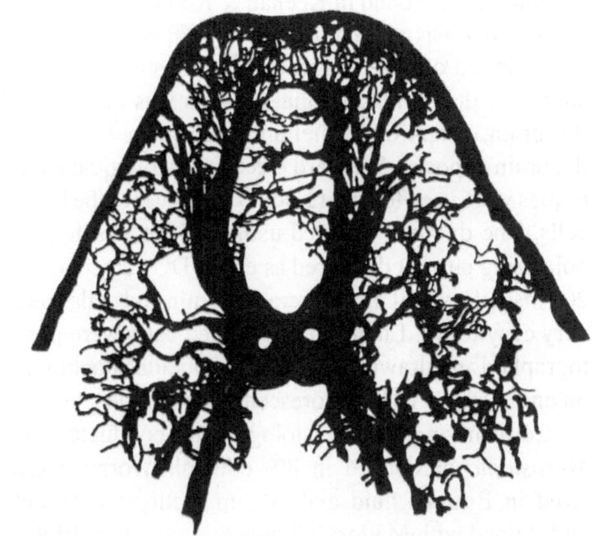

Fig. 1. Serial reconstruction of the brain and anterior nerve plexus of *Mesostoma ehrenbergii* from 40 mμ light microscopic sections. The white spots in the brain represents the position of the two eye spots inside the brain.

1989). Reports of this type of nerve cell in other flatworm clades are rare (Bullock & Horridge, 1965; Hanström, 1926). Most reports indicate a majority of unipolar cells (e.g. Baguñá & Romero, 1981) that are typical of other protostomous invertebrates. The differences between polyclads and other flatworm taxa could either be due to histological artifacts or real differences in the make-up of their nervous systems. This can be answered by comparing the polyclad nervous system with that of another clade. The fresh water rhabdocoel, *Mesostoma ehrenbergii* is of similar size to *Notoplana acticola* and provides a useful species for comparison. These two examples act as outgroups for each other, belonging to different clades of the Turbellaria and similarities between the two can then be taken as synapomorphies or plesiomorphies to indicate features that might be primitive or general features of flatworm nervous systems. Here we report on the results from such a comparison using injections of Lucifer Yellow into live neurons in intact living worms to reveal their fine anatomy.

Materials and methods

Notoplana acticola were collected from the intertidal zone at San Onofre, California and maintained in the laboratory in finger bowls of sea water until used. The *Mesostoma ehrenbergii* were originally collected in Ponoka, Central Alberta, Canada and cultured in our laboratory. Methods for handling and filling *Notoplana* neurons can be found in Keenan & Koopowitz (1984). *Mesostoma* was anesthetized using 8% ethanol. They were pinned out on Sylgard resin dishes using cactus spines. A dorsal longitudinal incision was made over the brain. Flaps of epithelium were pinned out and the brain exposed. Standard electrophysiological techniques were used to inject Lucifer Yellow into the brain cells. The dye was injected using 1–10 nA of hyperpolarizing current delivered as either DC current or as 200 ms pulses. Fill times averaged 5 min and cells were very easy to fill. Lucifer Yellow filled cells were photographed and drawn using a camera lucida attachment on an Olympus BH-2 fluorescent microscope.

Conventional wax histology was conducted on *Mesostoma* narcotized in 8% ethanol. Worms were fixed in Bouin's fluid and 40 μm sections were cut and stained with Mallory's Triple Stain (Pantin, 1960). Serial sections of the brain and nervous system were drawn using a camera lucida. Cerebral cell nuclei were also drawn for each section. These were then tallied for whole brains to get entire cerebral cell counts.

Results and discussion

Peripheral nervous system

The nervous system in Turbellaria ranges from diffuse networks to complex arrangements (Rieger *et al.*, 1991). The polyclad, *N. acticola*, for example has several plexes. The two major longitudinal nerve trunks form part of the ventral plexus and have a series of commissures that run between them resembling the rungs of a ladder. Within the plexus, nerve trunks join and anastomose to form a reticulation and a finer network of nerve fibers is to be found between that meshwork of plexus trunks. In *Mesostoma*, most illustrations of the nervous system show branching nerves but not an anastomosing net work as is so obvious in the polyclads. Using silver stains Bresslau & von Voss (1913) found a reticulation of fine fibers which they suggested were part of a nerve net. Using conventional light microscopy we have found an anastomosing network (Fig. 1) which is similar in many respects to the submuscular plexes seen in *Notoplana* but is much less well developed. Between the large branches of this plexus are finer fibers reminiscent of those described by Bresslau and von Voss. This network is not apparent in living material. In both animals we found cell somata embedded in meshes of the plexus.

In the region of the pharynx, there is a single commissure between the two main longitudinal trunks that emerge from the ventral part of the brain. Within the pharynx is a ring of five putative dopaminergic cells (J. Romero, unpublished). Whether or not these cells are those that comprise the pharyngeal nerve net reported by Göltenboth & Heitkamp (1977) is uncertain.

Central nervous system

Polyclads possess a single large central ganglion that is protected by a well developed but acellular proteinaceous sheath. The brain contains large swirls of neuropil. Golgi staining (Keenan *et al.*, 1981) and intracellular fills with the fluorescent stain Lucifer Yellow (Keenan & Koopowitz, 1984) showed that many of the cells in the brain of *Notoplana acticola* were heteropolar multipolar neurons. In addition to multipolar cells, a few unipolar cells similar to conventional invertebrate neurons were found and there were also a small number of isopolar bipolar cells. While most somata

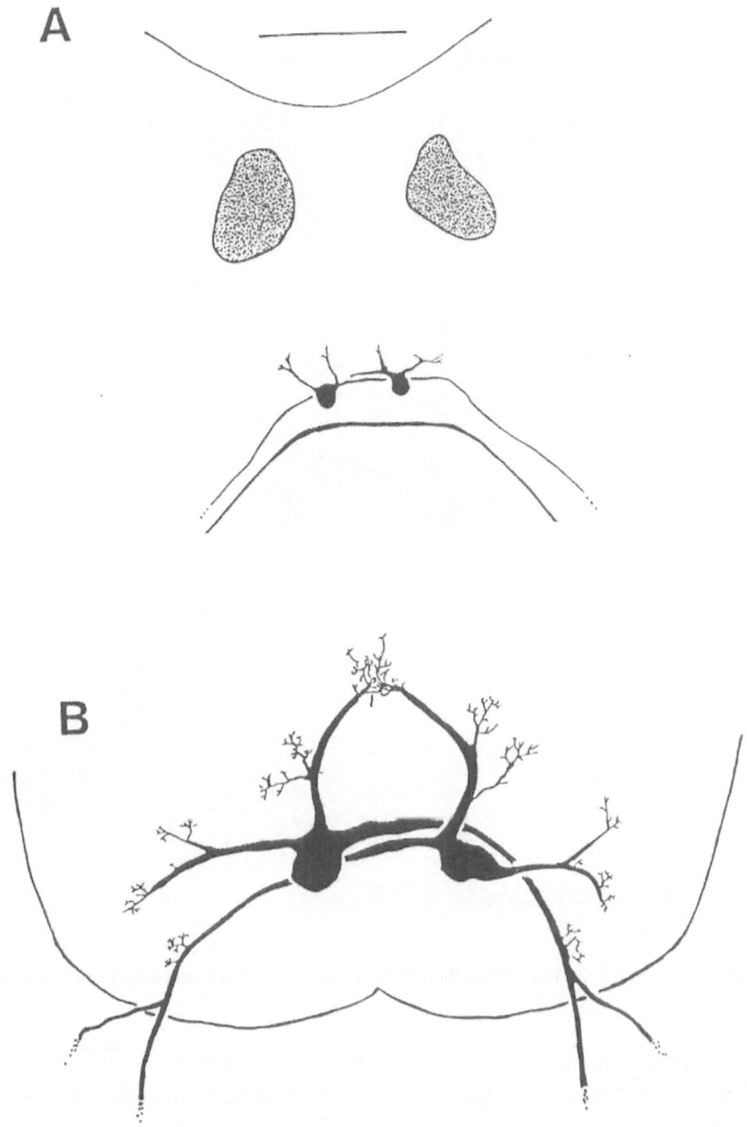

Fig. 2. Dye-coupled multipolar heteropolar cells pairs in the posterior part of the brain of *Mesostoma* (A) and *Notoplana* (B). The anterior and posterior margins of the brain are indicated in all of the *Mesostoma* drawings. The top of the page represents the anterior of the animal in all figures and all isobars represent 100 μm, the position of the eye spots are indicated by the dotted areas.

were arranged in a rind around the neuropil in typical invertebrate fashion, some *Notoplana* cerebral neurons had cell bodies buried deeply within the neuropil.

Unlike the polyclads, but similar to many other Turbellaria, *Mesostoma* brains lack a surrounding ganglion sheath. Whether or not this is the primitive condition or a simplification correlated with reductions in body size is unclear. The presence of glial wrappings around cell somata is known from *Notoplana* (Koopowitz, 1989) but ultrastructural investigations of *Mesostoma* nervous system are yet to be performed. Like *Notoplana*, *Mesostoma* has both an apparent rind of somata around the brain as well as many sunken somata bearing well developed neurites that occur within the neuropil.

Neuronal architecture

Lucifer Yellow fills of *Mesostoma* show a wide array of cell types many of which are similar to those that are also found in the polyclads. Among these are dye-

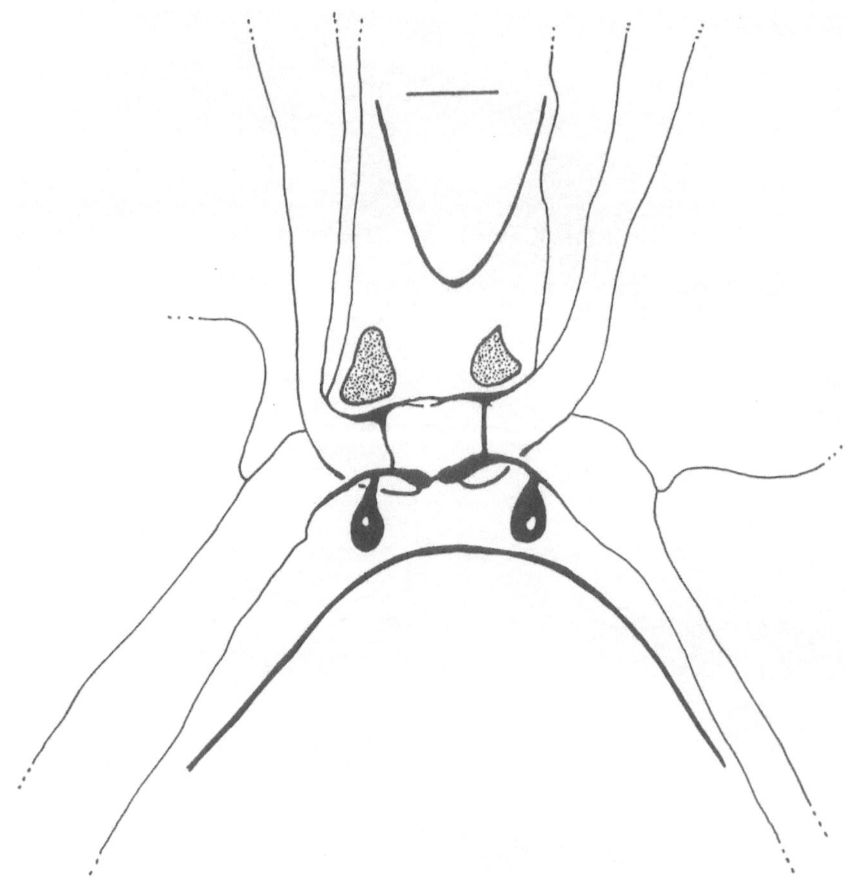

Fig. 3. A unipolar pair of dye-coupled cells in Mesostoma. Note that this pair of cells have very few short neurite branches.

coupled multipolar heteropolar cells pairs in the posterior part of the brain. One pair of these cells (Fig. 2) closely resembles the heteropolar multipolar BRA cells found in both of the polyclads, *Alloeoplana californica* (see Solon & Koopowitz, 1982) and *Notoplana acticola* (see Keenan & Koopowitz, 1984). Other dye-coupled neuron pairs, in *Mesostoma*, include smooth unipolars (Fig. 3), as well as some pairs of cells with more complex configurations.

In addition, there are a variety of bipolar cells some of which are isopolar (Fig. 4a) and others of which are heteropolar bipolars (Fig. 4b and c). Bipolar cells appear to be much more common in the central nervous system of *Mesostoma* than in the brain of *Notoplana*. In the rhabdocoel, many of the isopolar bipolars are single median cells with their somata positioned in the center of the brain. Cells like that have not been found in the polyclads. The median isopolar bipolars tend to occupy a central-medial to central-ventral cerebral position in *Mesostoma* and a similar tract occurs in the ventral central region of the brain of *Notoplana*. In the polyclad the cells tend to have their somata to the side and axons are extremely wide (Keenan *et al.*, 1981). In Göltenboth & Heitkamp's (1977) drawings, of the brain of *Mesostoma*, these tracts also extend through the ventral central region of the worm. It is possible in fact that similar cells are yet to be found in both worms. In both animals the tracts can be likened to commissures linking opposite sides of the brain or body and in both cases the somata appear to be attached to 'giant' axons.

We find that the preponderance of cell types in the two flatworm brains are heteropolar, but unipolar cells reminiscent of generalized invertebrate neurons also occur. Some of the unipolar neurons in the vicinity of the main posterior longitudinal nerve trunks have 'giant' cell bodies, 43.76 μm in diameter, which gradually taper over a distance of 100 μm (Fig. 5). In contrast to our findings, most earlier workers describe the cells in the brain of flatworms as being mainly

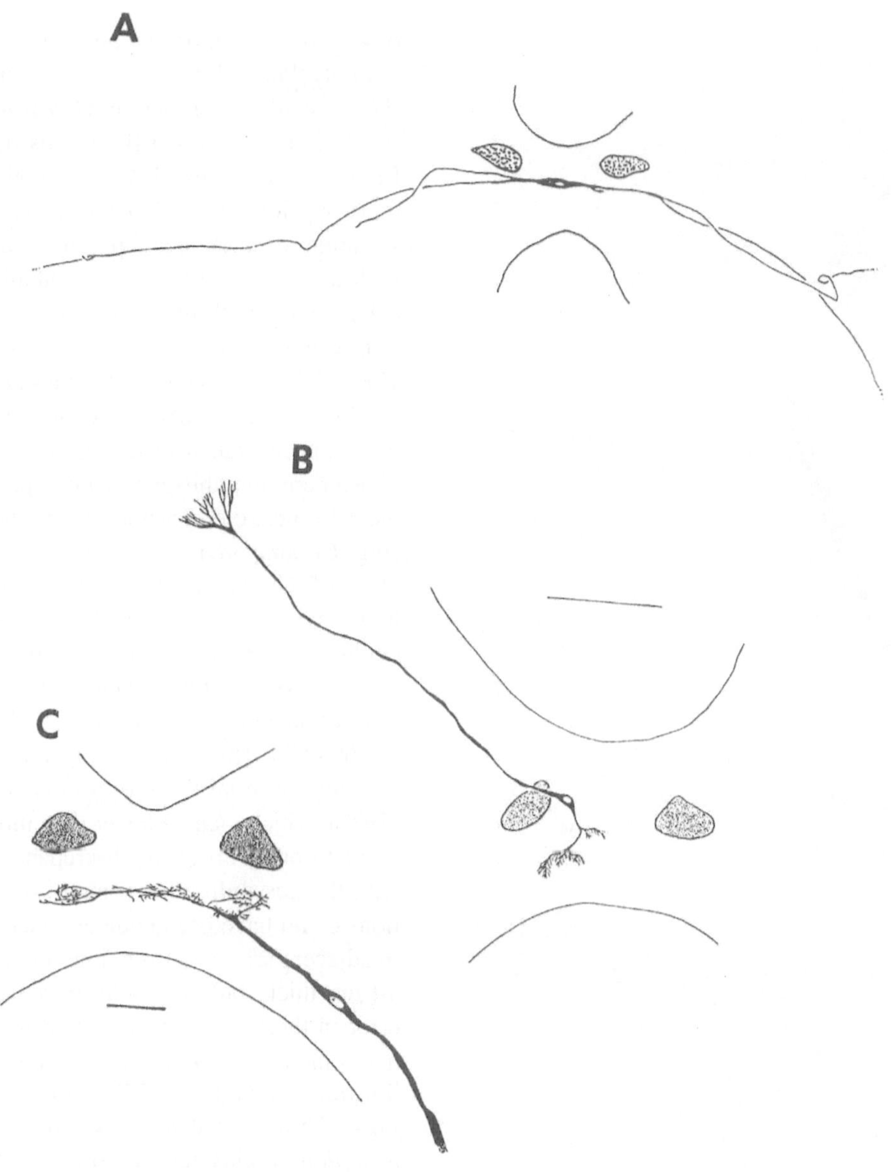

Fig. 4. Bipolar cells in the brain of *Mesostoma*. *A*) A medially located isopolar bipolar neuron. *B*) A heteropolar bipolar cell with an anterior projection. *C*) A heteropolar bipolar neuron with a major posterior projection. Both heteropolar bipolar cells have short fine branches which are presumably dendritic in the area of posterior neuropile.

unipolar cells (Bullock & Horridge, 1965). This was also true of both species that we have investigated, *Notoplana* (syn. *Leptoplana acticola* (Turner, 1946)) and *Mesostoma* (see Göltenboth & Heitkamp, 1977). The apparent abundance of unipolar cells in traditionally fixed light microscopic sections of flatworm brains can be explained in two ways. Firstly, tissue shrinkage during fixation and histological preparation could lead to a rounding of the cytoplasm and severance of fine neurites attached to the soma. Secondly, investigators expected to see unipolar somata and did not observe the cell bodies closely enough to find neurites coming off of them or when seen they were dismissed as artifacts. We find that careful observation usually reveals a few somata with fine branches and on fortunate occasions we have had preparations where the majority of somata appeared to have branches.

In both *Mesostoma* and *Notoplana* repeated fills of the same cell in different individuals were easy to obtain. Identified neurons with constancy in morphol-

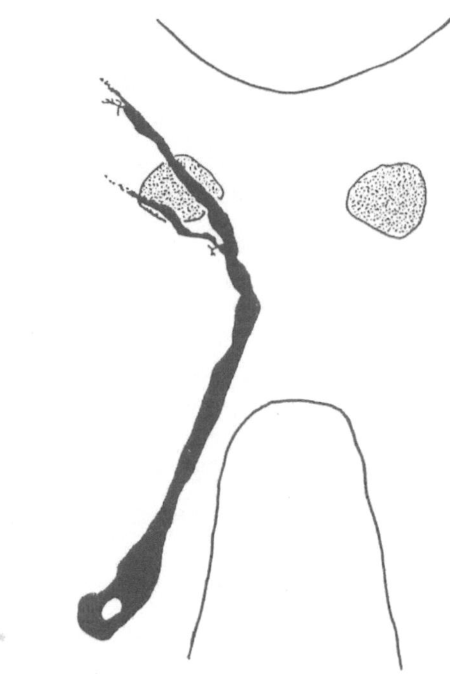

Fig. 5. A unipolar neuron, from the brain of *Mesostoma* in the vicinity of the origin of the main posterior longitudinal nerve trunk which has a 'giant' cell bodies over 40 μm in diameter that gradually tapers over a distance of 100 μm.

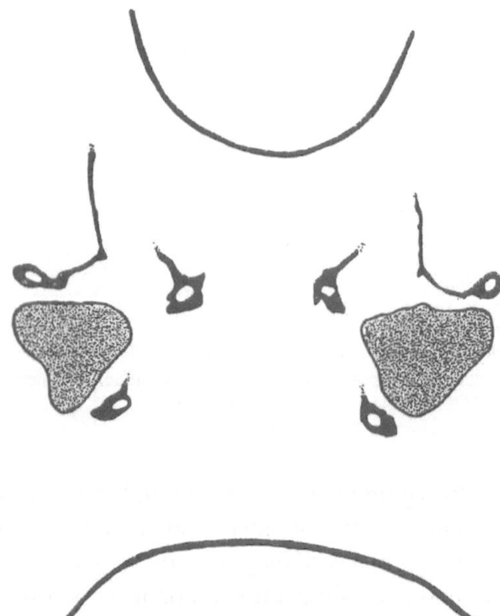

Fig. 6. The three pairs of putative dopaminergic cells found in *Mesostoma* brain situated around the eye spots (drawing from a photograph taken by J. Romero, unpublished).

ogy, cellular position and physiology are known from the polyclad and probably also occur in *Mesostoma*. We have filled neurons in *Mesostoma* brains which resemble the identified BRA cells of *Notoplana* (see Elvin & Koopowitz, 1994) but nothing is known of their physiology in *Mesostoma*. There is also some variation in function and possibly number of neurons with time in *Notoplana* where putative dopaminergic cells change both their number and position in relation to the size of the animal (Hauser & Koopowitz, 1987). When the same techniques are attempted with *Mesostoma*, the number of dopaminergic cells appears to remain constant irrespective of the age of the animal (J. Romero, unpublished). Only 3 pairs of dopaminergic cells, near the eye spots, were found in *Mesostoma* (Fig. 6) compared to the 20–28 pairs in *Notoplana*. None of the dopaminergic cells in *Mesostoma* can be homologized with those of *Notoplana* but once again could be interpreted as identified neurons.

Counts of the total cell nuclei, taken from serial sections of the brain of *Mesostoma*, yielded approximately 560 cells. This can be compared with counts from drawings of serial sections by Göltenboth & Heitkamp (1977), which yielded fewer than three hundred putative neurons. Part of the discrepancy may have to do with the possibility that some of the cells in our sections could have been counted twice if they appeared in adjacent sections. Both their and our sections were 40 μm thick, but one might have expected a similar error in their work. We also find that *M. ehrenbergii* has some cells that are very large but these were not illustrated in Göltenboth & Heitkamp's (1977) drawings of brain serial sections. The number of cells in *Notoplana* brains appears to be much greater than in *Mesostoma* but total brain neuron counts have not been made for *Notoplana*.

Conclusions

Until recently, it was not known if the neuronal types found in polyclads were peculiar to that order or if they reflected a general turbellarian condition. It is now clear that the heteropolar condition found in polyclads may be common to the Rhabditophora clade or at least those branches uniting the polyclads and rhabdocoels. Comparisons between *Mesostoma* and *Notoplana* show both similarities and differences between the two clades, in their brain structure. The similarities between the two include the presence of uni-, bi- and multipolar cells with a large percentage of these cell types being heteropolar. In addition both clades possess a similar arrangement of commissural tracts. The presence of very similar identified neurons such as the BRA cells argues very strongly for a common ancestral organism possessing that feature as well (Croll, 1986). Where there are differences, such as in the arrangements and numbers of dopaminergic cells, it is not clear what possible evolutionary relationships, if any, between the two situations could exist.

Previous work on polyclad neurons, pointed to the common occurrence of heteropolar neurons of both multi- and bi-polar types in the flatworms and the deuterostome phyla. Other similarities between acoel 'Turbellaria' and the deuterostomes, based on ciliary fine structure have been discussed (Rieger *et al.*, 1991). The range of cell types found in *Mesostoma* and *Notoplana* also argue for ties with the protostome phyla and perhaps the 'Turbellaria' should be sited at the base of the Bilateria similarly to the position of the flatworm-like common ancestor advocated by Barnes *et al.* (1988). If the flatworms actually straddled the phylogenetic tree before the divergence of protostomes and deuterostome lines, then one can assume that the protostomes eventually lost the preponderance of heteropolar bi- and multi-polar central neurons which now seems to be a property of turbellarian systems.

References

Baguñá, J. & R. Romero, 1981. Quantitative analysis of cell types during growth, degrowth, and regeneration in the planarians *Dugesia mediterrania* and *Dugesia tigrina*. In: Schockaert, E. R. and I. R. Ball, 'Turbellaria', Proc. Third Int. Symp. Hydrobiol. 84: 181–194.

Barnes, R. S. K., P. Calow & P. J. W. Olive, 1988. The Invertebrates: A new synthesis. Blackwell Scientific Publications, Oxford, 582 pp.

Bullock, T. H. & G. A. Horridge, 1965. Structure and Function in the Nervous Systems of Invertebrates, Vol. I. W. H. Freeman and Co., San Francisco, 798 pp.

Bresslau, E. & H. von Voss, 1913. Das Nervensystem von *Mesostoma ehrenbergii*. Zool. Anz. 43: 260–263.

Croll, R. P., 1986. Identified neurons and cellular homologies. In M. A. Ali (ed.), Nervous Systems in Invertebrates. Plenum, New York: 41–59.

Elvin, M. & H. Koopowitz, 1994. Neuroanatomy of the rhabdocoel flatworm *Mesostoma ehrenbergii* (Focke, 1983) I: Neuronal diversity in the brain. J. Comp. Neurol. (in press)

Göltenboth, F. & U. Heitkamp, 1977. *Mesostoma ehrenbergii* Plattwürmer (Strudelwürmer) Biologie, mikroskopische Anatomie und Cytogenetik. Grosses Zoologisches Praktikum, Gustav Fischer Verlag, Stuttgart, 6a: 1–60.

Hanström, B., 1926. Über den feineren Bau des Nervensystem der Tricladen Turbellarian auf Grund von Untersuchungen an *Bdelloura candida*. Acta zool., Stockh. 7: 101–115.

Hauser, M. & H. Koopowitz, 1987. Age-dependent changes in flourescent neurons in the brain of *Notoplana acticola*, a polyclad flatworm. J. exp. Zool. 241: 217–225.

Hori, H., A. Muto, S. Osawa, M. Takai, K. Lue & M. Kawakatsu, 1988. Evolution of turbellaria as deduced from 5S ribosomal RNA sequences. Prog. Zool. 36: 163–167.

Keenan, C. L. & H. Koopowitz, 1984. Ionic bases of action potentials in identified flatworm neurons. J. Comp. Physiol. 155: 197–208.

Keenan, C. L., R. Coss & H. Koopowitz, 1981 Cytoarchitecture of primitive brains: golgi studies in flatworms. J. Comp. Neurol. 195: 697–716.

Koopowitz, H., 1986. On the evolution of central nervous systems: implications from polyclad turbellarian neurobiology. Hydrobiologia 132 (Dev. Hydrobiol. 32): 79–87.

Koopowitz, H., 1989. Polyclad neurobiology and the evolution of central nervous systems. In A. V. Anderson (ed.), Evolution of the First Nervous Systems. Plenum, New York: 315–328.

Pantin, C. F. A., 1960. Notes on microscopic technique for zoologists. Cambridge University Press, Cambridge, 77 pp.

Solon, M. H. & H. Koopowit, 1982. Multimodal interneurons in the polyclad flatworm *Alloeoplana californica*. J. Comp. Physiol. 147: 171–178.

Rieger, R. M., S. Tyler, J. P. S. Smith & G. E. Rieger, 1991. Platyhelminthes: Turbellaria. In Harrison, F. W. & B. J. Bogitsh (eds), Microscopic Anatomy of Invertebrates, Vol. 3: Platyhelminthes and Nemertinea, Wiley-Liss Inc., New York: 7–140.

Hydrobiologia **305**: 135–139, 1995.
L.R.G. Cannon (ed.), *Biology of Turbellaria and some Related Flatworms.*

Glyoxylic acid induced fluorescence in the nervous system of *Gyratrix hermaphroditus* (Kalyptorhynchia, Polycystididae)

Elena A. Kotikova
Zoological Institute of the Russian Academy of Sciences, 199034, University emb. 1, St. Petersburg, Russia

Key words: nervous system, catecholamines, Kalyptorhynchia, Gyratricinae, Plathelminthes

Abstract

Catecholamines (CAs) are found in the neuropile of the brain, in 3 pairs of longitudinal nerve cords, in the transverse ventral commissure, in anterior ventral and dorsal nerves, in two pharyngeal nerve rings and in 24 neurons in the nervous system of *Gyratrix hermaphroditus*. The CA distribution pattern is compared with those of other neuroactive substances. Homology of neurons in the family of Polycystididae and in Plathelminthes in general is discussed.

Introduction

Three components: cholinergic, peptidergic and aminergic has been revealed in the nervous system of Plathelminthes (for reviews see Gustafsson & Reuter, 1992). The glyoxylic-acid-induced-fluorescence (GAIF) method has proven the most sensitive in the investigation of catecholamines (CAs) (Joffe & Kotikova, 1989a, b; Reuter & Eriksson, 1991; Gustafsson & Eriksson, 1991). Considering the principle of initial morphological diversity developed by Mamkaev (1986), the comparison of representatives of both distant systematic groups and closely related ones such as a family and a subfamily remains realistic.

The present aim was to locate CA-positive structures in the nervous system of *Gyratrix hermaphroditus* (family Polycystididae, subfamily Gyratricinae) and to analyse the morphological peculiarities of the distribution of nervous elements with different neurotransmitters in Gyratricinae and in Polycystididae as a whole.

Material and methods

Specimens of *G. hermaphroditus* were collected from the Jaroslavl Region (Russia) in a shallow swampy artificial reservoir. CAs were revealed by the GAIF method according to a slightly modified conventional method (Kabotyankyj, 1985) with a fivefold concentration of glyoxylate ions in the incubation medium. Preparations were examined in a fluorescence microscope LUMAM R-3. Photographs were made on RF-3 film.

Results

Green fluorescence indicates the presence of CA in *G. hermaphroditus*. The pattern of CA-ergic nervous system is distinct (Fig. 1). The brain is found in the eye region and has the shape of an arch with its edges dropped down. The central part of the brain is twice as thick as its marginal parts. Inside the brain there is a dense network of nervous fibres oriented both in longitudinal and transverse directions. The number of brain neurons is ten (Fig. 2A). They are all unipolar, spindle-shaped, 8–9 μm long. The neurons are distributed fan-like. Three pairs of neurons lie in front of the brain, and two pairs at the level of the neuropile. A small number of nerve fibres pass through every pair of the longitudinal cords. Along these cords many varicosities are observed. The largest of these are probably small neurons not exceeding 4 μm in size. The ventral cords originate from the lateral parts of the brain. Behind the pharynx the ventral cords bend sharply towards the body centre, forming a V-like structure (Fig 2 B). At this level they are connected with a single transverse

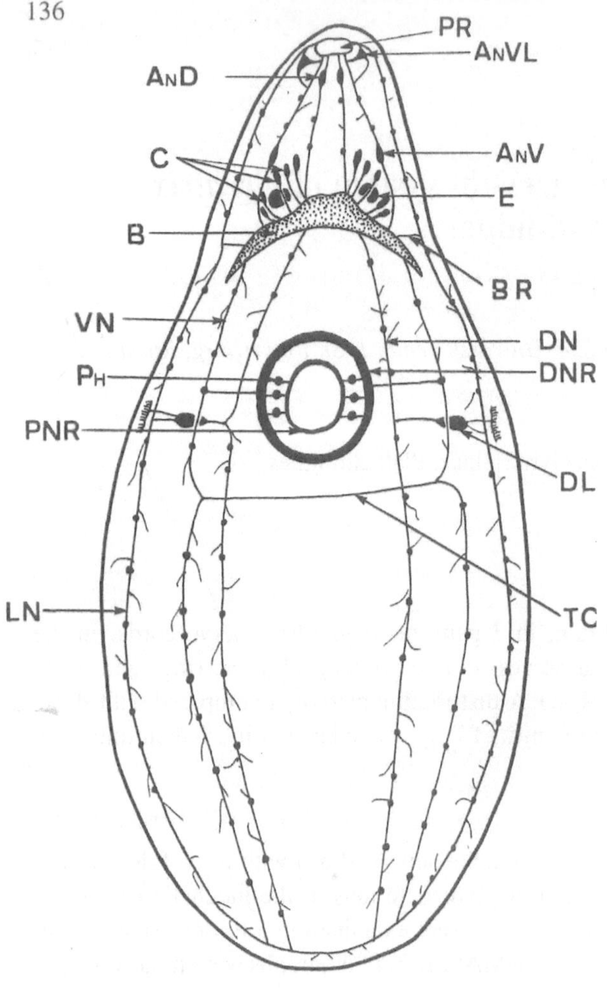

Fig. 1. Schematic drawing of the catecholaminergic nervous system revealed by GAIF method in *Gyratrix hermaphroditus*. View from the ventral side. Anterior neurons (AnD, AnV, AnVL); brain (B); brain rootlet (BR); brain cells (C); neuron at pharynx level (DL); dorsal nerve cord (DN); distal pharyngeal nerve ring (DNR); eye (E); lateral nerve cord (LN); pharyngeal neurons (Ph); proximal pharyngeal nerve ring (PNR); proboscis nerve ring (PR); transverse commissure (TC); ventral nerve cord (VN).

commissure. The lateral cords are connected with the brain by a thin brain rootlet, whereas the dorsal cords are independent and pass above the brain (Fig. 1). At the posterior end of the body the lateral cords are interconnected. At the anterior end they reach the thin ring nerve surrounding the proboscis sheath. A pair of ventral nerves containing one bipolar round neuron (*AnV*) branch off in an anterior direction from the anterior-lateral parts of the brain. These perikarya lie at the level of the three anterior pairs of brain neurons, they have a round shape pointed at the ends, their size is 7–9 μm (Fig. 2 C,D). The pharynx is innervated with two nerve rings: a strong distal one 4–5 μn thick and a weak proximal one up to 1.5 μm thick. Between these rings lie 6 bipolar round neurons 5–6 μm in size (Fig. 2 E). Lateral parts of the distal pharyngeal ring are connected with the longitudinal ventral cords through a pair of short pharyngeal nerves.

Near the anterior end of the body a pair of *AnVL* neurons forming an equilateral triangle 6–8 μm high is detected (Fig. 2 F). Two processes of these neurons pass to the ventral nerve, whereas the third one more often extends to the proximal part of the lateral cord, and very seldom to the thin nerve ring of the proboscis sheath. The anterior dorsal nerves deviating from the median brain region also have one bipolar spindle-like perikaryon (*AnD*) 8 μm long and 3 μm wide lying near the anterior end of the body. The dendrites of neurons *AnV* and *AnD* branch off at the anterior part of the body, forming a dense network of CA - positive fibres, with numerous small varicosities (Fig. 2 C,D). At the level of the pharynx a pair of lateral droplike DL neurons 7–8 μm long is found. Their main projections enter the dorsal cords, while the second ones join the lateral cords and the short, third ones end with numerous terminals coming ventrally from the lateral cords (Fig. 2 B). Many thin nerve fibres branch off from the longitudinal nerve cords (Fig. 2 G) forming a submuscular plexus. No neurons were revealed in the plexus, corroborating EM investigations (Reuter & Lindroos, 1979).

According to their shape, chemistry and location, brain neurons should be classified as the IV type of ganglion cells, whereas pharynx neurons - as the II type (Reuter & Lindroos, 1979).

Discussion

The GAIF-method was shown to be suitable for the whole-mount studies of the neuroanatomy of the flatworms (Kotikova & Joffe, 1988b; Orido, 1989; Buchman & Prento, 1989; Gustafsson & Eriksson, 1991; Joffe & Kotikova, 1991). It was in the representatives of the Polycystididae - *Gyratricella attemsi* (Gyratricinae) and *Acrorynchides robustus* (Acrorhynchinae) that the first data on the distribution of CA recorded by GAIF method in the Kalyptorhynchia were obtained (Joffe & Kotikova, 1989a; 1991). In two independent studies, made by the same method, brief notes of the distribution of CA-ergic elements in *G. hermaphroditus* are made (Reuter & Eriksson, 1991; Kotikova, 1991a). In the present study additional structures have

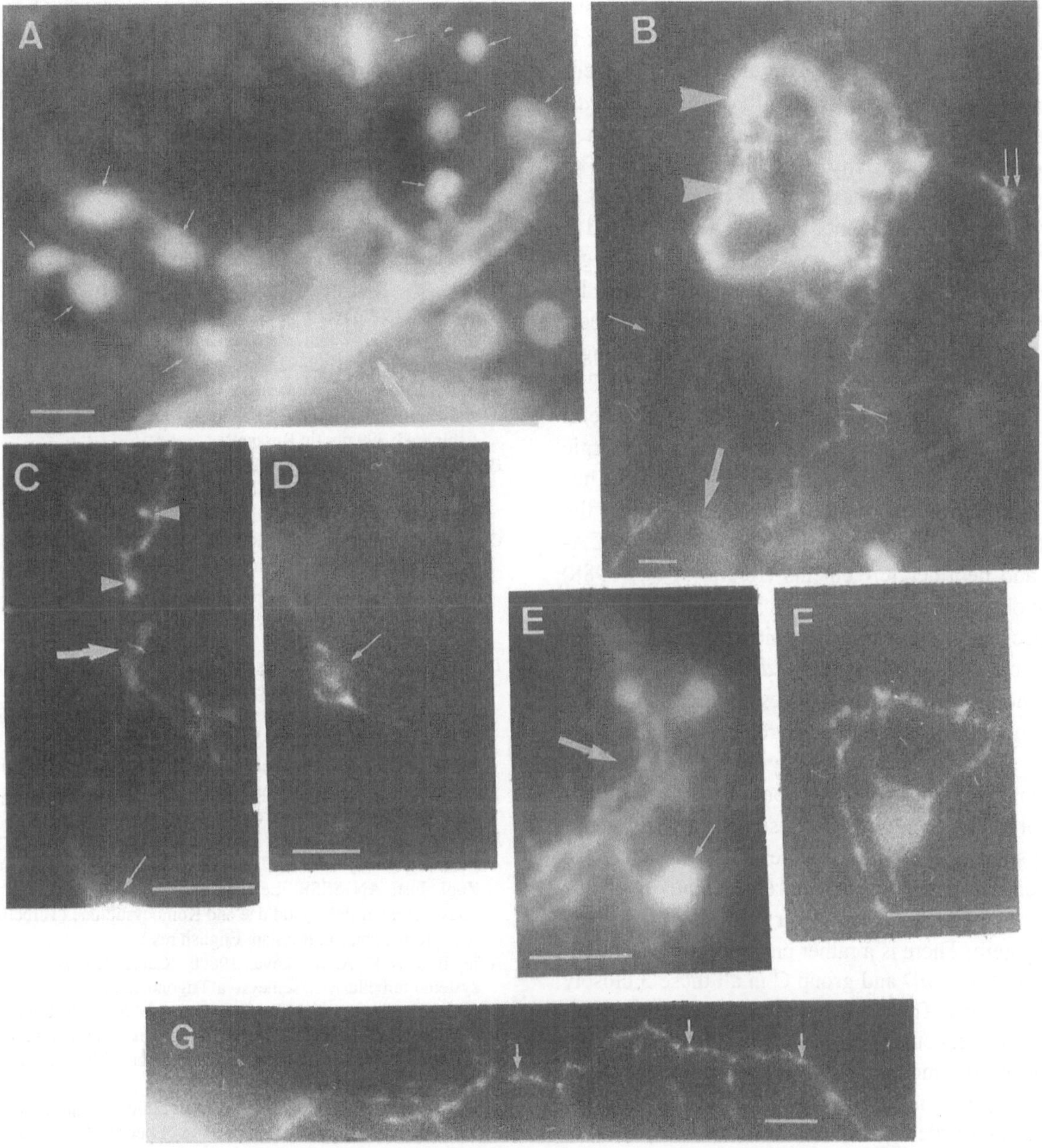

Fig. 2. *Gyratrix hermaphroditus*, whole - mounts, GAIF. Scale bars: 10 μm. (*A*) Detail of brain; brain cells (small arrows); neuropile (large arrow); (*B*) Ventral nerve cords with varicosities (small arrows) at the region of pharynx, transverse commissure (large arrow), distal pharyngeal nerve ring (arrow head), terminals of neuron DL (double arrows); (*C,D*) Detail of fluorescing ventral nerve (large arrow) with varicous thickenings (arrow heads) and neuron AnV (small arrow); (*E*) Detail of pharyngeal nervous system; neuron Ph (small arrow), distal pharyngeal nerve ring (large arrow); (*F*) Neuron AnVL; (*G*) Lateral nerve cord with varicosities (small arrows).

been discovered which allow comparison with the Polycystididae species studied previously.

An attempt to elaborate phylogenetic relationships within the Polycystididae was undertaken by Evdonin (1977). In Evdonin's opinion, most of other the subfamilies including Gyratricinae have originated from ancestral forms of Achrorynchinae. This phylogenetic relationship is also expressed in the structure of the Ch-ergic nervous system: the existence of two rather similar types of orthogons — irregular (*Acrorynchides*) and concentrated (*Gyratrix* and *Gyratricella*) ones (Kotikova, 1991b). The orthogon of *Achro-*

rynchides is at a lower stage of evolution of the commissural system, whereas the orthogon of *Gyratrix* is characterized by a distinct maximum level of concentration of conductive pathways. In *G. hermaphroditus* only 24 neurons have been identified which are characterized by a strictly defined localization. The brightness of their luminescence can be variable. In *A. robustus* green luminescence has been demonstrated in 22 cells, but in *G. attemsi*, only in 18. Most CA-ergic fibres in the nervous system of *G. hermaphroditus* lie parallel to the Ch-ergic ones, though there is a mediatorial differentiation of several parts along a single commissure. Of the 3 brain rootlets the strongest CA-ergic activity is observed in the rootlets of the lateral cords, while the two others are composed of Ch-ergic fibres. Coexistence of Ch- and CA-ergic patterns has been recorded in many other flatworms. Among the peptides, only some overlapping in the patterns of CA-ergic and peptidergic elements (Reuter *et al.*, 1988) can be observed.

Special attention should be paid to the neuron *LV* of *Gyratricella*, *L1* of *Achrorynchides* and *DL* of *Gyratrix*. Though the position of these neurons and their two processes is partially different, their third dendrite is always short and densely branched, ending in numerous terminals, disposed ventrally behind the lateral cord. The homology of these neurons within the Polycystididae is unquestionable. Sincer they have so far been described only in the representatives of this group, they can serve as marker neurons of their nervous system. There is a rather clear homology of neurons *Ph*, *AnV*, *AnD* and group C in all these 3 closely related species. The comparative morphological analysis of the nervous system of all basic groups of the flatworms convincingly demonstrates an independent, multiple appearance of the same type of orthogon in some distant orders (Kotikova, 1986; 1991b; Kotikova & Joffe, 1988a; Joffe, 1990). For example the irregular orthogon was described in such groups as Prolecithophora, Kalyptorynchia, Typhloplanoida and Dalyellioida. On the other hand a direct dependence of the orthogon type on the body shape is quite evident. To give one more example, the Polycladida, Temnocephalida, Monogenea and Trematoda with a leaflike body shape are characterized by a network orthogon (Kotikova, 1991). These peculiarities of the nervous system of the Plathelminthes must not be left out of consideration when establishing the homology of neurons.

Acknowledgments

I am grateful to our Finish collegues, especially Dr M. Reuter for the opportunity to take part in the VII ISBT. This work has been supported by the Russian Basic Research Foundation grant N 93-04-6601.

References

Buchman, K. & P. Prento, 1989. Cholinergic and aminergic elements in the nervous system of *Pseudodactylogyrus bini* (Monogenea) Dis. aqual. Org. 6: 89–92.

Evdonin, L. A., 1977. Turbellarii. Khobotkovye resnichnye chervi Kalyptorhynchia fauny SSSR i sopredelnych stran. Nauka, Leningrad: 399 pp. (in Russian).

Gustafsson, M. K. S. & K. Eriksson, 1991. Localization and identification of catecholamines in the nervous system of *Diphyllobothrium dendriticum* (Cestoda). Parasitol. Res. 77: 498–502.

Gustafsson, M. K. S. & M. Reuter, 1992. The map of neuronal signal substances in flatworms. In R. Naresh Singh (ed.), Nervous systems principles of design and function. Wiley Eastern Limited, New Delhi, Bangalore, Bombay, Calcutta, Guwahati, Hyderabad, Lucknow, Madras, Pune.: 165–188.

Joffe, B. I., 1990. Morfologicheskie zakonomernosty evolutzii nervnoi sistemy ploskih chervei: anatomicheskie varianty ortogona i ich svjaz s formoi tela. Trudy Zool. Inst. AN SSSR, Leningrad, 221: 87–125. (Morphological regularities in the evolution of the nervous system in the Plathelminthes: anatomical variants of the orthogon and their dependence upon the body form. In Russian, English res.)

Joffe, B. I. & E. A. Kotikova, 1989a. O nervnoi systeme Polycystididae i Koinocystididae (Turbellaria, Kalyptorhynchia). Trudy Zool. Inst. AN SSSR. Leningrad. 195: 47–69. (On the nervous system of Polycystididae and Koinocystididae (Turbellaria, Kalyptorhynchia. In Russian, English res.)

Joffe, B. I. & E. A. Kotikova, 1989b. Katecholaminy v nervnoi systeme turbellariy iz semeystva Trigonostomidae (Neorhabdocoela, Typhloplanoida). Trudy Zool. Inst. AN SSSR, Leningrad, 195: 70–83. (Catecholamines in the nervous system of turbellarians from the family Trigonostomidae (Neorhabdocoela, Typhloplanoida. In Russian, English res.)

Joffe, B. I. & E. A. Kotikova, 1991. Distribution of catecholamines in the turbellarians (with a discussion of neuronal homologies in the Platheminthes). J. Neurosci. In D. A. Sakharov & W. Winlow (eds), Simpler nervous systems. Manchester University Press, Manchester: 77–112.

Kabotyanakyj, E. A., 1985. Glioksilatnyj metod vyyavleniya kletochnykh monoaminov na totalnykh preparatakh nervnoj sistemy bespozvonovhnykh. In D. A. Sakharov (ed.), Prostye nervnye sistemy. Kazansk. Universitet. Kazan: 81–83 (in Russian).

Kotikova, E. A., 1986. Comparative characterization of the nervous system of the turbellaria. Hydrobiologia 132 (Dev. Hydrobiol. 32): 89–92.

Kotikova, E. A., 1991a. Catecholamines in the nervous system of *Gyratrix hermaphroditus* (Turbellaria, Kalyptorhynchia). In D. A. Sakharov (ed.), Simpler nervous systems. Abstracts ISIN, Nauka, Minsk: 45.

Kotikova, E. A., 1991b. Ortogon ploskikh chervej i osnovnye puti ego evolutsii. Trudy Zool. Inst. RAN. St. Petersburg. 241: 88–

112. (The orthogon of the Plathelminthes and main trends of its evolution. In Russian, English res.)

Kotikova, E. A. & B. I. Joffe, 1988a. On the nervous system of the dalyellioid turbellarians. Fortschr. Zool. 36: 191–194.

Kotikova, E. A. & B. I., Joffe, 1988b. Katecholaminy i cholinesterasy v nervnoi sisteme turbellarii *Provortex karlingi*. Zhurn. evol. byochim. Fiziol. 24: 142–148 (Catecholamines and cholinesterases in the nervous system of the turbellarian *Provortex karlingi*. In Russian, English res.)

Mamkaev, Y. V., 1986. Initial morphological diversity as a criterion in deciphering turbellarian phylogeny. Hydrobiologia 132 (Dev. Hydrobiol. 32): 31–33.

Orido, Y., 1989. Histochemical evidence of the catecholamines - associated nervous system in certain schistosome cercariae. Parasitol. Res. 76: 146–149.

Reuter, M. & K. Eriksson, 1991. Catecholamines demonstrated by glyocylic-acid-induced-fluorescence and HPLC in some microturbellarians. Hydrobiologia 227 (Dev. Hydrobiol. 69): 209–219.

Reuter, M. & P. Lindroos, 1979. The ultrastructure of the nervous system of *Gyratrix hermaphroditus* (Turbellaria, Rhabdocoela). I The brain. Acta Zool. (Stockh.) 60: 139–152.

Reuter, M., M. Lehtonen & M. Wikgren, 1988. Immunocytochemical evidence of neuroactive substances in flatworms of different taxa – a comparison. Acta Zool. (Stockh.) 69: 29–37.

Hydrobiologia **305**: 141–143, 1995.
L.R.G. Cannon (ed.), *Biology of Turbellaria and some Related Flatworms.*

The importance of the fixation of colour, pattern and form in tropical Pseudocerotidae (Platyhelminthes, Polycladida) *

L. J. Newman & L. R. G. Cannon**
Worms, Queensland Museum, P.O. Box 3300, South Brisbane, Queensland 4101, Australia
** *Present address: Zoology Department, University of Queensland, St. Lucia, Queensland 4072, Australia*

Key words: Polycladida, Pseudocerotidae, *Pseudoceros*, fixation, preservation, colour pattern

Abstract

At last a fixation method that ensures tropical pseudocerotid polyclads are fixed flat, preserved for histological preparation and which also retains their colour pattern has been developed. FCA-PGPP (Formaldehyde Calcium Acetate - Propylene Glycol, Propylene Phenoxetol) fixative is frozen and worms are coaxed onto filter paper which is then laid on the frozen fixative. As a consequence, over 230 species have been documented from the southern Great Barrier Reef and eastern Papua New Guinea (Newman & Cannon, 1994; in press). It was determined that species diagnoses need to be based on colour pattern, general morphology of living animals and serial reconstructions of the male anatomy.

Introduction

Tropical pseudocerotids are conspicuous and often flamboyant inhabitants of coral reefs being commonly found under ledges of the reef slope or under boulders at the reef crest (Newman & Cannon, 1994). These polyclads have usually been encountered in low numbers, their colours and patterns have not been adequately documented and fixation has often resulted in specimens without colour, which can be hopelessly curled, tangled, or simply broken into pieces. Such preparation has made further study on these delicate worms extremely difficult: consequently tropical polyclad diversity is poorly understood, animals are not readily identifiable and anatomical studies have been neglected.

Hyman (1954, 1959) and Prudhoe (1985) concluded that the only way to separate *Pseudoceros* spp. (the most speciose genus from tropical waters) was on the basis of colour pattern. This, however, has not been adequately recorded previously and is lost with fixation. Hence species identification has often not been possible for preserved animals. Prudhoe (1989) suggested that while colour may be variable, markings are consistent. He advocated that a closer examination of general features were needed to assess the diagnostic value of morphological characters which might be derived from living worms.

From 1989 to 1993, 230 pseudocerotids were collected from the southern Great Barrier Reef and eastern Papua New Guinea (Newman & Cannon, 1994; in press; Newman, unpubl.). It was obvious from the start that a new fixation method was needed to fix pseudocerotids intact, flat for histological preparation and to preserve their colour pattern. This method, developed by trial and error, is presented here.

Materials and methods

Tropical polyclads were hand collected from under ledges, coral rubble or under boulders at the reef crest. Animals were coaxed into sample containers with soft brushes and kept separate to reduce stress from the mucus of other species. In the laboratory, animals were kept alive in 2 L containers with clean seawater, unaerated to avoid damaging these delicate worms.

Polyclads were photographed either *in situ* with Nikonos cameras and Aqua Sea strobes or in the laboratory with a Canon T90, TTL flash and 50 mm macro lens and extension tubes. Polyclad colour pattern and

* Christensen Research Institute Contribution No. 82

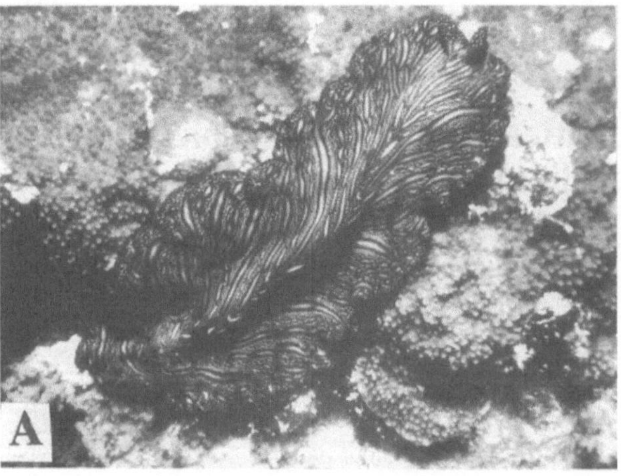

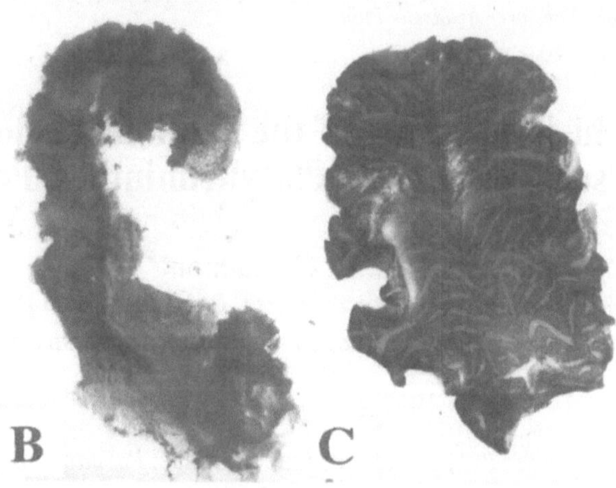

Fig. 1. Pseudobiceros bedfordi (Laidlaw, 1903); (*A*) live, Heron Island, southern Great Barrier Reef; (*B*) a museum specimen (of *P. bedfordi*?) fixed with formaldehyde; (*C*) fixed with FCA-PGPP.

behaviour were also recorded by video on Super VHS using a Sony Colour Video Camera DXX - 151.

The fixative, Formalin Calcium Acetate - Propylene Gylcol, Propylene Phenoxetol (FCA - PGPP) contains seawater 90 ml, formaldehyde (40%) 10 ml, calcium acetate 1.8 g, propylene glycol (propane 1, 2 - diol) 4.5 ml and propylene phenoxetol (phenoxy- propan-2-ol) 0.5 ml. For use, 1–2 cm depth of fixative was frozen in the bottom of a container. Polyclads were fixed by placing/coaxing the animal onto absorbent paper (e.g. filter paper) to restrict their movement and then placing this paper onto the frozen fixative. A thin layer of cold FCA-PGPP was added to just cover the worm and to facilitate melting of the fixative. A soft brush was used to ensure the worm remained under the fixative and remained flat.

After fixation for a minimum of 24 hours, the fixative was replaced by 70% ethanol for long term preservation and histological preparation. Whole mounts were prepared by regressively staining with Mayer's haemalum, dehydrating in graded alcohols, clearing in xylene and mounting in canada balsam. Longitudinal serial sections of the reproductive regions were prepared from paraffin wax (Paraplast, 56 °C), cut at 6–8 μm, and stained with Mayer's haemalum and eosin.

Results and discussion

Living pseudocerotids (Fig. 1A) are large, but extremely delicate and fragile. Traditional fixation methods (see Prudhoe, 1985) and deposition in museum collections has too often resulted in specimens being virtually unrecognisable (Fig. 1B). FCA-PGPP (modified from FCC-Steedman, see Winsor, 1991) proved excellent for morphological studies and long term storage since it stabilised the epithelium and minimised pigment loss (Fig. 1C). Substitution of sea water for tap water was used to minimise adverse osmotic effects in these for marine animals. Cobaltous nitrate (an ingredient of FCC) was found to inhibit staining of whole mounts by Mayer's haemalum and its omission does not appear to alter short term fixation. It is vital that FCA-PGPP is frozen to ensure quick fixation. If FCA-PGPP is not available, then 4% formaldhyde in seawater can be used but the results are not as consistent and delicate animals may fall apart.

Morphological characters such as shape of the pseudotentacles, degree of marginal ruffling, size and shape of the pharynx, pharyngeal folding, distribution and number of pseudotentacular and cerebral eyes (dorsally and ventrally) and position of the sucker were examined in living animals prior to fixation. Species descriptions need examination of morphological characters of living animals and colour pattern as well as reconstruction of the reproductive anatomy from longitudinal serial sections of mature animals.

An identification guide with colour photographs and brief descriptions of the polyclad fauna of the southern Great Barrier Reef is lodged at the Heron and One Tree Island Research Stations and at the Christensen Research Institute, Madang, Papua New

Guinea. All specimen material and colour transparencies are lodged at the Queensland Museum.

Without the development of a proper fixation method for the extremely delicate pseudocerotid polyclads, documentation of the diverse fauna of the southern Great Barrier Reef and eastern Papua New Guinea would not be possible. We consider that the procedures outlined in this paper will prove reliable for the study of these fascinating animals.

Acknowledgments

We wish to thank Mr L. Winsor for his histological advice, Mrs C. Lee and Ms Z. Khalil for histological preparations, Mr A. Flowers for assistance with collecting and photography, Mr J.-M. Ouin for suggesting the use of filter paper for polyclad fixation, the Director and staff of the Heron Island Research Station and the Christensen Research Institute and the Managers of the One Tree Island Research Station. Funding was provided by the Australian Research Council Grant AD# 9031806 (to L.R.G.C.), the Christensen Research Institute (to L.N.) and the P & O Pty. Ltd. – Heron Island Research Station Reef Fellowship (to L.N.).

References

Hyman, L. H., 1954. The polyclad genus *Pseudoceros*, with special reference to the Indo-Pacific region. Pac. Sci. 8: 219–225.

Hyman, L. H., 1959. Some Australian polyclads. Rec. Aust. Mus. 25: 1–17.

Newman, L. J. & L. R. G. Cannon, 1994. Biodiversity of tropical polyclad flatworms (Platyhelminthes, Polycladida) from the Great Barrier Reef, Australia. Mem. Qd Mus. 36: 159–163.

Newman, L. J. & L. R. G. Cannon, (in press). *Pseudoceros* and *Pseudobiceros* (Platyhelminthes, Polycladida, Pseudocerotidae) from eastern Australia and Papua New Guinea. Mem. Qld. Mus.

Prudhoe, S., 1985. A Monograph on Polyclad Turbellaria. British Museum (Natural History), Oxford Univ. Press, Oxford, 259 pp.

Prudhoe, S., 1989. Polyclad turbellarians recorded from African waters. Bull. Brit. Mus., N.H. (Zool.) 55: 47–96.

Winsor, L., 1991. Method for taxonomic and distributional studies of terrestrial flatworms (Tricladida: Terricola). Hydrobiologia 227 (Dev. Hydrobiol. 69): 349–352.

Hydrobiologia **305**: 145–150, 1995.
L.R.G. Cannon (ed.), Biology of Turbellaria and some Related Flatworms.

Structure and maintenance of the epidermis in *Friedmaniella* sp. (Prolecithophora)

Irina M. Drobysheva & Yurij V. Mamkaev
Laboratory of Evolutionary Morphology, Zoological Institute, Russian Academy of Sciences, 199034 St. Petersburg, Russia

Key words: Turbellaria, Platyhelminthes, Prolecithophora, epidermis, ultrastructure, physiological regeneration

Abstract

The epidermis of *Friedmaniella* sp. has been studied using light and electron microscopy. Three main morphological features characterize its cells, namely (1) DNA bodies in the nuclei, (2) an extensive Golgi apparatus with a well-developed system of transport vesicles, (3) clusters of centrioles mainly in the basal cytoplasm and axonemes and rootlets in the middle and apical cell parts. These peculiarities may indicate continuous physiological regeneration within the cell at the level of cell organelles. DNA bodies may prove to be a taxonomic feature distinguishing the Prolecithophora from other turbellarians.

Introduction

Recent phylogenetic schemes of the Plathelminthes have been constructed on the basis of molecular, developmental, ultrastructural and nerve lightmicroscopical studies (see e.g., Rieger, 1981; Hendelberg, 1986; Rieger *et al.*, 1991; Boyer, 1995; Ehlers, 1995; Rohde *et al.*, 1995). Special attention has been paid to tissue renewal, growth and origin. It has been shown, for example, that somatic cells of plathelminths do not divide, and that maintenance as well as tissue growth occur at the expense of recruitment of undifferentiated cells kept in reserve in the parenchyma (Lange, 1983; Morita & Best, 1984; Drobysheva, 1986; Baguñà *et al.*, 1989, for review; Palmberg, 1990, 1991; Ehlers, 1992).

As a rule, physiological regeneration (*i.e.* the continuous regeneration of tissues maintained in living animals) of the turbellarian epidermis is associated with migration of the precursor cells from the parenchyma, as has been demonstrated in specimens of various turbellarian orders (Ehlers, 1985; Drobysheva, 1991; Palmberg, 1990, for references). As regards the prolecithophorans, replacement of their epidermis has not been studied yet; the morphological descriptions available do not give a complete picture of regeneration (Friedmann, 1933; Karling, 1940; Ehlers, 1985; Timoshkin, 1986).

It has recently been found (Drobysheva, 1991) that after colchicine treatment of the prolecithophorans *Friedmaniella* sp. and *F. bargusinica*, mitotic figures accumulate in the parenchyma and gastrodermis but not in the epidermis. As was expected, differentiated epidermal cells cannot proliferate in these animals and no undifferentiated cells are present in the epidermis. However, migration of epidermal-cell precursors in the prolecithophoran *Friedmaniella* sp. has not been observed; that is, no nuclei are seen passing through the basement membrane. This evidence stimulated our interest in characterizing the morphology of the epidermis of this turbellarian through light and electron microscopy.

Materials and methods

Friedmaniella sp. was collected from Lake Baikal. Preparations for light microscopy were made by conventional histological methods: paraffin sections were stained with Mayer's haematoxylin, gallocyanin-eosin, and Feulgen. For electron microscopy, samples were fixed in 2.5% glutaraldehyde, followed by 2% osmium tetroxide. All fixatives were buffered to pH 7.4

with 0.05 M cacodylate. Dehydration was performed in graded concentrations of ethanol, and samples were embedded in an Epon-Araldite mixture. Thin sections were stained with uranyl acetate and lead citrate.

Results

The epidermis of *Friedmaniella* sp. could be seen to be a monolayered cellular, ciliated epithelium (Figs 1, 2, 4). A well-developed basement membrane separated the epidermis from the inner tissues (Figs 2, 6). Between the highly differentiated epidermal cells, numerous parenchymal gland-cell necks and single sensory-cell processes were found. Some of the parenchymal processes passed through the epidermal cell channels (Fig. 2) reaching the body surface.

The apical cell surface was covered with cilia and microvilli. Two rootlets, one horizontal pointing rostrally and one vertical, diverged from the basal body of each cilium. A terminal web appeared to be entirely absent.

The most prominent component of the cytoplasm of each epidermal cell was the highly developed Golgi apparatus, represented by numerous stacks of usually very flattened cisternae (Figs 2, 3). They were located mainly in the basal part of the epidermal cell between the nucleus and the basement membrane.

In spite of its intensive development, the Golgi appeared to produce no secretory granules for exocytosis. The usual rhabdites or ultrarhabdites were not observed and the rough endoplasmic reticulum was only slightly developed. On the other hand, swarms of small transport vesicles always surrounded the dictyosomes (Fig. 3).

Another remarkable feature of the epidermal cells was the presence of intracytoplasmic ciliary elements: centrioles, axonemes and rootlets. The centrioles were scattered throughout the cell, but many of them lay in the basal cytoplasm close to the basement membrane (Figs 5, 6) or in the vicinity of the dictyosomes of the Golgi complex. They occured singly or in groups, clusters might number 5. Frequently paired centrioles at right angles to each other were found (Fig. 2). At the same time, we never observed clusters of centrioles in the apical cytoplasm close to the ciliated surface: here only single centrioles occured occasionally. We believe that the centrioles were the presumptive basal bodies, but, following Tyler (1981: 233) we 'will still refer to them as centrioles as this is the convention in the literature'.

We have not observed any centrioles with the cartwheel-like structure which is considered inherent in mitotic centrioles; but even all centrioles found in parenchymal cells lacked that structure. In contrast to the commonly seen centrioles, axonemes and rootlets occured rarely and were never observed deep in the cytoplasm in close proximity to the basement membrane. While the rootlets always lay in the cytoplasm (Figs 7, 8), parts of the axonemes seen in the sections were within cytoplasmic vacuoles (Figs 7, 9). Inside a cell, the axonemes were arranged randomly, *i.e.*, they were at different angles to cell axes and, moreover, were typically slightly curved when in large vacuoles.

All epidermal nuclei contained prominent, more or less spherical electron dense bodies (Figs 2, 7). In sections of different nuclei the number of these bodies varied from 1 to 6. They were Feulgen-positive (Fig. 4), which implied that they were of heterochromatic nature, hence our reference to them as DNA bodies. True nucleoli appeared to be absent from epidermal nuclei of *Friedmaniella* sp.

The DNA bodies seemed to consist of thin fibrillar material packed extremely densely. They appeared highly lobulated. They had no envelope and the spaces within them opened directly to the nucleoplasm (Fig. 7). Short irregular processes or rays appeared to stem from the DNA body (Fig. 2), apparently as a result of peripheral fragmentation.

Discussion

According to published reports, the Golgi is the branch point of transport proteins to diverse membranes and organelles (see e.g., Rothman & Lenard, 1984; Zavarzin *et al.*, 1992, for references). Three processing paths pass through the Golgi: plasma membrane, secretory, and lysosomal proteins. The extensive development of the Golgi apparatus and the associated vesicular transport system as well as the absence of secretory granules may suggest that in the epidermal cells of *Friedmaniella* sp. the Golgi is active mainly in supplying constitutive proteins and hydrolytic enzymes. Location of centrioles near dictyosomes suppose that the Golgi complex could contribute to the development of centriole precursor structures as it is believed to be the case in two Dalyellioida (Cifrian *et al.*, 1992) and in many Vertebrata (Sorokin, 1968; Lemullois *et al.*, 1988; Dirksen, 1991, for review). Abundance of vacuoles with the debris of unidentifiable structures and

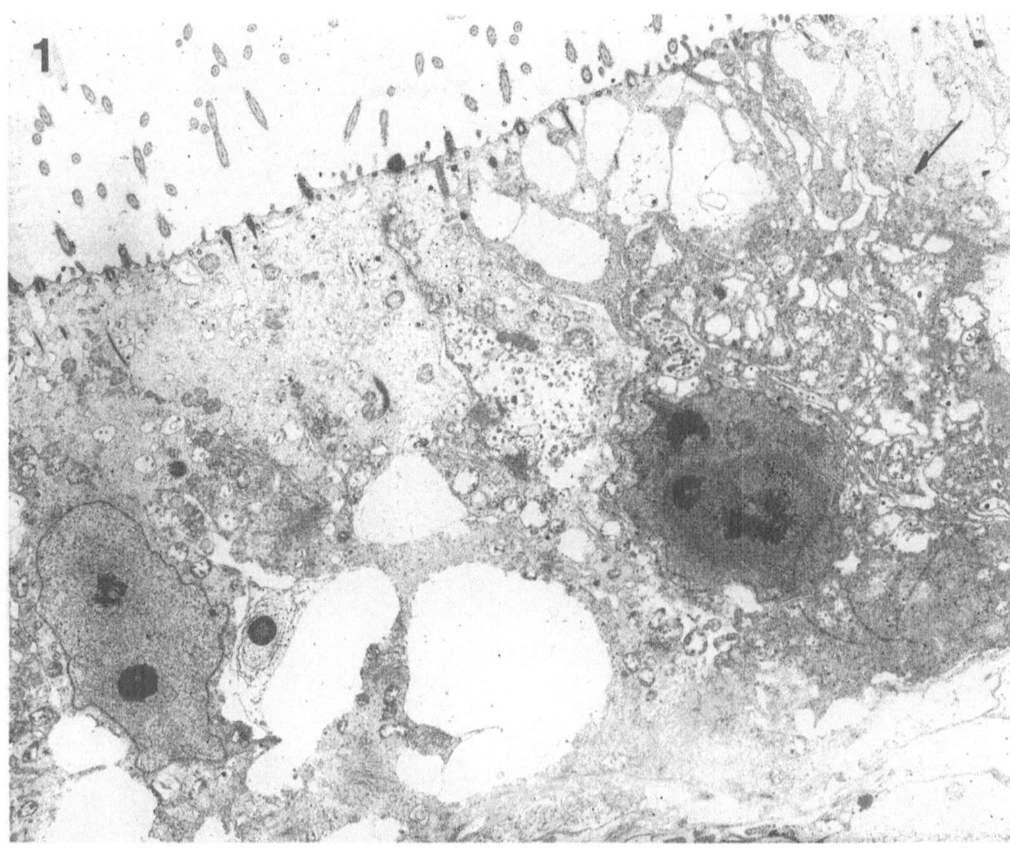

Plate 1. (Fig. 1) Overview of epidermis of *Friedmaniella* sp. Note position of the dictyosomes (basal and middle part of the cells), peculiar morphology of the epidermal nuclei (see text), oblique section through centriole (arrow) in upper cytoplasm, as well as differences in the cytoplasm with respect to vacuolization. × 6000

lamellar bodies supports the assumption that intensive autophagic processes occur in the epidermal cells.

Longitudinal sections through the centrioles show that the sides of a single centriole may be of different length (Fig. 5). These observations are consistent with the evidence of Tyler (1981) as regards the ciliogenesis in *Macrostomum hystricinum*, in which the presumptive basal bodies are also characterized by different lengths of their sides, whereas the centrioles of the mitotic apparatus have comparatively larger sizes and sides of equal length.

The presence of axonemes and rootlets in the cytoplasm might be interpreted as indicating retraction of worn-out cilia in the process of disintegration. But cases of retraction of ciliary axonemes into the cytoplasm reported in the literature show the axoneme intact and not separated from the surrounding cytoplasm by any membrane (Bloodgood, 1974). Intracytoplasmic axonemes surrounded by membranes, as observed in *Friedmaniella* sp., are indicative of nascent cilia (Sorokin, 1968). The similar axonemes in membrane sheaths as well as those without sheaths were observed in epidermal cells of turbellarians *Syndesmis echinorum* and *Paravortex cardii* (Cifrian *et al.*, 1992). The authors assume that these intracytoplasmic ciliary elements 'are generated by ciliogenesis rather than by reabsorption of cilia on the surface of the cell', for most of them are located in epidermal cells of *P. cardii* embryos and in non degenerating epidermal cells of *S. echinorum* and *P. cardii* adults. Unfortunately, the origin of the axonemes and the rootlets in epidermal cells of *Friedmaniella* sp. still remain obscure.

The ultrastructure of the DNA bodies appears identical to that of chromatin bodies in nurse and follicle cells of insects (Gruzova, 1974, 1985; Gaginskaya & Gruzova, 1975; Gruzova *et al.*, 1990). Chromatin clumps were also described in polyploid somatic cells of females in many butterfly species and for this reason were considered sex chromatin (Traut & Mosbacher, 1968). By means of autoradiography (for RNA and

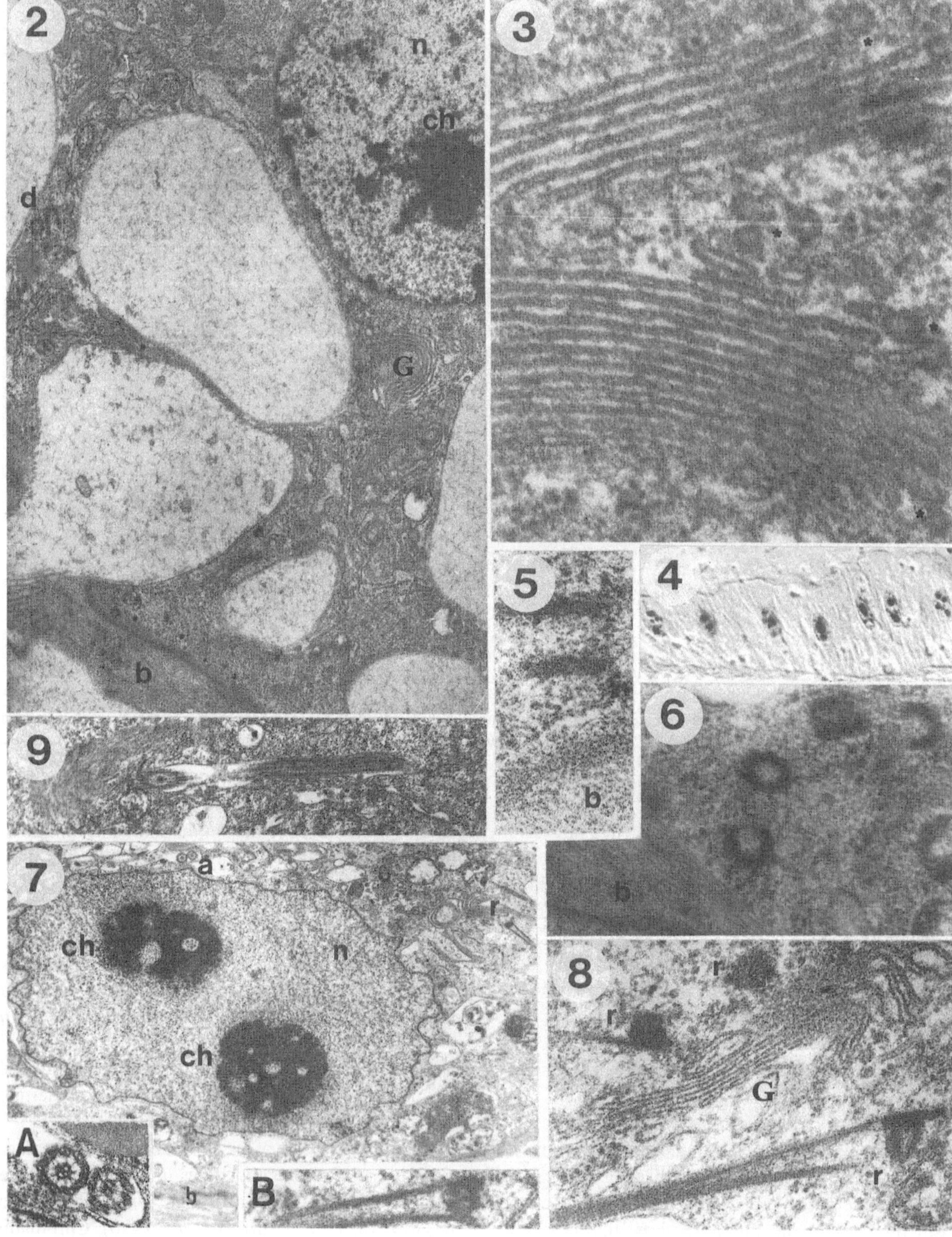
2
n
ch
d
G
b
3
5
b
4
6
b
9
7
a
r
ch
n
ch
8
r
r
G
r
A
b
B

DNA synthesis) and molecular hybridization *in situ*, it was shown that DNA bodies in trophocytes and follicle cells of insects contain repeating genes of ribosomal RNA and represent functioning nucleolus chromatin (Gruzova, 1974, 1985; Gaginskaya & Gruzova, 1975; Gruzova *et al.*, 1990). Such DNA bodies actively participate in ribosomal RNA synthesis and thus may be derivative of true nucleoli which may be absent (for references, see Gruzova, 1985). It is believed that the large size of DNA bodies is due to both the endopolyploidization of the DNA and amplification of the ribosomal DNA (Gruzova, 1985). Thus, in insects, the DNA bodies are the morphological expression of a particular functional state of intensive RNA synthesis.

The striking resemblance of DNA bodies of *Friedmaniella* sp. and those of insects give us every reason to assume that epidermal cells of this turbellarian are polyploid. Moreover, this surprising morphological similarity suggests that active production of ribosomal RNA also takes place in these epidermal nuclei. High numbers of ribosomes in the basal cytoplasm of epidermal cells also reflect this.

Conclusion

Three principal characteristics are prominent in the epidermal cells of *Friedmaniella* sp.: (1) DNA bodies in the nuclei, (2) extensive Golgi apparatus with a well-developed system of transport vesicles, and (3) clusters of centrioles close to the basement membrane and axonemes and rootlets in the middle and apical cell parts.

Friedmann (1933) described numerous sunken epidermal cells or parenchymal elements with processes in the epidermal layer in the prolecithophoran *Baicalarctia gulo*. She also noted that the nuclei were situated at the level of the basal lamina, an indication that, for its renewal, cell migration may be taking place from the inner part of the body into the epidermis.

In *Friedmaniella* sp. we did not observe nuclei passing through the basement membrane; by contrast, such cell migration appears to be common to polyclads (Drobysheva, 1988) and triclads (Skaer, 1961; Hori, 1978). This fact does not exclude the possibility of cell migration from the parenchyma where undifferentiated precursors of epidermal cells might be stored because cell migration may occur rarely and rapidly.

The absence of migrating nuclei suggests that the autapomorphy Ehlers (1986) suggests for the taxon Plathelminthes ('no mitosis in somatic cells') also applies to this prolecithophoran, but, instead of having stem cells outside the epidermis as do other turbellarians, the regenerative system consists of renewing organelles within the cell. Thus, *Friedmaniella* sp. represents another way of compensating for lack of epidermal mitoses.

The DNA bodies in the nuclei were observed not only in epidermal nuclei of *Friedmaniella* sp. but also in those of *Friedmaniella bargusinica* (Drobysheva, unpublished data). Those in *F. bargusinica* occur at twice the number (12–13 per section of a nucleus) than in *Friedmaniella* sp.. Moreover, these bodies occur in parenchymal and gastrodermal cells of these two species. Thus, the DNA bodies may prove to be a useful taxonomic feature for the Prolecithophora. This could be critical, given that the position of the Prolecithophora among turbellarians still remains debatable (Smith *et al.*, 1986; Ehlers, 1988).

Acknowledgments

We thank Dr Oleg Timoshkin for his assistance in collecting Baikal prolecithophorans, Dr Olga Raikova for fixing specimens for electron microscopy and Mrs Gunilla Henriksson for the technical assistance. We are also grateful to Dr M. N. Grusova, whose comments on the DNA bodies and related cytological topics were very useful. We are deeply indebted to Dr A. D. Kharazova and Dr S. A. Karpov for their advice and moral support. We would like to express our sincere gratitude to Dr Maria Reuter for the possibility to participate in 7 ISBT. Financial support was provided by the International Scientific Fund 'Cultural

Plate 2. (Fig. 2) Basal part of epidermal cell: the cytoplasm is perforated by parenchymal cell processes, one of which penetrates the basement membrane (b). × 17400. G. Golgi apparatus; d, diplosome; n, nucleus; ch, chromatin body. (Fig. 3) Golgi apparatus with transport vesicles (asterisks). -× 96000. (Fig. 4) Paraffin section of the epidermis stained by Feulgen. × 500. (Fig. 5, 6) Centrioles (presumptive basal bodies) close to the basement membrane (b). Fig. 5, × 84000; Fig. 6, × 42000. (Fig. 7) Part of epidermal cell with nucleus (n), axonemes (a) and rootlets (r). × 9500. b, basement membrane; ch, chromatin body. Insets: A, axonemes and B, nascent (?) rootlet at a higher magnification. A, × 50000; B, × 57000. (Fig. 8) Nascent (?) rootlets (r) near to Golgi apparatus (G). Another section of the cell in Fig. 7 at a higher magnification. × 57000. (Fig. 9) Axoneme in the apical cytoplasm. × 19400.

Initiative' and the Russian Academy of Natural Sciences.

References

Baguñà, J., E. Saló & M. C. Auladell, 1989. Regeneration and pattern formation in planarians. 111. Evidence that neoblasts are totipotent stem cells and the source of blastema cells. Development 107: 77–86.

Bloodgood, R. A., 1974. Resorption of organelles containing microtubules. Cytobios 9: 143–161.

Boyer, B. C., 1995. What studies of turbellarian embryos can tell us about the evolution of developmental mechanisms. Hydrobiologia 305 (Dev. Hydrobiol. 108): 217–222.

Cifrian, B., P. Garcia-Corrales & S. Martinez-Alos, 1992. Intracytoplasmic ciliary elements in epidermal cells of Syndesmis echinorum and Paravortex cardii (Platyhelminthes, Dalyellioida). J. Morphol. 213: 147–157.

Dirksen, E. R., 1991. Centriole and basal body formation during ciliogenesis revisited. Biol. Cell 72: 31–38.

Drobysheva, I. M., 1986. Physiological regeneration of the digestive parenchyma in Convoluta convoluta and Oxyposthia praedator (Turbellaria, Acoela). Hydrobiologia 132 (Dev. Hydrobiol. 32): 189–193.

Drobysheva, I. M., 1988. An autoradiographic study of the replacement of epidermis in polyclad turbellarians. Fortschr. Zool. 36: 97–101.

Drobysheva, I. M., 1991. Kambial'nost' epidermisa u turbellarij. Tr. Zool. Inst. AN SSSR 241: 53–87. [In Russian]

Ehlers, U., 1985. Das phylogenetische System der Plathelminthes. G. Fischer, Stuttgart, New York, 317 pp.

Ehlers, U., 1986. Comments on a phylogenetic system of the Platyhelminthes. Hydrobiologia 132 (Dev. Hydrobiol. 32): 1–12.

Ehlers, U., 1988. The Prolecithophora – a monophyletic taxon of the Plathelminthes? Fortschr. Zool. 36: 359–365.

Ehlers, U., 1992. No mitosis of differentiated epidermal cells in the Plathelminthes: mitosis of intraepidermal stem cells in Rhynchoscolex simplex Leidy, 1851 (Catenulida). Microfauna Mar. 7: 311–321.

Ehlers, U., 1995. The basic organization of Plathelminthes. Hydrobiologia 305 (Dev. Hydrobiol. 108): 21–26.

Friedmann, G. M., 1933. Anatomicheskoe stroenie Baicalarctia gulo Fr. i polozhenie eyo v sisteme Turbellaria. Tr. Bajkal. limnol. st. 5: 179–256. [In Russian]

Gaginskaya, E. R. & M. N. Gruzova, 1975. Vyyavlenie amplifitsirovannoj rDNK v kletkakh yaichnikov nekotorykh nasekomykh i ptits metodom gibridizatsii nukleinovykh kislot na preparatakh. Tsitologiya 17: 1132–1137. [In Russian]

Gruzova, M. N., 1974. Yadernye struktury v ovariolakh babochki Laspeyresia pomonella. O polovom khromatine v trofotsitakh. Ontogenez 5: 623–633. [In Russian]

Gruzova, M. N., 1985. Polovoj khromatin i evolyutsiya nukleolyarnogo apparata v trofotsitakh zhuka Blaps lethifera. Tsitologiya 27: 368–375. [In Russian]

Gruzova, M. N., F. Segnal & Yu. I. Gukina, 1990. Osobennosti organizatsii politrofnykh ovariol voskovoj moli Galleria melonella. Tsitologiya 32: 677–683. [In Russian]

Hendelberg, J., 1986. The phylogenetic significance of sperm morphology in the Platyhelminthes. Hydrobiologia 132 (Dev. Hydrobiol. 32): 53–58.

Hori, I., 1978. Possible role of rhabdite-forming cells in cellular succession of the planarian epidermis. J. Electron Microsc. 27: 89–102.

Karling, T. G., 1940. Zur Morphologie und Systematik der Alloeocoela Cumulata und Rhabdocoela Lecithophora (Turbellaria). Acta. Zool. Fenn. 26: 1–260.

Lange, C. S., 1983. Stem cells in planarians. In C. S. Potten (ed.), Stem cells. Churchill Livingstone, Edinburgh: 28–66.

Lemullois, M., E. Boisvieux-Ulrich, M.-Ch. Laine, B. Chailley & D. Sandoz, 1988. Development and functions of the cytoskeleton during ciliogenesis in metazoa. Biol. Cell 63: 195–208.

Morita, M. & J. B. Best, 1984. Electron microscopic studies of planarian regeneration. IY. Cell division of neoblasts in Dugesia dorotocephala. J. exp. Zool. 229: 425–436.

Palmberg, I., 1990. Stem cells in microturbellarians. An autoradiographic and immunocytochemical study. Protoplasma 158: 109–120.

Palmberg, I., 1991. Differentiation during asexual reproduction and regeneration in a microturbellarian. Hydrobiologia 227 (Dev. Hydrobiol. 69): 1–10.

Rieger, R. M., 1981. Morphology of the Turbellaria at the ultrastructural level. Hydrobiologia 84: 213–229.

Rieger, R. M., S. Tyler, J. P. S. Smith III & G. E. Rieger, 1991. Platyhelminthes, Turbellaria. In F. W. Harrison & B. J. Bogitsh (eds), Microscopic Anatomy of Invertebrates. Vol. 3. Platyhelminthes & Nemertinea. Wiley-Liss N.Y.: 7–140.

Rohde, K., A. M. Johnson, P. R. Baverstock & N. A. Watson, 1995. Aspects of the phylogeny of Platyhelminthes based on 18S ribosomal DNA and protonephridial ultrastructure. Hydrobiologia 305 (Dev. Hydrobiol. 108): 27–35.

Rothman, J. E. & J. Lenard, 1984. Membrane traffic in animal cells. Trends Biochem. Sci. 9: 176–178.

Skaer, R. J., 1961. Some Aspects of the Cytology of Polycelis nigra. Quart. J. Micr. Sci. 102: 295–317.

Smith, J. P. S., S. Tyler & R. M. Rieger, 1986. Is the Turbellaria polyphyletic? Hydrobiologia 132 (Dev. Hydrobiol. 32): 13–21.

Sorokin, S. P., 1968. Reconstructions of centriole formation and ciliogenesis in mammalian lungs. J. Cell. Sci. 3: 207–230.

Timoshkin, O. A., 1986. Osobennosti stroeniya i sistematicheskoe polozhenie Prolecithophora Bajkala (Turbellaria). Zool. zhurn. 65: 16–27. [In Russian]

Traut, W. & G. C. Mosbacher, 1968. Geschlechtschromatin bei Lepidopteren. Chromosoma 25: 343–356.

Tyler, S., 1981. Development of cilia in embryos of the turbellarian Macrostomum. Hydrobiologia 84: 231–239.

Zavarzin, A. A., A. D. Kharazova & M. N. Molitvin, 1992. Biologiya kletki: obshchaya tsitologiya. Izd-vo SPb. universiteta, Sankt-Peterburg, 320 pp. [In Russian]

Hydrobiologia **305**: 151–158, 1995.
L.R.G. Cannon (ed.), *Biology of Turbellaria and some Related Flatworms.*

A scanning electron microscope study of *Craspedella* sp. from the branchial chamber of redclaw crayfish, *Cherax quadricarinatus*, from Queensland, Australia

Kim B. Sewell[1] & Lester R. G. Cannon
The Queensland Museum, Cultural Centre, PO Box 3300, South Brisbane, Queensland 4101, Australia
[1]*(a) The Department of Parasitology and (b) The Department of Anatomical Sciences, The University of Queensland, Brisbane, Queensland 4072, Australia*

Key words: Temnocephalidae, *Craspedella*, ectosymbiotic flatworms, scanning electron microscopy, epidermis, methods of fixation

Abstract

Epidermal topography was examined, including papillate ridges, grooves and ciliated sensory papillae of *Craspedella* sp. from the branchial chamber of redclaw crayfish, *Cherax quadricarinatus*, from Queensland, Australia. Rhabdites were observed to discharge from ducts opening mainly in a small distal region of the ventral epidermis of the three central (of five) tentacles. These regions, devoid of ciliated sensory papillae, serve to adhere the anterior end of the worms during locomotion. Secretions from glands associated with the posterior attachment organ were observed to discharge from pores on the outside region of the ventral surface of the disc.

A comparison of various scanning electron microscopy (SEM) fixation techniques showed that (1) hot fixatives at 90 °C provide most information on the largest number of epidermal structures and (2) different fixation regimes highlight different epidermal features.

Introduction

Only four studies (on one New Zealand and four Australian species of *Temnocephala*) have used the scanning electron microscope (SEM) to elucidate the general features of the epidermis of temnocephalan worms (Williams, 1978, 1982, 1991, 1992). Damborenea (1992) concentrated on the ciliated sensory structures of the epidermal surface of six species of *Temnocephala* from Argentina. Studies on temnocephalans by Williams (1981), Matjašič (1990), Cannon (1991) and Jennings, Cannon & Hick (1992) used the SEM as an adjunct to other studies.

Craspedella spenceri Haswell, 1893 from the gills of *Cherax destructor* is the only described species within the genus, but other undescribed species of *Craspedella* are known to live in the branchial chamber of crayfish (Cannon & Sewell, in preparation). No SEM examination has been published on either *C. spenceri* or any undescribed species of *Craspedella*. We used specimens of an undescribed species of *Craspedella*, an ectosymbiont from the commercial redclaw crayfish, *Cherax quadricarinatus*, for (a) observation of surface structures and (b) experimentation on fixation procedures. We sought an optimum fixation regime for SEM of *Craspedella*, after routine methods of fixation by cold 3% glutaraldehyde proved inadequate for the resolution required. This paper forms part of a Ph.D. thesis by the senior author on the biology and functional morphology of *Craspedella* sp. from redclaw.

Materials and methods

Live *Cherax quadricarinatus* (=redclaw) were obtained from the University of Queensland Aquaculture Facility, Veterinary Science Farm, Pinjarra Hills, Brisbane. To obtain living *Craspedella* sp., redclaw were pithed, the gills excised and placed in a shallow glass vessel containing clean water.

For scanning electron microscopy (SEM) examination and trials of fixatives, live *Craspedella* sp. were fixed using the methods described in Appendix 1. Fixation was achieved either by immersion or flooding (Appendix 1). When fixation was by flooding, the water in which the specimens were held was reduced in volume to the minimum required to cover the worms immediately prior to fixation. After fixation, with the exception of freezing in liquid nitrogen (see Appendix 1), worms were washed for 30 sec to remove surface contamination in a 20% solution of Decon 90 detergent (Decon 90, Decon Laboratories Ltd., United Kingdom) and rinsed 6 times in filtered distilled water. Worms were then dehydrated for 10 minutes in each of a graded series of methanol, critical point dried, mounted on stubs, coated with gold, and examined with a Hitachi S-530 SEM operating at 20 KV.

Results

Three prominent transverse and four posterior but radially orientated ridges were observed on the posterior dorsal surface of *Craspedella* sp. (Fig. 1, A–B). The transverse ridges are neither lamellated nor divided into lobes, and the posterior ridges are not further divided into several minor elevations (for comparison with *C. spenceri* see Haswell, 1893, Plate XV; 3). Epidermal regions are delineated by a series of grooves and/or ridges (Fig. 1C–F; 2A, B). These may be used to divide the epidermis in 6 regions (after Williams, 1982): (1) the tentacles, delineated on the ventral surface by a transverse groove (Fig. 1C), and dorsally by (2) the dorsal saddle (Fig. 1D), (3) the dorsal body surface and (4) ventral body surface separated by lateral ridges on each side which extend also around the posterior region and link together (Fig. 1E), (5) the peduncle and dorsal surface of the disc of the posterior attachment organ delineated by a groove on the ventral body surface which circumscribes the peduncle (Fig. 1F) and (6) the disc ventral surface delineated by a peripheral ridge and associated groove around the perimeter of the disc (Fig. 2A, B). Pores of gland ducts perforate the epidermis and discharge specialised secretory products in two regions: (1) the 3 central tentacles, used to attach temporarily during locomotion, each have a small cavity on the distal ventral surface from which discharge concentrated intact rhabdites and/or rhabdite contents (Fig. 2E–G) and (2) the ventral surface of the disc except for the central region and the peripheral lip (Fig. 1F, 2A, B, I).

The entire epidermal surface, with the exception of selected regions described below, is covered with sparsely distributed conical papillae. Each papilla contains 1 to 10 sensory cilia, 1–3 μm in length. These ciliated sensory papillae are largest, and possess the most cilia, on the tentacles where they are arranged in 8 longitudinal rows (Fig. 2C, D). The papillae are also concentrated dorsally on the tip of each of the transverse, posterior and lateral ridges (Fig. 1B, E), and immediately anterior to the posterior groove of the dorsal saddle (Fig. 1D). Ventrally the papillae are arranged in 6 transverse rows on the ventral surface between the mouth and gonopore (Fig. 2H). What are presumed to be ciliated structures are regularly arranged over the entire ventral surface of the disc of the posterior attachment organ (Fig. 2I). No ciliated sensory papillae are present in the centre of the distal concavities of the tentacles (Fig. 2F) or on the ventral epithelium of the body delineated by the epidermal groove encircling the peduncle, on the peduncle (Fig. 1F), or on the dorsal surface of the disc. No elongate or locomotory cilia are present.

Considerable variation occurs in the shape and surface details of worms fixed by each of the various methods. However, a number of consistent results occur. A summary of the results of a trial of fixation regimes is presented in Appendix 1.

Fixation by freezing in acetone/dry ice and liquid nitrogen are not useful to demonstrate the resting shape of worms due to strong lateral curling and distortion. Most intact rhabdites are discharged to the surface of worms fixed by freezing in dry ice/acetone (Fig. 2G) and liquid nitrogen.

Cold 3% glutaraldehyde severely wrinkles the epidermis of worms and hides the pattern of the ciliated sensory papillae on the body, epidermal grooves and transverse and posterior ridges (Fig. 2C & E). Ciliated sensory papillae, however, are most prominent on the tentacles of worms fixed in this way (Fig. 2C, F). The distal concavities of the three central tentacles are usually deepest and most obvious in worms fixed in cold 3% glutaraldehyde (Fig. 2E) and the pore openings there of rhabdite tracts are best demonstrated this way (Fig. 2F).

Fixation at ambient temperature produces inferior results both for the general shape of the body and for individual surface structures. Formalin used at ambient temperature strongly curls the worms. Glacial acetic acid and Berland's Fluid fix worms in a resting shape but cause severe epidermal damage including sloughing and bubbling. The epithelium of the tentacles and dorsal saddle are consistently damaged with the saddle often almost completely lifted off.

Hot fixatives best preserve the natural resting shape of live worms. Hot water is adequate for low magnifi-

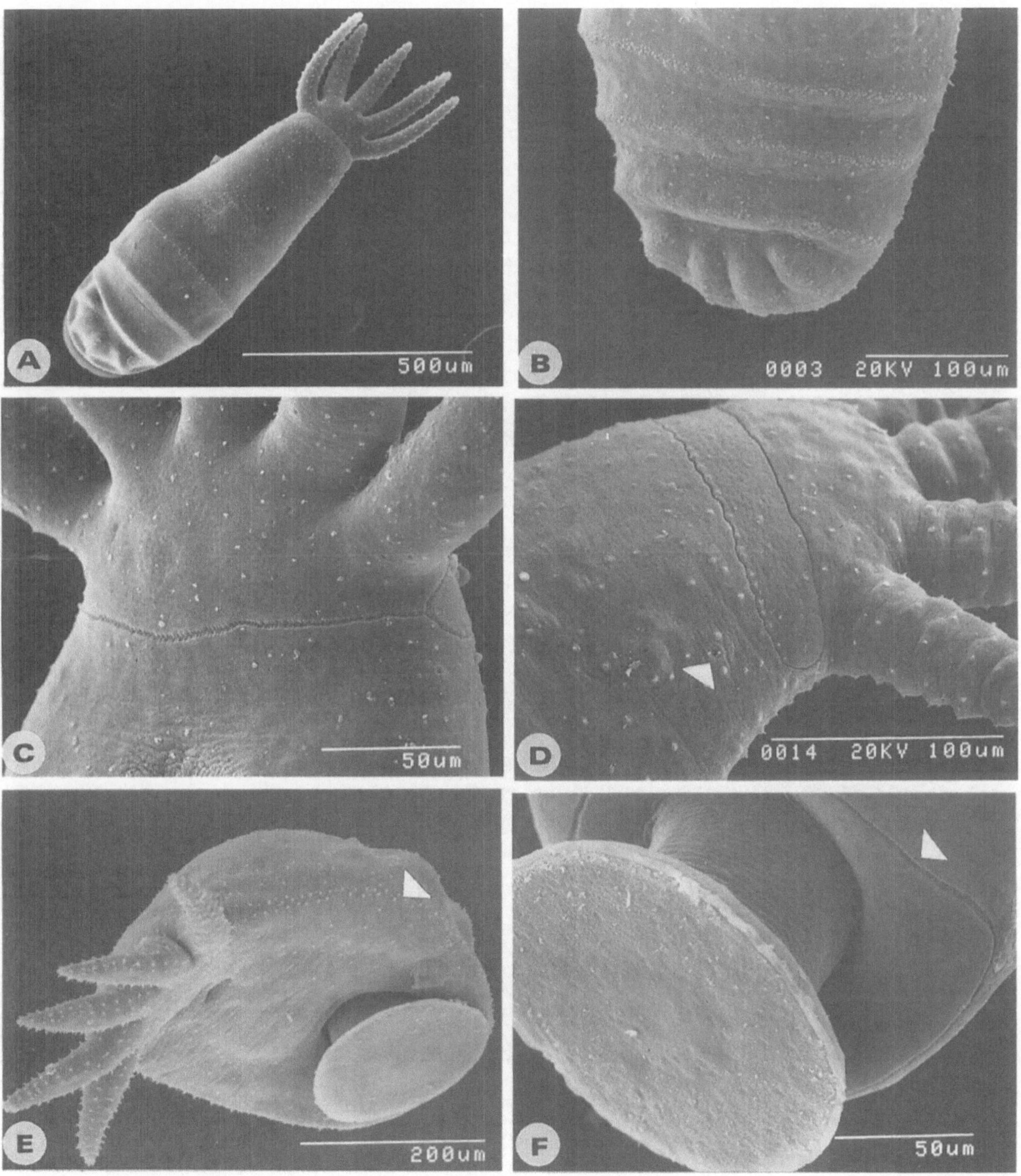

Fig. 1. (A) Dorsal view of *Craspedella* sp. fixed in hot 3% glutaraldehyde. (B) Dorsal view of posterior extremity of *Craspedella* sp. fixed in hot 3 % glutaraldehyde. (C) Anterior ventral surface of *Craspedella* sp. fixed in hot water showing the transverse epidermal groove at the bases of the tentacles. (D) Dorsal surface of *Craspedella* sp. fixed in hot water showing saddle-shaped region bordered by epidermal grooves at the bases of the tentacles. Arrow indicates a nephridiopore. (E) Ventral view of *Craspedella* sp. fixed in hot 10% formalin. Arrow indicates the line of ciliated sensory papillae on the lateral ridge. (F) The posterior attachment organ of *Craspedella* sp. fixed in hot water showing the epidermal groove on the ventral body surface which circumscribes the peduncle (arrow), and the central region of the ventral surface of the disc where pores are absent.

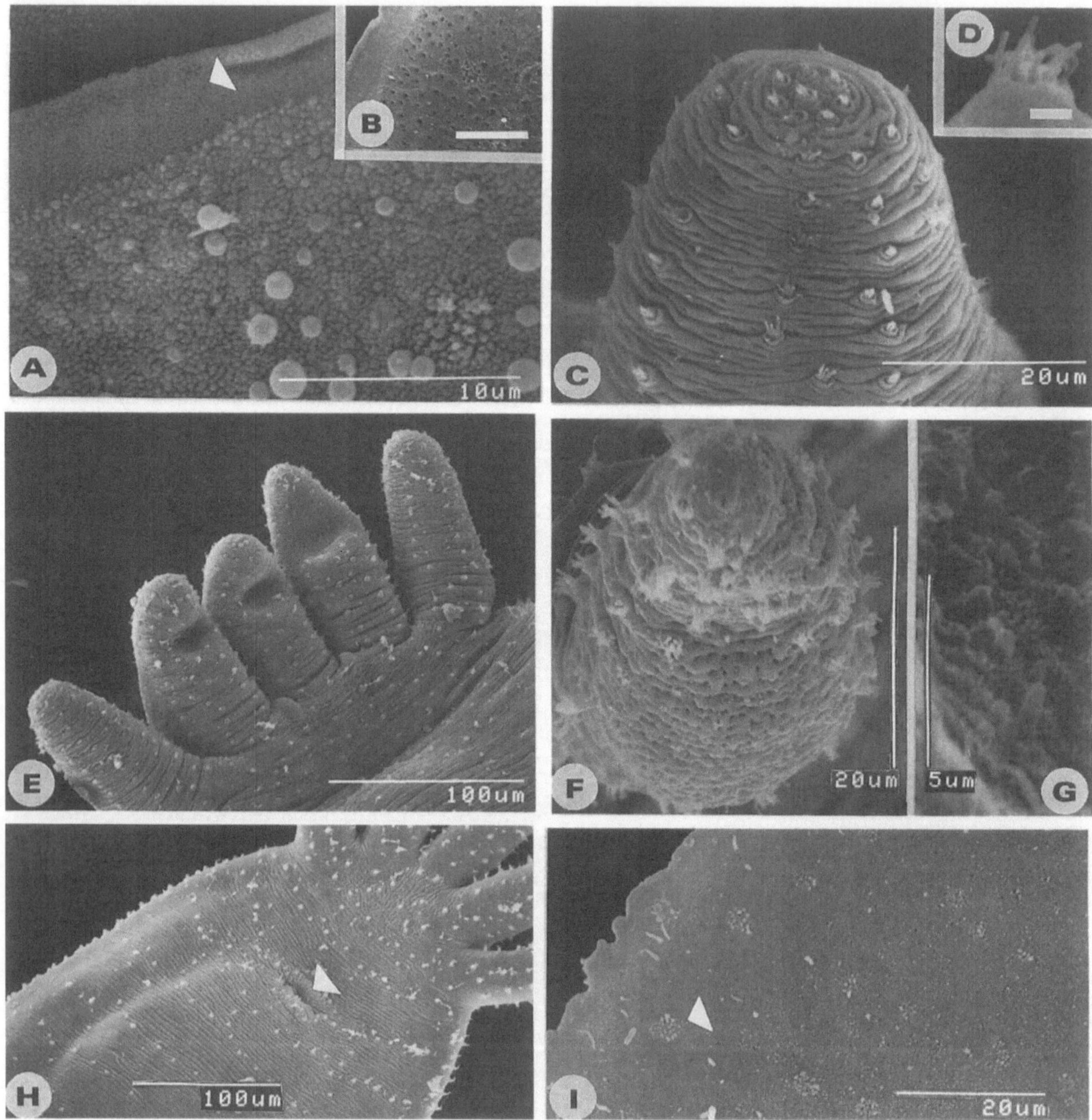

Fig. 2. (A) Edge of ventral surface of disc of *Craspedella* sp. fixed in hot 10% formalin) showing peripheral ridge (arrow) and surface covered with microvilli. (B) Edge of ventral surface of disc of *Craspedella* fixed in hot 3% glutaraldehyde showing pores, the openings of glands (scale bar = 10 μm). (C) Tentacle of *Craspedella* sp. fixed in cold 3% glutaraldehyde showing the arrangement of the multiciliated sensory papillae. (D) Multiciliated sensory papilla from tentacle fixed in hot 3% glutaraldehyde (scale bar = 1 μm). (E) Ventral view of the anterior end of *Craspedella* sp. fixed in cold 3% glutaraldehyde showing concavities in the distal region of the ventral epidermis of the central three tentacles. (F) Area of the tentacle concavity of *Craspedella* sp. fixed in cold 3% glutaraldehyde after narcotisation in 7% ethanol showing the opening of the rhabdite ducts and the absence of multiciliated sensory papillae. (G) Undischarged rhabdites protruding through the epithelium in the tentacle concavity of *Craspedella* sp. fixed by freezing in dry ice/acetone. (H) The ventral surface of *Craspedella* sp. fixed in hot 10% formalin showing the transverse rows of ciliated sensory papillae. Arrow indicates the mouth. (I) Disc of *Craspedella* sp. fixed in hot 10% formalin showing the presumed stunted cilia (arrow) on the surface of the disc.

Appendix 1

PROCEDURE

FREEZE FIXATION

(1) liquid nitrogen/ 3% glutaraldehyde in 100% methanol

Worms flooded with liquid nitrogen at -196°C while attached to the base of a plastic container. Half of the fluid was replaced with 3% glutaraldehyde in 100% methanol and the worms fixed for 24 hrs at -20°C. Worms were removed to room temperature (approx 25°C) and the fluid replaced with fresh fixative and stored for 24 hrs. Worms were then transferred to 100% methanol for 3 changes of 10 min duration at room temperature before they were critical point dried and processed as described for the other fixation methods.

(2) acetone/dry ice

Worms attached by their posterior attachment organ to a wooden point were immersed in a mixture of acetone/dry ice at -80°C. The solution was then allowed to return to room temperature.

COLD FIXATION

(3) 3% glutaraldehyde

Worms flooded with 3% glutaraldehyde buffered to pH 7.2 with 0.1M phosphate buffer at 4°C.

(4) 3% glutaraldehyde after narcotisation

Worms narcotised in 7% ethanol for 15 min then flooded with 3% glutaraldehyde buffered to pH 7.2 with 0.1M phosphate buffer at 4°C.

AMBIENT FIXATION

(5) glacial acetic acid

Worms flooded with 100% glacial acetic acid.

(6) Berland's Fluid

Worms flooded with Berland's Fluid (1 part formalin, 19 parts glacial acetic acid).

(7) 10% formalin

Worms flooded with 10% formalin phosphate buffered to pH 7.2.

HOT FIXATION

(8) tap water

Worms flooded with hot (approx. 90°C) water held for 30 sec then transferred to cold (4°C) 10% formalin phosphate buffered to pH 7.2.

(9) 10% formalin

Worms flooded with hot (approx. 90°C) 10% formalin phosphate buffered to pH 7.2 then cooled to room temperature.

(10) 3% glutaraldehyde

Worms flooded with hot (approx 90°C) 3% glutaraldehyde phosphate buffered to pH 7.2 and left to cool to room temperature.

Appendix 1. Continued

RESULTS

(1)

Rapid killing, strong lateral contortion of the body and twisting of the tentacles, most but not all worms well extended, epidermal fine detail well preserved, ciliated papillae prominent but pattern difficult to trace due to distortion of worms body and tentacles. Many intact rhabdites secreted from distal concavities of the central 3 tentacles.

(2)

as above except that distortion of the body and tentacles more severe.

(3)

Strong curling of body and tentacles with distortion, epidermis shrunk but fine detail well preserved, ciliated papillae prominent but pattern often difficult to trace due to distortion of worm's body and tentacles, epidermal grooves not obvious, distal concavities of the central 3 tentacles very obvious.

(4)

same as for 3% glutaraldehyde except curling of body and tentacles slightly less severe.

(5)

rapid killing, worms extended to resting position, epidermis swollen, epidermal damage very severe and often split along anterior epidermal grooves, epidermal fine detail extremely poor.

(6)

as for glacial acetic acid.

(7)

strong curling of worms, epidermis shrunk but fine detail well preserved, ciliated papillae prominent but pattern difficult to trace due to distortion of worm's body and tentacles, epidermal grooves not obvious.

(8)

kills rapidly, worms extended to resting position, epidermis swollen and ciliated sensory papillae obscured, epidermal fine detail poor, epidermal grooves very obvious, considerable contamination of epidermis by glandular secretions.

(9)

kills rapidly, worms extended to resting position, ciliated papillae prominent, epidermal grooves visible, distal concavities of tentacles visible, epidermal fine detail very good.

(10)

kills rapidly, worms extended to resting position, dorsal and ventral epidermis of body mildly shrunk into transverse folds, epidermal grooves visible, distal concavities of tentacles visible, epidermal fine detail excellent with ciliated papillae prominent.

cation work and best highlights the epidermal grooves but swells the epithelium so that the ciliated sensory receptors and microvilli cannot be observed clearly (Fig. 1C, D and F). Hot 10% formalin and particularly hot 3% glutaraldehyde is superior to hot water for the preservation of fine detail, including that of the ciliated sensory receptors and microvilli (Figs 1A & E; 2A, D, H & I). Pores on the disc are most obvious in worms fixed in hot 3% glutaraldehyde (Fig. 2B) For hot fixation, epidermal damage is less for worms fixed at approximately 90 °C than for those fixed at 100 °C. Lower temperatures increase curling of the worms and loss of the resting shape.

Discussion

Rohde (1987) stated that the multiciliated papillae of *Craspedella* sp. (= *C. spenceri*?) from *Cherax destructor* were sensory receptors which consisted of bundles of dendrites in a depression of the epidermis. Such papillae in temnocephalans, according to Williams (1982), may be mechanoreceptors or chemoreceptors. In *Craspedella*, the prominent transverse and posterior ridges presumably assist the sensory function of the ciliated sensory papillae by raising the papillae above the body surface thus increasing their chance of exposure to external stimuli.

Epidermal grooves have only previously been recorded for *Temnocephala novaezealandiae*, *T. dendyi*, *T. geonoma* and *T. fasciata* by Williams (1982, 1991, 1992), for in her study of species of *Temnocephala* from Argentina, Damborenea (1992) did not report grooves. Williams (1982) interpreted the grooves as the surface manifestations of lateral membranes with associated junctional synapses, and suggested that they divide the epithelium into a patchwork of structurally diverse regions. The position and pattern of the epidermal grooves which delineate the patchwork may prove useful as taxonomic characters. For example, the dorsal surface of *Temnocephala dendyi* has a pair of triangles on the dorsal surface linked by a single groove in place of the saddle described for *T. novaezealandiae* and *T. fasciata* (see Williams, 1982, 1992), and the nephridipore lies within the saddle of *T. novaezealandiae* and *T. fasciata* according to Williams, 1982, 1992, but outside the saddle of *Craspedella* sp.. Williams (1982) apparently did not observe with SEM an epidermal groove around the peduncle of *T. novaezealandiae*. She did report, however, (Williams, 1982), that transmission electron microscopy (TEM) revealed a well defined epidermal groove on the body surface near the stalk of the disc of *T. novaezealandiae*. We suggest that this groove is homologous to that observed on *Craspedella* sp. to circumscribe the ventral insertion of the peduncle. Similarly, we did not observe epidermal grooves associated with the lateral ridges of *Craspedella* as were reported for *T. novaezelandiae* by Williams (1982). We suggest that TEM examination may reveal epidermal grooves to occur in association with the lateral ridges of *Craspedella* sp.

The distal concavities of the three central tentacles, in which there is a concentration of rhabdite discharge, have not been described previously for any temnocephalan. The shape of the concavities suggests that during locomotion, worms may use a suctorial action to assist the attachment of the tentacles, perhaps generated by increasing the size of the concavity once the tentacle has been positioned (Nachtigall, 1974). If the concavities do function as suckers, the secretion may act (1) to seal the rim of the sucker and thereby enhance suction (Nachtigall, 1974), (2) to adhere the tentacles (Haswell, 1893) or (3) be protective (Williams, 1980; Williams & Ingerfield, 1988; Jennings *et al.*, 1992) by reducing abrasion in areas of the tentacle epidermis which contact the substrate during locomotion.

The distribution on the disc of the pores and their associated secretions, and in particular their absence from the centre of the disc is consistent with the roof of the organ being raised to create a partial vacuum to attach the worms to a substrate (Nachtigall, 1974; Williams, 1981). The peripheral ridge at the edge of the posterior attachment organ may serve as an aid to any suction, by sealing the rim (Nachtigall, 1974). A similar function has been ascribed to the marginal valve of the haptor of the monogenean *Entobdella solae* by Kearn (1964). The functional morphology of the attachment organs of *Craspedella* sp. from redclaw is the subject of another study (Sewell & Whittington, in press).

The precise nature of what we presumed were ciliated structures on the ventral surface of the disc of *Craspedella* sp. was not able to be determined. Damborenea (1992) observed 'bouquet-like' structures on the periphery of the ventral surface of the disc of *T. microdactyla*. She assumed these bouquets to be associated with the duo-gland system described by Tyler (1976). The structures we observed are more likely to be related to the short cilia observed on the ventral surface of the disc of *T. novaezealandiae* by Williams (1982). The structures are much less prevalent on the disc surface of *Craspedella* sp. than are the pore openings of the glands associated with the organ, and are present on the central region of the disc where pores and secretions are absent. We presume these structures are sensory.

Conclusion

If the primary goal of fixation is to preserve the specimen in a state that is as close to life as possible (Wischnitzer, 1981), then for SEM of the surface, hot fixatives best serve this purpose for *Craspedella*. We did not investigate, however, the likely detrimental effect of hot fixation on the internal structure of the worms. Hot fixatives were best for fixing worms in a shape close to that observed for resting live worms and performed optimally at approximately 90 °C. Fixation can be achieved safely by immersing a small vial of fixative into a container of near boiling water to heat the fixative to 90 °C.

For high resolution and fine detail, hot 3% glutaraldehyde and 10% formalin provided the best fixation while maintaining an excellent representation of the overall three dimensional appearance of live worms. Conversely, some of the less optimum fixatives were most useful to demonstrate the presence and/or nature of particular structures including the epidermal grooves, the distal concavities and associated rhabdite discharge of the tentacles.

Acknowledgments

The support of the Queensland Museum and the University of Queensland is gratefully acknowledged. We especially thank the Queensland Museum Photography Section. We thank Dr Malcolm Jones of the Centre for Microscopy and Microanalysis, The University of Queensland for assistance with the liquid nitrogen freezing of worms and for reading a draft of this manuscript. We also thank Dr Ian Whittington of the Department of Parasitology, University of Queensland for his comments on an early draft. This work was funded by an Australian Biological Resources Study grant 88/5909 awarded to LRGC.

Appendix

The details and results of various SEM fixation procedures on *Craspedella* sp.

References

Cannon, L. R. G., 1991. Temnocephalan symbionts of the freshwater crayfish Cherax quadricarinatus from northern Australia. Hydrobiologia 227 (Dev. Hydrobiol. 69): 341–347.

Damborenea, M. C., 1992. Scanning electron microscope study of epidermal surface of six ectosymbiotic Temnocephala species (Platyhelminthes) from Argentina. Hydrobiologia 246: 111–118.

Haswell, W. A., 1893. A monograph of the Temnocephaleae. Proceedings of the Linnean Society of New South Wales (Macleay Memorial Volume): 99–152.

Jennings, J. B., L. R. G. Cannon & A. J. Hick, 1992. The nature and origin of the epidermal scales of Notodactylus handschini – an unusual temnocephalid turbellarian ectosymbiotic on crayfish from northern Queensland. Biol. Bull. 182: 117–128.

Kearn, G. C., 1964. The attachment of the monogenean *Entobdella soleae* to the skin of the common sole. Parasitology 54: 327–325.

Matjašič, J., 1990. Monography of the family Scutariellidae (Turbellaria, Temnocephalidea). Academia Scientiarum et Artium Slovenica Classis IV: Historia Naturalis 28. Ljubljana, 1990, 167 pp.

Nachtigall, W., 1974. Biological Mechanisms of Attachment. The comparative morphology and bioengineering of organs for linkage, suction and adhesion. Translated by M. A. Biedermann-Thorson. Springer-Verlag, New York, Heidelberg, Berlin, 194 pp.

Rohde, K., 1987. Ultrastructural studies of epidermis, sense receptors and sperm of Craspedella sp. and Didymorchis sp. (Platyhelminthes, Rhabdocoela). Zool. Scr. 16: 289–295.

Sewell, K. B. & I. D. Whittington, (in press). A light microscope study of the attachment organs and their role in locomotion of *Craspedella* sp. (Platyhelminthes: Rhabdocoela: Temnocephalidae), an ectosymbiont from the branchial chamber of *Cherax quadricarinatus* (Crustacea: Parastacidae) in Queensland, Australia. J. Nat. Hist.

Tyler, S., 1976. Comparative ultrastructure of adhesive systems in the Turbellaria. Zoomorphologie 84: 1–76.

Williams, J. B., 1978. Studies on the epidermis of Temnocephala III. Scanning electron microscope study of the epidermal surface of Temnocephala dendyi. Aust. J. Zool. 26: 217–224.

Williams, J. B., 1980. Morphology of a species of Temnocephala (Platyhelminthes) ectocommensal on the isopod Phreatoicopsis terricola. J. Nat. Hist. 14: 183–199.

Williams, J. B., 1981. Classification of the Temnocephaloidea (Platyhelminthes). J. Nat. Hist. 15: 277–299.

Williams, J. B., 1982. Studies on the epidermis of Temnocephala VI. Epidermal topography of T. novaezealandiae and other Australasian temnocephalids, with notes on microvillar and coated vesicle function and the evolution of a cuticle. Aust. J. Zool. 30: 375–390.

Williams, J. B., 1991. Scanning electron microscope study of the epidermal surface of Temnocephala geonoma and an adherent diatom, with notes on the affinities of the Temnocephaloidea (Platyhelminthes). New Zeal. J. Zool. 30: 375–390.

Williams, J. B., 1992. Scanning electron microscopical study of Temnocephala fasciata (Platyhelminthes: Temnocephaloidea). J. Submicrosc. Cytol. Pathol. 24: 205–213.

Williams, J. B. & M. Ingerfield, 1988. Cells in the parenchyma of Temnocephala: rhabdite secreting cells of Temnocephala novaezealandiae (Temnocephalidae: Platyhelminthes). Int. J. Parasit. 18: 651–659.

Wischnitzer, S., 1981. Introduction to Electron Microscopy. Pergamon Press, New York, 405 pp.

Hydrobiologia **305**: 159, 1995.
L.R.G. Cannon (ed.), Biology of Turbellaria and some Related Flatworms.

An electron microscope study of primary epidermis formation in freshwater planarian embryos

Takashige Sakurai & Saburo Ishii
Division of Cell Science, Central Research Laboratory, Fukushima Medical College, Fukushima 960-12, Japan

Key words: Turbellaria, Tricladida, primary epidermis, ultrastructure, embryonic development

Abstract

Early development in freshwater planarians is generally considered to be highly modified to the point of being unique. A careful examination by TEM, however, suggests that the primary epidermis (Skaer, 1965) is formed in a rather regular manner but is partially inverted with respect to the definitive body axes. After formation of the yolk cell syncytium, the blastomeres enclosed within it increase in number in the central area. Some of these blastomeres then move peripherally as a group and fuse to form another syncytium, the primordium of the primary epidermis. This primordium contacts the surface of the yolk-cell syncytium at the place where the primordium will subsequently flow out. The primordium spreads to the opposite pole through the spaces among the syncytial and non-syncytial yolk cell masses.

We have shown that the formation of the primary epidermis in planarian embryos involves an epiboly-like movement of cells similar to that described for other neoophoran groups (Thomas, 1986). This suggests that triclad development may not have diverged as markedly from that of the other turbellarians as previously thought and that epidermal epiboly in the triclads might be homologous to gastrulation in the archoophora (Kato, 1940), though the direction of cell migration is reversed.

References

Kato, K., 1940. On the development of some japanese polyclads. Jap. J. Zool. 8: 537-573.

Skaer, R. J., 1965. The origin and continuous replacement of epidermal cells in the planarian *Polycelis tuenuis* (Iijima). J. Embryo. Exp. Morph. 13: 129-139.

Thomas, M. B., 1986. Embryology of the Turbellaria and its phylogenetic significance. Hydrobiologia 132 (Dev. Hydrobiol. 32): 105-115.

Hydrobiologia **305**: 161–165, 1995.
L.R.G. Cannon (ed.), *Biology of Turbellaria and some Related Flatworms.*

Ultrastructure of the epidermis of *Meara stichopi* (Platyhelminthes, Nemertodermatida) and associated extra-epidermal bacteria

Kennet Lundin & Jan Hendelberg
Department of Zoology, University of Göteborg, Medicinareg. 18, S-41390 Göteborg, Sweden

Key words: Nemertodermatida, epidermis, ciliary rootlets, glands, symbiosis, bacteria

Abstract

The ultrastructure of the epidermis of *Meara stichopi* Westblad, an endosymbiont in the holothurian *Parastichopus tremulus* (Gunneri) was studied. The entire body surface of the worm is ciliated. The columnar epithelial cells interdigitate with each other. The epidermal cells are not insunk and have a prominent terminal web underlined by mitochondria. Each cilium has a main, anteriorly pointing, cross-striated rootlet with a knee and a posterior rootlet dividing into two lateral fibre bundles. These bundles descend to contact the two neighbouring rootlets in the row behind. The basal bodies contain glycogen granules. Epidermal glands containing secretion globules are common. Some epidermal glands contain rhabdoids.

Between the epidermal cilia we found elongated procaryotic organisms, evidently symbiotic gram-negative bacteria, anchored with a tapering end in the glycocalyx.

The observations imply that the ultrastructure of the epidermis of *M. stichopi* is similar to that described earlier from *Nemertoderma* and *Flagellophora* (Nemertodermatida). Thus these three genera share some systematically interesting ultrastructural character states, especially the pattern of the ciliary rootlet system.

Introduction

Meara stichopi is a small flatworm with many primitive features. When describing the species, Westblad (1949) referred it to the taxon Nemertodermatida Karling 1940 within the 'Turbellaria archoophora'. The species is characterized by the two features regarded by Karling (1974) to be apomorphic character states for the Nemertodermatida, *i.e.* a nemertine-like epidermis and a statocyst with two statoliths. Electron microscopy has made it possible to use a number of new characters for evaluating the relationships of nemertodermatids, acoels and other flatworms, see for example Tyler & Rieger (1977), Hendelberg (1977, 1986), Tyler (1979), Smith (1981), Smith *et al.* (1982, 1986), Rieger (1981), Rieger *et al.* (1991), Ehlers (1985, 1992a, 1992b), Rohde *et al.* (1988) and Mamkaev (1991). Some of the significant characters used by these authors relate to the epidermal structure of the worms. The intention of this study was to find out if the ultrastructure of the epidermis of *M. stichopi* could be used to attain a better basis for evaluating the relationships between this species and other acoelomorphs. In the following we will give a description of the general ultrastructural morphology of the epidermis of *M. stichopi*, and discuss the phylogenetic implications of our findings.

In contrast to species of other genera referred to the Nemertodermatida which are freeliving, *M. stichopi* is an endosymbiont, found mainly in the intestine of a holothurian, *Parastichopus tremulus*. When studying the epidermis of *M. stichopi* we searched for features which could be interpreted as adaptations to the parasitic life. We then found bacteria, evidently well adapted to living between the epidermal cilia of the worm, anchored in the glycocalyx. A short account of the observations of these bacteria is given in this paper.

Materials and methods

Specimens of *Parastichopus tremulus* (Gunneri) were collected in October 1992 by dredging from a depth of

120–130 m in the Fanafjord and the Raunefjord west of the marine biological station at Espegrend, Bergen, on the west coast of Norway. The holothurians were kept in aquaria with running deepwater. Holothurians were dissected within a few hours or up to one day after capture. The intestine and the coelomic cavity of the holothurians were examined for parasitic flatworms under a dissecting microscope. Living specimens of *Meara stichopi* Westblad were easily identified by the statocyst which could be clearly seen near the anterior end. The worms were transferred to dishes with coelomic fluid from the holothurian. While swimming in this for a few seconds they cleaned themselves from debris from the intestine of the host. They were then immediately immersed in fixing fluid. To be certain of the identification of the species we studied squash preparations of some specimens under a light microscope with phase contrast optics. Comparision was made with specimens collected in the same region in Nov. 1976 and in Sept. 1989.

Two methods of fixation for electron microscopy were used. (1) Fixation in 3% glutaraldehyde in 0.1 M sodium cacodylate buffer pH 7.2 in filtered seawater for 4–24 h at about 6 °C followed by rinsing in buffer and postfixation for 3–4 h in 1% OSO_4 in buffer at 6 °C. (2) Fixation for 3–5 h at about 6 °C in 1% OSO_4 in filtered seawater. After fixation the specimens were stored at 4 °C for one or more weeks in buffer (1) or in filtered seawater (2). For transmission electron microscope (TEM) studies, specimens were dehydrated in an ethanol series and then embedded in Epon^{TM}. Ultrathin sections (±60 nm) were stained with uranyl acetate and lead citrate and examined in a Zeiss CEM 902 TEM. Uncontrasted comparatively thick sections (±0.5 μm) were examined in the TEM using electron spectroscopic imaging (ESI) as a method to get rid of inelastically scattered electrons which would otherwise interfere with the image of such thick sections. For scanning electron microscopy (SEM) studies, the specimens were fixed according to method (2), dehydrated in an acetone series, and critical point dried. They were examined in a Zeiss DSM 950 SEM.

Results

Specimens of *M. stichopi* were found most often in the foremost part of the intestine of *P. tremulus*, 1–2 cm behind the mouth, but were present in the whole intestine and sometimes in the coelomic cavity. The worms studied in vivo in the host during dissection were always seen swimming freely and were capable of rapid movement.

SEM studies showed the epidermis to be wholly covered with locomotory cilia. According to TEM studies the epidermis is cellular. The two fixation methods used gave the same results as far as the structures are reported here. The ciliated epithelial cells are columnar and somewhat interdigitating (Fig. 1). The basal parts are not insunk, *i.e.* the nucleus-bearing portion of the cell does not extend below the musculature. There is an outer layer of circular muscle cells and an inner layer of longitudinal muscle cells. It is noteworthy that *M. stichopi* has more strongly developed body-wall muscles on the dorsal side of the body. Therefore, when fixed, the worm curls towards the dorsal side. The extracellular matrix below the epithelial cells is poorly developed and indistinct.

The epidermal, locomotory cilia are about 5–6 μm long and have the 9+2 axonemal pattern typical of eukaryotes. Observations in longitudinal sections of the distal parts of the cilia indicate that some of the microtubular doublets do not reach the tip of the cilium. The remaining microtubular doublets continue to the distal tip where they end in a dense cap. The basal bodies of the cilia were often found to contain granules of the type described as glycogen granules in other acoelomorphs (for example Silveira, 1972; Hendelberg, 1976, 1981; Ehlers, 1985) (Fig. 4). The rootlets of the cilia are interconnected in a system as follows. Each cilium has an apically oriented, striated, main rootlet with a knee-like bend and a posterior rootlet (Fig. 3). The latter splits into two fibre bundles that descend to connect to the two neighbouring main rootlets in the row next behind (Fig. 4). Below the ciliary rootlets there is a prominent, thick terminal web (Figs 2–4) which is underlined by mitochondria (Fig. 2).

Epidermal gland cells are abundant. Most of them contain spherical to ovoid electrondense secretion globules about 0.5–1.0 μm across (Fig. 1). Some glands contain elongated rhabdoids, about 5 μm long and 0.4–0.5 μm in diameter (Fig. 2), with a homogeneous fine-grained substructure albeit with a slightly denser core. Individual rhabdoids lack a surrounding membrane; instead, several rhabdoids occupy a membrane-bounded compartment. The rhabdoid glands appear quite scarce in the epidermis.

Sparsely occurring noncollared monociliary receptor cells with single, straight, striated rootlets were found to be dispersed in the epidermis.

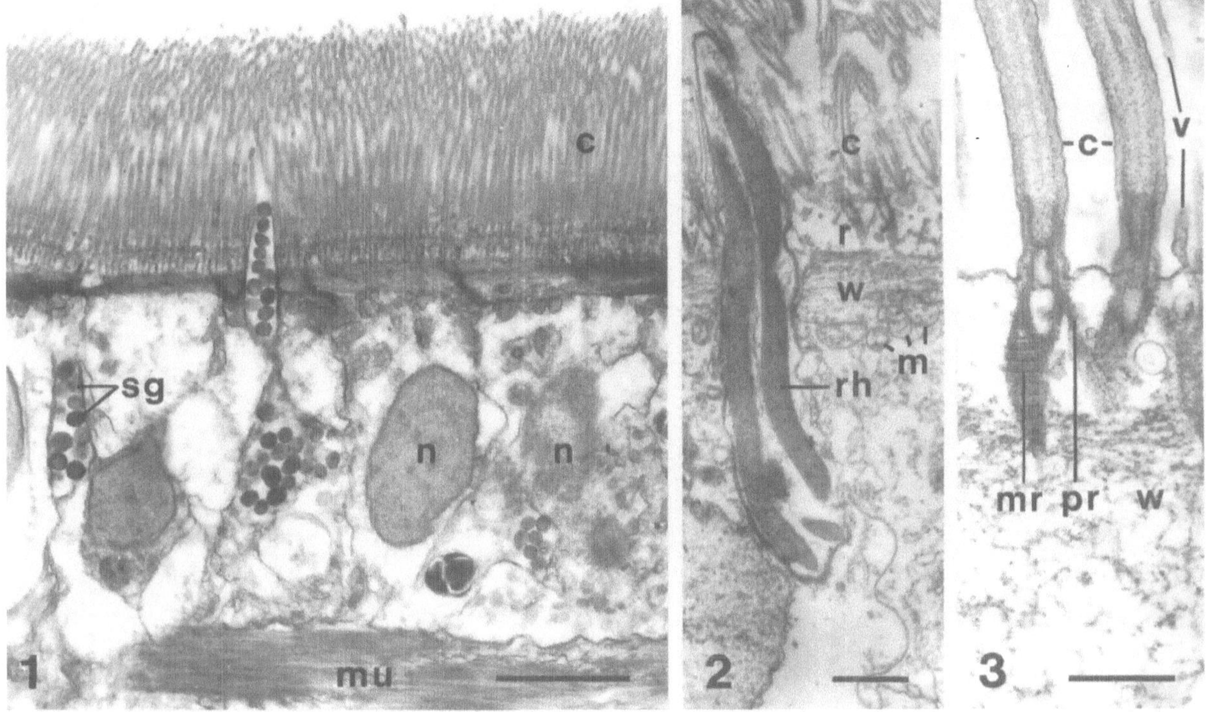

Figs 1–3. Transmission electron micrographs of the epidermis of *M. stichopi.* (*1*) Semithin (0.5 μm), unstained cross section showing dorsal body wall; scale bar 3 μm. (*2*) Ultrathin (as the following) nearly longitudinal section of epidermis with rhabdoids; scale bar 1.0 μm. (*3*) Nearly longitudinal section; scale bar 0.5 μm. – c, cilia; m, mitochondria; mr, main rootlet; mu, circular muscle cell; n, nucleus; pr, posterior rootlet; r, ciliary rootlets; rh, rhabdoid; sg, secretion globules; v, microvilli; w, terminal web.

Bacteria of about the same length and diameter as the epidermal cilia were found on the surface of the epidermis. They are mingled with and parallel to the cilia (Figs 5–6). However, they do not make contact with the epithelial surface. Instead the ends of the bacteria bend off close to the cell membrane (Fig. 4) and, while tapering, winds between the bases of the cilia. Thus each bacterium is anchored in the glycocalyx. However, occasionally one end of a bacterium was found in a deep, narrow pit in the epidermis. The substructure of individual bacteria is characterized by two successive membranes surrounding an electron-dense outer part, the ribosome-rich cytoplasm, and a lucent inner part with a central filamentous strand, the DNA (Fig. 5). No bacterial flagella were found.

In the specimens of *M. stichopi* studied in this respect, the bacteria were found to be most abundant on the ventral side of the worm. In this region we found in three worms studied mean values of about 1–2 bacteria/μm^2. Studies of material collected earlier, in 1976 and 1989, showed the same kind of bacteria to occur in the same way as was found in the material collected for the present study.

Discussion

Two general patterns of interconnection between the rootlets of epidermal locomotory cilia have been found in the flatworm group Acoelomorpha. In the Acoela the rootlets have been found to be interconnected at two levels. One is most often near the middle of the main rootlets, where two fibre bundles from the posterior rootlets make contact with the main rootlets in the row next behind, not easy to reveal, but observed in some of the species studied; another is the level of the tips of the main and lateral rootlets, where two lateral rootlets starting from the main rootlet make contact with the two main rootlets in the row next behind (e.g. Dorey, 1965; Bedini & Papi, 1974; Hendelberg & Hedlund, 1973, 1974; Tyler, 1979; Hendelberg, 1981; Ax, 1984; Ehlers, 1985; Smith *et al.*, 1986; Rohde *et al.*, 1988; Rieger *et al.* 1991; Ehlers, 1992a). In the few

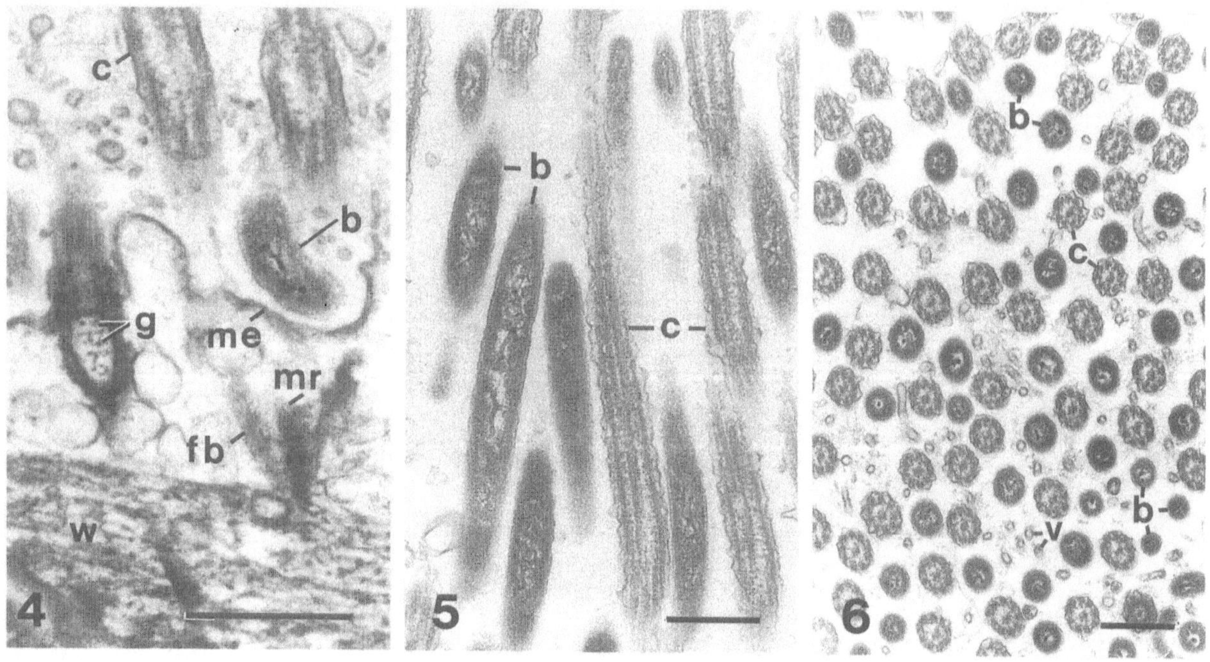

Figs 4–6. Transmission electron micrographs of the epidermis of *M. stichopi*. Scale bar 0.5 μm. (*4*) Part of cross section showing how two fibre bundles (from posterior rootlets) connect with a main rootlet in the row behind, and how a bacterium bends off near the cell membrane. (*5*) Part of cross section showing bacteria and cilia in longitudinal section. (*6*) Part of tangential section showing bacteria and cilia in cross section. –b, bacterium; c, cilium, cilia; fb, fibre bundle; g, glycogen granules; me, cell membrane; mr, main rootlet; v, microvilli; w, terminal web.

free-living species of Nemertodermatida studied earlier, connections were found on the first of the two levels mentioned for the Acoela. Thus, in *Nemertoderma* two fibre bundles extending from the posterior rootlet make contact with the knee-like bend of the main rootlets of adjacent cilia (Tyler & Rieger, 1977; Rieger *et al.*, 1991). The micrographs of the rootlet system of *Flagellophora* demonstrated by Smith *et al.* (1986, Figs 5 and 12) apparently shows the same kind of connections. Lateral rootlets emerging from the main rootlets are absent from these nemertodermatids. In *M. stichopi* we found an arrangement of the ciliary rootlets similar to that found in the mentioned, free-living nemertodermatids, that is, no lateral rootlets emerging from the main rootlets were found. If this general type is plesiomorphic for the Acoelomorpha, as suggested by Ehlers (1985), it means that the character states of the ciliary rootlet system do not contradict referal of *M. stichopi* to the Nemertodermatida. However, the possibility exists that the two kinds of ciliary rootlet patterns found in Acoela and Nemertodermatida respectively, evolved independently from a more primitive type, cf. Rohde *et al.* (1988). If so, the rootlet pattern demonstrated here in *M. stichopi* might be an autapomorphy for the Nemertodermatida. More data are needed before the phylogeny of the Nemertodermatida can be evaluated.

A character state specific for *M. stichopi* is the unusually thick terminal web in the epidermal ciliated cells. Another is the occurrence of rhabdoids, which have not been reported from any other nemertodermatid, but are present in acoels and other flatworms. The absence of an individual rhabdoid membrane and the aggregation of several rhabdoids in a common membrane-bounded compartment are characters also found in some acoels (see Smith *et al.*, 1982). However, the rhabdoids of *M. stichopi* lack the filamentous structures found in the acoel rhabdoids.

Our observations indicate but do not clearly show that there is a shelf near the tip of the locomotory cilia like that described earlier for free-living Nemertodermatida (Tyler & Rieger, 1977; Rieger *et al.*, 1991). A three-dimensional reconstruction is wanted to prove such a resemblance.

In spite of the life of *M. stichopi* as an endosymbiont, no conspicuous morphological adaptations to this mode of life could be found, such as a partial reduction of the ciliary coverage, found in many other endosymbiotic flatworms.

The extraepidermal bacteria found among the cilia clearly show a symbiotic relationship with the worm. That their presence was not a matter of temporary infection is evident from our observations of the same kind of bacteria on specimens of *M. stichopi* collected earlier, in 1976 and 1989. In fact, the size and position of the bacteria being so reminiscent of the cilia, indicate a far reaching adaptation to their life on the surface of the worm. The two surrounding membranes, the outer membrane and the cytoplasmic membrane, are diagnostic of gram-negative bacteria. Bacteria associated with the epidermis of turbellarian flatworms have been reported from proseriates (B. Ehlers, 1977; Sopott-Ehlers, 1989). In these worms they were found, however, in areas without cilia.

Acknowledgments

The directors and staff of the Institute for Marine Biology, Bergen, provided working space, equipment and help with collection of material. Ulf Jondelius provided material collected in 1989 for comparision. Ms Inger Holmqvist helped with preparations. Financial support was received from the Swedish Natural Science Research Council.

References

Ax, P., 1984. Das phylogenetische System. Systematisierung der lebenden Natur aufgrund ihrer Phylogenese. G. Fischer, Stuttgart, New York, 349 pp.

Bedini, C. & F. Papi, 1974. Fine structure of the turbellarian epidermis. In N. W. Riser & M. P. Morse (eds), Biology of the Turbellaria, McGraw Hill, New York, St. Louis, etc.: 108–147.

Dorey, A. E., 1965. The organization and replacement of the epidermis in acoelous turbellarians. Q. J. Microsc. Sci. 106: 147–142.

Ehlers, B. 1977. 'Trematoden-artige' Epidermisstrukturen bei einem freilebenden proseriaten Strudelwurm (Turbellaria, Proseriata). Acta zool. Fenn. 154: 129–136.

Ehlers, U., 1985. Das phylogenetische System der Plathelminthes. G. Fisher, Stuttgart, New York, 317 pp.

Ehlers, U., 1992a. On the fine structure of Paratomella rubra Rieger & Ott (Acoela) and the position of the taxon Paratomella Dörjes in a phylogenetic system of the Acoelomorpha (Plathelminthes). Microfauna Marina 7: 265–293.

Ehlers, U., 1992b. Frontal glandular and sensory structures in Nemertoderma (Nemertodermatida) and Paratomella (Acoela): ultrastructure and phylogenetic implications for the monophyly of the Euplathelminthes (Plathelminthes). Zoomorphology 112: 227–236.

Hendelberg, J., 1976. Granules of glycogen beta-particle type demonstrated in epidermal ciliary rootlets of acoelous turbellarians. J. Ultrastruct. Res. 54: 491.

Hendelberg, J., 1977. Comparative morphology of turbellarian spermatozoa studied by electron microscopy. Acta zool. Fenn. 154: 149–162.

Hendelberg, J., 1981. The system of epidermal ciliary rootlets in Turbellaria. Hydrobiologia 84: 240.

Hendelberg, J., 1986. The phylogenetic significance of sperm morphology in the Platyhelminthes. Hydrobiologia 132 (Dev. Hydrobiol. 32): 53–58.

Hendelberg, J. & K.-O. Hedlund, 1973. Electron microscope studies of the junctions between ciliary rootlets and the orientation the ciliary axonemes in the epidermis of acoelous turbellarians. J. Ultrastruct. Res. 44: 440.

Hendelberg, J. & K.-O. Hedlund, 1974. On the morphology of the epidermal ciliary rootlet system of the acoelous turbellarian Childia groenlandica. Zoon. 2: 13–24.

Karling, T. G., 1974. On the anatomy and affinities of the turbellarian orders. In: N. W. Riser & M. P. Morse (eds), Biology of the Turbellaria. New York: McGraw-Hill; 1–16.

Mamkaev, Yu. V., 1991. On the phylogenetic significance of sagittocysts and copulatory organs in acoel turbellarians. Hydrobiologia 227 (Dev. Hydrobiol. 69): 307–314.

Rieger, R. M., 1981. Morphology of the Turbellaria at the ultrastructural level. Hydrobiologia 84: 213–229.

Rieger, R. M., S. Tyler, J. P. S. Smith & G. E. Rieger, 1991. Platyhelminthes: Turbellaria. In F. W. Harrison (ed.), Microscopic Anatomy of Invertebrates, 3. Wiley-Liss, N.Y.: 7–140.

Rohde, K., N. Watson & L. R. G. Cannon, 1988. Ultrastructure of epidermal cilia of Pseudactinoposthia sp. (Platyhelminthes, Acoela); implications for the phylogenetic status of the Xenoturbellida and Acoelomorpha. J. Submicrosc. Cytol. Pathol.: 20: 759–767.

Silveira, M., 1972. Association of polysaccharid material with certain flagellar and ciliary structures in turbellarian flatworms. Revue Microsc. Electron. 1: 96.

Smith, J. P. S., 1981. Fine-structural observations on the central parenchyma of Convoluta sp. Hydrobiologia 84: 259–265.

Smith, J. P. S., S. Tyler & R. Rieger, 1986. Is the Turbellaria polyphyletic? Hydrobiologia 132 (Dev. Hydrobiol. 32): 13–21.

Smith, J. P. S., S. Tyler, M. B. Thomas & R. M. Rieger, 1982. The morphology of turbellarian rhabdites: phylogenetic implications. Trans. am. Microsc. Soc. 101: 209–228.

Sopott-Ehlers, B. 1989. Zur Ultrastruktur von Bulbotoplana acephala Ax (Proseriata, Plathelminthes). Microfauna Marina 5: 219–226.

Tyler, S., 1979. Distinctive features of cilia in metazoans and their significance for systematics. Tissue Cell 11: 385–400.

Tyler, S & R. M. Rieger, 1977. Ultrastructural evidence for the systematic position of the Nemertodermatida (Turbellaria). Acta Zool. Fenn. 154: 193–207.

Westblad, E., 1949. On Meara stichopi (Bock) Westblad, a new representative of Turbellaria archoophora. Ark. Zool. 1: 43–57.

Hydrobiologia **305**: 167, 1995.
L.R.G. Cannon (ed.), Biology of Turbellaria and some Related Flatworms.

Ultrastructure and differentiation of ciliated pits in *Microstomum lineare* (Turbellaria, Macrostomida)

Irmeli E. Palmberg
Department of Teacher Education, Åbo Akademi University, P.O. Box 311, 65101 Vasa, Finland

Key words: ultrastructure, ciliated pits, differentiation, immunoreactivity, Turbellaria

The microturbellarian *Microstomum lineare* bears two pairs of glandulo-sensory organs, *i.e.* epidermal pigmented eyespots and ciliated pits. The ciliated pits are bottle-shaped grooves which open to the lateral margins of the worm. The walls of each pit consist mainly of two kinds of cells. The first, dominant cell type is placed postero-laterally along the wall. It is characterized by several cilia with prominent rootlets, apical mitochondria and a well-developed Golgi-complex with amorphous vacuoles of different sizes. The second cell type has numerous microvilli, in addition to the cilia, abundant microtubules and apically placed, electron-dense granules. A few multiciliary receptors lie between these cells, characterized by abundant neurotubules, mitochondria and vacuoles (Fig. 1).

The ciliated pits begin as invaginations of the epidermis in asexually reproducing worms and continue as a pair of cell aggregations posterior to the brain primordium. Differentiating nerve cells, gland cells, and cells with numerous centrioles in the cytoplasm are observed in these cell aggregations which simultaneously are abundantly innervated by fibres from the lateral nerve cords, and later also from the brain. Immunoreactivity to the neuropeptides FMFR/RFamide, NPF and substance P are observed in these fibres and in some cells in the vicinity of the pits, but not in any of the cells forming the pits.

Because of their ultrastructure, the function of the ciliated pits is argued to be chemoreceptive and similar to the olfactory organs of fishes. Their close connection to the protonephridium indicates a possibility that they are also involved in osmoregulation.

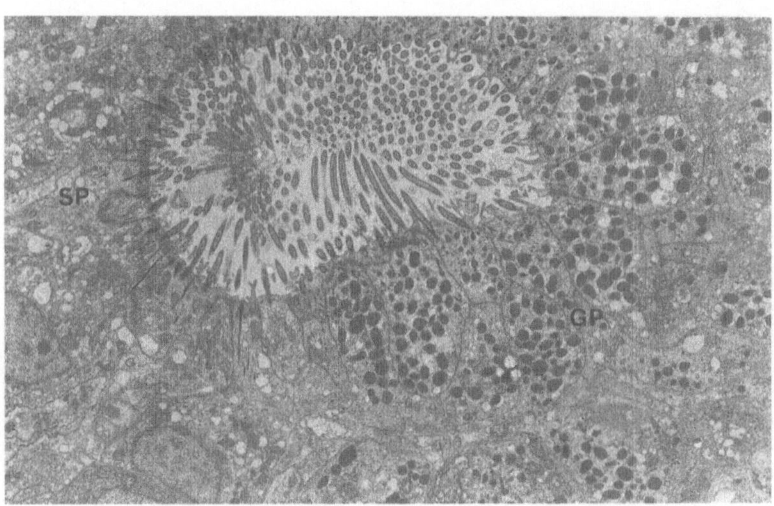

Fig. 1. Overview of ciliated pit showing glandular part (GP) and sensory part (SP) of wall in *M. lineare* (× 6000).

Hydrobiologia **305**: 169–175, 1995.
L.R.G. Cannon (ed.), Biology of Turbellaria and some Related Flatworms.

Ultrastructure of the ciliary pits in the *Geocentrophora* group (Platyhelminthes, Lecithoepitheliata)

Olga I. Raikova[1], Maria Reuter[2], Inger Böckerman[2] & Oleg A. Timoshkin[3]
[1]*Zoological Institute of the Academy of Sciences of Russia, 199034 St. Petersburg, Russia*
[2]*Department of Biology, BioCity, Åbo Akademi University, Fin-20521 Åbo, Finland*
[3]*Limnological Institute of Siberian Branch of Academy of Sciences of Russia, 664033 Irkutsk, Russia*

Key words: Lecithoepitheliata, ultrastructure, ciliary pit, sensilla, protonephridia, osmoregulation

Abstract

The ultrastructure of the paired lateral ciliary pits in several endemic species of *Geocentrophora* from Lake Baikal and in one cosmopolitan species, *G. baltica*, has been compared and the possible functional significance is discussed. The pit is composed of two distinctive parts; the bottom of the pit is an extensive sensitive area, filled with uni-and biciliary sensory receptors with reduced rootlets and numerous neurotubules. The walls of the pit are formed by several large 'dark cells', characterized by a dark cytoplasm with numerous mitochondria, a large nucleus, intracellular canaliculi, basal infoldings of the cell membrane, glycogen granules and a varying number of cilia. A protruding, densely ciliated ridge occurs along the anterior wall of the pit. The cilia have a strengthened rootlet system and seem to provide a strong water current into the pit. Dark cell processes penetrate the basement membrane of the pit and come into the vicinity of large cells with a cytoplasm similar to that of the 'dark cells' of the pit. These large cells in their turn come close to the terminal parts of the protonephridial canals, containing a weir. Smaller protonephridial capillaries without a weir seem to open directly into the pit lumen. The morphological data obtained suggest that the ciliary pit in not only a sensory structure, but plays a part in osmoregulation and ion exchange as well.

Introduction

Geocentrophorans are a turbellarian group, 50% of which are endemics of Lake Baikal. Because of the random and incomplete data on turbellarians belonging to the taxon Lecithoepitheliata, more data are needed for determination of their taxonomic position and phylogenetic relationships (Timoshkin, 1991). The presence of two lateral ciliary pits located ventrally at the anterior end of the animal is characteristic for the family Prorhynchidae. The pits have been described only by light microscopy (Kepner & Taliaferro, 1916, Reisinger, 1968) and are thought to be sensory structures. In the present study the fine structure of the ciliary pits of several representatives of *Geocentrophora* is compared and the possible functions are discussed.

Material and methods

Five species of the genus *Geocentrophora* from Lake Baikal: *G. porfirievae*, *G. wagini*, *G. wasiliewi*, *G. levanidorum* and *G. intersticialis* and the cosmopolitan species *G. baltica* were fixed according to Böckerman *et al* (1994). The best results were obtained when animals were flat fixed on ice in 2% glutaraldehyde buffered with 0.05 M Na-cacodylate buffer pH 7.4 at 4°C for 1.5 h, washed three times in the same buffer, postfixed for 1 h in cold buffered 1% OsO_4, dehydrated in acetone and embedded in Epon 812. Ultrathin sections mounted on Formvar-coated grids, stained with uranyl acetate and lead citrate were examined under a JEOL JEM 100 SX electron microscope.

Results

Paired ciliary pits are situated in the anterior body region. The ultrastructural features of the pits in all the investigated *Geocentrophora* spp. are similar. The pits open at the lateral body margins, the pit itself extending backwards from the opening. Each ciliary pit (Fig. 1) is composed of two distinctive parts: the bottom of the pit is a sensitive area, penetrated by numerous sensilla. The walls of the pit are formed by large cells with a dark cytoplasm, here called 'dark cells'. The lumen of the pit contains numerous cilia, those of the sensory cells arise from the bottom and do not reach the opening of the pit, while the cilia of the dark cells protrude from the pit above the body surface. The pit is underlined by a thick basement membrane.

The sensitive area at the bottom of the pit is composed of closely packed sensory receptors, bearing one (sometimes two) cilia (Fig. 2a,b). The distal parts of the cilia are deformed, having a dark matrix and axonemes lacking dyneine arms (Fig. 2c). The sensilla are characterized by a reduced rootlet arising from the basal body and numerous neurotubules (Fig. 2d). The receptor processes pass through the epithelial cells with a highly vacuolized cytoplasm and then through the basement membrane. Neuronal vesicles of different sizes and densities are observed in the sensilla and in a neuropilar area immediately beneath the pit (Fig. 2b). In this area synapses are seen.

The walls of the pit are formed by several (5-8) dark cells underlined by a basement membrane. These cells are unevenly ciliated, sometimes the ciliary density is very low (Figs. 1, 3a) and in some places it is higher than on the body surface. The anterior wall of the pit forms a longitudinal cytoplasmic ridge, protruding into the pit lumen (Fig. 3a). The ridge is densely ciliated with cilia having a strengthened rootlet system (Fig. 3b). Both the anterior and the vertical rootlets are split into two, sometimes into three, long striated rootlets. Derivates of the anterior rootlet are oriented towards the pit opening, thus the water current, provided by the movement of the ridge cilia, must go into the pit. On the opposite side of the pit the dark cells bear long straight microvilli and few cilia with normal rootlets.

Large nuclei of dark cells are situated above the basement membrane of the pit (Figs. 1, 3, 4). The apical cytoplasm of the dark cells contains numerous vacuoles or canaliculi, sometimes filled with flocculent material, sometimes opening at the surface (Fig. 3c, d). Occasional ciliary structures are seen in the canaliculi of the cytoplasm (Fig. 3d). Many basal infoldings of the cell membrane are observed (Fig. 3e), and a considerable number of mitochondria with a dark matrix and irregular cristae occur both in the apical and basal parts of the cell (Fig. 3d, e). In addition, glycogen granules and Golgi complexes are observed (Fig. 3c, e). Processes of the dark cells filled with numerous mitochondria pass down through the basement membrane (Fig. 4a). Below the basement membrane they approach large cells with a cytoplasm similar to that of the dark cells at the surface of the pit (Figs 1, 4a).

In the vicinity of the ciliary pit, numerous protonephridial canals and capillaries are observed. Some of them have a ciliary weir inside, but the small ones lack ciliary structures and have only thin cytoplasmic walls surrounding the canal lumen. No complete set of serial sections was made, but the examination of certain section reveal, that at least some of the smaller capillaries are continuous with the flame bulbs. The large cells on the proximal side of the basement membrane make contact with the terminal parts of protonephridial canals (Fig. 4a, b). In the cytoplasm of the protonephridia there are closely packed mitochondria with few cristae but no intracellular canaliculi (Fig. 4c). The smaller capillaries come close to the pit lumen and seem to open into it (Figs 1, 4d). They have walls of granular cytoplasm with dense homogenous granules. Similar capillaries open on the epidermal surface of the anterior end of the body (Fig. 4e).

Discussion

Ciliary pits occur in several turbellarian taxa: Catenulida, Macrostomida, Prolecithophora and Lecithoepitheliata. They have generally been considered as chemoreceptive sense organs (Rieger *et al.*, 1991). To date only the ciliary pits of Catenulida have been studied by electron microscopy (Reuter *et al.*, 1993). In the ciliary pits of *Stenostomum leucops*, three ultrastructurally different types of sensilla have been found. None of them shows structural homology with the type of sensillum found in the ciliary pit of geocentrophorans. The 'microvillous' type with a dark cytoplasm, numerous microvilli and abundant mitochondria instead resembles the microvilli bearing dark cells. Any interpretation of the sensory modality of the sensilla is premature. The construction of the walls of the pit in geocentrophorans, however, indicates that environmental information may be brought to the sensilla at the bottom of the pit in a very effective mode. Cilia of the pro-

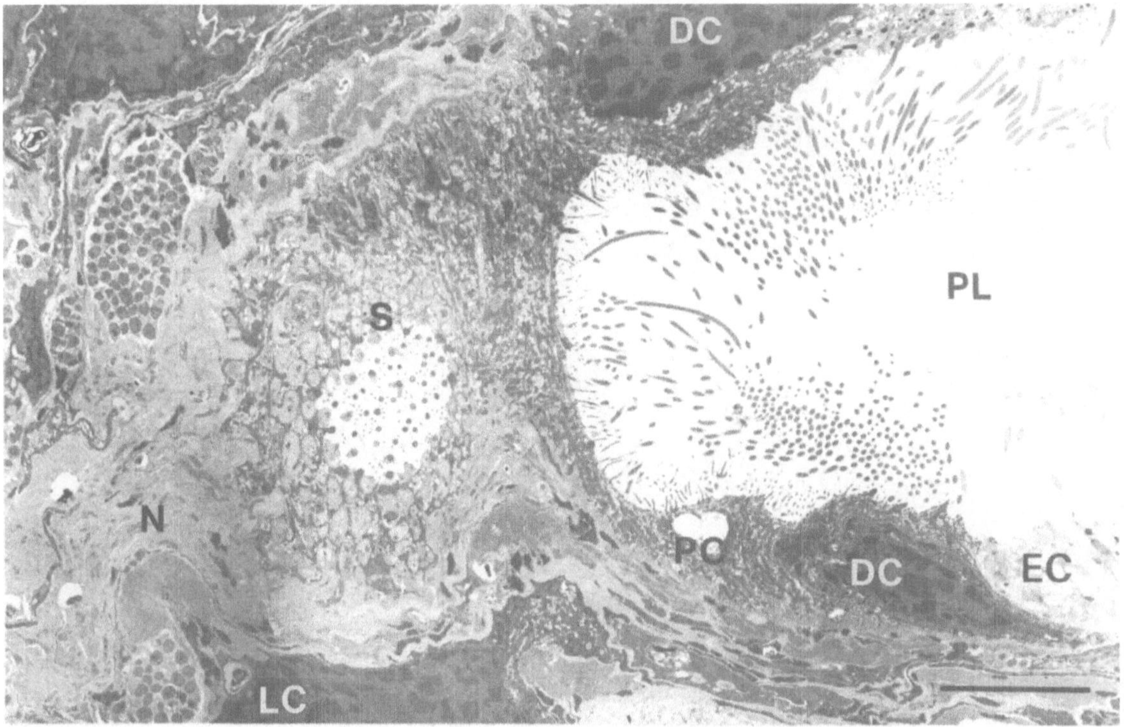

Fig. 1. *Geocentrophora levanidorum*, ciliary pit, longitudinal section. Epidermal cells (EC), dark cells (DC), large cells around the pit (LC), nerve processes (N), protonephridial capillaries (PC), pit lumen (PL), sensillae at the bottom of the pit (S). Scale 5 μm.

truding ridge in the pit have multiple rootlets strongly anchoring the cilium. It suggests high activity of these cilia provides a strong water current into the pit.

In Catenulida, no connection between the pit and protonephridia was detected (Reuter *et al.*, 1993), while in the geocentrophorans, neighbouring protonephridial canals and small capillares were observed. In the lecithoepitheliate *Xenoprorhynchus* paranephrocytes engaged in the excretory process appear along the protonephridial canals (Reisinger, 1968). Ultrastructurally the dark cells as well as the large cells below the pit resemble the parenephrocytes. Other similar features are the intracellular lacunae, numerous mitochondria and surface enlargement by microvilli, features typical of transporting epithelia. In *Daphnia magna*, gill epithelial cells with elaborate tubular systems, surface infoldings and a rich supply of mitochondria, are suggested to play an important role in osmoregulation (Kikuchi, 1983). Similar morphological features of the gill epidermal cells of the annelid *Diopatra neapolitana* are interpreted in accordance with an osmoregulatory function (Mendez *et al.*, 1984). The infoldings of the basal cell membrane, the numerous mitochondria and the canaliculi in the dark cells, are features that closely resemble invertebrate gill epithelial cells presumed to be involved in ion transport and osmoregulation (for review, see Mendez *et al.*, 1984). Thus in *Geocentrophora*, the ciliary pit seems to be a complex organ which might perform not only sensory but also osmoregulatory functions.

Acknowledgments

The authors are very grateful to Prof. Y. V. Mamkaev, Dr D. Bogoliubov and Dr B. I. Joffe for providing us with live material. We also want to thank Mrs G. Henriksson and Mr E. Nummelin for technical assistance and Åbo Akademi University for financial support. The work was partly carried out with the financial support of the Russian Basic Research Foundation under the project No. 93-04-21226.

References

Böckerman I., M. Reuter & O. Timoshkin, 1994. Ultrastructural study of the central nervous system of endemic Geocentrophora

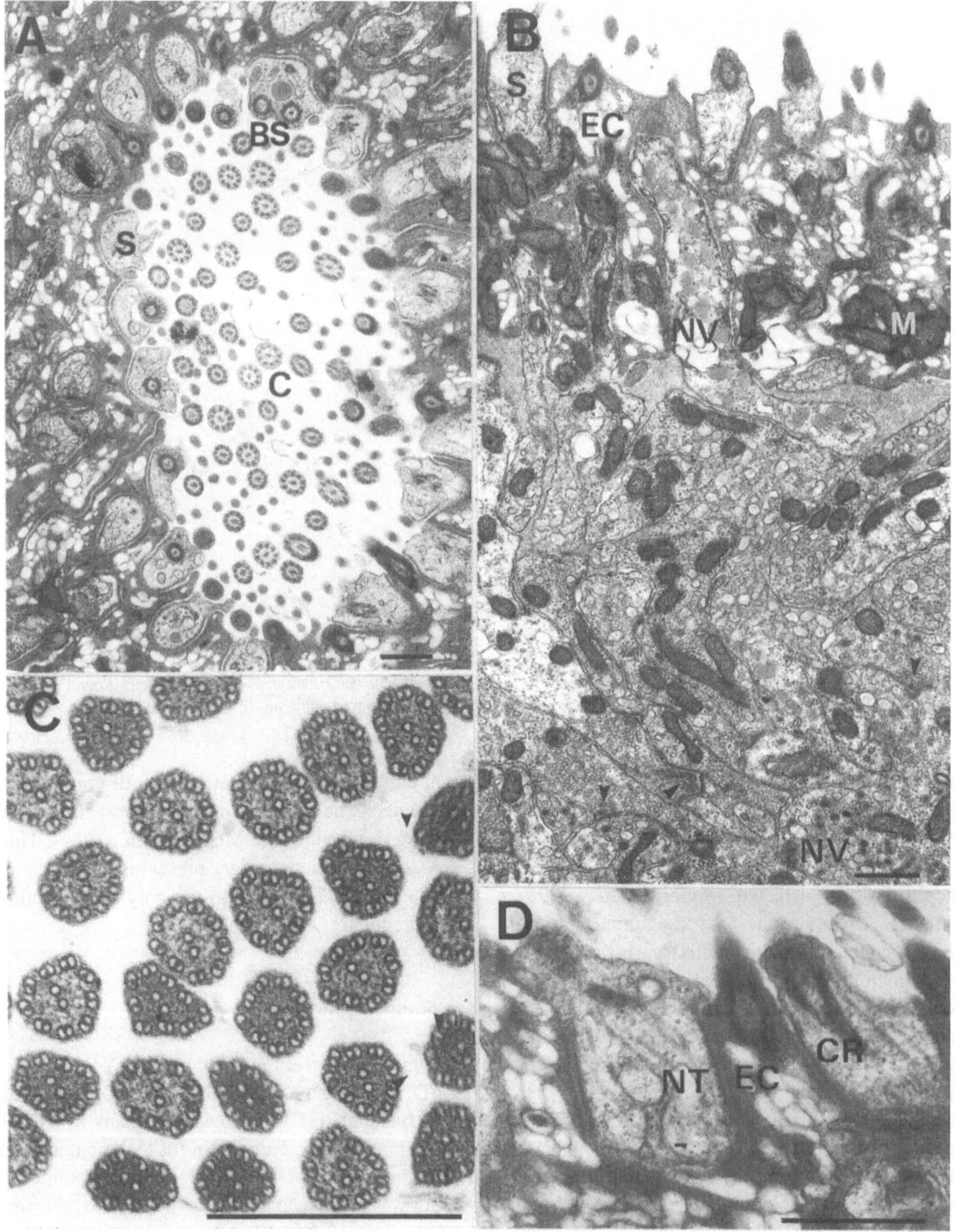

Fig. 2. Sensory part of the ciliary pit in *Geocentrophora* spp. Scale 0.5μm. A. *G. porfirievae*, cross section of the deeper region of the pit. B. *G. baltica*, sensory endings arising from the neuroplilar area below the pit. C. *G. baltica*, cross section of distal parts of sensory cilia in the pit. D. *G. porfirievae*, sensilla with reduced rootlet and numerous neurotubules in the pit. Biciliary sensillae (BS), cilia (C), ciliary rootlet (CR), epidermal cell (EC), mitochondria (M), neurotubules (NT), neuronal vesicles (NV), sensillae (S), synapses (arrowheads).

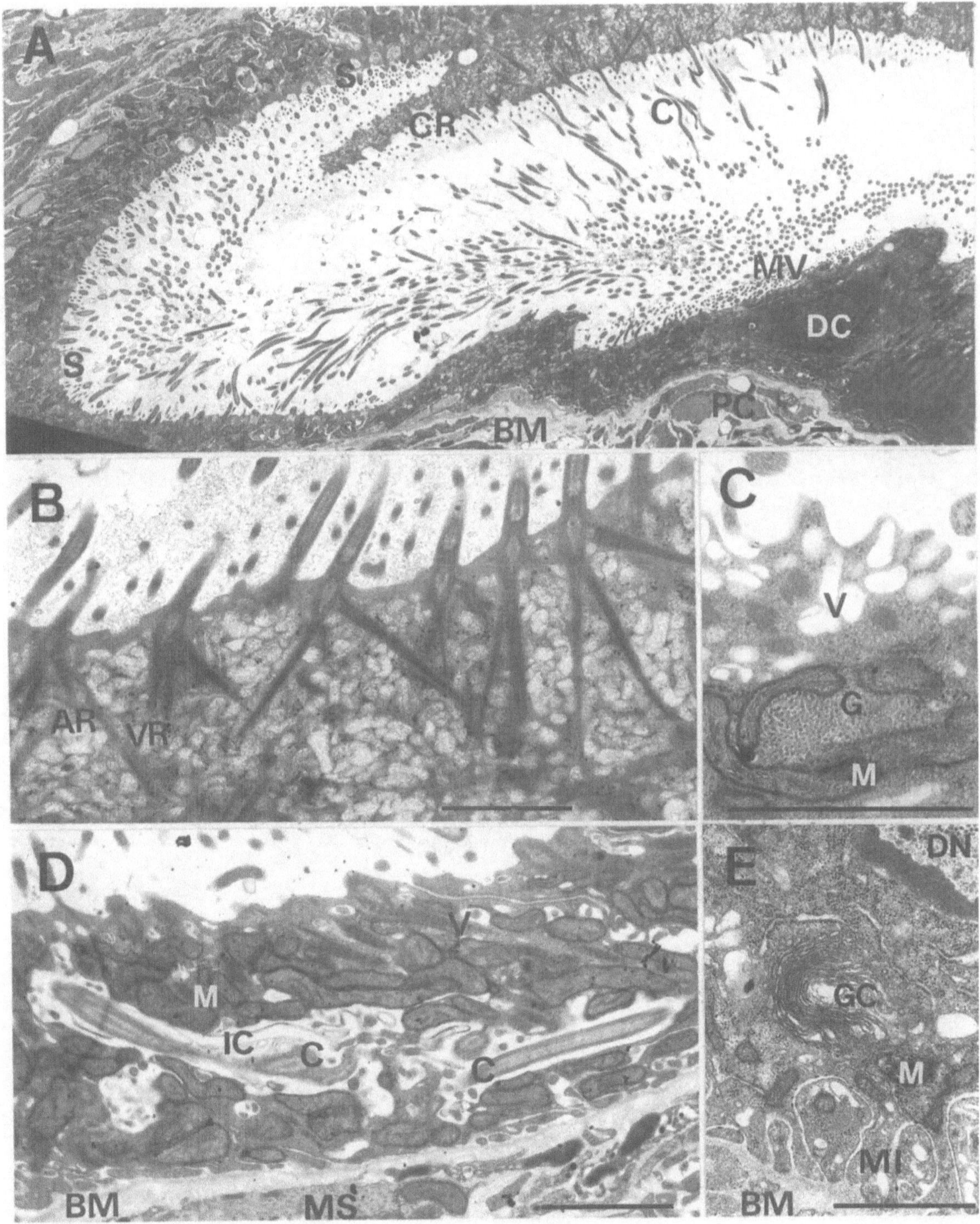

Fig. 3. Middle part of the ciliary pit in geocentrophorans. Scale 1 μm. A. *G. wasiliewi*, longitudinal section through the ciliary pit. B. *G. wasiliewi*, strengthened rootlet system of the cilia on the protruding cytoplasmic ridge. C. *G. baltica*, glycogen granules surrounded by mitochondria in the apical cytoplasm of the dark cells. D. *G. baltica*, ciliary axonemes in the intracellular canaliculi of the dark cells; E. *G. baltica*, basal infoldings of the cell membrane of the dark cell. Anterior rootlet derivates (AR), basement membrane (BM), cilia (C), cytoplasmic ridge (CR), dark cells (DC), dark cell nucleus (DN), glycogen granules (G), Golgi complex (GC), intracellular canaliculi (IC), mitochondria (M), membrane infoldings (MI), muscle fibres (MS), microvilli (MV), protonephridial capillaries (PC), sensilla (S), vacuoles (V), vertical rootlets (VR).

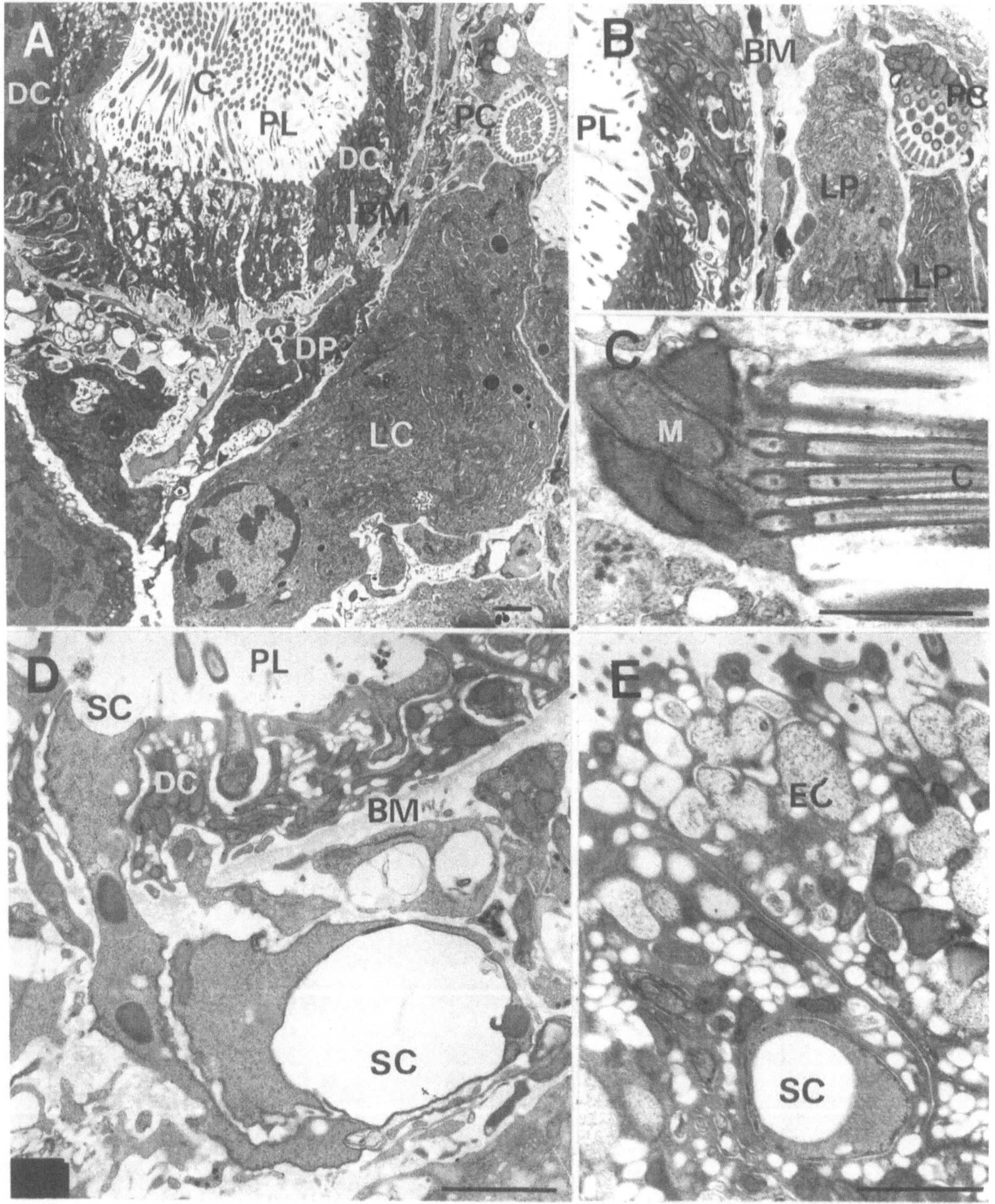

Fig. 4. *G. baltica*, pit-protonephridia relationships. Scale 1 μm. A. Dark cell projection passing (arrowhead) through the basement membrane to a large cell coming close to the protonephridial canal. The whole area is penetrated by smaller capillaries. B. Serial section from the same region as in A. Note terminal part of protonephridial canal in contact with processes of a large cell. C. Terminal part of a protonephridial canal in the vicinity of the pit. D. Protonephridial capillary close to the pit lumen. E. Protonephridial capillary close to the body surface. Basement membrane (BM), cilia (C), dark cells (DC), dark cell projections (DP), epidermal cell (EC), large cell (LC), large cell projection (LP), mitochondria (M), protonephridial canal (PC), pit lumen (PL), smaller protonephridial capillary (SC).

(Prorhynchidae, Plathelminthes) from Lake Baikal. Acta Zool. (Stockholm) (in press).

Kepner W. A. & W. H. Taliaferro, 1916. Organs of special sense of Prorhynchus applanatus Kennel. J. Morphol. 27: 163–177.

Kikuchi, S., 1983. The fine structure of the gill epithelium of a freshwater flea Daphnia magna (Crustacea: Phyllopoda) and changes associated with acclimation to various salinities. Cell Tissue Res. 299: 253–268.

Mendez, A., J. L. Arias, D. Tolivia & M. Alvarez-Uria, 1984. Ultrastructure of gill epithelial cells of Diopatra neapolitana (Annelida, Polychaeta). Zoomorphology 104: 304–309.

Reisinger, E., 1968. Xenoprorhynchus - ein Modellfall für progressiven Funktionswechsel. Z. Zool. Syst. Evolutionsforsch. 6: 1–55.

Reuter, M., B. Joffe & I. Palmberg, 1993. Sensory receptors in the head of Stenostomum leucops. II. Localization of catecholaminergic histofluorescence - ultrastructure of surface receptors. Acta Biol. Hungarica 44: 125–131.

Rieger, R. M., S. Tyler, J. P. S. Smith III & G. E. Rieger, 1991. Platyhelminthes, Turbellaria. In F. W. Harrison & B. J. Bogitsh (eds) Microscopic Anatomy of Invertebrates, 3. Wiley-Liss. Inc, New York, Chichester, Brisbane, Toronto, Singapore: 7–140.

Timoshkin, O. A. 1991. Turbellaria Lecithoepitheliata: morphology, systematics, phylogeny. Hydrobiologia 227 (Dev. Hydrobiol. 69): 323–332.

Hydrobiologia **305**: 177–182, 1995.
L.R.G. Cannon (ed.), Biology of Turbellaria and some Related Flatworms.

A new type of photoreceptor in *Anthopharynx sacculipenis* (Plathelminthes, Solenopharyngidae)

Beate Sopott-Ehlers
II. Zoologisches Institut und Museum der Universität Göttingen, Berliner Strasse 28, D-37073 Göttingen, Germany

Key words: Solenopharyngidae, photoreceptors, lamellate organelles, Plathelminthes

Abstract

The unpaired eye of *A. sacculipenis* consists of two sensory cells and one pigmented mantle cell. The light-sensing organelles are formed by flattened and rolled lamellae, which are not of ciliary origin but derive from the surface membrane of the sensory cells. The pigment of the mantle cell appears black in living animals, but electron-lucent in the EM picture. The significance of these special features is discussed.

Introduction

Considering the total number of species comprising the Solenopharyngidae, species with pigmented eyes are quite rare. Moreover, these light-sensing organs exhibit diversity, as paired and unpaired pigmented eye spots (cf. Ehlers 1972, p. 21), or showing variations in the nature of the pigment (Karling 1940, p. 118).

The primary purpose of this study was to determine the construction of the unpaired eye spot in *Anthopharynx sacculipenis* Ehlers, 1972 and to obtain information on the ultrastructure of the receptor organelles formed by its sensory cells.

Material and methods

The animals were collected from sand samples taken at the type locality (see Ehlers, 1972). The specimens were extracted from the sediment by the sea water ice method. Fixation, dehydration and staining followed the procedure described by Sopott-Ehlers (1993). Series of transverse and sagittal sections were examined using the Zeiss electron microscopes EM9, EM900 and EM10B.

Results

The unpaired and bilobed eye of *Anthopharynx sacculipenis* Ehlers, 1972 appears as a microscopic black spot in living animals (Ehlers, 1972, p. 38 fig. 2 A–C). This eye is medially located in close proximity to the brain, which lacks a capsule. The light- sensing organ of *A. sacculipenis* consists of only three cells: a single pigmented mantle cell forming the cup and two bilaterally arranged sensory cells.

Sensory cells

Both sensory cells are roughly mushroom-shaped, and their apical parts penetrate the pigment cup from the left and the right dorso- lateral direction (Fig. 1A). The light-sensitive organelles of the cells are formed by evaginations of the cell membrane. In cross sections these elaborations of the plasmalemma extend almost from one side of the cell to the other (Fig. 2A). In sagittal sections, however, they appear as closely packed, curved or flat lamellae (Fig. 1B, 2B; Fig. 3). These whorls are not developmentally derived from cilia and vestigial cilia were not seen. Processes of nerve cells can be observed near the lamellar whorls (Fig. 2B). The cytoplasm between the different lamellae contains voluminous mitochondria, small profiles

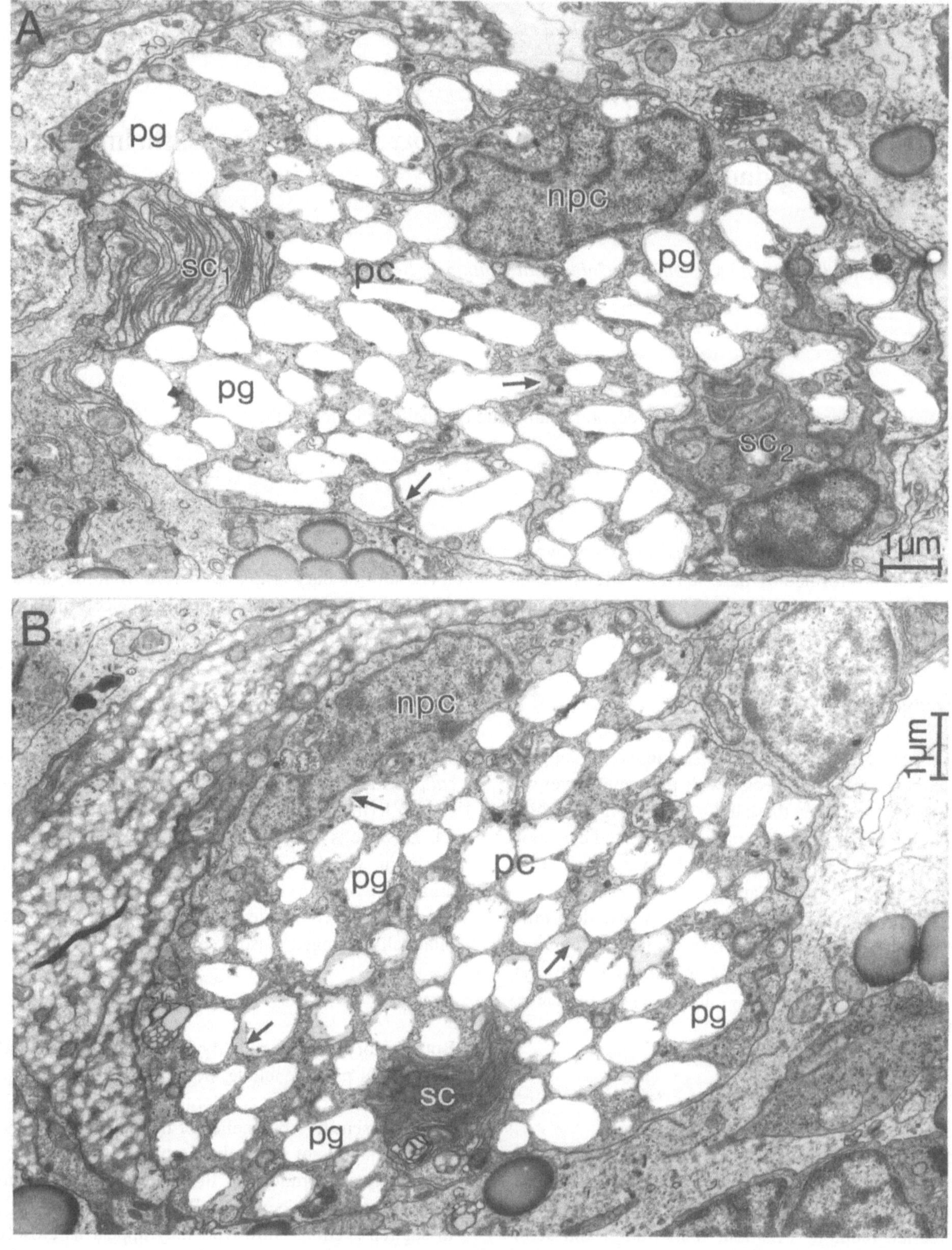

Fig. 1. Eyespot of *Anthopharynx sacculipenis* (A) Transverse section. (B) Sagittal section. The arrows in (B) point to remnants of the contents of pigment granules. *npc*, nucleus of the pigment cell; *pc*, pigment cell; *pg*, pigment granule; *sc, sc1 sc2*, sensory cell.

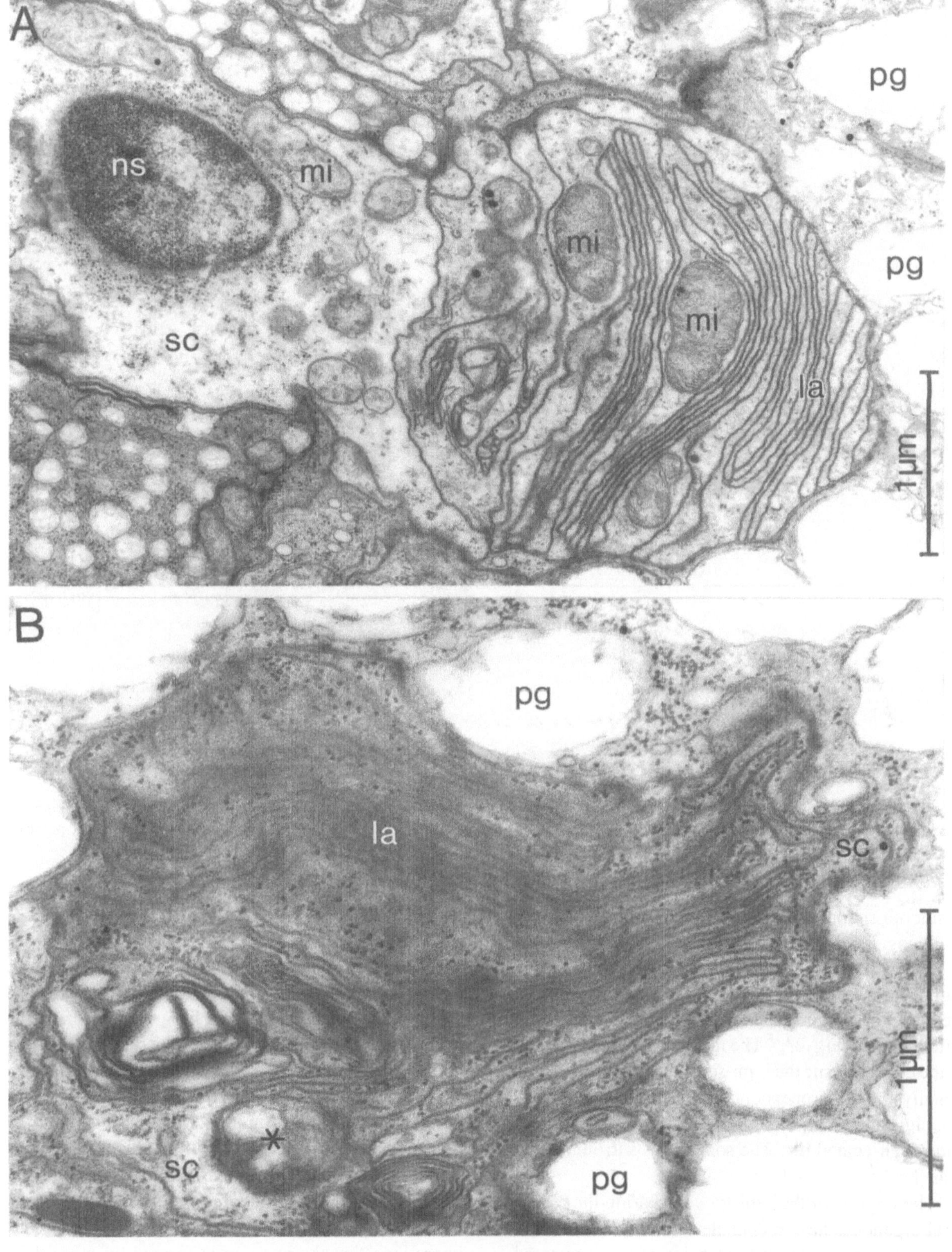

Fig. 2. (A) Sensory cell with receptoral organelles from a transverse section. (B) Sensory cell from a sagittal section showing the rolled lamellae. The asterisk marks the branch of a nerve cell lying in close proximity to the lamellae. *la*, lamellae; *mi*, mitochondrion; *ns*, nucleus of the sensory cell; *pg*, pigment granule; *sc*, sensory cell.

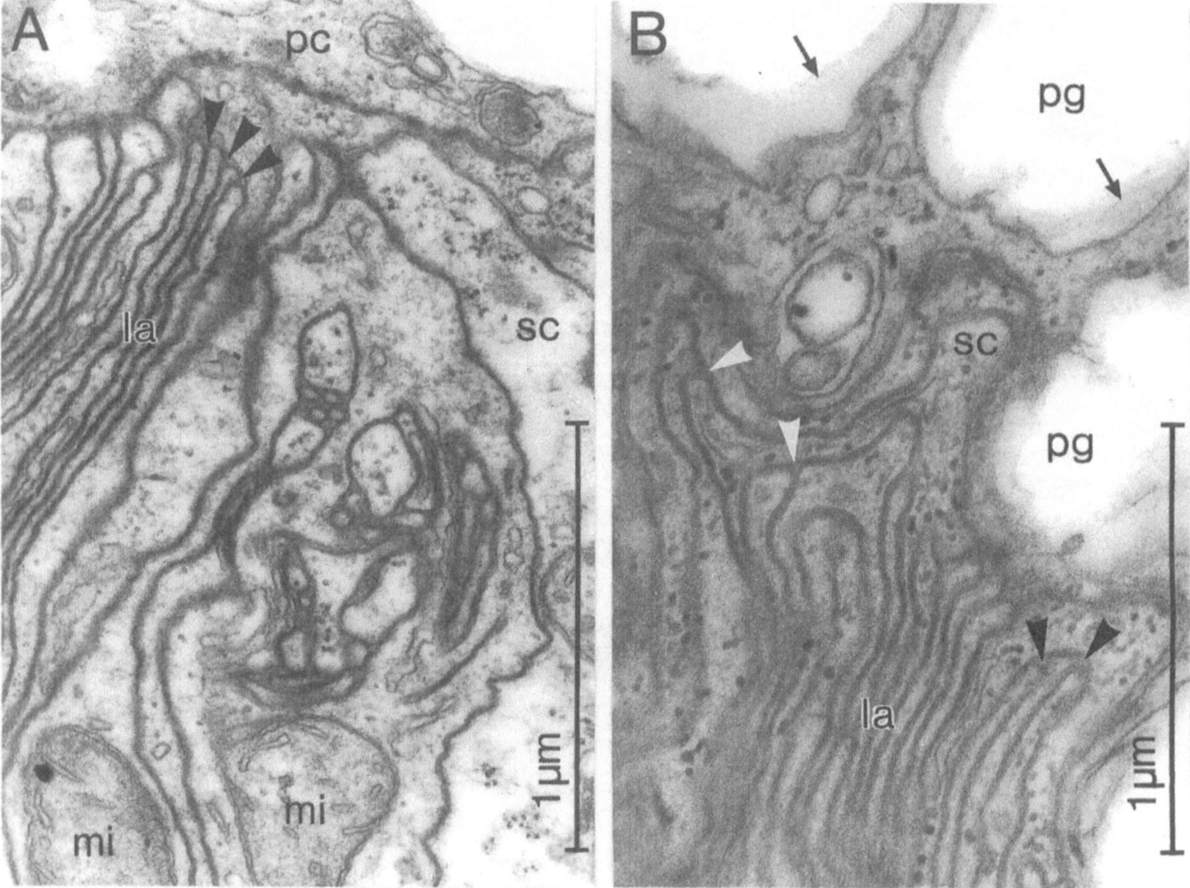

Fig. 3. (A) (B) Bases of the membranous lamellae (arrow heads) from a transverse (A) and a sagittal (B) section. At the right in (B) pigment granules showing remnants of their electron-lucent content (small arrows). *la*, lamellae; *mi*, mitochondrion; *pc*, pigment cell; *pg*, pigment granule; *sc*, sensory cell.

of smooth endoplasmic reticulum, and a large quantity of glycogen deposits (Fig. 2; Fig. 3).

The somata of the sensory cells, lying outside the pigment cup, are stout. Closely beneath the bases of the receptor processes, there is a layer of spherical mitochondria (Fig. 2A). The nucleus lies in the middle of the cell, opposite the light-sensing organelles. Additionally, the cell proper contains free ribosomes, several Golgi centres, glycogen rosettes, some microtubules and rough-coated ER. The somata are surrounded by branches of nerve cells.

The regions of the sensory cell bearing the receptoral organelles have a diameter of about 2.8 μm and are small compared to the extension of the pigmented cup cell.

Pigment cup

The pigment cup consists of a single bilobular mantle cell. It has a diameter of about 12–14 μm. Most of the cellular lumen is occupied by granules which are oval to elongate and surrounded by a bordering membrane (Fig. 1A, B). The contents of the granules were almost totally dissolved, apart from a small peripheral layer of electron-lucent material (arrows in Figs. 1A, B; Fig 3B). The nucleus of the mantle cell is situated medially between the two lobes. The cell membrane is strongly folded and penetrates deeply into the lumen of the cell. Thus, at the first glance, one might get the impression that the eye cup is formed by more than one cell. However, series of transverse and sagittal sections show that only one nucleus exists. Aside from nucleus and pigment granules, there are a few mitochondria, short profiles of smooth and rough ER, and some dictyosomes.

Discussion

In more than one aspect the eye spot of *Anthopharynx sacculipenis* represents an organ differing from the 'typical' pigmented rhabdomeric photoreceptors well-known for many species of the Rhabdocoela. 1.It is an unpaired 'cyclopean' eye. 2. The photoreceptive organelles are of the epigenous type sensu Burr (1984). That is to say, a lamellar type of photoreceptor organelle without any vestigial cilia, formed by the cell membrane proper and not by microvillar (rod-shaped) elaborations of the surface membrane. 3. The pigment granules appear to be neither of melanin nor of carotenoid nature.

Although we have no physiological evidence for a photoreceptor, and despite its special construction, the function of the black spot in *A. sacculipenis* as a light-sensing organ is indicated by its morphological picture. It possess an extensive membranous surface, which is enclosed by a shading device; many mitochondria lie nearby this membranous surface, and the cell from which the membranes originate has an identifiable connection with the nervous system.

EM-findings on unpaired median 'cyclopean' eyes in free-living Plathelminthes have not been reported previously. Eyes formed by a single mantle cell and two sensory cells also exist in the kalyptorhynch *Polycystis naegelii* (see Lanfranchi & Bedinii, 1982) and in the prolecithophoran species *Pseudostomum quadrioculatum* (see Sopott-Ehlers, 1988), but these species possess paired eyes.

Light-sensing organelles formed by flattened and rolled membranes have been found in the eyes of polyclad larvae (Ruppert, 1978; Eakin & Brandenburger, 1981; Lanfranchi *et al.*, 1981; Lanfranchi & Bedini, 1986). However, in all these instances the lamellae derive developmentally from cilia. This is also true for the presumed photoreceptors called ciliary lamellate bodies (Ehlers & Ehlers, 1977; Sopott-Ehlers, 1986, 1991) existing in representatives of the Proseriata Parotoplaninae.

Because the lamellae in the eye spot of *A. sacculipenis* derive from the surface membrane of the sensory cell proper, they are not homologous to the ciliary lamellae in larval polyclad eyes or to those of the lamellate ciliary bodies in parotoplanid species. Since within the free-living Plathelminthes studied at the EM-level to date the rhabdomeric type, which consists of rod- shaped microvilli, is the dominant photoreceptor element, the lamellae in *A. sacculipenis* are hypothesized to be an evolutionary novelty.

As far as pigmented eyes exist with the Plathelminthes, pigment granules of melanin or carotenoid nature are the prevailing shading device. Although the eye spot in *A. sacculipenis* appears black in living animals, the pigment granules are obviously of a different nature. Based on light-microscopical observations granules of a similar consistency have already been mentioned by Karling (1940, p. 118) for some other solenopharyngid species.

With the electron-microscope the granules of *A. sacculipenis* appear similar to the platelets in the eyes of some Acoela at first glance (Popova & Mamkaev 1985; Yamasu 1991). However, those platelets are bound together in a vacuole. Furthermore, most of the acoel species possess, in addition, typical black pigment granules.

Beyond acoel species, reflective platelets rectangular in shape have been found in the eyes of larvae of *Polystoma intergerrimum* (see Fournier & Combes, 1978) and in larval *Kronborgia isopodicola* (see Williams, 1991; Watson *et al.*, 1992). But, in both instances those platelets are arranged in regular concentric rows concentrated on this side of the pigment cup which faces the sensory cells.

Not the shape, nor the position within the cup cell, nor the contents seen by EM, give any reason to conclude that the platelets in the eyes of the species mentioned above are identical to the oval shaped granules in the eye spot of *A. sacculipenis*.

No other EM investigations on photoreceptors in Solenopharyngidae exist, but the ultrastructure of the eyes of other species need to be determined before a conclusion can be drawn as to whether the lamellar organelles and the pigment granules are autapomorphic features of *Anthopharynx sacculipenis* or of a taxon comprising several species of the Solenopharyngidae, respectively.

Acknowledgments

Financial support was provided by the Akademie der Wissenschaften und der Literatur, Mainz. Thanks are due to Mrs K. Lotz and Mrs E. Hildenhagen-Brüggemann for technical assistance.

References

Burr, A. H., 1984. Evolution of eyes and photoreceptor organelles in the lower phyla. In M. A. Ali (ed.), Photoreception and vision in invertebrates. Plenum Press, London, New York: 241– 288.

Eakin, R. M. & J. L. Brandenburger, 1981. Fine structure of the eyes of Pseudoceros canadensis (Turbellaria. Polycladida). Zoomorphology 98: 1–16.

Ehlers, U., 1972. Systematisch-phylogenetische Untersuchungen an der Familie Solenopharyngidae (Turbellaria, Neorhabdocoela). Mikrofauna Meeresboden 11: 1–78.

Ehlers, B. & U. Ehlers, 1977. Die Feinstruktuur eines ciliären Lamellarkörpers bei Parotoplanina geminoducta Ax (Turbellaria, Proseriata). Zoomorphologie 87: 65–72.

Fournier, A. & C. Comber, 1978. Structure of photoreceptors of Polystoma intergerrimum (Platyhelminthes Monogenea). Zoomorphologie 91: 145–155.

Karling, T. G., 1940. Zur Morphologie und Systematik der Alloeocoela Cumulata und Rhabdocoela Lecithophora (Turbellaria). Acta Zool. Fenn. 26: 3–260.

Lanfranchi, A. & C. Bedini, 1982. The ultrastructure of the sense organs of some Turbellaria Rhabdocoela. I. The eyes of Polycystis naegelii Kölliker (Eukalypterhynchia Polycystididae). Zoomorphology 101: 95–102.

Lanfranchi, A. & C. Bedini 1986. Electron microscopic study of larval eye development in Turbellaria Polycladida. Hydrobiologia 132 (Dev. Hydrobiol. 32): 121–126.

Lanfranchi, A., C. Bedini & E. Ferrero, 1981. The ultrastructure of the eyes in larval and adult polyclads (Turbellaria). Hydrobiologia 84: 267–275.

Popova, N. V. & Yu. V. Mamkaev, 1985. Ultrastructure and primitive features of the eyes of Convoluta convoluta (Turbellaria Acoela). Dokl. Akad. Nauk SSSR 283: 756–759.

Rupport, E. E., 1978. A review of metamorphosis in turbellarian larvae. In F. S. Chia & M. E. Rice (eds), Settlement and metamorphosis of marine invertebrate larvae. Elsevier, New York: 65–81.

Sopott-Ehlers, B., 1986. Die Feinstruktur der Sehkolben und der Lamellarkörper von Parotoplana capitata (Plathelminthes, Proseriata). Zoomorphology 106: 44–48.

Sopott-Ehlers, B., 1988. Fine structure of photoreceptors in two species of the Prolecithophora. Fortschr. Zool.36: 221–227.

Sopott-Ehlers, B., 1991. Comparative morphology of photoreceptors in free-living plathelminths - a survey. Hydrobiologia 227 (Dev. Hydrobiol. 69): 231–239.

Sopott-Ehlers, B., 1993. Ultrastructural features of the pigmented eye spot in Pseudomonocelis agilis. Microfauna Marina 8: 77–88.

Watson, N. A., J. B. Williams & K. Rohde, 1992. Ultrastructure and development of the eyes of larval Kronborgia isopodicola (Platyhelminthes, Fecampiidae). Acta Zool. (Stockh.) 73: 95–102.

Williams, J. B., 1991. Ultrastructural studies on Kronborgia (Platyhelminthes: Fecampiidae): observations on the encapsulated larva of K. isopodicola. New Zeal. J. Zool. 18: 251–265.

Yamasu, T., 1991. Fine structure and function of ocelli and sagittocysts of acoel flatworms. Hydrobiologia 227 (Dev. Hydrobiol. 69): 273–282.

Hydrobiologia **305**: 183–188, 1995.
L.R.G. Cannon (ed.), *Biology of Turbellaria and some Related Flatworms.*

Ultrastructure of the nerve cells and sensilla of *Geocentrophora baltica* (Platyhelminthes, Lecithoepitheliata) and the surface sensilla in the *Geocentrophora* group

Inger Böckerman[1], Olga I. Raikova[2], Maria Reuter[1] & Oleg Timoshkin[3]
[1]*Åbo Akademi University, Department of Biology, BioCity, Artillerigatan 6, FIN-20520, Åbo, Finland*
[2]*Zoological Institute of the Academy of Sciences of Russia, Universitetskaya nab. 1, 199034 St. Petersburg, Russia*
[3]*Limnological Institute, Siberian Division of the Russian Academy of Sciences, Box 4199, 664033 Irkutsk, Russia*

Key words: ultrastructure, nervous system, sensilla, cilia, Lecithoepitheliata, Lake Baikal

Abstract

Two types of nerve cell could be distinguished ultrastructurally in the central nervous system of *Geocentrophora baltica* (Prorhynchida, Lecithoepitheliata). Both show invaginations in the plasma membrane, but they differ in the character of the cytoplasm (light or densely stained) and the distribution of the neuronal vesicles (evenly or in groups). Different kinds of vesicles and neuronal release sites are observed. Special features of the synapses are pronounced local thickenings of the presynaptic membrane connected to paramembranous densities. In *G. baltica* and five endemic *Geocentrophora* spp. from Lake Baikal six types of surface sensillum were observed at the epidermal surface: 1. those with a long thin rootlet; 2. a short, balloon-shaped cilium with an aberrant axoneme and a reduced rootlet; 3. a rootlet branching into many striated bundles; 4. a thick rootlet; 5. a reduced rootlet and numerous neurotubules; and 6. collared sensilla each with one cilium in a deep pit surrounded by a collar of 11 to 12 microvilli. The variable number of microvilli in the collared sensillum is considered plesiomorphic relative to the stable number of eight microvilli known in sensilla of the Prolecithophora, Proseriata, and Rhabdocoela. The ultrastructure of the collar sensillum indicates that the Lecithoepitheliata is only distantly related to the Prolecithophora and higher turbellarians.

Introduction

In recent years many papers have been published on the ultrastructure of the nervous system and ciliary sensilla of several turbellarian orders (for references see Rieger, *et al.*, 1991). But a gap still exists in our knowledge of the ultrastructure of the nervous system and sensilla of Lecithoepitheliata.

The phylogeny and the taxonomic position of Lecithoepitheliata is still open to discussion (see Karling, 1968; Ehlers, 1985; Timoshkin, 1991). The order is therefore regarded as *incerta sedis* (Ehlers, 1985). The Prorhynchidae, the major family of the Lecithoepitheliata, has a conservative morphology in most organ systems (Timoshkin, 1991). Electron microscopic data on these animals are scarce and incomplete (see Timoshkin, 1991 for review).

The endemic *Geocentrophora* spp. (Prorhynchida, Lecithoepitheliata) from Lake Baikal probably originated from an archetype resembling two cosmopolitan species, namely *G. baltica* and *G. sphyrocephala*. The present study concerns the ultrastructure of the brain neurons and synapses of *G. baltica*. The central nervous system of the endemic species of *Geocentrophora* from Lake Baikal has been previously studied (Böckerman *et al.*, 1994). On the basis of that study, the ultrastructure of the nerve cells and synapses of *G. baltica* is discussed. In addition, the sensilla of *G. baltica* and five endemic *Geocentrophora* spp., are studied.

Material and methods

Specimens of *Geocentrophora intersticialis*, *G. levanidorum*, *G. porfirievae*, *G. wagini*, and *G. wasiliewi* (Order Lecithoepitheliata) were collected from Lake Baikal, Russia. Specimens of *G. baltica* were collected in Sosnovka Park in St. Petersburg, Russia. Two different fixation procedures were used: *G. intersticialis*, *G. levanidorum*, *G. wagini* and *G. wasiliewi* were fixed in the same way as *G. intersticialis* in Böckerman *et al.* (1994), and *G. baltica* and *G. porfirievae* were fixed as *G. porfirievae* in Böckerman *et al.* (1994). The fixed specimens were dehydrated in a series of acetone, except for specimens of *G. levanidorum* which were dehydrated in a series of ethanol, and embedded in Epon 812. Ultrathin sections, stained with uranyl acetate and lead citrate, were examined with JEM 100 SX electron microscope.

Results

Two types of brain nerve cell and a putative glial cell can be distinguished in *Geocentrophora baltica*. Both types of nerve cell contain dense-core vesicles of different densities and sizes, have irregularly shaped nuclei, and invaginations of the cell membrane filled with extracellular material (ECM). One type (Fig. 1A) measures, 7×13 μm and shows a large (6×11 μm) heterochromatic nucleus, and a high nucleocytoplasmic ratio with densely staining cytoplasm. The second type of nerve cell (Fig. 1B) is large (13×28 μm) and has a low nucleocytoplasmic ratio with light cytoplasm. The putative glial cell (Fig. 1C), measuring 2×11 μm, has a high nucleocytoplasmic ratio with a strongly heterochromatic nucleus measuring 1.5×8.5 μm.

In the neuropile (Fig. 1D), various kinds and numbers of vesicles occur: large dense vesicles, dense-core vesicles of varying densities, small lucent vesicles and pleiomorphic vesicles of intermediate densities. The nerve fibres form both single and shared synaptic contacts. Some synapses show a paramembranous density inside the presynaptic terminal. At the presynaptic membranes dense material is concentrated in two or three thicker areas along the membrane, and in some, strings connects these areas to the paramembranous density. Nonsynaptic contacts (omega profiles), contacts between nerve terminals and basement membrane, and neuromuscular contacts are also observed.

In all six species studied, five types of surface sensilla were observed. One additional type was observed in *G. porfirievae*. In all types, dendrites of the cells of sensilla penetrated through channels in the epidermal cells (Fig. 2A, B). Distal parts of the dendrites formed septate junctions with the epidermal cell. Processes of the sensilla contained mitochondria, neurotubules, and either large (irregularly shaped) lucent or dense-core vesicles. Cell bodies of the sensilla were not observed; they were situated below the basement membrane in the parenchyma. Sensilla usually occurred in groups, sometimes of different types. The six types could be distinguished largely by morphology of the ciliary rootlet as follow. 1. Sensilla with a long thin rootlet (Fig. 2A) were seen scattered on the body surface in all species studied. They had a single cilium with a normal (9 + 2) axoneme and a long thin cross-striated rootlet penetrating deep into the dendrite. The neck of the dendrite slightly protruded above the epidermal surface. 2. Sensilla with a balloon-shaped cilium (Fig. 2A) were rarely seen and occurred along anterior lateral body margins. The axoneme of the cilium was short and showed single microtubules closely packed in a dense matrix. The ciliary rootlet was reduced or practically lacking. 3. Sensilla with a branching rootlet (Fig. 2A–C) occurred together with the first mentioned sensilla with a long thin rootlet. They each had one cilium with a normal axoneme and a short vertical rootlet branching into many striated bundles. Five to seven bundles arose from the posterior surface of the basal body. Others arose more basally through branching of the main rootlet (Fig. 2A–E). 4. Sensilla with a thick rootlet and proximally widened dendrite (Fig. 3A) were observed only in *G. porfirievae* and were rare even in this species. This type may represent a variation of sensilla with a long thin rootlet. 5. Sensilla with a reduced rootlet and numerous neurotubules (Fig. 3B) were common in the pharyngeal cavity and in the ciliary pit, but were rare on the body surface. The dendrite had one normal cilium and sometimes thin microvilli and was filled with vesicles and neurotubules. 6. Collared sensilla (Fig. 3C, D) occurred on the body surface in all species studied but were rare. They each had one cilium in a deep pit surrounded by a collar of 11 to 12 microvilli. Below the epidermal surface, basal parts of the microvilli formed vertical ridges lining the pit. Only tips of microvilli protruded above the epidermal surface.

In locomotory cilia of the epidermis, the microtubule doublet number 2 appeared shorter than the other eight (Fig. 3F). In other locomotory cilia doublets on

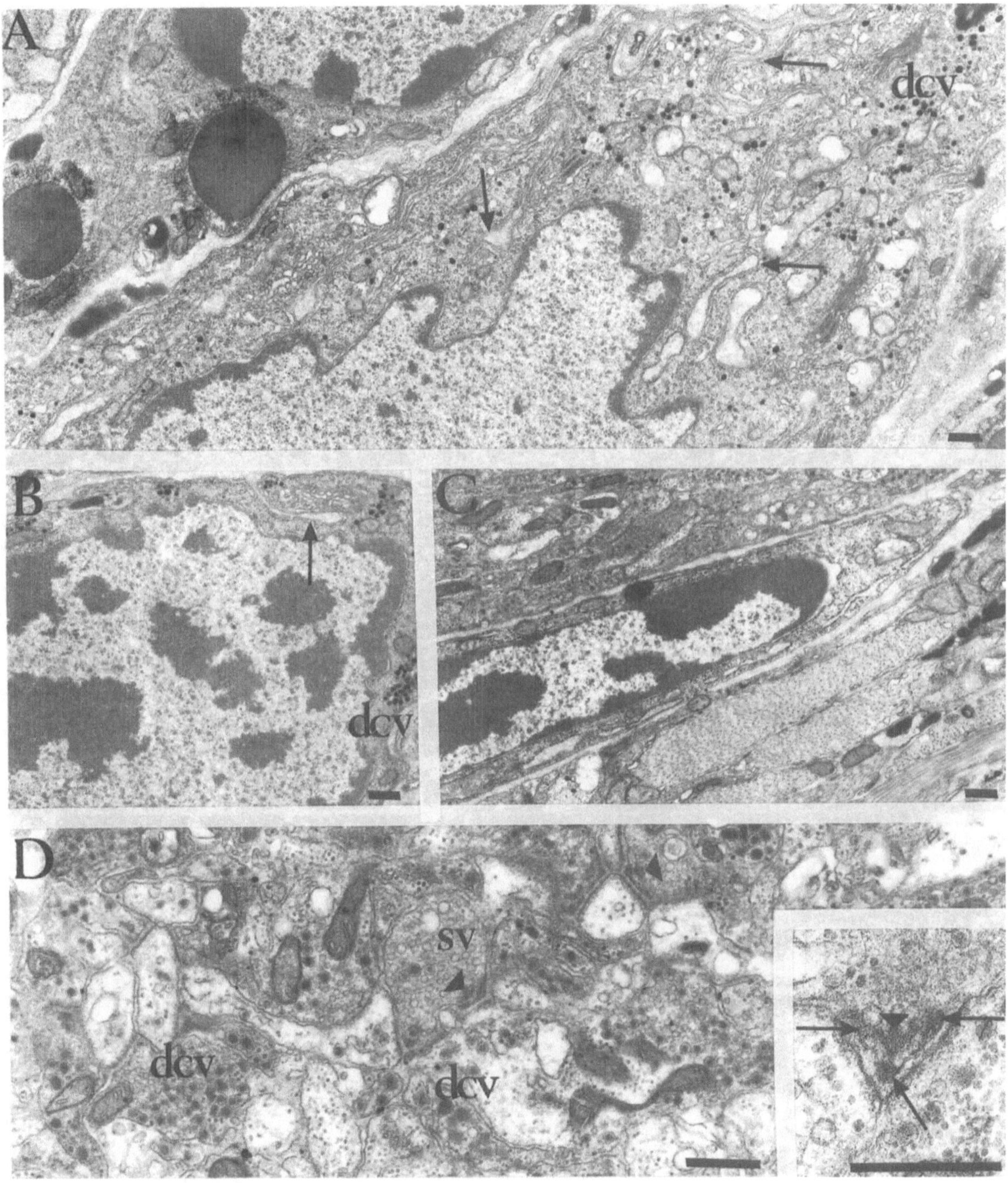

Fig. 1. Geocentrophora baltica. A) – Nerve cell type I showing characteristic dark cytoplasm, ECM-filled invaginations of plasma membrane (*arrow*), dense-core vesicles (*dcv*). *B)* – Nerve cell type II showing characteristic light cytoplasm, ECM-filled invaginations of plasma membrane (*arrow*), dense-core vesicles (*dcv*) of various densities. *C)* – Putative glial cell. *D)* – Nerve fibres of neuropile with dense-core vesicles (*dcv*) of different sizes and small lucent vesicles (*sv*), conventional synapses and paramembranous density (*arrowhead*). Inset shows synapse with local thickenings (arrow) of presynaptic membrane connected to same paramembranous density (*arrowhead*). Scale = 0.4 μm.

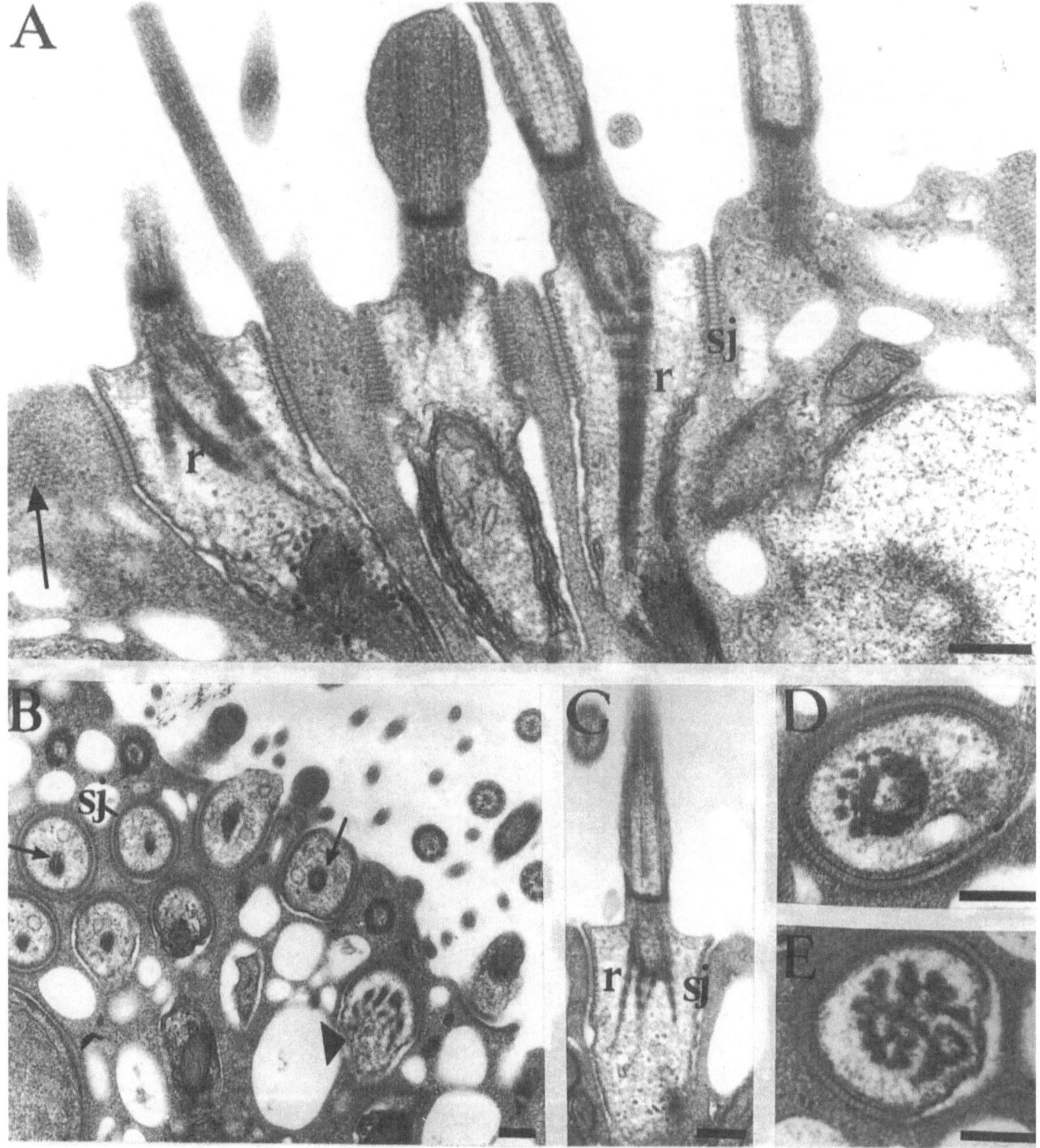

Fig. 2. Surface sensilla in *Geocentrophora baltica. A*) – Group of sensilla on the lateral body margin. Sensilla, one with a long thin rootlet, one with a balloon-shaped cilium, and one with a branching rootlet. Tangential section of a septate junction gives a honeycomb structure (*arrow*). *B*) – Tangential section of a cluster of sensilla, one with long thin (*arrow*) and one with branching rootlets (*arrowhead*), penetrating an epidermal cell. *C*) – Sensillum with branching rootlet. *D, E*) – Cross sections through a branching rootlet. Rootlet (*r*), septate junctions (*sj*). Scale = 0.2 μm.

one side of the cilium (numbers 2–5) became singlets at a certain distance from the tip (Fig. 3G), possibly due to ciliary bend. Most ciliary tips appeared rather symmetrical (Fig. 3E). At the extreme tip of the cilium, all nine doublets terminated as singlets, and an electron dense rod joined the central microtubules (Fig. 3H). In some cilia peripheral rods appeared under the ciliary membrane. The axoneme terminated in a dense terminal cap.

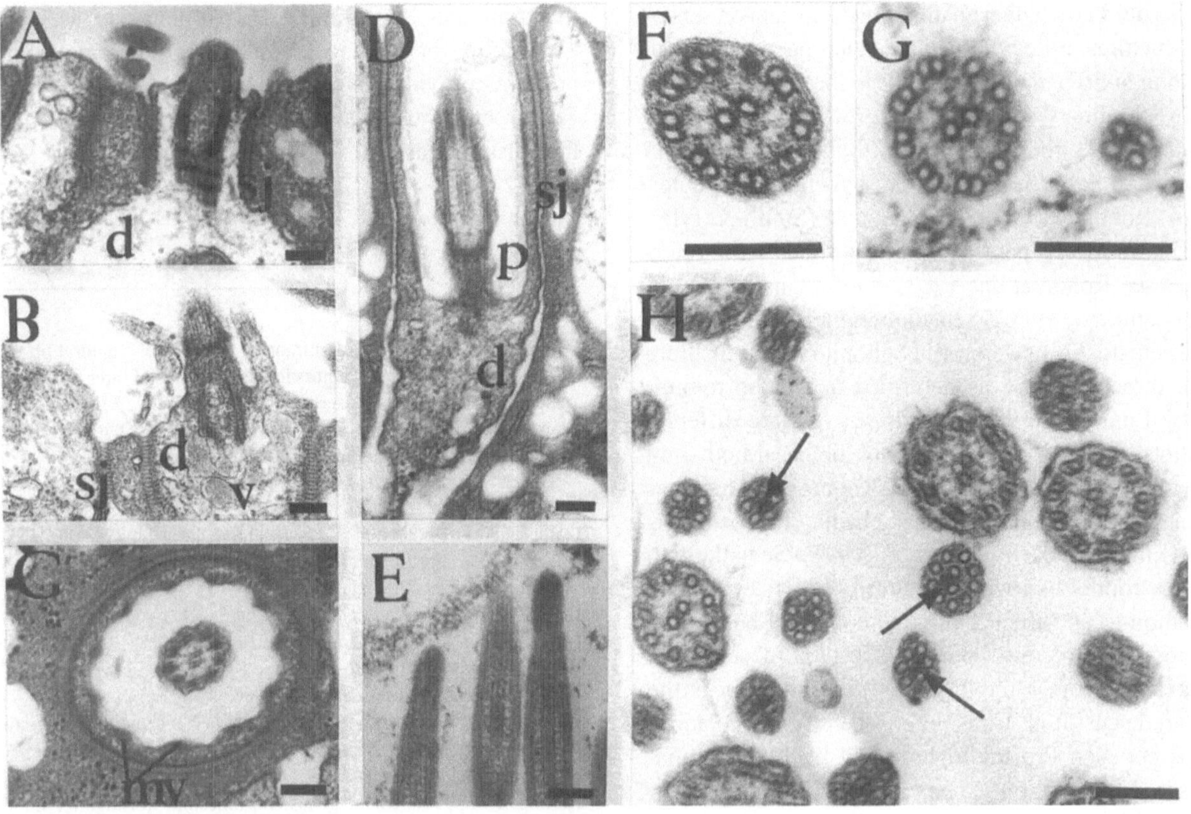

Fig. 3. A.–D. Types of sensilla in *Geocentrophora* spp. *A*) – Thick rootlet sensillum in *G. porfirievae*. *B*) – Sensillum with reduced rootlet in the sensory pit of *G. baltica*. *C*) – Transverse section of a collared sensillum (with 11 microvilli), penetrating the epidermis of *G. intersticialis*. *D*) – Collared sensillum in the epidermis of *G. baltica*. E.–H. *G. baltica*, details of ciliary tip termination. *E*) – Tapering tips of epidermal cilia. *F*) – Cross-section of a distal region of a cilium with a short no.-2 doublet. *G*) – Cross-section of a cilium with doublet numbers 2–5 as singlets and an extreme tip of a cilium with only three remaining microtubules. *H*) – Extreme ciliary tips with electron-dense supporting rods (*arrow*) and a reduced number of microtubules. Dendrite (*d*), microvilli (*mv*), pit (*p*), septate junctions (*sj*), vesicles (*v*). Scale =0.15 μm.

Discussion

Böckerman *et al.* (1994) described the structure of the central nervous system (CNS) of Baikal *Geocentrophora* spp. Our present study of the cosmopolitan *G. baltica* reveals differences in the structure of its CNS and that of the Baikal species.

Classification of nerve cells into types is difficult because no clearly distinct characters distinguish them. As Lentz (1967) has assumed, this could be because the different forms of nerve cells simply reflect different functional states of the same cell type. The two features that characterize the neurons of endemic *Geocentrophora* spp., namely ECM-filled invaginations in the cell membrane and local specializations of the presynaptic membrane (Böckerman *et al.*, 1994), are also typical of *G. baltica*. Noticeable differences are that nerve cells in *G. baltica* sometimes have an irregular nucleus, as is also characteristic of ganglion cells of planarians (Lentz, 1967) and neurosecretory cells in the rhabdocoel *Gyratrix hermaphroditus* (see Reuter & Lindroos, 1979) and that omega profiles do not occur as frequently as in the endemic *Geocentrophora* spp. The second category of cells found in the nervous system of some flatworms, called glial cells, are also found in both endemic (Böckerman *et al.*, 1994) and cosmopolitan *Geocentrophora*. This cell type lacks vesicles, has a thin cytoplasm with very few organelles and the main mass of cytoplasm concentrated in branches (Golubev, 1988). These cells are believed to form supporting, isolating and metabolic functions. The few differences between the CNS of endemic and cosmopolitan geocentrophorans, and the high degree of ultrastructural specialization agree with the conclusion that Prorhynchidae is an advanced taxon of living Platyhelminthes (Böckerman *et al.*, 1994).

To our knowledge, no ultrastructural data on sensory structures in Lecithoepitheliata has been published. Among six types of sensilla, described in the present study, the first (with a long thin rootlet) occurs practically in all turbellarian groups. Balloon-shaped cilia have been described in Kalyptorhynchia (see Reuter, 1975; Rohde *et al.*, 1988), in Acoela (Smith & Tyler, 1985), in many parasitic platyhelminths (for references see Rohde *et al.*, 1988), and in Gnathostomulida (Lammert, 1986). Sensilla with a branching rootlet seem to be new for Rhabditophora; a sensillum with three rootlets was described in Gnathostomulida by Lammert (1986), but these are three different rootlets, deriving separately from the basal body, not one rootlet branching into parts. Rootlet branching has been described in monociliary sensilla of Acoelomorpha (Ehlers, 1985; Rieger *et al.*, 1991). Sensilla with a thick rootlet and with a reduced rootlet seem to be common to all turbellarians and could not be used in phylogenetic reconstructions. The type of sensillum interesting from a phylogenetic point of view is the collared sensillum. Considering the possible relationships between Prolecithophora and Lecithoepitheliata (see Timoshkin, 1991) and the similarities in protonephridial structure in Lecithoepitheliata and Rhabdocoela (Rohde, 1991), it is interesting to compare collared sensilla in these groups. Collared sensilla in Prolecithophora (Ehlers, 1977) have two cilia, each surrounded by a collar of eight microvilli. Monociliary sensilla with eight microvilli are common to Proseriata and Rhabdocoela (Bedini *et al.*, 1975; Ehlers, 1977; Ehlers & Ehlers, 1977) and Ehlers (1985) concluded that eight microvilli is an apomorphic character for Prolecithophora, Proseriata, and Rhabdocoela. As for geocentrophorans, their collar sensilla have 11 to 12 microvilli, presumably a plesiomorphic condition. Thus, the opinion that the Lecithoepitheliata is closely related to Prolecithophora and higher turbellarians is not supported by the present observation of the collar sensilla.

This study show that the *Geocentrophora* group is very conservative also concerning the ultrastructure of the central nervous system and the sensilla, although they, the endemic and cosmopolitan species, have been geographically separated for a very long time.

Acknowledgments

We wish to thank Dr Yuri Mamkaev, Dr Dmitriy Bogoliubov and Dr Boris Joffe for collecting living material, and Mrs Gunilla Henriksson and Mr Esa Nummelin for technical assistance. Financial support was provided by Åbo Akademi University, The Research Institute of the Åbo Akademi Foundation and the Russian Basic Research Foundation grant No 93-04-21226.

References

Bedini, C., E. Ferrero & A. Lanfranchi, 1975. Fine structural observations on the ciliary receptors in the epidermis of three otoplanid species (Turbellaria, Proseriata). Tissue & Cell 7: 253–266.

Böckerman, I., M. Reuter, & O. Timoshkin, 1994. Ultrastructural study of the central nervous system of endemic *Geocentrophora* (Prorhynchida, Platyhelminthes) from Lake Baikal. Acta zool. 75: 47–55.

Golubev, A. I., 1988. Glial and neuroglia relationships in the central nervous system of the Turbellaria (electron microscopic data). Fortschritte der Zoologie 36: 31–37.

Ehlers, U., 1977. Vergleichende Untersuchungen über Collar-Rezeptoren bei Turbellarien. Acta zool. fenn. 154: 137–148.

Ehlers, U., 1985. Das Phylogenetische System der Platyhelminthes. Gustav Fischer Verlag, Stuttgart, New York, 317 pp.

Ehlers, U. & B. Ehlers, 1977. Monociliary receptors in interstitial Proseriata and Neorhabdocoela (Turbellaria, Neoophora). Zoomorphologie 86: 197–222.

Karling, T. G., 1968. On the genus *Gnosonesima* Reisinger (Turbellaria). Sarsia 33: 81–108.

Lammert, V., 1986. Vergleichende Ultrastruktur-Untersuchungen an Gnathostomuliden und die phylogenetische Bewertung ihrer Merkmale. Ph. D. thesis, Göttingen, 218 pp.

Lentz, T. L., 1967. Fine structure of nerve cells in a planarian. J. Morph. 121: 323–338.

Reuter, M., 1975. Ultrastructure of the epithelium and the sensory receptors in the body wall, the proboscis and the pharynx of *Gyratrix germaphroditus* (Turbellaria, Rhabdocoela). Zool. Scr. 4: 191–204.

Reuter, M. & P. Lindroos, 1979. The ultrastructure of the nervous system of *Gyratrix hermaphroditus* (Turbellaria, Rhabdocoela). I. The brain. Acta zool. 60: 139–152.

Rieger, R. M., S. Tyler, J. P. S. Smith III & G. E. Rieger, 1991. Platyhelminthes: Turbellaria. In F. W. Harrison & B. J. Bogitsh (eds), Microscopic anatomy of invertebrates, 3. Wiley-Liss. Inc, New York: 7–140.

Rohde, K., 1991. The evolution of protonephridia of the Platyhelminthes. Hydrobiologia 227 (Dev. Hydrobiol. 69): 315–321.

Rohde, K., L. R. G. Cannon & N. Watson, 1988. Sense receptors with electron-dense supporting structures, and centrioles in the nerve fibres of Gieysztoria and Rhinolasius (Platyhelminthes, Rhabdocoela). J. Submicrosc. Cytol. Pathol. 20: 153–160.

Smith, J. P. S. & S. Tyler, 1985. Fine-structure and evolutionary implications of the frontal organ in Turbellaria Acoela. I. *Diopisthoporus gymnopharyngeus* sp.n. Zool. Scr. 14: 91–102.

Timoshkin, O. A., 1991. Turbellaria Lecithoepitheliata: morphology, systematics, phylogeny. Hydrobiologia 227 (Dev. Hydrobiol. 69): 323–332.

Hydrobiologia **305**: 189–196, 1995.
L.R.G. Cannon (ed.), *Biology of Turbellaria and some Related Flatworms.*

Structure and function of the reticular cell in the planarian *Dugesia dorotocephala*

Michio Morita
Department of Anatomy and Neurobiology, Colorado State University, Fort Collins, Colorado 80523, USA

Key words: planaria, reticular cell, ultrastructure, phagocytosis, carcinogenesis

Abstract

Structural and functional characteristics of the reticular cell in the planarian *Dugesia dorotocephala* were studied by light and electron microscopy. Since the reticular cells have numerous glycogen granules, lipid droplets and some lysosomes in their cytoplasm, they are easily distinguishable from other cell types. They migrate into the injured tissue, cover the injured mesenchyme, and also phagocytize debris of degenerating cells. The reticular cells also recognize foreign invaders such as bacteria. The larger aggregates of killed bacteria are encapsulated by reticular cells and eliminated into the intestine, whereas small aggregates are phagocytized by reticular cells. When cell wall extract of bacteria was inserted into the planarian body before insertion of killed bacteria, reticular cells were found to respond more quickly and vigorously to subsequent insertion of killed bacteria, indicating that the reticular cell has an immune response memory. When planarians were treated with 0.3 ppm cadmium sulfate and 0.01 ppm TPA, reticuloma tumors were induced in 76% of exposed planarians, indicating the similarity to blood cell diseases in mammals such as leukemia or lymphoma which are also induced by TPA. When these tumors were transplanted into normal hosts, the tumor cells were attacked by host reticular cells. These observations indicate that planarian reticular cells are primitive blood cells, playing important roles in nutrient transportation, homeostatic control of cells, and in defence and immune surveillance systems.

Introduction

In previous papers (Morita & Best, 1974, 1984) we have stated that planarians have a specific type of mesenchymal cell which plays an important role in nutrient transportation and phagocytosis in the body. We have named this type of cell the 'reticular cell'. This cell type is similar to the 'fixed parenchymal cell' described by Pedersen (1961) and Ishii (1965). However, we have demonstrated that the reticular cell is capable of migrating into the injured area, covering the cut-surface of the mesenchymal tissue, and phagocytizing debris of damaged or degenerating cells. These observations indicate that the reticular cell is not a 'fixed' type of cell but has the mobility similar to white blood cells such as lymphocytes and macrophages in higher animals. When heat-killed bacteria were inserted into the planarian body, they were phagocytized or encapsulated by the reticular cells (Morita, 1991), indicating that the reticular cell can recognize foreign invaders such as bacteria. Furthermore, we have demonstrated that tumors can be induced in the planarian body by treatment with mammalian carcinogens (Hall *et al.*, 1986a, b). In fact, a type of malignant tumor was induced by treatment with cadmium and 12-O-tetradecanoylphorbol-13-acetate (TPA). Since the tumor consists of a type of cell similar to the reticular cell but transformed, we have named this type of tumor the 'reticuloma'. Interestingly, the reticuloma seems to be similar to leukemia or lymphoma of mammals which have also been induced by TPA treatment (Berenblum & Lonai, 1970; Armuth & Berenblum, 1974). It becomes clear from the results of these observations that the reticular cell acts like coelomocytes or blood cells of higher animals, although planarians have not yet evolved the vascular system in their body. The present study was designed to provide more evidence for the blood cell theory of the reticular cell.

Materials and methods

Asexual planarians of the species *Dugesia dorotocephala* were used for this study. Healthy planarians, approximately 20–22 mm long, were selected from a laboratory colony that had been maintained in aged tap water and fed twice a week with raw beef liver.

In the first experiments, healthy planarians were decapitated by transverse section just behind the auricles, and the heads were discarded. The decapitated planarians were then maintained in aged tap water without food. The anterior portions of the bodies were dissected 6, 8, 12, 16, 20, 24 and 48 h after decapitation, respectively and prepared immediately for electron microscopy.

In the second experiments, each planarian was immobilized on cold, moist filter paper on a metal plate cooled to below $-0\,^{\circ}$C, and an incision was made just behind the right eye, and then heat-killed bacteria (*Mycobacterium tuberculosis H37Ra*) were inserted into the incision. Treated planarians were maintained in aged tap water, collected 2, 4, 6, 8, 10, 12, 24 and 48 h after insertion of bacteria, respectively and then prepared for electron microscopy. In the other groups of planarians, a small amount of bacterial cell wall extract (*M. bovis*: RIBI Immuno-Chem. Res. Inc.) was inserted into the incision made just behind the right eye of each planarian. After 5 days of maintenance in normal aged tap water, a small aggregate of heat-killed bacteria was inserted into the incision made in the same site of the previous operation in each planarian. These treated planarians were maintained in aged tap water for 2, 4, 6, 8, 10, 12, 16, 24 and 48 h, respectively and then prepared for electron microscopy.

In the third experiments, planarians were treated with 0.3 ppm cadmium sulfate and 0.01 ppm 12-0-tetradecanoylphorbol-13-acetate (TPA) for 2 weeks and then maintained in normal aged tap water for another 2 weeks. As reported in previous papers (Hall *et al.*, 1986a, b), malignant tumors were induced in about 76% of exposed planarians. Some tumors were transplanted into normal planarian bodies. The host tissues including transplanted tumors were collected a week after transplantation and prepared immediately for electron microscopy.

For electron microscopy, all experimental planarians described above were fixed for 1 h in a mixture of 2.5% glutaraldehyde and 0.5% formaldehyde in 0.1 M phosphate buffer (pH = 7.4). After aldehyde fixation, the target tissues were dissected from the body and postfixed for 1.5 h with 1.0% buffered osmium textroxide. These specimens were then dehydrated through increasing concentrations of ethanol, replaced with propylene oxide and embedded in Poly/Bed 812 resin. Thick sections (about 0.5 to 1.0 μm in thickness) were cut with glass knives and stained with 1.0% toluidine blue O for light microscopy. Thin sections, cut using a diamond knife, were stained with uranyl acetate and lead citrate (Sato, 1968) and observed using a JEOL electron microscope, 2000EXII.

Results

Glycogen-rich cells, which we named 'the reticular cell', are observable everywhere in the mesenchyme of the planarian body. These reticular cells are multipolar, each extending 2 or 3 cytoplasmic processes to form reticular networks among neighboring cells. Their nucleus usually has an irregular shape with frequent convex protrusions which relate morphologically to areas where cytoplasmic processes are extended. The cytoplasm of these reticular cells contains glycogen granules, lipid droplets and lysosomes in addition to the ordinary cellular organelles (Fig. 5). A large number of reticular cells are usually seen in the vicinity of the intestine (Figs 1, 2). Since the intestinal boundary is organized loosely by muscle fibers and primitive connective tissue, either intestinal cells or reticular cells are seen extruding across the tissue and associating closely with each other (Figs 6, 7). In some cases, the debris of degenerating cells, which are encapsulated by reticular cells, were seen near the intestinal boundary (Fig. 7). Light microscopy revealed that large pieces of encapsulated debris appear to be extruded from the mesenchyme tissue and eliminated into the intestine. On the other hand, in the mesenchyme, the cytoplasmic processes of reticular cells are in close association with one another and form gap junctions between them (Fig. 8).

When the planarian body is injured, the reticular cells appear on the damaged tissue 6–8 h after the injury (Fig. 9); some reticular cells appear to extend their cytoplasmic processes and cover the surface of the damaged mesenchymal tissue (Figs 10, 11). More importantly, they phagocytize debris of damaged or degenerating cells. These results indicate that the reticular cell can migrate freely like lymphocytes or macrophages in higher animals (Fig. 3). On the other hand, it is clear that the reticular cell can also recognize foreign invaders such as bacteria. When aggregates of heat-killed bacteria were inserted into the planari-

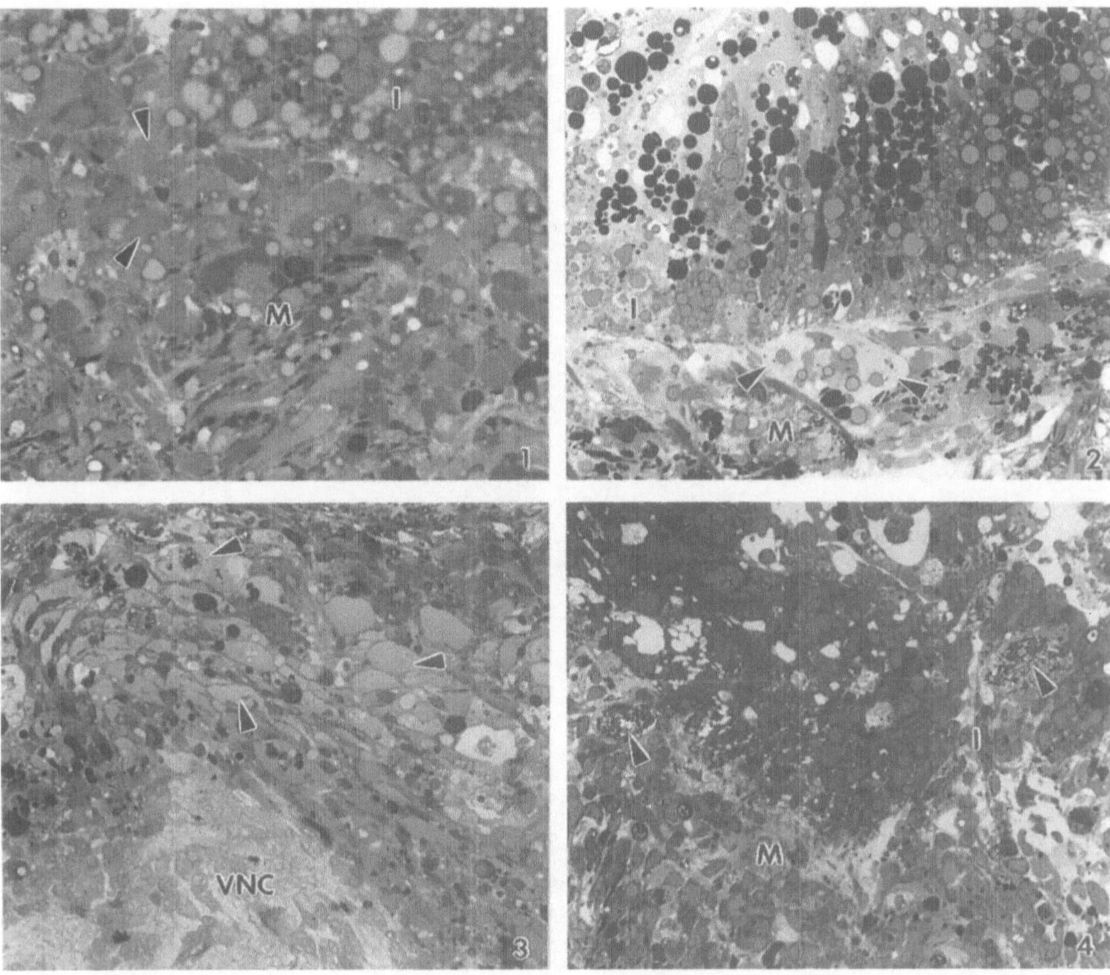

Figs 1–4. Light micrographs showing localization of reticular cells (arrowheads) in the mesenchyme (*M* near the intestine (I) (Figs 1 & 2) and the migratory process of the reticular cells (arrowheads) near the ventral nerve cord (*VNC*) (Fig. 3). In (Fig. 4) encapsulated bacteria (arrowheads) are also seen in the mesenchyme (*M*) as well as in the intestine (*I*). Figs 1–3, × 480: Fig. 4, × 260.

an body, the larger aggregates of killed bacteria were observed to be encapsulated by the cytoplasmic processes of reticular cells 12 h after insertion of killed bacteria (Fig. 12) and by 24 h, they were eliminated into the intestine (Fig. 4). In some cases, a few bacteria were seen within the cytoplasmic processes of reticular cells which formed the encapsulating wall (Fig. 13). However, small aggregates of bacteria were often observed in the phagosomes of reticular cells 12 h after insertion. These results indicate that the reticular cell can recognize differences between 'self' and 'non-self'. In addition, when the cell wall extract of bacteria were inserted into the target area of the planarian body 5 days before insertion of killed bacteria, reticular cells were observed to phagocytize bacteria more vigorously and quickly. In fact, reticular cells were seen phagocytizing or encapsulating bacteria as earlier as 6 h after insertion of killed bacteria. These results indicate that the reticular cell is capable of expressing an immune response memory similar to those of vertebrates.

When planarians were treated with 0.3 ppm cadmium sulfate and 0.01 ppm TPA, simultaneously, tumors were induced in 76% of exposed planarians 2 weeks after exposure to carcinogens. Histopathological observations indicate that large, distinct tumor formations appear in the mesenchyme, outside of the intestine and the ventral nerve cords, in the postpharyngeal region of the body. These tumors consist mainly of a type of cell similar to the reticular cell but are transformed abnormally. Immature cells, which are similar

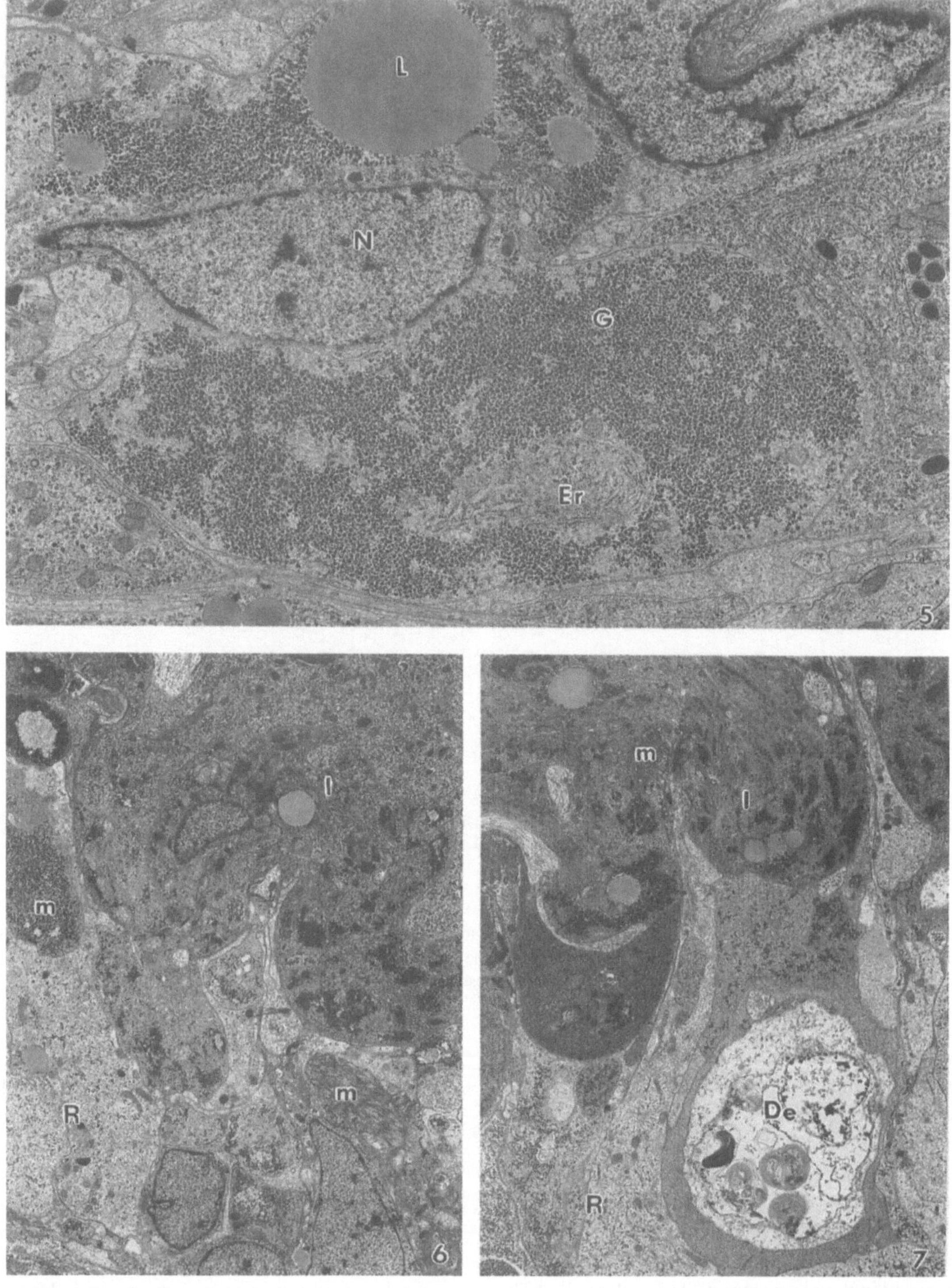

Fig. 5. An electron micrograph showing a typical structure of the reticular cell which as irregular shapes of nucleus (*N*) and cytoplasm containing numerous glycogen granules (*G*) and lipids (*L*). *Er*: endoplasmic reticulum. × 10 000. *Figs 6–7.* Electron micrographs showing many reticular cells (*R*) in the vicinity of the intestine (*I*). In some cases, debris (*De*) of cell encapsulated by reticular cell processes are observable. *m*: muscles. × 4000.

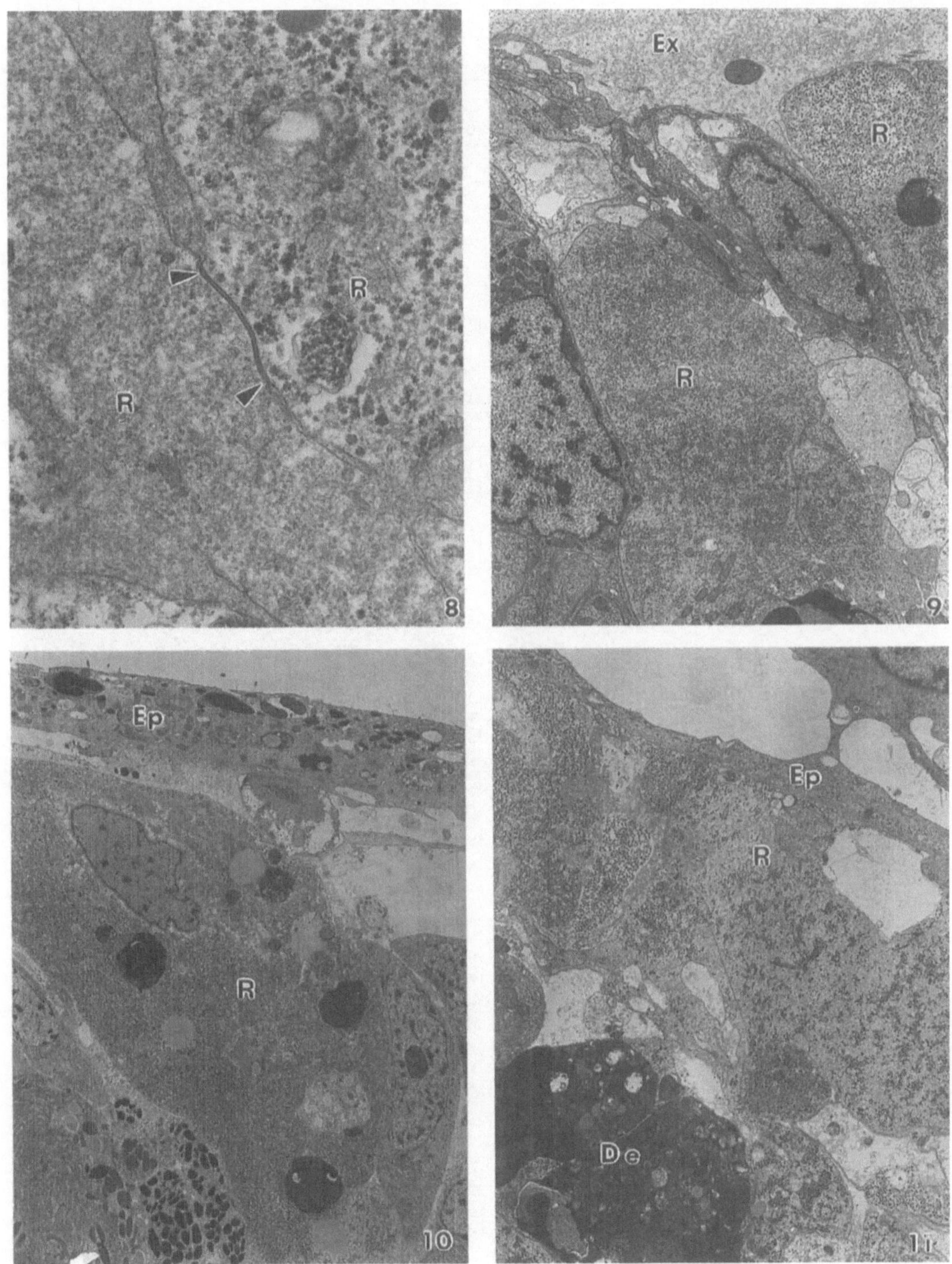

Fig. 8. Electron micrograph showing the reticular cells (*R*) are in close association with each other to form gap junctions (arrowheads) between them., × 15 000. *Figs 9–11.* Electron micrographs show the reticular cells (*R*) appeared 4 h (Fig. 9) after decapitation and covered the cut surface 6 h (Fig. 10) and 8 h (Fig. 11). *Ex*: the cut-surface, *Ep*: epithelium, *De*: debris. Figs 9 & 11, × 6400, Fig. 10, × 3200.

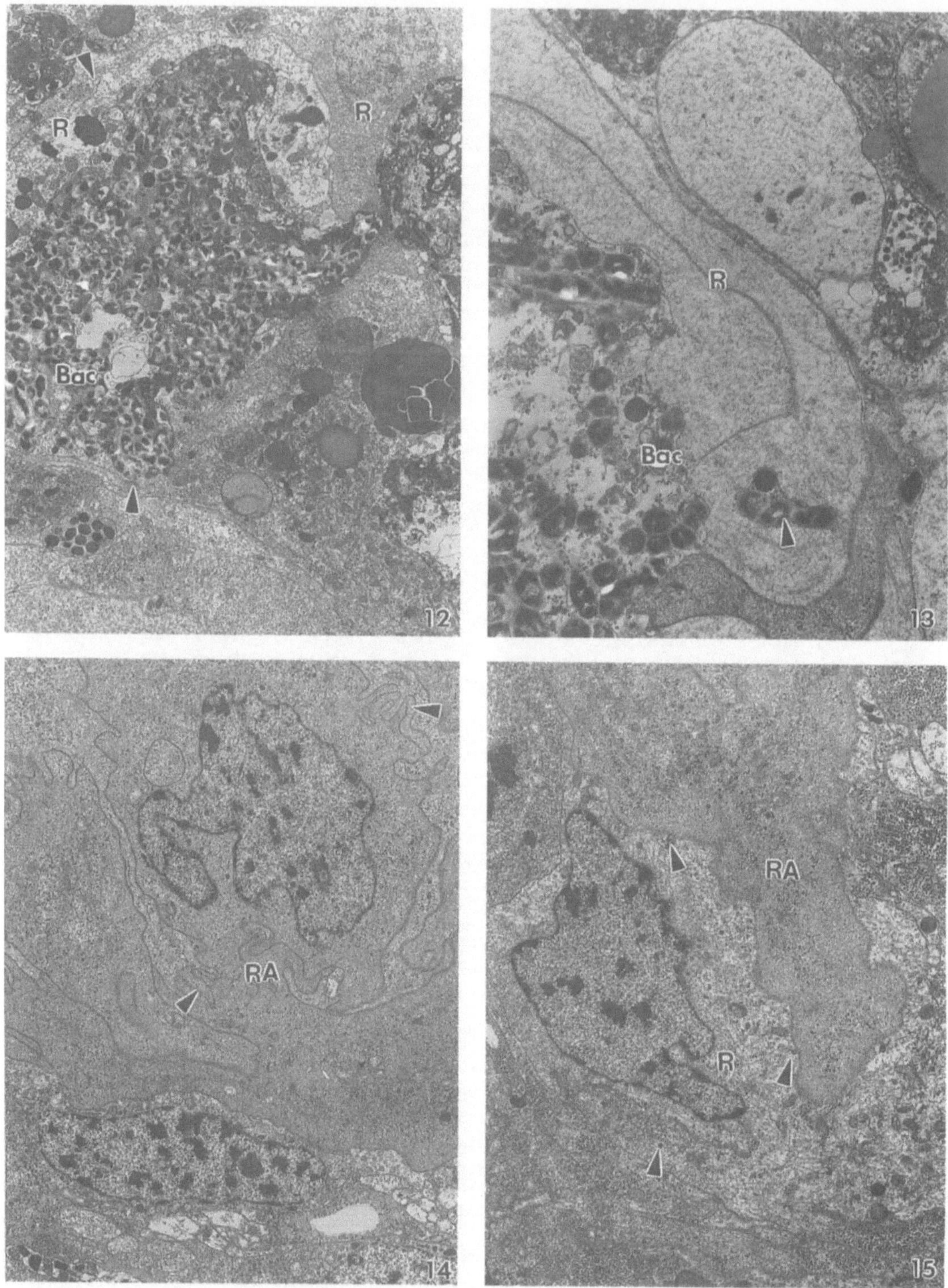

Figs 12–13. When heat-killed bacteria (*Bac*) are inserted into the planarian body, they are observed being encapsulated (Fig. 12) or phagocytized (Fig. 13) by cytoplasmic processes (arrowheads) of reticular cells (*R*). Fig. 12, × 5800; Fig. 13, × 9600. *Fig. 14.* An electron micrograph showing the reticuloma cell (RA), a transformed reticular cell, induced by cadmium and TPA. The tumor cell has an irregular nucleus, extensive cytoplasmic interdigitations with adjacent cells an numerous filopodial extrusions (arrowheads). × 6900. *Fig. 15.* Some host reticular cells (*R*) appear to enfold the transplanted tumor cell (*RA*). Small electron-dense vesicles (arrowheads) are seen in the cytoplasm of this reticular cell near the contact area. ×7300.

to the neoblast in their phenotype, are also seen in the vicinity of the tumor cell assembly. Since many tumor cells have nuclear satellite material which is a typical structure of the neoblast, it is clear that these tumor cells are differentiated from these immature but transformed neoblasts. The tumor cells named as reticuloma cells display irregular nuclei an abnormal number of mitochondria, fibrous cortical cytoplasms, extensive cytoplasmic interdigitations with adjacent cells and numerous filopodial extrusions (Fig. 14). On the other hand, when tumor tissue is transplanted into a normal planarian body, these transplanted tumor cells are found in close contact with host reticular cells. Interestingly, some reticular cells appear to enfold the tumor cell as if the reticular cell is attacking the tumor cell (Fig. 15). Those reticular cells have many small electron-dense vesicles in their cytoplasm; especially in the cytoplasm which is near the contact region of the tumor cell. Although the contents of these vesicles are not identified yet, their size and topographical and temporal location appear similar to vesicles in the stimulated cytotoxic T cells of higher animals. These results indicate that the reticular cells function like lymphocytes or macrophages in higher animals, and certainly have structured characteristics of these cells.

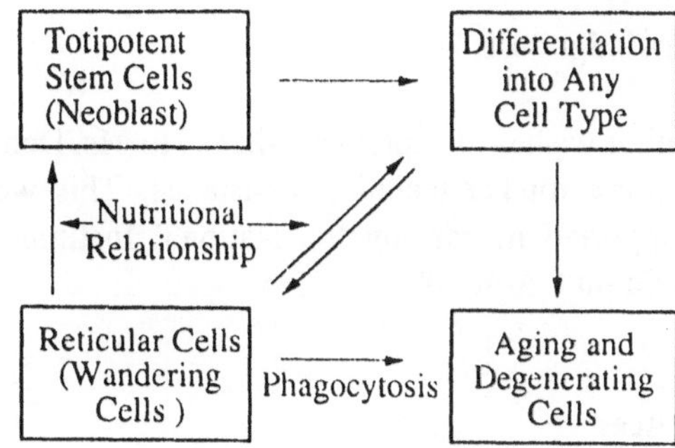

Figs 16. A proposed hypothesis: Homeostatic control of cells in planaria.

Discussion

At the electron microscopic level, Pedersen (1959, 1961) first reported two types of fundamentally important cells in the mesenchyme of the planarian body; the 'free' neoblast and the 'fixed' parenchymal cell. He described that the fixed parenchymal cell has a large cytoplasm containing numerous glycogen granules, which exhibit a strong PAS-reaction in light microscopy, lipids and lysosome-like structures. He also stated that although the fixed parenchymal cells extend their cytoplasmic processes to form reticular networks, the polymorphic shape of the nucleus and the lack of desmosomes at the surface of these cells may be related to a presumed high degree of their mobility. In conclusion, he stated that there is rather strong indirect evidence for an active participation of the fixed parenchymal cell in nutrient transport and intracellular digestive processes. It seems that he tried to find characteristics of primitive blood cells in the fixed parenchymal cell. Hori (1991) also suggested that the fixed parenchymal cell has functions of energy storage, mobility and phagocytosis as well as contact guidance for migration of regenerative cells.

In animals without a true circulatory system, wandering cells or migratory cells can be regarded as progenitors of coelmocytes and blood cells (Andrew, 1965). However, despite some of the morphological similarities, it is difficult to establish a close parallel between the functions of primitive wandering cells and of blood cells. From the results of our observations, it is clear that a specific type of cell shows characteristics of the blood cell such as nutrient transportation, protection of injured mesenchymal tissue and phagocytosis of debris of degenerating cells in the body. These functional characteristics of the cell are similar to those of the fixed parenchymal cell described by Pedersen (1961). However, since this type of cell has a high degree of mobility and characteristics of blood cells observed in higher animals, we have called this cell the 'reticular cell' instead of the 'fixed parenchymal cell'.

As described previously, the planarian neoblast is the totipotent stem cell which differentiates into any type of cell in the body (Morita, 1967; Morita *et al.*, 1969; Morita & Best, 1974, 1984). However, as observed here the reticular cell functions in nutrient transportation, protection of the mesenchymal tissue, phagocytosis of degenerating cells, recognition of foreign invaders and in immune response memory. It seems that the totipotent stem cells and reticular cells are fundamental, important cells in the planarian body. As shown in the Fig. 16, I hypothesize that the planarian body consists of only the renewing population of cells, and that the reticular cell plays some kind of regulatory role in the homeostatic control of the total cellular population. That is, the totipotent stem cell (neoblast) renews any type of cell in the planarian body, and the reticular cell plays an important role in the nutritional support and physiological surveillance of cells.

Acknowledgments

The author wishes to express his thanks to Ms. Denise Fay-Guthrie for her technical assistance. This work was supported in part by the National Institute of Health Grant CA-46409.

References

Andrew, W., 1965. Comparative aspects of blood invertebrates. In W. Andrew (ed.), Comparative Hematology. Greene and Stratton, New York: 62–68.

Armuth, V. & I. Berenblum, 1974. Promotion of mammary carcinogenesis and leukemogenic action by phorbol in virgin female Wistar rats. Cancer Res. 34: 2704–2707.

Berenblum, I. & V. Lonai 1970. The leukemogenic action of phorbol. Cancer Res. 30: 2744–2748.

Hall, F. L., M. Morita & J. B. Best, 1986a. Neoplastic transformation in the planarian. I. Cocarcinogenesis and histopathology. J. exp. Zool. 240: 211–227.

Hall, F. L., M. Morita & J. B. Best, 1986b. Neoplastic transformation in the planaria. II. Ultrastructure of malignant reticuloma. J. exp. Zool. 240: 229–244.

Hori, I., 1991. Role of fixed parenchyma cells in blastema formation of the planarian *Dugesia japonica*. Int. J. dev. Biol. 35: 101–108.

Ishii, S., 1965. Electron microscopic observations on the planarian tissue. II. The intestine. Fukushima J. med. Sci. 12: 67–87.

Morita, M., 1967. Observations on the fine structure of the neoblast and its cell division in the regenerating planaria. Sci. Rep. Tohoku Univ. Biol. 33: 339–406.

Morita, M., 1991. Phagocytic response of the planarian reticular cells to heat-killed bacteria. Hydrobiologia 227 (Dev. Hydrobiol. 69): 193–199.

Morita, M. & J. B. Best, 1974. Electron microscopic studies of planarian regeneration. II. Changes in epidermis during regeneration. J. exp. Zool. 187: 345–374.

Morita, M. & J. B. Best, 1984. Electron microscopic studies of planarian regeneration. III. Degeneration and differentiation of muscles. J. exp. Zool. 229: 425–436.

Morita, M., & J. B. Best & J. Noel, 1969. Electron microscopic studies of planarian regeneration. I. Fine structure of neoblasts in *Dugesia dorotocephala* J. ultrastruc. Res., 27: 7–23.

Pedersen, K. J., 1959. Cytological studies on the planarian neoblast. Z. Zellforsch. mikroskop. Anat. 50: 700–817.

Pedersen, K. J., 1961. Studies on the nature of planarian connective tissue. Z. Zellforsch. mikroskop. Anat. 53: 569–608.

Sato, T., 1968. A modified method for lead staining of thin sections. J. electron Microsc., 17: 158–159.

Hydrobiologia **305**: 197, 1995.
L.R.G. Cannon (ed.), *Biology of Turbellaria and some Related Flatworms.*

Fine structure of the astrocyte-like neuroglial cell in the planarian *Dugesia dorotocephala*

Michio Morita[1] & Jay Boyd Best[2]
[1] *Department of Anatomy and Neurobiology*
[2] *Environmental Health, Colorado State University, Fort Collins, Colorado 80523, USA*

Key words: planarian, neuroglial cell, ultrastructure

Abstract

Astrocyte-like neuroglial cells in the planarian central nervous system were studied by electron microscopy. The neuroglial cell observed in *Dugesia dorotocephala* is multipolar, extending 3 or 4 major cytoplasmic processes, each with many smaller daughter branches. The nucleus is often irregular, which seems to be related to the terminals of the cytoplasmic processes. Many glycogen granules, mitochondria, granular endoplasmic reticulum, lysosomes and phagosomes are characteristically observed in the cytoplasm of the neuroglial cells. Within phagosomes, debris of nerve or neurosecretory fibers appear in various degrees of disintegration. These neuroglial cells are closely associated with bundles of nerve fibers, nerve cells, neurosecretory cells, muscle fibers and other neuroglial cells. They also form gap junctions with each other in a manner similar to astrocytes of higher animals. A comparison of the planarian neuroglial cell with the migratory, phagocytic reticular cell reveals several similarities but also some important differences. That is, both cells have functions of phagocytosis, nutrient transportation and junctional communication, but their shapes and locations as well as phagocytized material are different from each other. From an evolutionary point of view, these observations suggest that the planarian neuroglial cell is not merely a reticular cell that has migrated from the mesenchyme, but represents a distinct cell type; e.g. a primitive astrocyte that transports nutrients for neurons and phagocytizes degenerating neurons, in addition to its role as a supporting element of the central nervous system.

Hydrobiologia **305**: 199–206, 1995.
L.R.G. Cannon (ed.), Biology of Turbellaria and some Related Flatworms.

Ultrastructural and cytochemical studies of the female gonad of *Prorhynchus* sp. (Platyhelminthes, Lecithoepitheliata)

Alessandra Falleni, Paolo Lucchesi & Vittorio Gremigni
Dipartimento di Biomedicina Sperimentale, Sez. Biologia e Genetica, via A. Volta 4, Università di Pisa, 56126 Pisa, Italy

Key words: Platyhelminthes, Lecithoepitheliata, female gonad, oocytes, vitellocytes, ultrastructure

Abstract

The female gonad of *Prorhynchus* is heterocellular (neoophoran organization) and consists of an unpaired, elongate germovitellarium enveloped by a finely granular extracellular lamina. It is composed of a posterior germinative area where early oocytes are randomly associated with differentiating vitellocytes and a growth area with follicular organization. In each follicle a single oocyte is surrounded by a layer of vitellocytes. By electron microscopy, the oocytes showed features typical of non-vitellogenic germ cells; they had chromatoid bodies, annulate lamellae, lipid droplets and R.E.R. and Golgi complexes producing small granules with a multilamellar pattern. Vitellocytes showed features typical of secretory cells with the R.E.R. and Golgi complex developed to a great extent and involved in the production of type A and type B globules, respectively. We speculate that type A globules are shell-globules and type B globules are yolk. The structure, composition and role of vitellocyte globules of *Prorhynchus* are compared with those of homologous inclusions from other Platyhelminthes.

Abbreviations

A	= type A globule
B	= type B globule
ECL	= extracellular lamina
GC	= Golgi complex
L	= lipid
RER	= rough endoplasmic reticulum
O	= oocyte
V	= vitellocyte

Introduction

Several recent studies have focused on the submicroscopic anatomy of Lecithoepitheliata (Rohde & Watson, 1991; Antoniazzi & Silveira, 1992; Watson & Rohde, 1992, 1993). However, very little is known about the ultrastructure of the female gonad, particularly concerning the origin, chemical composition, and function of oocyte and vitellocyte structures. Limited data on the germovitellarium of *Gnosonesima* (see Gremigni, 1988) and *Geocentrophora* (see Bogolyubov & Timoshkin, 1993) and on vitellocytes of *Prorhynchus* (see Gremigni & Falleni, 1991, 1992) are available, while a number of papers have been published to date on the ultrastructure of the female gonad of other taxa of free-living and parasitic Platyhelminthes (see Davis & Roberts, 1983; Gremigni, 1983, 1988; Smyth & Halton, 1983; Coil, 1991; Fried & Haseeb, 1991; Rieger *et al.*, 1991 for reviews).

The female gonad of the Lecithoepitheliata might represent an intermediate stage between the archoophoran and neoophoran organization, or more appropriately an early stage of neoophoran organization (Steinböck, 1927; Bresslau, 1928–33; Karling, 1940; Hyman, 1951; Gremigni, 1983; Timoshkin, 1991). Therefore, ultrastructural investigations on this organ are of interest and the discovery of new characteristics from gonadal cells may prove useful in clarifying the uncertain phylogenetic relationships of Lecithoepitheliata with other turbellarians (Ehlers, 1985; Smith *et al.*, 1986; Ax, 1987).

This paper reports a study of the germovitellarium of *Prorhynchus* by conventional and cytochemical electron microscopy and compares the present findings with those obtained from other platyhelminths.

Materials and methods

Some adult specimens of *Prorhynchus* sp. were collected from springs near Verona (Italy); other specimens came from a pond on the campus of the University of New England, Armidale, NSW (Australia) and were given to us by Dr Rohde and Dr Watson. Sexually mature specimens were fixed in 2.5% glutaraldehyde in 0.05–0.1 M cacodylate (or phoshate) buffer, pH 7.2, for 1 h at 4 °C. Specimens were rinsed in buffer and then postfixed in 1% osmium tetroxide in the same buffer for 2 h at room temperature. They were then dehydrated in ethanol and embedded in Epon-Araldite or Spurr resin. Ultrathin sections were cut with a diamond knife on a Reichert-Jung Ultratome E, stained with uranyl acetate and lead citrate and observed with a JEOL 100 SX T.E.M..

Test for polyphenolic substances. Whole specimens, fixed in 3% glutaraldehyde, were treated according to the methenamine silver method of Locke & Krishnan (1971). Ultrathin sections were observed without further staining.

Enzymatic protein extraction. Ultrathin sections obtained from blocks used for morphological studies were treated with 2% H_2O_2 in water for 15 min at 37 °C and then incubated in a solution of 0.5% protease (Pronase E, Sigma) adjusted to pH 7.5 for 3–6 h.

Test for polysaccharides and glycoproteins. Ultrathin sections obtained from blocks used for morphological studies were treated according to the Thiéry (1967) method.

Control specimens or sections were prepared for each cytochemical test.

Results

The germovitellarium of *Prorhynchus* sp. is an unpaired, elongate organ located ventral to the gut. It consists of a posterior germinative area where young germ cells are randomly associated with early differentiating vitellocytes (Fig. 1A) and an anterior growth area with follicular organization (Fig. 1B, C). Usually from five to seven follicles forming a ventral string are distributed caudo-cranially. They have been classified as stage I or stage II follicles according to ultrastructural features. Each follicle is composed of a single oocyte surrounded by a layer of vitellocytes. The germovitellarium is enveloped by a narrow, finely granular extracellular lamina-like layer (Fig. 2) and no accessory cells have been observed within follicles.

Germinative area

The germinative area occupies the posterior portion of the germovitellarium. It is composed of early differentiating oocytes and vitellocytes not organized in well-defined follicles. Oocytes are oblong cells measuring approximately 10×20 μm. The large nucleus contains small patches of condensed chromatin and a well-developed nucleolus (Fig. 3). The cytoplasm is weakly differentiated, packed with free ribosomes and contains a few mitochondria and lipid droplets. Vitellocytes are rounded cells measuring approximately 7×10 μm. The nucleus is ovoid and contains large clumps of heterochromatin and a prominent nucleolus (Fig. 3). The cytoplasm contains large amounts of free ribosomes and some mitochondria.

Growth area

Stage I and stage II follicles occupy the median and anterior portion of the germovitellarium; each consists of one oocyte surrounded by a layer of vitellocytes. The growing oocyte has a large nucleus whose envelope is rich in pores. It contains diffuse chromatin and a large nucleolus that, in more developed oocytes, has lost its compact structure and shows enlarged areas of low electron-density (Fig. 4). Several fibrillar or finely granular aggregates of presumed nuclear origin, the so-called chromatoid bodies, are visible in the perinuclear cytoplasm (Fig. 5). Stacks of annulate lamellae are present in the cytoplasm, particularly in proximity to the nucleus (Fig. 6). Other cytoplasmic features that characterize this stage are the increase in number of mitochondria and lipid droplets and the appearance of rough endoplasmic reticulum (R.E.R.) and Golgi complexes (Fig. 7). The Golgi complex appears to be involved in the formation of small vesicles that contain a material of low electron-density. The vesicles appear to give rise by fusion to small granules (0.4–0.6 μm in diameter) with a multilamellar pattern that remains scattered throughout the ooplasm (Fig. 8).

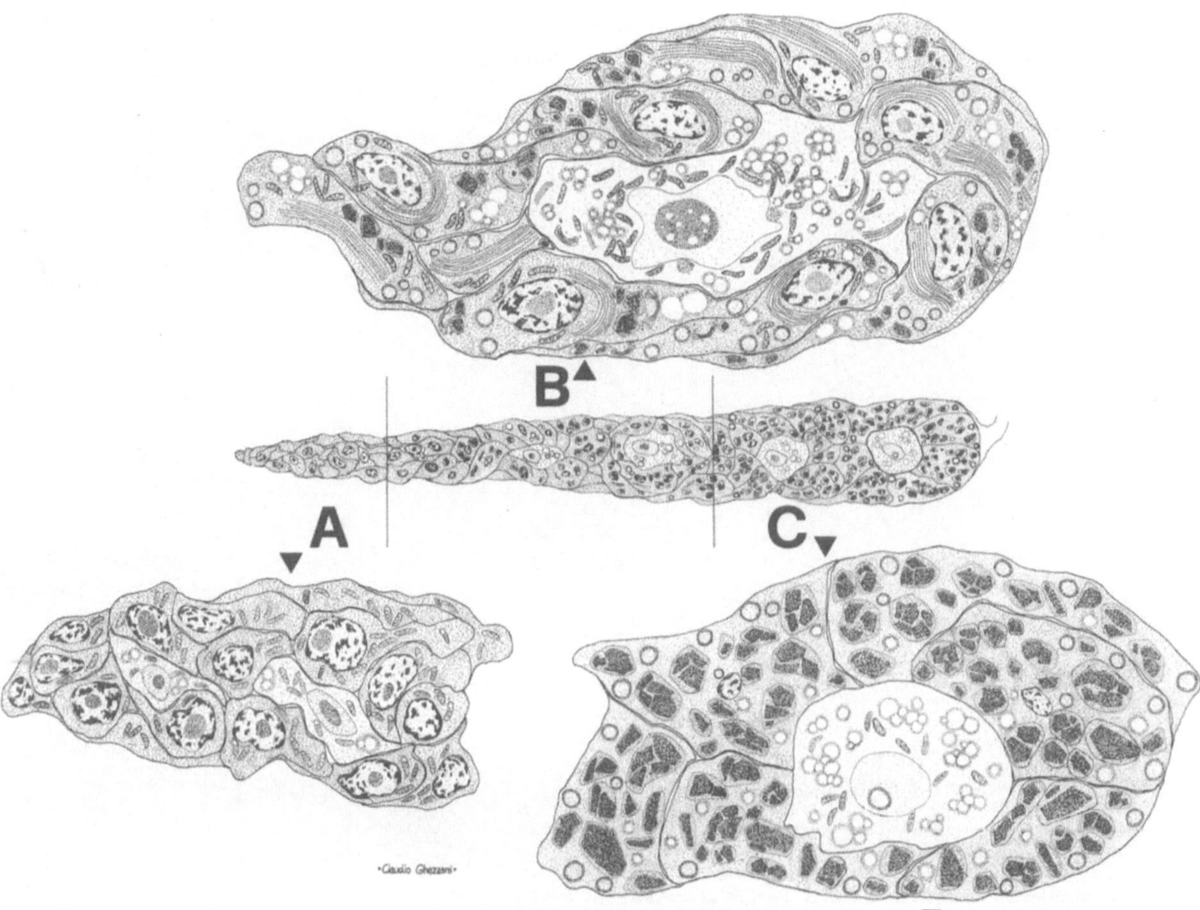

Fig. 1. Diagram of the germovitellarium of *Prorhynchus*. A, detail of the germinative area, B, C, details of the growth area indicating a stage I and a stage II follicle respectively.

Vitellocytes surrounding the oocyte have a rounded nucleus with large clumps of chromatin and a well-developed nucleolus (Fig. 9). The main features of differentiation are long cytoplasmic processes projecting from the outer part of the cell beneath the extracellular lamina (Fig. 2), giant profiles of R.E.R., Golgi complexes, lipid droplets, and two types of membrane-bounded inclusions which we called type A and type B globules. Type A globules appear first, are relatively scarce and are characteristically enveloped by a granular membrane derived from the R.E.R. (Fig. 10). Mature type A globules have a round shape, measure about 2.5 μm in diameter, consist of a narrow electron-dense cortex and a less electron-dense granular core, and are localized in the peripheral cytoplasm (Fig. 14). Their content tested negative for polyphenols, was digested by protease and showed a light positive reaction to the Thiéry test (Fig. 11). Type B globules appear to be formed by fusion of vesicles derived from the Golgi complex, contain electron-dense material and are enveloped by a smooth membrane (Fig. 12). At an intermediate stage of maturation, type B globules are oblong and contain a central, electron-dense area surrounded by a granular, less electron-dense material. Large irregularly shaped globules up to 5×7 μm appear to have arisen by coalescence of smaller globules. Mature type B globules are composed of one or more electron-dense areas, sometimes showing a paracrystalline pattern, embedded in a finely granular, less electron-dense material (Fig. 15). They are numerous and randomly scattered in the cytoplasm of mature vitellocytes (Figs 13, 14). The granular component showed a light positive reaction while the central component tested negative for polyphenols (Fig. 16). The central component showed a fine silver precipitate with the Thiéry test (Fig. 17) and was extracted by protease (Fig. 18).

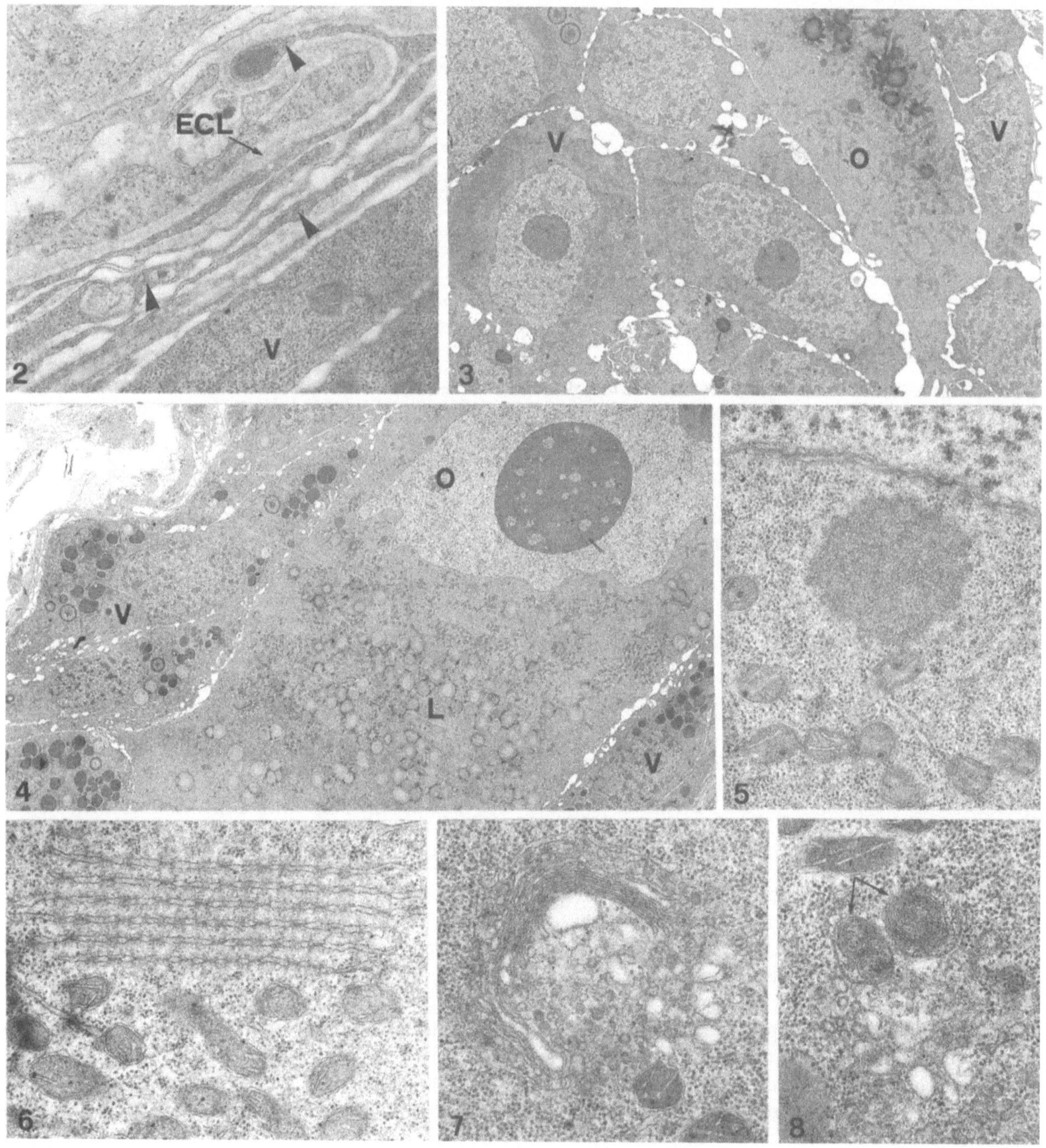

Figs 2–8. (Fig. 2) Processes of the peripheral cytoplasm (arrowheads) of a vitellocyte. × 20 000. (Fig. 3) A young oocyte and some early vitellocytes from the germinative area. × 26 000. (Fig. 4) Portion of a stage I follicle. × 1500. (Fig. 5) Chromatoid body in the perinuclear cytoplasm of a growing oocyte. × 24 000. (Fig. 6) Annulate lamellae in a growing oocytes. × 23 000. (Fig. 7) Golgi complex producing some weakly electron-dense vesicles. × 28 000. (Fig. 8) Two small granules (arrows) adjacent to Golgi-derived vesicles. × 30 000.

In stage II follicles, the mature oocyte is still elongate and reaches a size of 40×80 μm. It has diffuse chromatin and no nucleolus. The cytoplasm contains some clustered lipid droplets and a few, randomly scattered, small granules. Vitellocytes are up to about five times their original diameter and are cuboidal. Their cytoplasm is packed with type A and type B globules (Fig. 13) and contains diffuse glycogen particles and a few lipid droplets.

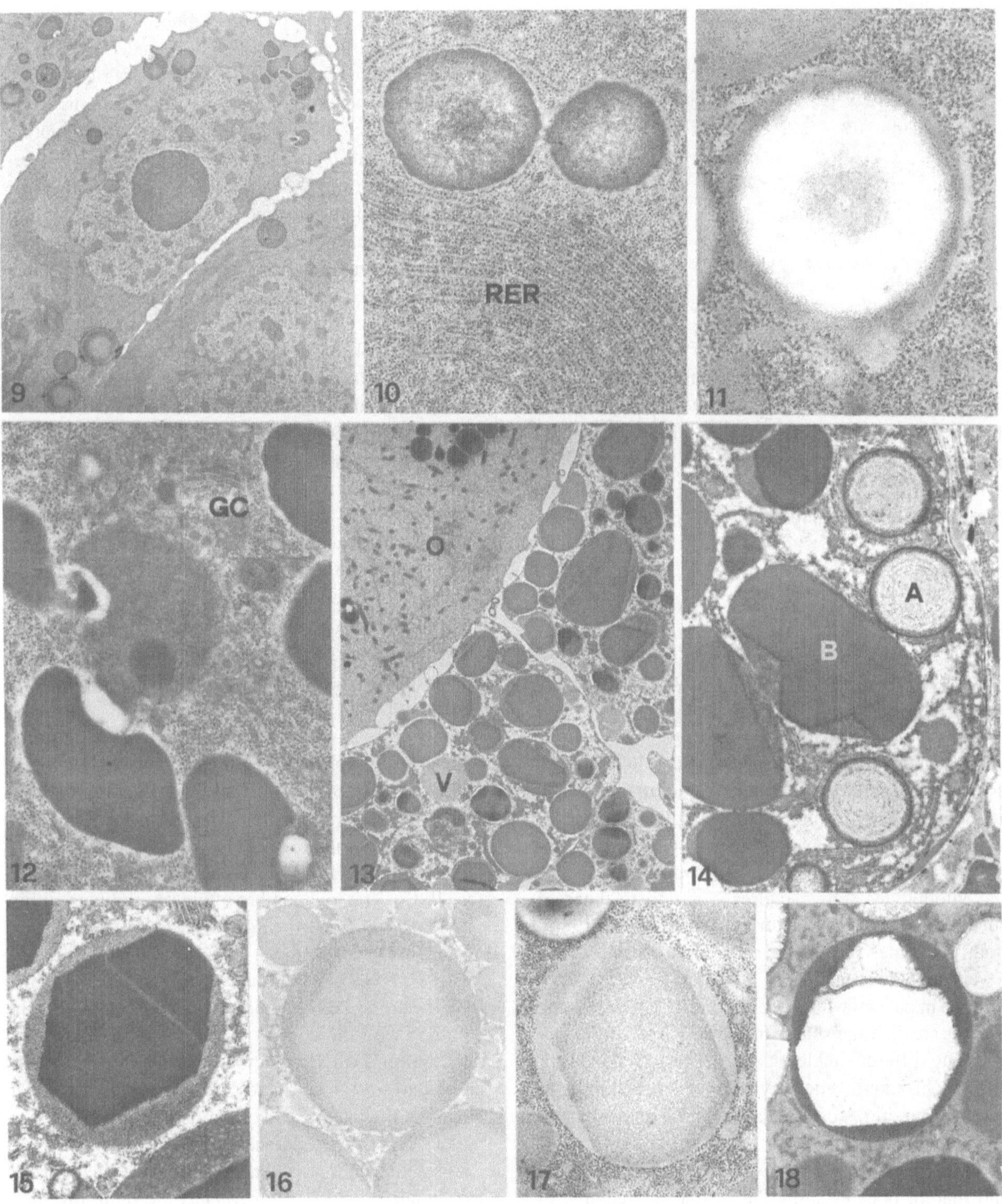

Figs 9–18. (Fig. 9) Differentiating vitellocytes from a stage I follicle. × 3000. (Fig. 10) Two nascent type A globules bounded by a granular membrane. Differentiating vitellocyte. × 19 000. (Fig. 11) Mature type A globule after Thiéry test. A fine silver precipitate indicating the presence of glycoproteins is visible on the globule cortex. × 24 000. (Fig. 12) Nascent type B globules. × 20 000. (Fig. 13) Portion of a stage II follicle showing part of two mature vitellocytes adjacent to the mature oocyte. × 3300. (Fig. 14) Distal part of a vitellocyte. Type A globules are located beneath the membrane, type B globules remain dispersed in the cytoplasm. × 6000. (Fig. 15) Type B globule. × 12 000. (Fig. 16) Type B globule. A light silver precipitate is visible only on the peripheral component. Locke & Krishnan test. × 5500. (Fig. 17) Type B globule. A fine silver precipitate is visible on the central component. Thiéry test. × 15 000. (Fig. 18) Type B globule. The peripheral component is unaffected after incubation with protease. × 8500.

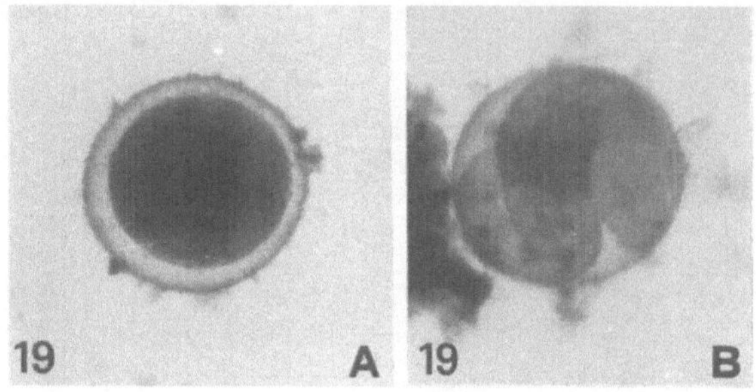

Figs 19. Light photomicrographs of capsules containing a cleaving embryo (A) and a worm just before hatching (B). Both capsules are bounded by a thin, translucent shell. × 110.

Laid capsules, containing a single embryo, are spherical, about 200 μm in diameter, and have a thin, translucent, flexible shell. A thin space is visible between the embryo and the shell (Fig. 19A, B). Just before hatching, the worm could be seen moving vigorously until the shell broke.

Discussion

The heterocellular female gonad of *Prorhynchus* is enveloped by an extracellular lamina and is devoid of a cellular tunica. On the other hand, the germovitellarium of the marine species *Gnosonesima mediterranea* appears to have an incomplete cellular tunica (Rieger *et al.*, 1991 citing personal communications by Schockaert). Unlike other neoophorans (Rieger *et al.*, 1991), *Prorhynchus* does not have accessory cells in the female gonad. Differentiating oocytes and vitellocytes show features common to those of other neoophoran platyhelminths (Gremigni, 1988). Both cell types have high nucleo/cytoplasmic ratio, a well-developed nucleolus and weakly differentiated cytoplasm packed with free ribosomes. Differentiating oocytes are distinguishable from vitellocytes principally by the different nuclear pattern. Oocytes have very diffuse chromatin; vitellocytes have large clumps of heterochromatin, mostly adjacent to the nuclear envelope.

In stage I and stage II follicles, differentiating and mature oocytes show the distinctive features of non-vitellogenic germ cells, *i.e.* the presence of chromatoid bodies, short R.E.R. profiles, annulate lamellae and lipid droplets and the absence of yolk globules. Oocytes also produce a few small granules of unknown function. Vitellocytes increase in size and assume the appearance of secretory cells with well-developed R.E.R. and Golgi complex involved in the production of two types of globules. Type A globules are round, characteristically bounded by a granular membrane and contain a central area consisting of proteins and a thin cortex of glycoproteins. They become located in the peripheral cytoplasm of mature vitellocytes and appear quite different from all other inclusions observed in the vitellocytes of either the lecithoepitheliate *Gnosonesima*, or other turbellarians. Type B globules are irregularly shaped, bounded by a smooth membrane and have one or more proteinaceous homogeneous or paracrystalline areas embedded in a granular component containing a small amount of polyphenols. They remain randomly scattered throughout the cytoplasm of mature vitellocytes and are similar to some inclusions observed in the vitellocytes of *Gnosonesima* (Gremigni, 1988).

Based on their location in mature vitellocytes, we suggest that type A globules are shell-globules and that type B globules are yolk. Unlike the vitellocytes of most neoophoran platyhelminths (Gremigni, 1983, 1988; Shinn, 1993), those of *Prorhynchus* do not have polyphenol-containing globules. This finding correlates with the observation that laid capsules of *Prorhynchus* are enveloped by a thin, translucent shell which appears to be proteinaceous and non-sclerotized. A similar translucent, non-sclerotized shell is also seen in other platyhelminths, namely in the Acoelomorpha which lack polyphenol-containing granules (Gremigni, 1988; Smith *et al.*, 1988; Falleni & Gremigni, 1989, 1990), in the typhloplanoid *Mesostoma* (subitaneous eggs only) (Domenici & Gremigni, 1977) and in some members of the parasitic Neodermata (Smyth & Halton, 1983). The low number of type A globules within vitellocytes of *Prorhynchus* correlates with the thinnes of the non-sclerotized egg shell. Moreover, secretions from two types of shell glands which open around the female pore in Prorhynchidae (Reisinger, 1968) might combine with the content of type A globules to produce the egg shell. Further investigations are necessary, particularly on freshly laid capsules of this and other lecithoepitheliates, to ascertain definitively to role of the two types of vitellocyte globules.

Present findings indicate that some features of the female gonad of *Prorhynchus* are quite different from those of *Gnosonesima*. In particular, (a) the germovitellarium of *Prorhynchus* is enveloped by an extracellular lamina, while the female gonad of *Gnosonesima* appears to be enveloped by an incomplete tunica; (b) unlike shell globules in the vitellocytes of *Gnosonesima*, those of *Prorhynchus* do not contain polyphenols. These aspects may support the hypothesis of a diphyletic origin of Lecithoepitheliata (Karling, 1968; Timoshkin, 1991) suggesting that Prorhynchidae and Gnosonesimidae are not closely related.

Furthermore, the structure and composition of both small granules in the oocytes and shell globules in the vitellocytes of *Prorhynchus* are quite different from those observed in the prolecithophorans *Plagiostomum*, *Vorticeros* (see Nigro & Gremigni, 1987; Gremigni, 1988) and *Acanthiella* (Rieger, 1981). Inclusions in the oocytes and vitellocytes of prolecithophorans studied to date are quite similar to homologous inclusions of the Rhabdocoela (Gremigni, 1988; Lucchesi *et al.*, 1995). These ultrastructural features do not support the speculation that Lecithoepitheliata (or at least Prorhynchidae) are closely related to Prolecithophora (Timoshkin, 1991), but correlate with the recent statement of Böckerman *et al.* (1995) that there is no close relationship between these two groups.

The position of Lecithoepitheliata within Platyhelminthes remains uncertain and further ultrastructural and molecular studies are necessary to find useful apomorphies that can clarify the phylogenetic relationships of Prorhynchidae and Gnosonesimidae with other taxa of Platyhelminthes-Turbellaria.

Acknowledgments

Our special thanks are due to Dr K. Rohde and Dr N. Watson for providing us with both live and fixed specimens of *Prorhynchus* sp.. We are also grateful to Mr C. Ghezzani for technical assistance and Mr J. Franceschina for English revision. The work was supported by a grant from the Consiglio Nazionale Ricerche (C.N.R.) of Italy.

References

Antoniazzi, M. M. & M. Silveira, 1992. Pharyngeal and gastrodermal ultrastructure of Prorhynchus stagnalis Schulze, 1851 (Turbellaria, Lecithoepiteliata). Acta zool. 73: 255–262.

Ax, P., 1987. The Phylogenetic system. The systematization of organisms on the basis of their phylogenesis. J. Wiley & Sons, Chichester, 249 pp.

Böckerman, I., O. I. Raikova, M. Reuter & O. Timoshkin, 1995. Ultrastructure of the nerve cells and sensilla of *Geocentrophora baltica* (Platyhelminthes, Lecithoepitheliata) and the surface sensillae in the Geocentrophora group. Hydrobiologia 305 (Dev. Hydrobiol. 108): 183–188.

Bogolyubov, D. S. & O. A. Timoshkin, 1993. Comparative characteristics of the female gonads of the Lecithoepitheliata (Plathelminthes) with remarks on the taxonomy of the order. Zool. Zh. 72: 17–26.

Bresslau, E., 1928–33. Turbellaria. In W. Kukenthal & T. Krumbach (eds), Handbuch der Zoologie. Vol. II, Part 1. Walter de Gruyter, Berlin: 52–304.

Coil, W. H., 1991. Platyhelminthes: Cestoidea. In F. W. Harrison and B. J. Bogitsh (eds), Microscopic Anatomy of Invertebrates. Vol. 3. Platyhelminthes and Nemertinea. Wiley-Liss, Inc., New York: 211–283.

Davis, R. E. & L. S. Roberts, 1983. Platyhelminthes-Eucestoda. In K. G. Adiyodi & R. G. Adiyodi (eds), Reproductive Biology of Invertebrates. Vol. 1, Oogenesis, oviposition and oosorption. J. Wiley & Sons, Chichester: 109–233.

Domenici, L. & V. Gremigni, 1977. Fine structure and functional role of the coverings of the eggs in Mesostoma ehrenbergii (Focke) (Turbellaria, Neorhabdocoela). Zoomorphology 88: 247–257.

Ehlers, U., 1985. Phylogenetic relationships within the Platyhelminthes. In S. Conway Morris, J. D. George, R. Gibson & H. M. Platt (eds), The origins and relationships of lower invertebrates. Oxford University Press, Oxford: 143–158.

Falleni, A. & V. Gremigni, 1989. Egg covering formation in the acoel Convoluta psammophila (Platyhelminthes, Turbellaria): an ultrastructural and cytochemical investigation. Acta Embryol. Morphol. Exper. n.s. 10: 105–112.

Falleni, A. & V. Gremigni, 1990. Ultrastructural study of oogenesis in the acoel turbellarian Convoluta. Tissue Cell 22: 301–310.

Fried, B. & M. Haseeb, 1991. Platyhelminthes: Aspidogastrea, Monogenea and Digenea. In F. W. Harrison & B. J. Bogitsh (eds), Microscopic Anatomy of Invertebrates. Vol. 3. Platyhelminthes and Nemertinea. Wiley-Liss, Inc., New York: 141-209.

Gremigni, V., 1983. Platyhelminthes-Turbellaria. In K. G. Adiyodi & R. G. Adiyodi (eds), Reproductive Biology of Invertebrates. Vol. 1. Oogenesis, oviposition and oosorption. J. Wiley & Sons, Chichester: 67–107.

Gremigni, V., 1988. A comparative ultrastructural study of homocellular and heterocellular female gonads in free-living Platyhelminthes-Turbellaria. Fortschr. Zool. 36: 245–261.

Gremigni, V. & A. Falleni, 1991. Ultrastructural features of cocoon-shell globules in the vitelline cells of neoophoran platyhelminths. Hydrobiologia 227 (Dev. Hydrobiol. 69): 105–111.

Gremigni, V. & A. Falleni, 1992. Mechanisms of shell-granule and yolk production in oocytes and vitellocytes of Platyhelminthes-Turbellaria. Anim. Biol. 1: 29–37.

Hyman, L. H., 1951. The Invertebrates. II. Platyhelminthes and Rhynchocoela. The Acoelomate Bilateria. Mc Graw-Hill, New York, 550 pp.

Karling, T. G., 1940. Zur Morphologie und Systematik der Alloecoela Cumulata und Rhabdocoela Lecithophora (Turbellaria). Acta zool. fenn. 26: 1–260.

Karling, T. G., 1968. On the genus Gnosonesima Reisinger (Turbellaria). Sarsia 33: 81–108.

Locke, M. & N. Krishnan, 1971. The distribution of polyphenoloxidases and polyphenols during cuticle formation. Tissue Cell 3: 103–126.

Lucchesi, P., A. Falleni & V. Gremigni, 1995. The ultrastructure of the germarium in some Rhabdocoela. Hydrobiologia 305 (Dev. Hydrobiol. 108): 207–212.

Nigro, M. & V. Gremigni, 1987. Ultrastructural features of oogenesis in a free-living marine platyhelminth, Vorticeros luteum. Tissue Cell 19: 377–386.

Reisinger, E., 1968. Xenoprorhynchus ein Modellfall für progressiven Funktionswechsel. Z. Zool. Syst. Evolutionsforsch. 6: 1–55.

Rieger, R. M., 1981. Morphology of the Turbellaria at the ultrastructural level. Hydrobiologia 84: 213–229.

Rieger, R. M., S. Tyler, J. P. S. Smith & G. E. Rieger, 1991. Platyhelminthes: Turbellaria. In F. W. Harrison & B. J. Bogitsh (eds), Microscopic Anatomy of Invertebrates. Vol. 3. Platyhelminthes and Nemertinea. Wiley-Liss, Inc., New York: 7–140.

Rohde, K. & N. Watson, 1991. Ultrastructure of the flame bulbs and protonephridial capillaries of *Prorhynchus* (Lecithoepitheliata, Prorhynchidae, Turbellaria). Zool. Scr. 20: 99–106.

Shinn, G. L., 199.3. Formation of egg capsules by flatworms (Phylum Platyhelminthes). Trans. am. Microsc. Soc. 112: 18–34.

Smith, J. P. S., M. B. Thomas, R. M. Chandler & S. Zane, 1988. Granular inclusions in the oocytes of Convoluta sp., Nemertoderma sp., and Nemertinoides elongatus (Turbellaria, Acoelomorpha). Fortschr. Zool. 36: 263–269.

Smith, J. P. S., S. Tyler & R. M. Rieger, 1986. Is the Turbellaria polyphyletic? Hydrobiologia 132 (Dev. Hydrobiol. 32): 13–21.

Smyth, S. D. & D. W. Halton, 1983. The Physiology of Trematodes. Cambridge University Press, Cambridge, 417 pp.

Steinböck, O., 1927. Monographie der Prorhynchidae (Turbellaria). Z. Morph. Okol. Tiere 8: 538–662.

Thiéry, J. P., 1967. Mise en évidence des polysaccharides sur coupes fines en microscopie électronique. J. Microscopie 6: 987–989.

Timoshkin, O. A., 1991. Turbellaria Lecithoepitheliata: morphology, systematics, phylogeny. Hydrobiologia 227 (Dev. Hydrobiol. 69): 323–332.

Watson, N. A. & K. Rohde, 1992. Ultrastructure of the pharynx of *Prorhynchus* (Platyhelminthes, Lecithoepitheliata). Zool. Scr. 21: 325–333.

Watson, N. A. & K. Rohde, 1993. Ultrastructure of spermiogenesis and spermatozoa in *Prorhynchus* sp. (Platyhelminthes, Lecithoepitheliata, Prorhynchidae). Invertebr. Reprod. Dev. 23: 215–223.

Hydrobiologia **305**: 207–212, 1995.
L.R.G. Cannon (ed.), Biology of Turbellaria and some Related Flatworms.

The ultrastructure of the germarium in some Rhabdocoela

Paolo Lucchesi, Alessandra Falleni & Vittorio Gremigni
Dipartimento di Biomedicina Sperimentale, Sez. Biologia e Genetica, via A. Volta 4, Università di Pisa, 56126 Pisa, Italy

Key words: Rhabdocoela, Platyhelminthes, germarium, ovary, oocyte, cortical inclusions, ultrastructure

Abstract

Ultrastructural features of the germarium (ovary) have been investigated in several species of Rhabdocoela. The gonad is usually unpaired, small and pear-shaped; it is enveloped by an extracellular lamina and contains oocytes at different stages of maturation. Elongate accessory cells surround germ cells in the peripheral zone of the gonad and can also fill the internal spaces between oocytes with their long, flattened processes. The main features observed during oocyte maturation were the appearance of chromatoid bodies, annulate lamellae, lipid droplets and glycogen particles, and the development of R.E.R. and Golgi complexes which appeared correlated with the production of small inclusions that became localized in the cortical ooplasm of mature germ cells. The inclusions exhibited a different structure in different taxa and contained variable amounts of polyphenols. Cortical inclusions of rhabdocoels most probably represent residual shell-granules and may participate in the capsule-shell formation.

Abbreviations

AC	= accessory cell
ECL	= extracellular lamina
GC	= Golgi complex
L	= lipid
NSG	= nascent shell-granule
O	= oocyte

Introduction

The germarium (ovary) of Platyhelminthes has been extensively studied by electron microscopy for the last two decades (see Gremigni, 1983, 1988; Davis & Roberts, 1983; Smyth & Halton, 1983; Guraya & Parshad, 1988; Rieger *et al.*, 1991; Cifrian *et al.*, 1993; Shinn, 1993 for reviews). However, only a recent comprehensive article (Falleni & Lucchesi, 1992) and a few sporadic observations on specific features of the oocyte structure (Domenici & Gremigni, 1977; Gremigni & Domenici, 1977; Sopott-Ehlers & Ehlers, 1986; Gremigni, 1988) have been published on Rhabdocoela, a group of turbellarians which is not yet well-defined from a systematic point of view and includes organisms possibly related to the parasitic forms of Platyhelminthes (Neodermata *sensu* Ehlers, 1985).

The aim of this work was to study comparatively the ultrastructure of the female gonad in different taxa of rhabdocoels (namely Typhloplanoida, Dalyellioida, Kalyptorhynchia and Temnocephalida) to obtain further insights on the differentiation of oocytes and possibly new characteristics that may clarify the uncertain relationships among the rhabdocoelan taxa and between them and other groups of Platyhelminthes.

Materials and methods

Freshly collected specimens of the following species were used: Typhloplanoida: *Castrada viridis*, *Mesostoma ehrenbergii*. Dalyellioida: *Dalyellia viridis*, *Castrella truncata*, *Gieysztoria diadema*. Kalyptorhynchia: *Cheliplana* sp., *Gyratrix hermaphroditus*. Temnocephalida: *Temnocephala* sp.

For conventional electron microscopy, specimens were fixed in 0.05–0.1 M phosphate buffered 3% glutaraldehyde for 1 h and postfixed in 1% osmium tetroxide in the same buffer for 2 h. Following dehydration, specimens were embedded in Epon-Araldite, sectioned with a diamond knife, stained with uranyl acetate and lead citrate and observed with a JEOL 100 SX TEM.

For cytochemical investigations, whole specimens were processed according to the methenamine silver method of Locke & Krishnan (1971) for polyphenols. Thin sections obtained from blocks used for conventional electron microscopy were either treated with 2% H_2O_2 and incubated in a solution of Pronase E for enzymatic protein extraction (Anderson & André, 1968) or treated according to the Thiéry (1967) method for polysaccharides and glycoproteins. Control specimens or sections were prepared for each cytochemical test.

Results

The germarium of most species studied is an unpaired, small and pear-shaped organ located posterior to the pharynx. It is enveloped by a fibrous, finely granular extracellular lamina-like layer of medium electron-density and contains several oocytes at different stages of maturation associated with accessory cells. It has a saccular structure and can be divided into a germinative and a growth area. The extracellular lamina appears very thin (0.1 μm) in Dalyellioida (Figs 1, 3) and Typhloplanoida, while it is much thicker (up to 0.5 μm) in *Temnocephala* (Fig. 2). The accessory cells can envelop each growing oocyte as in *Gieysztoria* or be confined to the outer portion of the gonad as in *Castrada*. The nucleus of accessory cells is usually located in the peripheral interspaces between two oocytes and is very elongate. It contains some patches of heterochromatin and a well-developed nucleolus. The cytoplasm has a central, enlarged portion surrounding the nucleus and long, flattened processes that completely envelop the gonad (*Castrada*) or even each germ cell (*Gieysztoria*). Accessory cells contain some mitochondria, short cisternae of the rough endoplasmic reticulum (R.E.R.) and Golgi complexes that produce small vesicles with a poorly electron-dense content (Fig. 3).

Oocytes display a similar basic pattern in different species. Early differentiating germ cells (10 × 15 μm) have a large nucleus with diffuse chromatin and a prominent nucleolus. The nuclear envelope is rich in pores. The cytoplasm is packed with free ribosomes and contains some mitochondria and a few cisternae of R.E.R. (Fig. 4). A common feature in the perinuclear ooplasm is the presence of chromatoid bodies and annulate lamellae that remain even in later stages of oogenesis when they can become scattered throughout the ooplasm (Figs 5, 6). Larger oocytes (30–40 μm in diameter) are characterized by an increase in the number of R.E.R. profiles and the appearance of Golgi complexes, clustered lipid droplets and glycogen particles (Fig. 7). Small vesicles containing either a granular content of medium electron-density or a homogeneous more electron-dense content arise from the Golgi complex (Figs 8, 9). Repeated coalescence of the vesicles produces small inclusions that show a similar structure in species of the same taxon and a quite different structure in species of different taxa (Figs 10, 11, 12). The content of such inclusions tests more or less positive for polyphenols and is usually poorly affected by incubation with protease (Figs 13, 14, 15, 16). In mature oocytes the electron-dense inclusions are located in the cortical ooplasm just beneath the oolemma where they form a discontinuous layer (Fig. 17).

Discussion

The germarium of rhabdocoels has a saccular structure and is enveloped by a fibrous, finely granular extracellular lamina that can vary in thickness from 0.1 μm as observed in Dalyellioida and Typhloplanoida, up to 0.5 μm as in *Temnocephala* sp.. Peripherally within the gonad some very elongated accessory cells surround germ cells. In Dalyellioida and Kalyptorhynchia accessory cells extend long, flattened processes to occupy the internal spaces between oocytes. The origin of the extracellular lamina and accessory cells has not yet been ascertained. The complex extracellular lamina + accessory cells is quite similar to that described as surrounding the gonads of some Neodermata (Xylander, 1987; Cifrian *et al.*, 1993; Martinez-Alõs *et al.*, 1993). It may function to protect oocytes and to transfer nutrient precursors from the parenchyma to growing oocytes as previously suggested for other turbellarians (Gremigni & Nigro, 1983, 1984; Nigro & Gremigni, 1987; Falleni & Gremigni, 1992). The cellular tunica enveloping the germarium usually described by light and electron microscope investigations (see Cannon, 1986; Rieger *et al.*, 1991) presumably corresponds to the complex extracellular lamina + accessory cells here described.

Growing oocytes of rhabdocoels show the typical features of non-vitellogenic female germ cells, including the presence of numerous free ribosomes, some chromatoid bodies, annulate lamellae, R.E.R. cister-

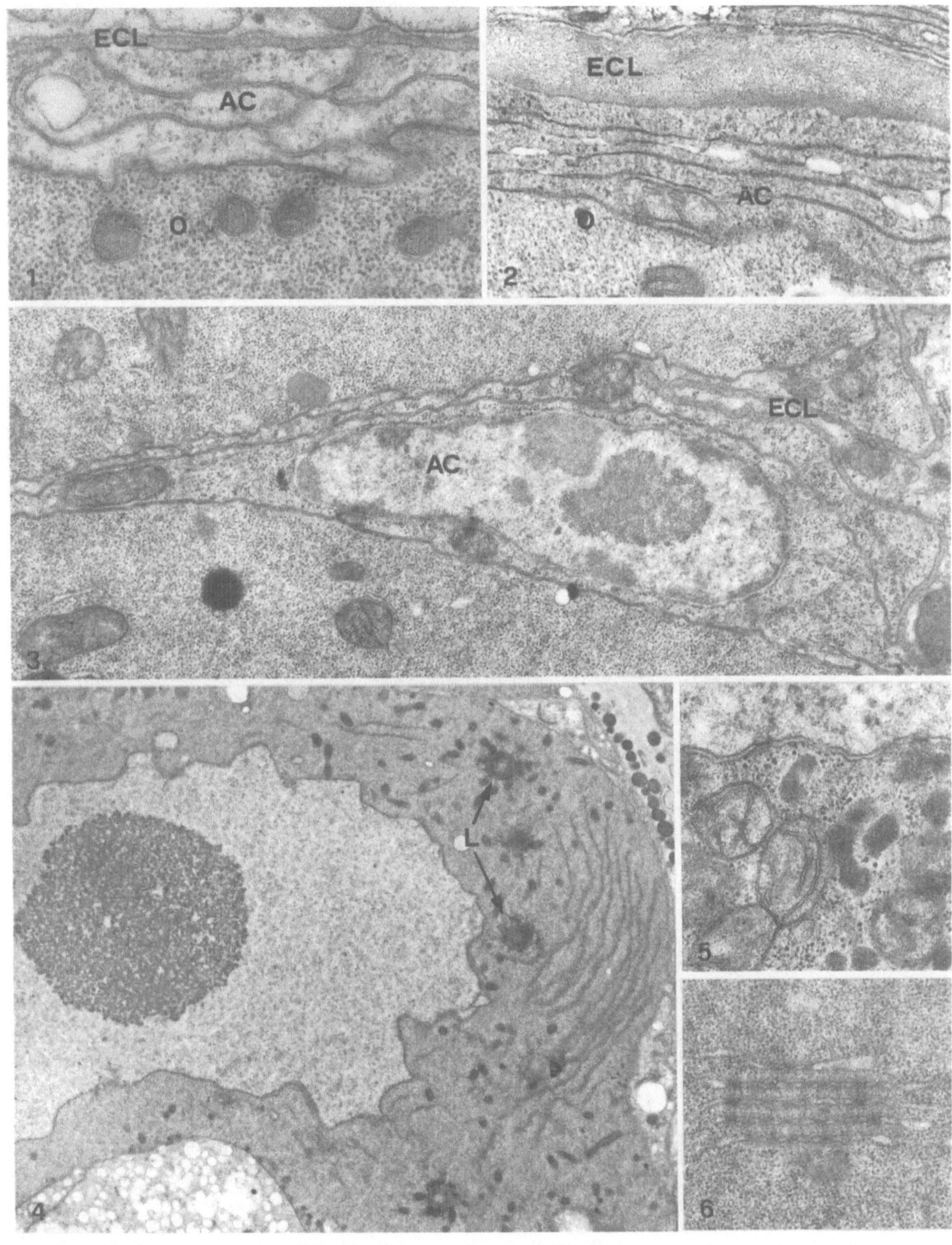

Figs 1–6. (Fig. 1) Outer portion of the germarium enveloped by a thin extracellular lamina. *Castrella truncata*. × 33 000. (Fig. 2) Outer portion of the germarium enveloped by a thick extracellular lamina. *Temnocephala* sp. × 30 000. (Fig. 3) An accessory cell between two growing oocytes. *Gieysztoria diadema*. × 21 000. (Fig. 4) Early differentiating oocyte. *Castrella truncata*. × 5000. (Fig. 5) Chromatoid bodies in the perinuclear cytoplasm of an early differentiating oocyte. *Dalyellia viridis*. × 30 000. (Fig. 6) Annulate lamellae in a growing oocyte. *Gieysztoria diadema*. × 18 000.

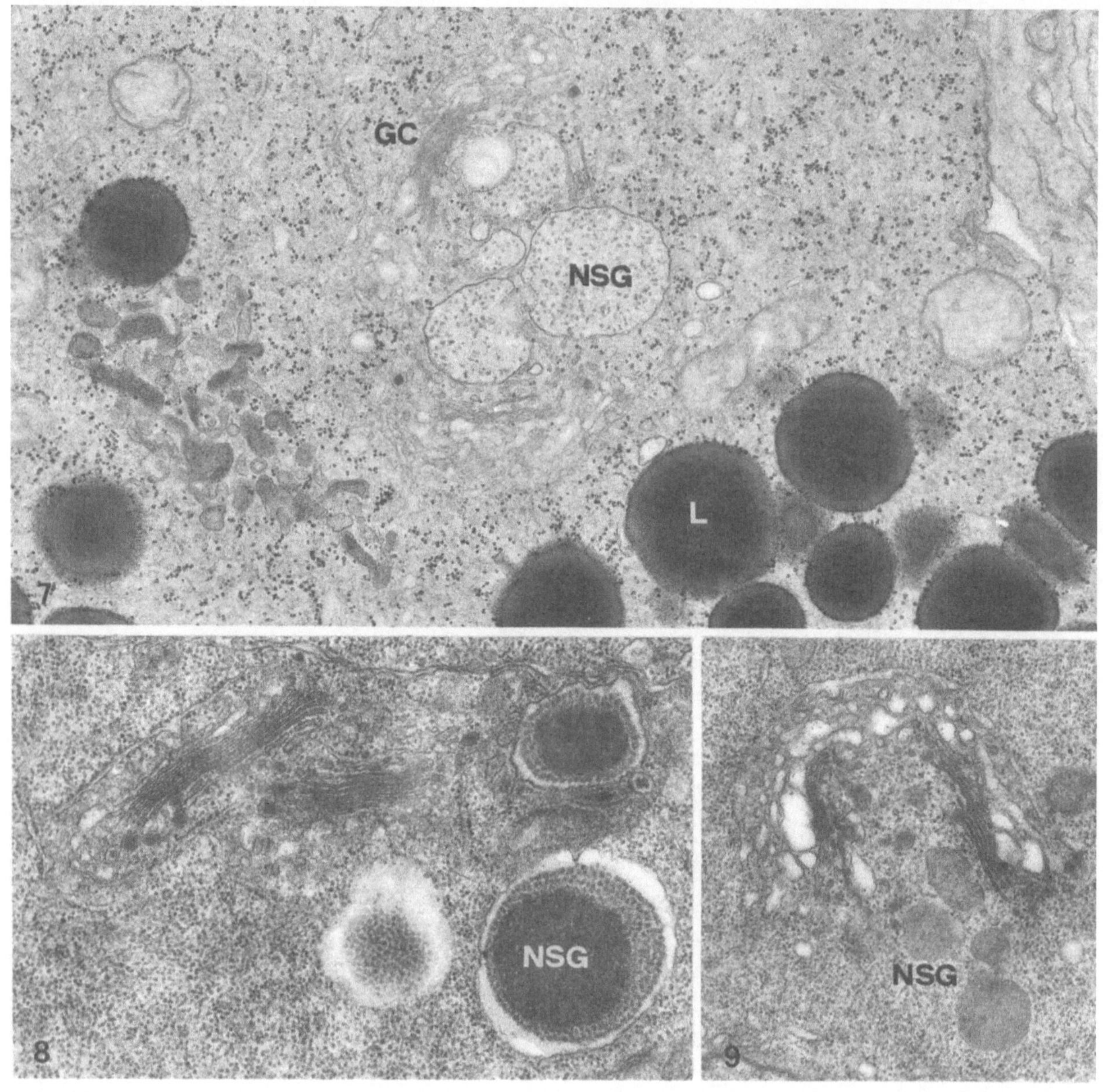

Figs 7–9. (Fig. 7) Part of a growing oocyte. Thiéry test. *Dalyellia viridis.* × 20 000. (Fig. 8) Two Golgi complexes and some nascent shell-granules in a growing oocyte. *Castrada viridis.* × 28 500. (Fig. 9) Golgi complexes and nascent shell-granules in a growing oocyte. *Gieysztoria diadema.* × 26 500.

nae and Golgi complexes. In addition, all mature oocytes studied have some small inclusions forming an incomplete layer in the cortical cytoplasm. These inclusions contain variable amounts of polyphenols and might therefore represent residual shell-granules. Shell-granules in oocytes are characteristic of turbellarian platyhelminths with archoophoran organization (Gremigni, 1988). Cortical inclusions of rhabdocoel oocytes could participate with the shell-globules produced by vitellocytes and the shell glands in the formation of the capsule-shell, in particular in the formation of hatching sutures, as suggested by Shinn & Cloney (1986).

The structure of cortical inclusions of rhabdocoels appears homogeneously granular in Dalyellioida, while two distinct components are observed in Typhloplanoida and Kalyptorhynchia, where the inclusions assume a different pattern, with a more electron-dense area containing polyphenols and a less electron-dense, proteinaceous area. A double structure, in particular

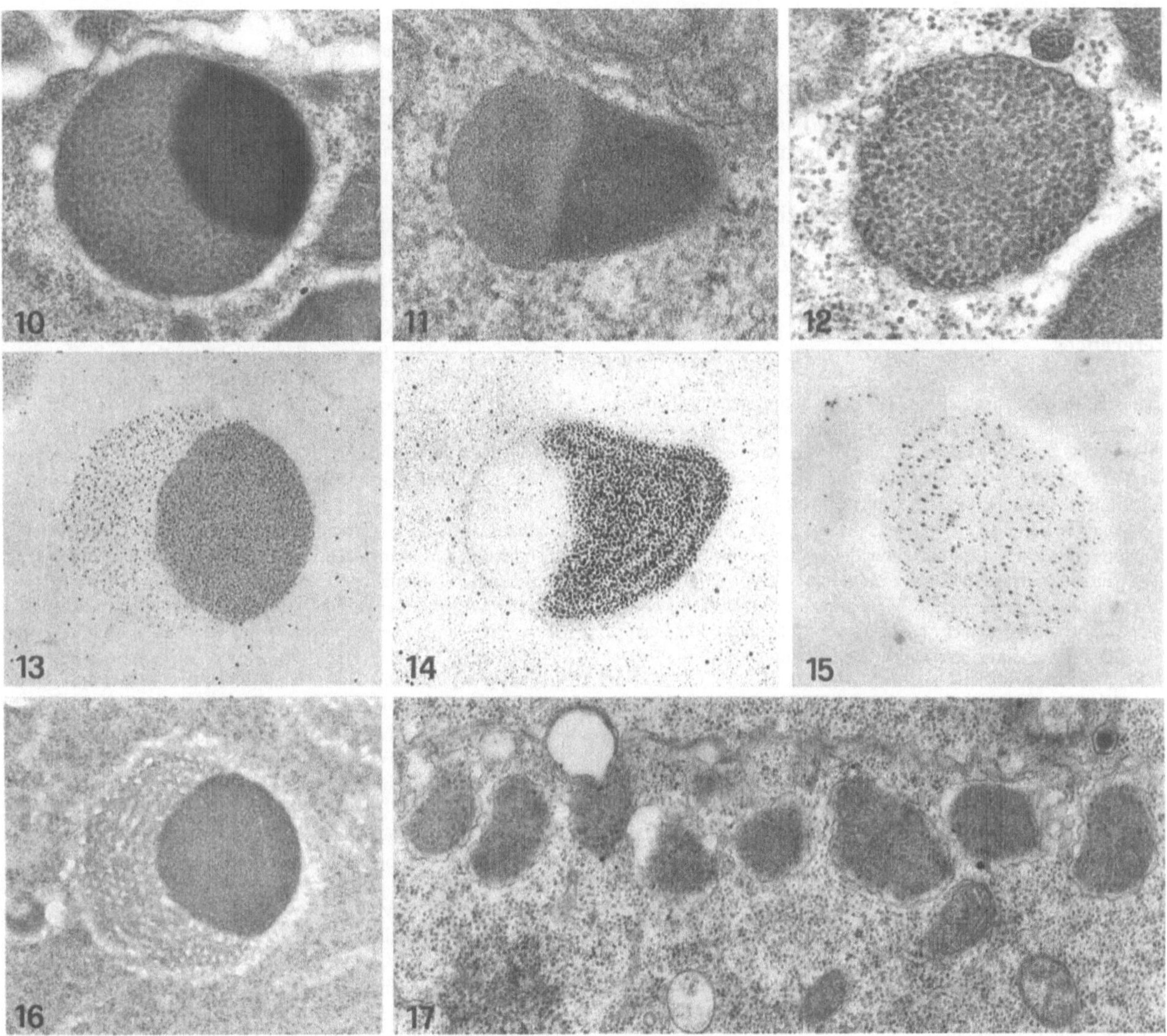

Figs 10–17. (Fig. 10) A mature double-structured shell-granule. *Castrada viridis*. × 33 000. (Fig. 11) A mature double-structured shell-granule. *Cheliplana* sp. × 69 000. (Fig. 12) A mature shell-granule. *Dalyellia viridis*. × 44 000. (Fig. 13) A shell-granule similar to that of Fig. 10. The silver precipitate is mainly concentrated on the subcentral area. Locke & Krishnan test. *Castrada viridis*. × 34 000. (Fig. 14) A shell-granule similar to that of Fig. 11. Locke & Krishnan test. *Cheliplana* sp. × 69 000. (Fig. 15) A shell-granule similar to that of Fig. 12. Locke & Krishnan test. *Dalyellia viridis*. × 45 000. (Fig. 16) A shell-granule similar to those of Figs 10 and 13 after 3 h of incubation with protease. *Castrada viridis*. × 25 000. (Fig. 17) Shell-granules in the peripheral cytoplasm of a mature oocyte. *Castrella truncata*. × 24 000.

similar to that observed in Typhloplanoida, is shared by homologous inclusions in the eggs of some Prolecithophora (Nigro & Gremigni, 1987; Gremigni, 1988). Small peripheral granules with complex structures have also been described in the oocytes of many parasitic Neodermata where they can have a different composition and are considered cortical granules (see Cifrian *et al.*, 1993 for references). Furthermore, shell-globules with a similar multigranular structure have been observed in the vitellocytes of Rhabdocoela, Neodermata and Prolecithophora (Gremigni, 1988; Gremigni & Falleni, 1992). The distinctive structures of both peripheral granules in the oocytes and shell-globules in the vitellocytes might represent synapomorphies for the Rhabdocoela (*sensu* Ehlers, 1985, including Neodermata) and Prolecithophora, and contribute to clarify the uncertain phylogenetic position of the latter group (Ehlers, 1985; Smith *et al.*, 1986; Ax, 1987).

Acknowledgments

We are indebted to Dr L. Cannon who provided us with embedded specimens of *Temnocephala*. Warm thanks are also due to Mr C. Ghezzani for technical assistance and Mr J. Franceschina for English revision. The work was supported by the Consiglio Nazionale Ricerche (C.N.R.) of Italy.

References

Anderson, A. & J. André, 1968. The extraction of some cell components with pronase and pepsin from thin sections of tissue embedded in Epon-Araldite mixture. J. Microscopie 7: 343–345.

Ax, P., 1987. The phylogenetic system. The systematization of organisms on the basis of their phylogenesis. J. Wiley & Sons, Chichester, 249 pp.

Cannon, L. R. G., 1986. Turbellaria of the world. A guide to families & genera. Brisbane: Queensland Museum, 131 pp.

Cifrian, B., S. Martinez-Alõs & V. Gremigni, 1993. Ultrastructural and cytochemical studies of the germarium of Dicrocoelium dendriticum (Plathelminthes, Digenea). Zoomorphology 113: 165–171.

Davis, R. E. & L. S. Roberts, 1983. Platyhelminthes-Eucestoda. In K. G. Adiyodi & R. G. Adiyodi (eds), Reproductive Biology of Invertebrates. Vol. 1, Oogenesis, oviposition and oosorption. J. Wiley & Sons, Chichester: 109–133.

Domenici, L. & V. Gremigni, 1977. Fine structure and functional role of the coverings of the eggs in Mesostoma ehrenbergii (Focke) (Turbellaria, Neorhabdocoela). Zoomorphology 88: 247–257.

Ehlers, U., 1985. Phylogenetic relationships within the Platyhelminthes. In S. Conway Morris, J. D. George, R. Gibson & H. M. Platt (eds), The origins and relationships of lower invertebrates. Oxford University Press, Oxford: 143–158.

Falleni, A. & V. Gremigni, 1992. An ultrastructural study of growing oocytes in Nematoplana riegeri (Platyhelminthes). J. Submicrosc. Cytol. Pathol. 24: 51–59.

Falleni, A. & P. Lucchesi, 1992. Ultrastructural and cytochemical aspects of oogenesis in Castrada viridis (Platyhelminthes, Rhabdocoela). J. Morphol. 213: 241–250.

Gremigni, V., 1983. Platyhelminthes-Turbellaria. In K. G. Adiyodi & R. G. Adiyodi (eds), Reproductive Biology of Invertebrates. Vol. 1. Oogenesis, oviposition and oosorption. J. Wiley & Sons, Chichester, 67–107.

Gremigni, V., 1988. A comparative ultrastructural study of homocellular and heterocellular female gonads in free-living Platyhelminthes–Turbellaria. Fortschr. Zool. 36: 245–261.

Gremigni, V. & L. Domenici, 1977. On the role of specialized, peripheral cells during embryonic development of subitaneous eggs in the turbellarian Mesostoma ehrenbergii (Focke): an ultrastructural and autoradiographic investigation. Acta Embryol. Morphol. Exper. 2: 251–265.

Gremigni, V. & A. Falleni, 1992. Mechanisms of shell-granule and yolk production in oocytes and vitellocytes of Platyhelminthes-Turbellaria. Anim. Biol. 1: 29–37.

Gremigni, V. & M. Nigro, 1983. An ultrastructural study of oogenesis in a marine triclad. Tissue Cell 15: 405–415.

Gremigni, V. & M. Nigro, 1984. Ultrastructural study of oogenesis in Monocelis lineata (Turbellaria, Proseriata). Int. J. Invert. Reprod. Devel. 7: 105–118.

Guraya, S. S. & V. R. Parshad, 1988. Platyhelminthes. In K. G. Adiyodi and R. G. Adiyodi (eds), Reproductive Biology of Invertebrates. Vol. 3. Accessory sex glands. J. Wiley & Sons, Chichester: 1–49.

Locke, M. & N. Krishnan, 1971. The distribution of polyphenoloxidases and polyphenols during cuticle formation. Tissue Cell 3: 103–126.

Martinez-Alõs, S., B. Cifrian & V. Gremigni, 1993. Ultrastructural investigations on the vitellaria of the digenean Dicrocoelium dendriticum. J. Submicrosc. Cytol. Pathol. 25: 583–590.

Nigro, M. & V. Gremigni, 1987. Ultrastructural features of oogenesis in a free-living marine platyhelminth, Vorticerods luteum. Tissue Cell 19: 377–386.

Rieger, R. M., S. Tyler, J. P. S. Smith & G. E. Rieger, 1991. Platyhelminthes: Turbellaria. In F. W. Harrison & B. J. Bogitsh (eds), Microscopic Anatomy of Invertebrates. Vol. 3. Platyhelminthes and Nemertinea. Wiley-Liss, Inc., New York: 7–140.

Shinn, G. L., 1993. Formation of egg capsules by flatworms (Phylum Platyhelminthes). Trans. am. Microsc. Soc. 112: 18–34.

Shinn, G. L. & R. A. Cloney, 1986. Egg capsules of a parasitic turbellarian flatworm: ultrastructure of hatching sutures. J. Morphol. 188: 15–28.

Smith, J. P. S., S. Tyler & R. M. Rieger, 1986. Is the Turbellaria polyphyletic? Hydrobiologia 132 (Dev. Hydrobiol. 32): 13–21.

Smyth, S. D. & D. W. Halton 1983. The Physiology of Trematodes. Cambridge University Press, Cambridge, 417 pp.

Sopott-Ehlers, B. & U. Ehlers, 1986. Differentiation of male and female germ cells in neoophoran Plathelminthes. In M. Porchet, J. C. Andries & A. Dhainaut (eds), Advances in Invertebrate Reproduction 4. Elsevier Science Publ., Amsterdam: 187–194.

Thiéry, J. P., 1967. Mise en évidence des polysaccharides sur coupes fines en microscopie électronique. J. Microscopie 6: 987–989.

Xylander, W. E. R., 1987. Ultrastructural studies on the reproductive system of Gyrocotylidea and Amphilidea (Cestoda). II. Vitellaria, vitellocyte development and vitelloduct of Gyrocotyle urna. Zoomorphology 107: 293–297.

Hydrobiologia **305**: 213–216, 1995.
L.R.G. Cannon (ed.), Biology of Turbellaria and some Related Flatworms.

Ultrastructure of oncospheral envelopes in Hymenolepidids (Cestoda) with aquatic life cycles

L. Chomicz[1], Z. Swiderski[2] & A. Czubaj[3]
[1]*Department of General Biology and Parasitology, Medical Academy, 02-004 Warsaw, Poland*
[2]*Laboratory of Comparative Anatomy and Physiology, University of Geneva, CH-1211 Geneva 4, Switzerland*
[3]*Department of Cytology, University of Warsaw, 00-972 Warsaw, Poland*

Key words: Cestoda, Hymenolepididae, oncospheral envelopes, ultrastructure, freshwater environment

Abstract

The ultrastructure of the oncospheral envelopes of five species of hymenolepidid cestodes, namely, *Fimbriaria fasciolaris, Dicranotaenia coronula, Sobolevicanthus gracilis, Diorchis elisae* and *Diorchis ovofurcata* are described. These cestodes are parasites of aquatic hosts and utilise aquatic invertebrates as intermediate hosts. The ultrastructure of the oncospheral envelopes was essentially similar among these species. However, the structure of the components of the inner oncospheral envelopes displayed significant structural variation among the five species examined. The small eggs of *D. coronula* and *S. gracilis* have a thin embryophore and oncospheral membrane. The eggs of these species settle at the bottom of the water body and are eaten by benthic crustaceans. Eggs of *F. fasciolaris, D. elisae* and *D. ovofurcata* have thick multilayered embryophores with specific ornamentation. Moreover, in *Diorchis* species the eggs are elongate, and the oncospheral membrane is thick and striated. Zones of electron-dense aggregates penetrate the polar filaments in these species. Eggs of *F. fasciolaris* float passively in water whereas those of *Diorchis* species are attached by long filaments to aquatic plants and are eaten by crustaceans in the littoral zone. The interrelationships among the ultrastructure of the oncospheral envelopes, cestode life cycles and habitat of crustacean intermediate hosts are drawn and discussed.

Introduction

The development and ultrastructure of oncospheral envelopes have been described mainly in three hymenolepidid cestodes with terrestrial life cycles: *Hymenolepis diminuta* (see Pence, 1970; Rybicka 1972), *H. microstoma* (see Swiderski, 1975) and *H. nana* (see Fairweather & Threadgold, 1981). The purpose of the present comparative studies is to describe and compare the ultrastructure of oncospheral envelopes of five species of hymenolepidids which infect aquatic birds and use crustaceans (Copepoda, Ostracoda) as intermediate hosts.

Material and methods

Adult specimens of five following hymenolepidids with aquatic life cycles: *Diorchis ovofurcata* Czaplinski, 1972 obtained from *Aythya fuligula* and *Diorchis elisae* (Skrjabin 1914) Spassky et Frese, 1961, *Sobolevicanthus gracilis* (Zeder, 1803), *Fimbriaria fasciolaris* (Pallas, 1781), *Dicranotaenia coronula* (Dujardin, 1845), all obtained from the same host, *Anas platyrhynchos*, were examined.Small fragments of gravid proglottides were fixed in 2.5–3% glutaraldehyde in 0.1 M cacodylate buffer (Ph 7.4) for 3 h at 4°C, washed in the same buffer and postfixed in 1% OsO_4 in 0.1 M cacodylate buffer for 2 h at 4°C. The samples were then processed for transmission electron microscopy using standard methods of dehydration and embedding. Ultrathin sections stained with uranyl acetate and lead citrate were examined in Jeol 100B electron microscope.

Table 1. Ultrastructure of oncospheral envelopes and life cycle characteristics of five hymenolepidids examined

Species of tapeworms	Location in water of: Infective eggs		Intermediate hosts	Ultrastructure of the oncospheral envelopes
Dichonotaenia coronula *Sobslevicanthus gracilis*	Small eggs at the bottom of the water reservoir		In the benthos zone	Thin, delicate embryopore, thin oncospheral membrane
Fimbriaria fasciolaris	Chains of eggs	Near the water surface	In the littoral zone	Thick secondary envelopes (with additional zone of electron-dense aggregates in diorchids)
Diorchis elisae *D. ovofurcata*	Elongated eggs with filaments			

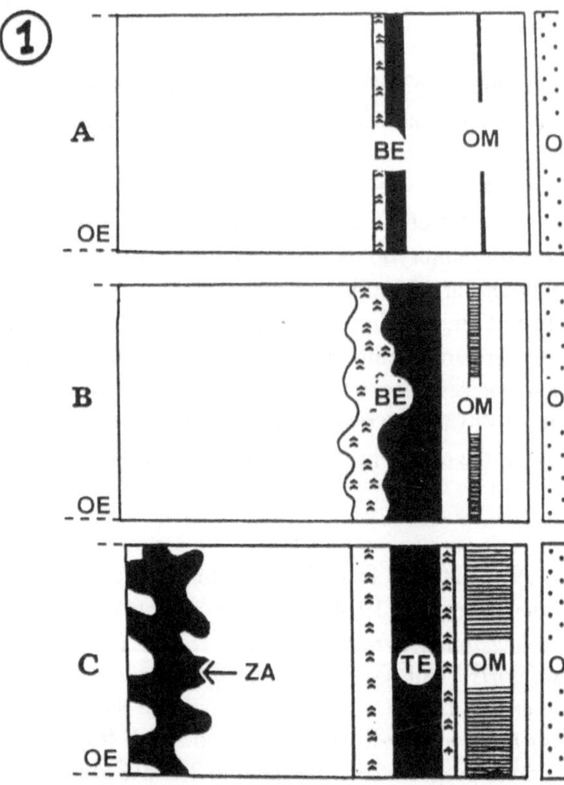

Fig. 1. Diagrams showing secondary oncospheral envelopes at the end of egg development in five hymenolepidids examined: *A*)- *Dicranotaenia coronula* and *Sobolevicanthus gracilis*; *B*)-*Fimbriaria fasciolaris*; *C*)-*Diorchis elisae* and *D. ovofurcata*. *BE*-bi-layered embryophore; *OM*-oncospheral membrane; *O*-oncosphere; *TE*-tri- layered embryophore; *ZA*-zone of electron-dense aggregates.

Results

Three primary envelopes (capsule, outer and inner envelopes) appear in the five hymenolepidid species examined in the early stage of embryonic development. The inner envelope forms two or three derivative layers: oncospheral membrane, embryophore, and, eventually, a zone of electron-dense aggregates, which is visible at the end of oncospheral development. In *D. coronula* and *S. gracilis* two secondary envelopes (embryophore and oncopheral membrane) can be distinguished (Fig. 1A and 2). In the both the latter species, the embryophores and oncospheral membranes are relatively thin, and they acquire their final shape simultaneously. The infective eggs of *D. coronula* and *S. gracilis* are small but heavy; they remain at the bottom of the water reservoir and are eaten by benthic ostracods (for example *Cypria ophthalmica*). In *F. fasciolaris*, two secondary envelopes (a thick, bi-layered embryophore and relatively thick, striated oncospheral membrane (Fig. 1B) undergo simultaneous differentiation from the inner envelope. The infective eggs of *F. fasciolaris* form long chains and are usually located near the water surface and are eaten by different copepods and ostracods in littoral zone. In *D. elisae* and *D. ovofurcata* (Fig. 1C) there are three secondary envelopes, all originating from the primary inner envelope. The eggs of *Diorchis* species have a thick, tri-layered embryophore, a thick striated oncospheral membrane and a characteristic zone of electron-dense aggregates situated just under the outer membrane of the inner envelope (Figs 1C and 3). The electron-dense aggregates penetrate into the polar filaments of the infective eggs, which are formed by outer and inner oncospheral envelopes. The elongate eggs of the diorchids attach to aquatic plants with the long polar filaments and thus remain near the surface of the water column. Here, the eggs are exposed to littoral copepods and ostracods, the intermediate hosts. The

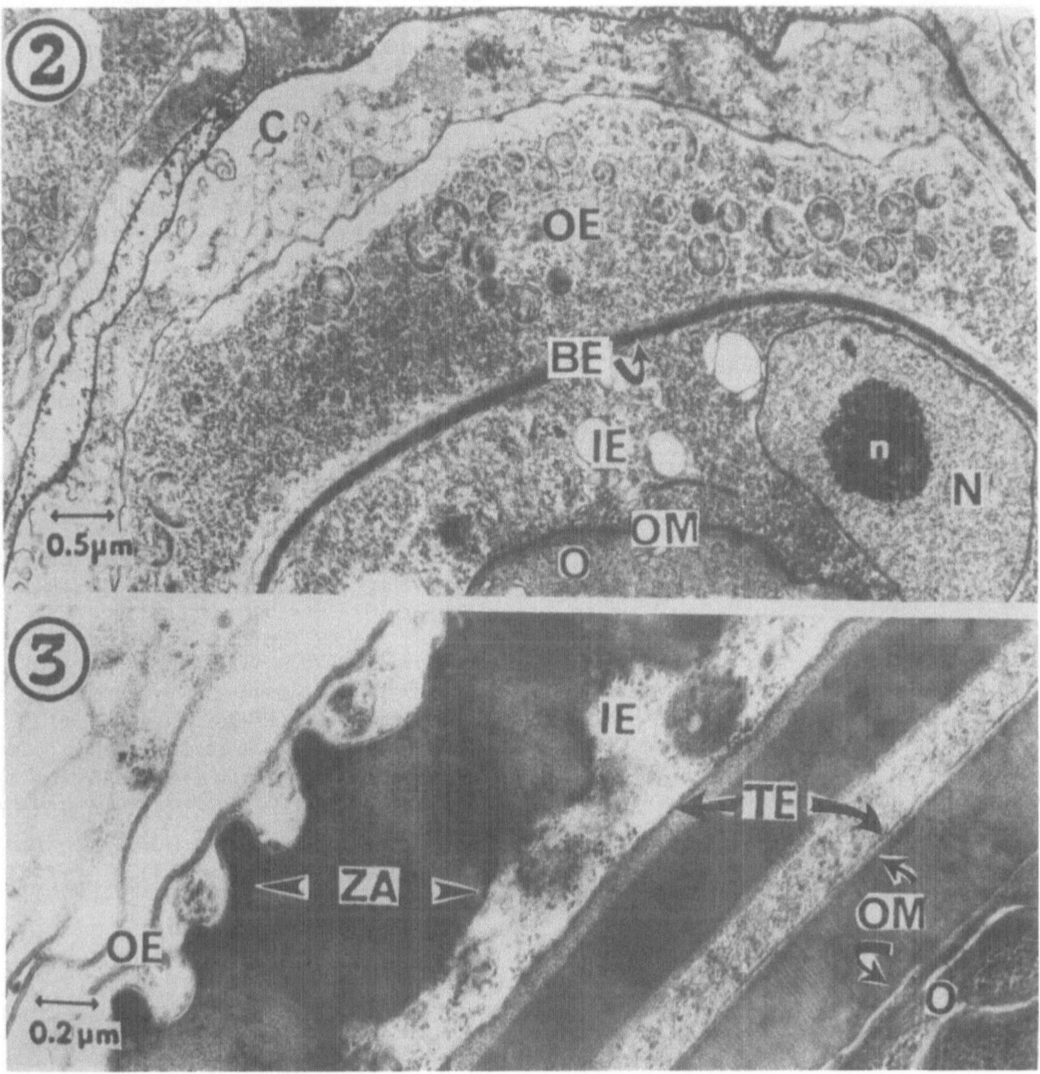

Fig. 2. Part of the fully developed oncosphere of *D. coronula* surrounded by five oncospheral envelopes. Note a prominent nucleus (*N*) with large, electron-dense nucleulus (*n*) in the cytoplasm of the inner envelope (*IE*). *C*-capsule; *BE*-bi-layered embryophore; *IE*-inner envelope; *O*-oncosphere; *OE*-outer envelope; *OM*- oncospheral membrane. (3) Section through the oncospheral envelopes in *D. ovofurcata*. Note presence of an additional zone of electron-dense aggregates (*ZA*) and striated oncospheral membrane (*OM*). *IE*-inner envelope; *O*-oncosphere; *OE* -outer envelope; *OM*-oncospheral membrane; *TE*-tri- layered embrophore.

data concerning the ultrastructure of the oncospheral envelopes in relation to life cycle characteristics of five hymenolepidids examined are summarized in Table 1.

Discussion

Comparison of the ultrastructure of the oncospheral envelopes in the five hymenolepidid species (see also Chomicz & Czubaj, 1991; Chomicz & Walski, 1991) confirms existence of a common pattern in the formation of the embryonic envelopes in Cestoda (Rybicka, 1966; Ulbelaker, 1980; Swiderski, 1981; Burt, 1986). There are, however, evident differences in the shape of the secondary envelopes in the five species of hymenolepidids examined here. Consideration of the results of our study with biological data on the life cycles and developmental conditions of the hymenolepidids in aquatic environment (Jarecka, 1961; Czaplinski & Szelenbaum, 1974) indicates close interrelationships among: (1) the biotope of the intermediate hosts, (2) location of infective eggs in the water reservoir, and (3) the ultrastructure of the oncospheral envelopes. Thus, thin secondary envelopes (embryophores and oncospheral membranes) are found in eggs of *D. coronula* and *S. gracilis*, which settle in benthic layers of water column. Thickened secondary envelopes are found in *F. fasciolaris, D. elisae* and *D. ovofurcata*. The interrelations may indicate that the structural and chemical composition of infective egg envelopes determine their position in different biotopes and/or layers of water reservoir. It facilites, therefore, a direct contact of infective eggs with littoral or benthic crustaceans, the intermediate hosts of hymenolepidids with aquatic life cycle.

References

Burt, M. D. B., 1986. Early morphogenesis in the Plathelminthes with special reference to egg development and development of cestode larvae. In: Howell, M. J. (ed.), Parasitology-Quo vadit? Proceedings of the Sixth International Congress of Parasitology. Australian Academy of Sciences (Canberra): 241–253.

Chomicz, L. & A. Czubaj, 1991. Transmission electron micrograph studies of developing oncospheral envelopes of *Fimbriaria fasciolaris* (Hymenolepididae). Parasitology Research 77: 503–508.

Chomicz, L. & M. Walski, 1991. Ultrastructure of oncospheral envelopes of *Diorchis elisae* (Skrjabin, 1914) Spassky et Frese, 1961 (Cestoda, Hymenolepididae). Parasitology Research 77: 550–552.

Czaplinski, B. & D. Szelenbaum, 1974. Morphological and biological differences between *Diorchis ransomi* Johri, 1939 and *Diorchis parvogenitalis* Skrjabin et Mathevosian, 1945 (Cestoda, Hymenolepididae). Acta Parasitol. Polon. 22: 113–132.

Fairweather, I. & L. T. T. Threadgold, 1981. *Hymenolepis nana*: the fine structure of the embryonic envelopes. Parasitology 82: 429–443.

Jarecka, L., 1961. Morphological adaptaptations of tapeworm eggs and their importance in the life cycles. Acta Parasitol. Polon. 9: 409–426.

Pence, D. B., 1970. Electron microscope and histochemical studies on the eggs of *Hymenolepsis diminuta*. J. Parasitol. 56: 84–97.

Rybicka, K., 1966. Embryogenesis in cestodes. Adv. Parasitol. 4: 107–186.

Rybicka, K., 1972. Ultrastructure of embryonic envelopes and their differentiation in *Hymenolepis diminuta* (Cestoda). J. Parasitol. 58: 849–863.

Swiderski, Z., 1975. Comparative fine structure of cestode embryos. Proceedings, 2nd European Multicolloquy of Parasitology, Trogir: 265–272.

Swiderski, Z., 1981. Reproductive and developmental biology of the cestodes. In: Clark, W. H. & Adams, T. S. (eds), Advances in invertebrate Reproduction. Elsevier/North Holland, New York, Amsterdam, Oxford: 364–367.

Ubelaker, J. E., 1980. Structure and ultrastructure of the larvae and metacestodes of *Hymenolepis diminuta*. In: Arai, H. P. (ed), Biology of the tapeworm *Hymenolepis diminuta*. Academic Press, New York: 59–156.

Hydrobiologia **305**: 217–222, 1995.
L.R.G. Cannon (ed.), Biology of Turbellaria and some Related Flatworms.

What studies of turbellarian embryos can tell us about the evolution of developmental mechanisms

Barbara C. Boyer
Department of Biology, Union College, Schenectady, NY 12308, USA (address for correspondence); Marine Biological Laboratory, Woods Hole, Mass., USA

Key words: Turbellaria, Polycladida, spiralians, development, D quadrant determination, cleavage patterns

Abstract

In spiralian embyros determination of the axes of bilateral symmetry is associated with D quadrant specification. This can occur late through equal cleavage and cell interactions (conditional specification) or by the four-cell stage through unequal cleavage and cytoplasmic localization (autonomous specification). Freeman & Lundelius (1992) suggest that in spiralian coelomates the former method is ancestral and the latter derived, with evolutionary pressure to shorten metamorphosis resulting in early D quadrant determination through unequal cleavage and appearance of adult features in the larvae. Because of the key phylogenetic position of the turbellarian platyhelminthes, understanding the method of axis specification in this group is important in evaluating the hypothesis. Polyclad development, with equal quartet spiral cleavage, is believed to represent the most primitive condition among living turbellarians and has been examined experimentally in *Hoploplana inquilina*. Blastomere deletions at the two and four-cell stage produce larvae that are abnormal in morphology and symmetry, indicating that early development is not regulative, and also establish that the embryo does not have an invariant cell lineage. Deletions of micromeres and macromeres at the eight-cell stage indicate that cell interactions are involved in dorso-ventral axis determination, with cross-furrow macromeres playing a more significant role than non-cross-furrow cells. The results support the idea that conditional specification is the primitive developmental mode that characterized the common ancestor of the turbellarians and spiralian coelomates. Evolutionary trends in development in polyclads and other turbellarian orders are discussed.

Introduction

Embryonic cell fate is established through one of two fundamentally different mechanisms, either autonomous or conditional specification. Autonomous specification occurs when a blastomere during cleavage inherits regions of egg cytoplasm containing localized determinants that activate particular developmental pathways. Organisms whose embryos become determined primarily autonomously by cytoplasmic localization are said to exhibit mosaic development. Conditional specification occurs when cells induce a specific developmental program in nearby cells, a mechanism that is associated with regulative development. Classically spiral cleavage and mosaic development have been linked with the protostomes and radial cleavage and regulative development with the deuterostomes. However it has long been known that some deuterostomes, specifically the tunicates, are highly mosaic, and more recently it has been shown that many inductive interactions occur in spiralian embryos. The question of whether autonomous or conditional specification is primitive remains unanswered. Because of the key position of the turbellarian Platyhelminthes in most phylogenetic schemes, study of the development of members of this group should provide further insights into this problem.

In considering the adaptive advantage of autonomous specification, Raff & Kaufman (1983) state, “One intriguing possibility for the role of mosaic development is that it allows the rapid production of specialized larvae from a limited number of embryonic cells.” They suggest that a trend toward direct development with the loss of specialized larvae in an evolutionary lineage was associated with a change in the relative contributions of the autonomous and conditional modes of specification. In an extensive discussion on the evolution of developmental mechanisms, Davidson (1991) also concludes that autonomous specification is the ancestral mode of development and that conditional specification is derived.

Freeman & Lundelius (1992) have been concerned specifically with one of the first major events in embryonic organization, the establishment of the axes of bilateral symmetry, which in coelomate spiralians occurs at the time of D quadrant determination. They conclude that in these phyla conditional specification is ancestral and the autonomous mode is derived. D quadrant specification has been most thoroughly examined in molluscs and can occur by either of these mechanisms. When the first two cleavages produce blastomeres that are unequal in size, dorso-ventral is determined at the four-cell stage by cytoplasmic localization. The best known examples are the embryos of animals such as *Ilyanassa* that form polar lobes. Other molluscs such as *Lymnaea* and *Patella* produce embryos with equal first and second cleavages and specify the D quadrant much later by cell interactions that occur between formation of the third and fourth quartets of micromeres.

Freeman & Lundelius (loc. sit.) propose an evolutionary sequence to explain this puzzling difference in such closely related species. They examine cleavage in coelomate protostomes and find that in those groups widely regarded as conservative, cleavage is equal, D quadrant specification occurs relatively late, and feeding larvae are produced (typically a trochophore) that are markedly different from the adult. Deletion of the first quartet of micromeres in these embryos produces larvae that remain radially symmetrical and do not form a D quadrant (van den Biggelaar & Guerrier, 1979). D quadrant specification through cytoplasmic localization, on the other hand, is characteristic of advanced groups, in which cleavage is unequal, the dorso-ventral axis is established at the four-cell stage, and either development is direct or the larvae exhibit adult characteristics. Deletion of the first quartet in this type of embryo produces larvae with bilateral symmetry and a D quadrant, and that lack first quartet derivatives (van Dam & Verdonk, 1982).

Freeman & Lundelius (loc. sit) suggest that there is a cause and effect relationship between the time of D quadrant determination and the developmental stage when an organism begins to function in its environment. They conclude that moving the time of D quadrant specification to an earlier developmental stage through unequal cleavage seems to accelerate subsequent developmental processes. The selective advantage for this evolutionary change might be that the transition from larva to adult is shortened when the development of adult organs is initiated earlier; this trend of incorporating more adult features early in development led to suppression of the larval stage and ultimately to direct development. Experimental creation of unequal blastomeres in the normally equally cleaving *Lymaea* embryo provides support for this hypothesis; once a macromere exceeds 35% of the embryo's volume it always becomes D, inheriting most of the vegetal region of the egg. This suggests that a shift to unequal cleavage could be a preadaptation that would facilitate the specification of the D quadrant via cytoplasmic localization, since any vegetal localizations would be inherited by the large D blastomere rather than divided among four equal size cells.

Though Freeman & Lundelius (loc. sit) refer to the Turbellaria, they do not use this key group, whose ancestors presumably gave rise to the coelomates with spiral cleavage, to develop their argument. Spiral cleavage, including the quartet pattern, is widespread in the Turbellaria, suggesting that this mode of cleavage is primitive for the class (Thomas, 1986). The canonical quartet pattern and the presence of a lobed larva – similar to the trochophore – in the polyclad turbellarians suggest the polyclad embryo as an appropriate system for examination of the evolutionary relationship between autonomous and conditional specification.

Polyclad development

A. Cleavage

Cleavage, which has been examined in a number of species of polyclads, is of the quartet spiral type. Although there is not total agreement among investigators on the relative sizes of the blastomeres at the two-cell stage, differences, when they exist, are slight. In examining cleavage in 11 species of polyclads, Kato (1940) found that the cells were usually of equal size, while Teshirogi *et al.* (1981) report that in the seven species studied (four of which were also examined by Kato) the CD blastomere is generally larger than AB. Surface (1907) observed a consistent though small size difference in the blastomeres of *Planocera* (*Hoploplana*) at two cells but my observations have not confirmed this. In fact usually the blastomeres are equal in size, though occasionally cleavage produces two noticably unequal cells in some of the embryos of a single individual.

B. Blastomere specification

Experimental studies of blastomere specification have been carried out only on the embryos of *Hoploplana inquilina* (see Boyer, 1986a, 1986b, 1987, 1988, 1989, 1992). Separation of blastomeres at the two-cell stage results in development of abnormal "half larvae" that have poor lobe development and abnormal symmetry, while deletions at the four-cell stage produce "three-quarter larvae" that are morphologically more normal, having a Müller's-like shape, though typically lobe development is not complete. Interestingly there is not an absolute correlation between a particular deletion and a subsequent abnormality, indicating that the blastomeres do not have the same fate in each embryo. However there is a significant tendency for the A and C quadrants to become left and right and B and D to become ventral and dorsal as in higher Spiralia.

C. Axis specification

Blastomere deletions at the eight-cell stage have been done in which all cells have been deleted singly and in various combinations. The resulting larvae can be divided into two categories based on body symmetry: 1. Larvae that established the normal axes of bilateral symmetry, and 2. Larvae that were either asymmetric or radially symmetrical and therefore had not formed the normal bilateral axes. Micromere deletions support the conclusion that specific blastomeres do not receive consistent cytoplasmic localizations (Boyer, 1989, 1992). There is a significant tendency for 1a and 1c to be lateral but 1b and 1d may be also, unlike the invariant lineage of molluscs. With increase in the number of micromeres deleted, greater deficiencies resulted, with a significant change from bilateral to radial or asymmetry when three were ablated. When all four micromeres were deleted very few embryos survived and those that did were extremely abnormal, never forming lobes, eyes or any differentiated tissues.

Deletion of macromeres at the eight-cell stage produced different results depending on whether cross-furrow (B & D) or non-cross-furrow (A & C) blastomeres were involved. Table 1 indicates a clear relationship between larval symmetry and presence of macromeres, with abnormal symmetry becoming increasingly common as the number of cells removed was increased, particularly when cross-furrow macromeres were included in the deletions. When one macromere was killed significantly more larvae were radially symmetrical if the deleted cell was a cross-furrow than a non-cross-furrow macromere ($\chi^2 = 5$, 1 d.f., $0.05>p>0.01$). In the two macromere deletion series, inclusion of one or both cross-furrow cells produced significantly more radially symmetrical larvae than when neither cross-furrow cell was removed ($\chi^2 = 12$, 1 d.f., $p<0.001$ and $\chi^2 = 19.4$, 1 d.f., $p<0.001$ respectively). There was no significant difference in symmetry abnormalities when the two deleted macromeres were one or both cross-furrow cells. Removal of three macromeres produced similar results, with over 80% of the larvae having radial symmetry, regardless of whether one or two cross-furrow cells were included in the deletion. With removal of all four macromeres, 100% of the larvae were abnormal in symmetry.

Table 1. Effect on larval symmetry of deleting macromeres from eight-cell stage embryos of *Hoploplana inquilina*

Deletion	Bilateral symmetry	Asymmetric or radial symmetry
−1A or 1C	56 (81%)	13 (19%)
−1B or 1D	43 (64%)	24 (36%)
−1A & 1C	31 (62%)	19 (38%)
−2 adjacent macromeres	13 (27%)	35 (73%)
−1B & 1D	10 (19%)	42 (81%)
−3 macromeres (A & C included)	6 (18%)	28 (82%)
−3 macromeres (B & D included)	4 (13%)	27 (87%)
−4 macromeres	0	22 (100%)

Eight-cell stage deletions suggest that micromere-macromere interactions play a role in axis specification in *Hoploplana.* Bilateral symmetry is dependent on the presence of at least two first quartet micromeres when all four macromeres are present. When macromeres are removed, three, or at least two if both are cross-furrow macromeres, must be present to produce a bilateral larva. Therefore cross-furrow macromeres play a more significant role in D quadrant determination, as in molluscs with equal cleavage, where B or D almost always becomes dorsal.

Discussion

When D quadrant specification in coelomates occurs through induction, any macromere at the four-cell stage has the potential to become D but in fact it is almost always one of the vegetal cross-furrow cells that does (van den Biggelaar & Guerrier, 1979). Apparently because of the packing arrangement of these macromeres, a single one makes contact with the greatest number of micromeres at the animal pole and is induced to become D. If this contact is prevented (Martindale *et al.*, 1985) or the first quartet micromeres are removed (van den Biggelaar & Guerrier, 1979), a D quadrant is not specified and the embryo remains radially symmetrical. Based on their extensive analysis of development in four coelomate phyla, Freeman & Lundelius (1992) conclude that equal cleavage and the inductive mode of D quadrant specification are ancestral. They suggest that among the morphological characteristics attributed to the early ancestors of the spiralian phyla should be added "an inductive mechanism for specifying bilateral symmetry mediated by equal first two cleavages." Thus their proposed ancestral coelomate would have had a larva that was subject to selective pressure to shorten metamorphosis, which could be accomplished through accelerating the appearance of adult characteristics. The specification of bilateral symmetry by cytoplasmic localization of vegetal determinants inherited by the larger of the two blastomeres at the two-cell stage is one way of accelerating development.

Though living polyclads have many advanced features, their development, including canonical quartet spiral cleavage and a lobed larva, is conservative (Galleni & Gremigni, 1989), and clearly links this group with the rest of the Spiralia. If the Freeman & Lundelius proposal is correct, polyclads should have equal cleavage and late D quadrant determination by induction, whereas if the ancestral mode of spiralian development was determination by cytoplasmic localization, present day polyclads might be expected to retain this mode of development. As examination of first and second cleavage in a number of different polyclad species and the experimental results presented above indicate, polyclad development seems to be most similar to that of the equally cleaving coelomates. The development of radialized larvae with removal of combinations of micromeres or macromeres plus the early arrest of development with removal of the entire first quartet in *Hoploplana* suggest that D quadrant determination occurs late by induction and lends support to the idea that axis specification by induction is ancestral. This concept is also consistent with the conclusions of Baguñà & Boyer (1990) that equal cleavage is ancestral and unequal cleavage and early blastomere determination are derived.

Other aspects of polyclad development also indicate that their mode of development is ancestral and suggest a possible sequence of evolutionary events leading to the invariant cell lineages and specification by cytoplasmic localization that characterize many of the higher spiralians. Blastomere deletion experiments at the two and four-cell stages of the *Hoploplana* embryo produce aberrant larvae indicating that the embryo is not regulative, but specific blastomeres may have different fates in different embryos. Although the polyclad cleavage pattern is of the invariant quartet type, the cell lineage is not invariant and probably represents a conservative condition. Moreover, removal of vegetal cytoplasm from fertilized eggs before first cleavage results in normal Müller's larvae (Boyer, 1988), indicating that either determinants are not localized vegetally or that localization occurs during cleavage; either of these possibilities probably is more likely to be the ancestral condition than the vegetal localization that typically occurs during oogenesis in higher spiralians.

Given the likelihood of a widespread evolutionary trend toward unequal cleavage accompanied by cytoplasmic localization and direct development in coelomate spiralians, it is relevant to look for similar trends within the Turbellaria. Though there is little consistent size difference between the blastomeres at the two-cell stage in polyclads, differences may occur in the individual eggs of a single organism, as I have observed in *Hoploplana inquilina* and is apparent in photographs of *Pseudostylochus* two-cell embryos where a larger CD cell is common (Teshirogi *et al.*, 1981). Moreover Teshirogi describes a cytoplasmic protrusion appearing during cleavage that is similar to a molluscan polar lobe and I have observed a similar protrusion in *Notoplana atomata*. These observations suggest a trend towards unequal cleavage in the polyclads.

Within the polyclads, the suborder Acotylea is considered to be derived and in this group direct development predominates, typically involving large eggs and a relatively long developmental time. However a number of species do exhibit indirect development, producing either a Müller's or a Götte's larva. Although most of the acotyleans with small eggs and rapid development that have been examined form a lobed larva, *Euplana gracilis* with 85–100 μm eggs and a 7 day developmental time (Christensen, 1971), *Pseudostylochus* sp. with 106–116 μm eggs and an 8 day developmental time (Rho, 1976) and *Stylochoplana pusilla* with 115 μm eggs and a developmental time of 8–9 days (Teshirogi *et al.*, 1981) all have direct development, reflecting an acceleration of the process. It is relevant to know if D quadrant specification occurs at an earlier stage in these embryos.

Spiral cleavage occurs in several other turbellarian orders, including the archoophoran acoels and macrostomids and the neoophoran lecithoepitheliates, rhabdocoels and proseriates. The duet form of spiral cleavage characteristic of the acoels is generally believed to be derived but in this group there has been an evolution toward predominantly conditional rather than autonomous specification (Boyer, 1971). Macrostomid early cleavage is described by Beauchamp (1961) as primarily quartet spiral but is modified after the 16-cell stage. The only other group known to have classical quartet spiral cleavage is the lecithoepitheliates. Reisinger (1970) describes cleavage in *Xenoprorhynchus steinboecki* as quartet spiral with a large D macromere. Cleavage in proseriates has been examined in *Monocelis fusca* by Giesa (1966), and *Minoma trigonopora* and *Otomesostoma auditivum* by Reisinger *et al.* (1974a, b). Early cleavages are of the quartet spiral type in which the blastomeres are equal or slightly unequal in size. Though traces of spiral cleavage have been observed in a few other neoophoran orders such as the Neorhabdocoela, the trend in this evolutionary line has been toward *Blastomeren-Anarchie* and appears to involve specialization in a direction different from the D quadrant cytoplasmic localization mechanism. The presence of unequal quartet spiral cleavage in some groups suggests that these might be fruitful organisms for experimental analysis of D quadrant specification. In addition, development of the nemertodermids and catenulids remains unexamined and is of considerable interest in the analysis of evolution of developmental mechanisms. Although many more studies need to be done to determine if unequal cleavage and early specification are an evolutionary trend in the turbellarians, the occurrence of modified spiral cleavage patterns including unequal early cleavage and direct development associated with a shortened developmental time in some groups, suggest such a trend may be widespread in all spiralians.

Conclusions

Polyclad turbellarian development provides support for the evolutionary scheme proposed by Freeman & Lundelius (1992) where equal cleavage and relatively late conditional specification of the D quadrant are hypothesized to be primitive. The invariant quartet cleavage of polyclads in association with a variable cell lineage and probable conditional specification of the D quadrant suggest that this condition is plesiomorphic and that present day polyclads probably exhibit a developmental program that is most similar to the ancestral worms that gave rise to the turbellarians as well as to the coelomates.

Acknowledgments

I wish to thank Yoshiko Nakagawa for help with translation. This work was supported by NSF grant DCB-8817760 and a grant from Research Corporation.

References

Baguñà, J. & B. C. Boyer, 1990. Descriptive and experimental embryology of the Turbellaria: Present knowledge, open questions and future trends. In H. Marthy (ed.), Experimental Embryology in Aquatic Plants and Animals, NATO ASI Series A: 195: 95–128.

Beauchamp, P. de, 1961. Classe des Turbellaries. In P. P. Grasse (ed.), Traite de Zoologie, IV. Masson, Paris: 35–212.

Boyer, B. C., 1971. Regulative development in a spiralian embryo as shown by cell deletion experiments on the acoel *Childia*. J. exp. Zool. 176: 96–105.

Boyer, B. C., 1986a. Experimental evidence for the origins of determinative development in the polyclad Turbellaria. Hydrobiologia 132 (Dev. Hydrobiol. 32): 117–119.

Boyer, B. C., 1986b. Determinative development in the polyclad turbellarian *Hoploplana inquilina*. Int. J. invert. Repro. Dev. 9: 243–251.

Boyer, B. C., 1987. Development of in vitro fertilized embryos of the polyclad flatworm *Hoploplana inquilina* following blastomere separation and deletion. Roux's Arch. dev. Biol. 196: 158–164.

Boyer, B. C., 1988. The effect of removing vegetal cytoplasm during the maturation divisions on the development of *Hoploplana inquilina* (Turbellaria, Polycladida). Fortschr. Zool. 36: 277–282.

Boyer, B. C., 1989. The role of the first quartet micromeres in the development of the polyclad *Hoploplana inquilina*. Biol. Bull. 177: 338–343.

Boyer, B. C., 1992. The effect of deleting opposite first quartet micromeres on the development of the polyclad *Hoploplana*. Biol. Bull. 183: 374–375.

Christensen, D. J., 1971. Early development and chromosome number of the polyclad flatworm *Euplana gracilis*. Trans. am. microsc. Soc. 90: 457–463.

Davidson, E. H., 1991. Spatial mechanisms of gene regulation in metazoan embryos. Development 113: 1–26.

Freeman, G. & J. W. Lundelius, 1992. Evolutionary implications of the mode of D quadrant specification in coelomates with spiral cleavage. J. evol. Biol. 5: 205–247.

Galleni, L. & V. Gremigni, 1989. Platyhelminthes–Turbellaria. In K. G. Adiyodi & R. G. Adiyodi (eds), Reproductive Biology of Invertebrates, IV. J. Wiley & Sons, N.Y.: 63–89.

Giesa, S., 1966. Die Embryonalentwicklung von *Monocelis fusca* Oersted (Turbellaria, Proseriata). Z. Morph. Okol. Tiere 57: 137–230.

Kato, K., 1940. On the development of some Japanese polyclads. Jap. J. Zool. 8: 537–573.

Martindale, M. Q., C. Q Doe & J. B. Morrill, 1985. The role of animal-vegetal interaction with respect to the determination of dorsoventral polarity in the equal cleaving spiralian, *Lymnaea palustris*. Roux's Arch. dev. Biol. 194: 281–295.

Raff, R. A. & R. C. Kaufman, 1983. Embryos, Genes and Evolution. Macmillan Publishing Co., Inc., New York, 395 pp.

Reisinger, E., 1970. Zur Problematik der Evolution der Coelomaten. Z. zool. Syst. Evolut-forsch 8: 81–109.

Reisinger, E., I. Cichocki, T. Erlach, & T. Szyskowitz, 1974a. Ontogenetische Studien an Turbellarien: ein Beitrag zur Evolution der Dotterverarbeitung im ektolecitalen Ei. I. Z. zool. Syst. Evolut.-forsch. 12: 161–195.

Reisinger, E., I. Cichocki, T. Erlach, & T. Szyskowitz, 1974b. Ontogenetische Studien an Turbellarien: ein Beitrag zur Evolution der Dotterverarbeitung im ektolecitalen Ei. II. Z. zool. Syst. Evolut.-forsch. 12: 241–278.

Rho, S., 1976. Studies o:n the polyclad Turbellaria of Korea II. Spawning and early development of *Stylochus ijimai* Yeri et Kat-uraki and *Pseudostylochus* sp. under laboratory conditions. Bull. Fish. Res. Dev. Agency 15: 125–140.

Surface, F. M., 1907. The early development of a polyclad, *Planocera inquilina*. Proc. Acad. nat. Sci. Philad. 59: 514–559.

Teshirogi, W., S. Ishida & K. Jatani, 1981. On the early development of some Japanese polyclads. Rep. Fukaura mar. biol. Lab. 9: 2–31.

Thomas, M. B., 1986. Embryology of the Turbellaria and its phylogenetic significance. Hydrobiologia 132 (Dev. Hydrobiol. 32): 105–115.

van Dam, W. I. & N. H. Verdonk, 1982. The morphogenetic significance of the first quartet micromeres for the development of the snail *Bithynia tentaculata*. Roux's Arch. dev. Biol. 191: 112–118.

van den Biggelaar, J. A. M. & P. Guerrier, 1979. Dorsoventral polarity and mesentoblast determination as concomitant results of cellular interactions in the mollusk *Patella vulgata*. Dev. Biol. 68: 462–471.

Hydrobiologia **305**: 223–224, 1995.
L.R.G. Cannon (ed.), Biology of Turbellaria and some Related Flatworms.

Cellular succession in the epithelium lining the pharyngeal cavity of planarians

Saburo Ishii
Division of Cell Science, Central Research laboratory, Fukushima Medical College, Fukushima 960-12, Japan

Key words: planarian, pharyngeal cavity epithelium, cellular succession, TEM & SEM

Abstract

Regardin cell renewal of planarian epithelial tissue, little is known at present except for the epidermis (Tyler, 1984). According to Skaer (1965), the definitive epidermal cells in planarian embryos originate in the parenchyma where they develop as precursor cells which then migrate into the epidermis. This process of development is also reported in regenerating planarians (Spiegelman & Dudley, 1973; Morita & Best, 1974; Hori, 1978). Recently the validity of Skaer's conclusion has clearly been confirmed by Ishii & Sakurai (1988) using SEM and TEM.

The present study deals with the origin and renewal of cells of another epithelium – that lining the pharyngeal cavity of a freshwater planarian *Dugesia japonica*. This epithelium is known to consist of cells with secretory granules of fingerprint-like structure (Ishii, 1966; Bowen & Ryder, 1973; Gamo & Garcia-Corrales, 1988; Asai, 1991). The characteristic feature of these specific granules, as well as epitheliosomes (Tyler, 1984) of the epidermis, is very useful as a cell marker for distinguishing cell types and following cell differentiation. Gamo & Garcia-Corrales (1988) claimed the cell populations of the pharyngeal epithelium of *Dugesia gonocephala* is morphologically and functionally heterogeneous. The sequence of morphological alteration offers the opportunity to distinguish at least three functional stages. These represent different adaptive forms of a single cell type, and indicate transformation from one to the next induced by cell renewal. It suggests a beta cell (neoblast) origin of the epithelial cells.

In the present study SEM observations revealed theat there are morphologically three distinct cell types which are considered to be juvenile, mature, and senile cells respectively, judging from the differences in cells and the many transitional forms among them (Fig. 1). When observed by TEM, the cytoplasmic patterns of these cell types are correspondingly different from one another, similarly the surface specializations of the plasma membranes of each cell type which facilitate their identification by TEM also vary. Apparently alterations of cell morphology are involved in ageing in a single cell type. Morphological features of each cell type are summarized. Juveniles (R) rarely occur, are small cells provided with prominent apical ruffles and considered to be phagocytic because the cells often contain newly formed phagosomes. Mature cells (F) predominate, are large, polygonal, flat cells, presumably with an intense secretory activity. The cells contain many profiles of granular ER and a small number of specific (secretory) granules. Senile cells (M) are intermittently distributed, irregularly contoured cells with heavy surface microvilli. The specific granules are dense in the apical cytoplasm, and profiles of granular ER decease in number.

The precursor epithial cells that are seen in the parenchyma exhibit a considerable degree of differentiation before their translocation into the lining. The primordial cells observed are neoblast-like, with several cohromatoid bodies and many free ribosomes (Morita *et al.*, 1969). These results obtained thus support the evidence and prediction of Gamo & Garcia- Corrales (1988). The observed mode of cell replacement appears to be the same as found in the epidermis (Lentz, 1967).

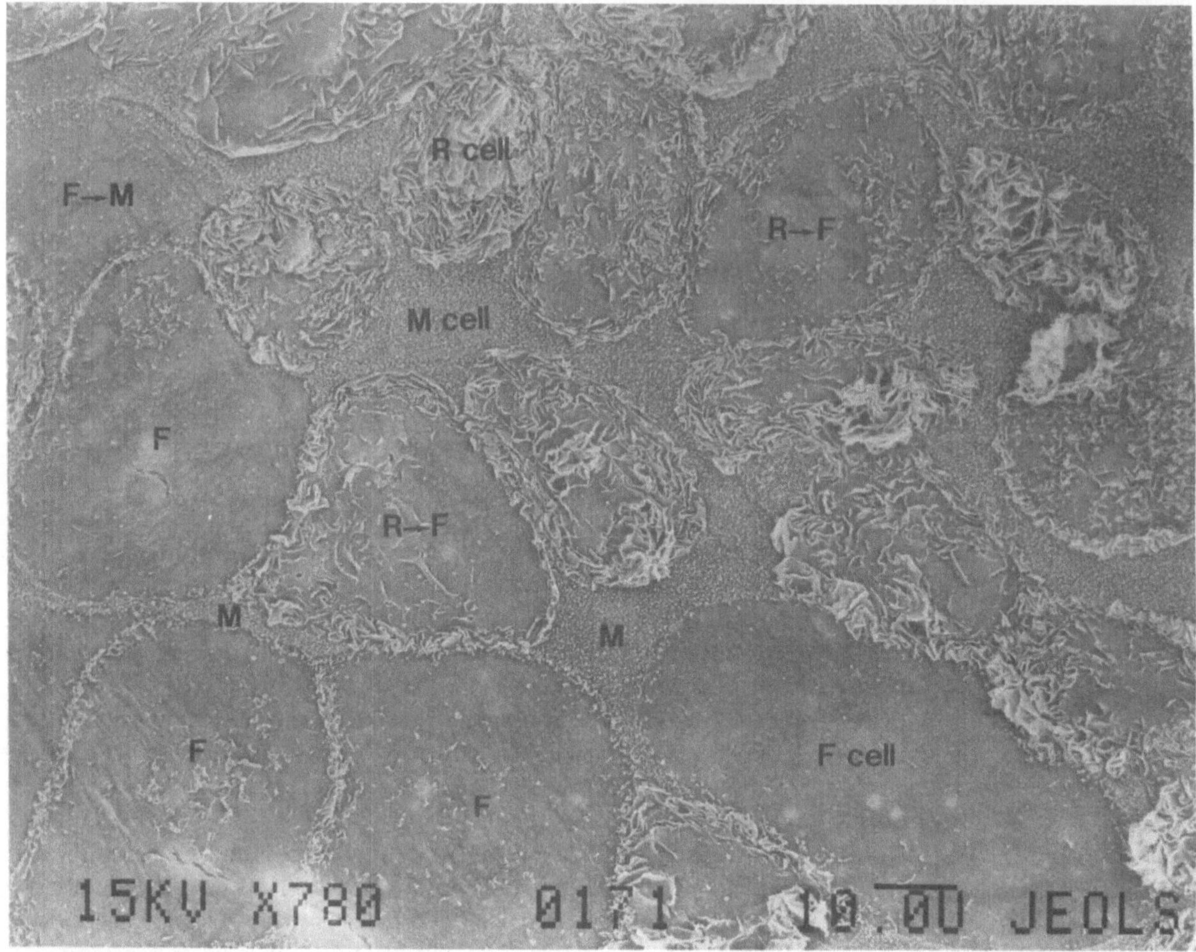

Fig. 1. SEM micrograph of the pharyngeal epithelium of *Dugesia japonica* showing the surface granulation characterising juvenile (R), mature (F) and senile (M) cells and transitions between them.

References

Asai, E., 1991. Regeneration of the pharynx in a freshwater planarians: An electron- microscopic study with special reference to the formation of the pharyngeal cavity and pharyngeal lumen. Zool. Sci. 8: 775–784.

Bowen, I. D. & T. A. Ryder, 1973. The fine structure of the planarian Polycelis tenuis Ijima. I. The pharynx. Protoplasma, 78: 223–241.

Gamo, J. & P. Garcia-Corrales, 1988. Ultrastructure of the pharyngeal cavity epithelium of Dugesia gonocephala (Tricladida, Paludicola). Fortschr. Zool. 36: 429–434.

Hori, I., 1978. Possible role of rhabdite-forming cells in cellular succession of the planarian epidermis. J. Electron Microsc. 27: 89–102.

Ishii, S., 1966. The ultrastructure of the insunk epithelium lining the planarian pharynx cavity. J. Ultrastruct. Res. 14: 345–355.

Ishii, S. & T. Sakurai, 1988. Scanning and transmission electron microscopic observations on epidermal development in the embryo of the freshwater triclad, Bdellocephala brunnea. Fortschr. Zool. 36: 291–295.

Lentz, T. L., 1967. Rhabdite formation in planaria: The role of microtubules. J. Ultrastruct. Res. 17: 114–126.

Morita, M. & J. B. Best, 1974. Electron microscopic studies of planarian regeneration. II. Changes in epidermis during regeneration. J. Exp. Zool. 187: 345–358.

Morita, M., J. B. Best & J. Noel, 1969. Electron microscopic studies of planarian regeneration. I. Fine structure of neoblast in Dugesia dorotocephala. J. Ultrastr. Res. 27: 7–23.

Skaer, R. J., 1965. The origin and continuous replacement of epidermal cells in the planarian Polycelis tenuis (Ijima). J. Embr. Exp. Morph. 13: 129–139.

Spiegelman, M. & P. L. Dudley, 1973. Morphological stages of regeneration in the planarian Dugesia tigina: a light and electron microscopic study. J. Morphol. 139: 155–184.

Tyler, S., 1984. Turbellarian Plathelminths. In J. Bereiter-Hahn, A. G. Matoltsy & K. S. Richards (eds), Biology of the Integument. I. Invertebrates. Springer-Verlag, Berlin: 112–131.

Hydrobiologia **305**: 225, 1995.
L.R.G. Cannon (ed.), Biology of Turbellaria and some Related Flatworms.

Differentiation of the body wall musculature in *Macrostomum* and *Hoploplana* (Turbellaria, Platyhelminthes)

R. M. Rieger[1], W. Salvenmoser[1], D. Reiter[1] & Barbara C. Boyer[2]
[1]*Institut für Zoologie, Universität Innsbruck, Austria*
[2]*Department of Biological Sciences, Union College, Schenectady, N.Y., USA*

Key words: Muscle differentiation, Macrostomida, Polycladida., fluorescence labelling technique

Abstract

Using the Phalloidin-Rhodamine fluorescence-labelling technique for F-actin, we have studied the development of the body wall musculature in *Macrostomum hystricinum marinum* and in the polyclad *Hoploplana inquilina*. The structure of the muscle grid in the freshly hatched *Macrostomum* (see also Rieger & Salvenmoser, 1991) and the young larva of *Holplana* served as reference systems for the embryonic development of the body wall musculature. In *Macrostomum* muscle fiber differentiation starts around 60% of developmental time between egg-laying and hatching, and in *Hoploplana* around 80% of embryonic development.

In *Macrostomum*, early stages show TV-antenna-like arrangements of one longitudinal and several circular fibers. In *Hoploplana* our preliminary results show a particularly large, longitudinal fiber on either side of the body. These primary longitudinal fibers may serve as a 'founder cell' for other longitudinal fibers and as spatial guides for the circular muscles. Similar 'founder cells' have been reported during early muscle differentiation in leeches (Jellies & Kristan, 1988; Jellies, 1990). In *Hoploplana*, a special muscle system is present at the outset under the apical organ. It consists of what seems to be a spirally arranged fiber – when seen in head-on view – and of two additional fibers crossing this spiral, from the later developing posterior to the anterior lobe.

TEM-studies of embryos of *Macrostomum* suggest that the longitudinal nerve cords represent an important guide during early differentiation of the pattern within the body wall musculature. Young stages of myoblasts can be identified along the main lateral nerve cord. Commonly, the myoblasts are seen to alternate with young neurons in their position along the nerve cord. Embryonic stages of *Macrostomum hystricinum marinum* were obtained from our cultures (Rieger *et al.*, 1988). Immediately prior to fixation (Paraformaldehyde, Stephanini's fixative) the eggshells were punctured with tungsten needles. We noted some variability of developmental time for certain embryonic stages, which we cannot explain. Developmental stages of *Hoploplana inquilina* were collected at the Marine Biological Laboratory, Woods Hole, MA, USA according to the procedure outlined in Boyer (1987) and Boyer (1989). They have been timed in relation to normal developmental time to an early Müller's larva at about 100 hours.

Acknowledgment

This work was supported by FWF grant P1638-BIO and NSF grant DCB-8817760.

References

Boyer, B. C., 1987. Development of in vitro fertilized embryos of the polyclad flatworm Hoploplana inquilina following blastomere separation and deletion. Roux's Arch. dev. Biol. 196: 158–164.

Boyer, B. C., 1989. The role of the first quartet micromeres in the development of the polyclad Hoploplana inquilina, Biol. Bull. 177: 338–343.

Jellies, J. & W. B. Kristan, 1988. Embryonic assembly of a complex muscle is directed by a single identified cell in the medicinal leech. J. Neurosci. 8: 3317–3326.

Jellies, J., 1990. Muscle assembly in simple systems. TINS 13: 126–131.

Rieger, R. M., M. Gehlen, G. Haszprunar, M. Homlund & A. Legniti, 1988. Laboratory cultures of marine Macrostomida (Turbellaria). Progr. Zool. 36: 523.

Rieger, R. & W. Salvenmoser, 1991. Demonstration of the muscle system in whole mounts of Macrostomum hystricinum marinum (Turbellaria, Macrostomida). Am. Zool. 31: 35.

Hydrobiologia **305**: 227, 1995.
L.R.G. Cannon (ed.), *Biology of Turbellaria and some Related Flatworms.*

Autoradiographic studies of spermatogenesis in *Dugesia bengalensis*, Kawakatsu

A. K. Aditya & P. N. Ghosh
Department of Zoology, School of Life Sciences, Visva-Bharati University, Santiniketan – 731 235, W. B. India

Key words: Turbellaria, tricladida, *Dugesia*, spermatogenesis, autoradiography.

Abstract

Spermatogenesis in *Dugesia bengalesis* Kawakatsu was studied by exposing worms to water containing 10 μci to H^3-Thymidine. Specimens were killed and sectioned at intervals from 4 h to 48 h.

Kodak AR-10 film strips were exposed to the histological sections (Ghosal *et al.*, 1983). Spermatogonia and spermatocytes became readily labelled within 4 h, but spermatozoa first displayed labelling at 35 h signifying that both meiosis and spermiogenesis were completed by then (Table 1).

Table 1. Appearance of radiographic labelling on the spermatozoa of *Dugesia bengalensis* Kawakatsu.

Duration (in hours) of exposure to H^3-thymidine	Most advanced stage of spermiogenesis labelled
4	Spermatogonia and pre-meiotic spermatocyte
11	meiotic spermatocyte
18	early spermatid
24	middle spermatid
30	late spermatid
35	mature spermatozoa
47	mature spermatozoa
48	mature spermatozoa

Reference

Ghosal, S. K., Ghosh, P., Aditya, A. K., Basu Roy, T., Sengupta, A., Dutta, G. & Dev. T., 1983. Estimation of total duration of meiosis and spermiogenesis in *D. bengalensis*. Ind. Sci. Cong. p. 83.

Hydrobiologia **305**: 229–233, 1995.
L.R.G. Cannon (ed.), Biology of Turbellaria and some Related Flatworms.

Never ending growth and a growth factor. II. Immunocytochemical evidence for the presence of epidermal growth factor in a tapeworm

Margaretha K. S. Gustafsson, Krister Eriksson & Annika Hydén
Department of Biology, Åbo Akademi University, FIN-20520 Åbo, Finland

Key words: EGF, immunocytochemistry, flatworm, *Diphyllobothrium*

Abstract

Flatworm growth patterns are interesting and the guiding principles behind them have long been sought. Epidermal growth factor immunoreactivity (EGF-IR) was detected in a dispersed population of nerve cells of the constantly growing adult gull-tapeworm *Diphyllobothrium dendriticum* (Cestoda, Pseudophyllidea). The EGF-IR cells are located close to regions characterized by active growth and the presence of mitotically competent cells. As no EGF-IR was observed in the non-growing plerocercoid larva a correlation with the growth rate of the worm and the presence of EGF-IR is suggested.

Introduction

Flatworm growth patterns are interesting; turbellarians regenerate, asexual reproduction takes place both among free-living and parasitic worms, and adult pseudophyllidean tapeworms never stop growing along their longitudinal axis. These are rare phenomena in the animal kingdom. The guiding principles behind planarian regeneration and the never-ending growth of adult tapeworms have long been sought (Baguñà *et al.* 1989; Gustafsson, 1990; Gustafsson & Eriksson, 1992). A stimulating effect of growth factors, such as epidermal growth factor (EGF) and substance P (SP), on cell proliferation in planarians has been demonstrated by Baguñà *et al.* (1989). Immunoreactivity (IR) to EGF receptor (EGFr) was localized in the planarian nervous system (Baguñà *et al.* 1990). In the nervous system of the microturbellarians *Microstomum lineare* and *Stenostomum leucops* IR towards EGF, EGFr and basic fibroblast growth factor (bFGF) was demonstrated by Reuter & Kuusisto (1992). The same authors also reported an inhibitory effect of anti-EGF, anti-EGFr and anti-bFGF on the asexual reproduction of *S. leucops*. In the gull-tapeworm *Diphyllobothrium dendriticum* the presence of bFGF-IR nerve cells was demonstrated by immunocytochemistry (ICC) and Western blot analysis indicated the presence of a 47 kDa bFGF-like molecule (Gustafsson & Eriksson, 1992). Recently SP-IR was detected in the nervous system of larval and adult *D. dendriticum* by Gustafsson *et al.* (1993) and in the nervous system of *M. lineare* and *S. leucops* by Reuter (1994).

In order to broaden the picture of growth factors in flatworms a study was undertaken, in which *D. dendriticum* was searched for EGF-IR cells with the ICC method. Generally *D. dendriticum* is well suited for growth studies; the growth rates of larval and adult worms differ greatly, presenting an opportunity to compare the presence and pattern of EGF-IR with growth rates.

Material and methods

EGF-IR in Diphyllobothrium dendriticum

Plerocercoid larvae of *D. dendriticum* Nitzsch, 1824 (Cestoda, Pseudophyllidea) were obtained from whitefish from Lake Pyhäjärvi, SW Finland. They were excised from cysts on the stomach wall. Some were fixed directly from the cysts, others were cultivated *in vitro* in Dulbecco's Modified Eagles Medium for 3 and 10 h at 38 °C. Adult worms were obtained from golden hamsters, infected experimentally according to the method described by Wikgren *et al.* (1970).

The worms were fixed, sectioned and stained with the immunoperoxidase-antiperoxidase method (Sternberger, 1974; Gustafsson *et al.*, 1986)), using anti-EGF (Ab-3) from Oncogene Science (conc 1:20–1:50) and anti-EGF from Biomedical Technologies (conc 1:50–1:200). Both antibodies are purified polyclonals raised in rabbits against recombinant human EGF, are specific for human EGF and show no cross reactivity with human transforming growth factor-α. Control sections were incubated with antiserum inactivated by addition of exess antigen.

Results

EGF-IR in D. dendriticum

To facilitate the interpretation of the results a diagram of the anatomy and histology of the adult tapeworm is included (Fig. 1) (Gustafsson *et al.*, 1986).

The antibodies from the two sources gave identical results. The control sections were negative (Fig. 2). In larvae fixed directly from whitefish no EGF-IR was found. After cultivation *in vitro* for 3 h at 38 °C a weak EGF-IR was detected in a few cells dispersed in the medullary and cortical parenchyma and in the layer of longitudinal muscles. Cultivation *in vitro* for 10 h revealed distinctly EGF labelled cells mainly located in the cortical parenchyma close to the outer border of the longitudinal muscle layer (Fig. 3). The EGF-IR occurred as distinct granules in the cytoplasm. EGF-IR was observed also in nerve fibres of the main and peripheral nerve cords.

In adult worms the EGF-IR was distinct, along the whole body, from head to tail. It occurred inside cell bodies in the form of large granules in the cytoplasm (Fig. 5, inset). The EGF-IR cell bodies were bi- to multipolar in shape and measured about 10 × 30 μm (6–15 × 15–50 μm). The following pattern was observed. EGF-IR occurred: 1. in nerve fibres along the two main nerve cords and in nerve cells surrounding them (Fig. 4), 2. in cell bodies associated to the peripheral nervous system (PNS), which is well developed in the cortical parenchyma and where the EGF-IR cells have projections towards the basal lamina of the surface tegument (Fig. 5), 3. in the PNS of the longitudinal muscle layer, especially along its outer border (Fig. 5), 4. in the PNS of the of transverse muscles (Figs 4, 6), 5. in the medullary parenchyma (Figs 4, 6), 6. in association with the primary genital anlage, especially in the periphery of the anlage (Fig. 6), 7. in the capsule

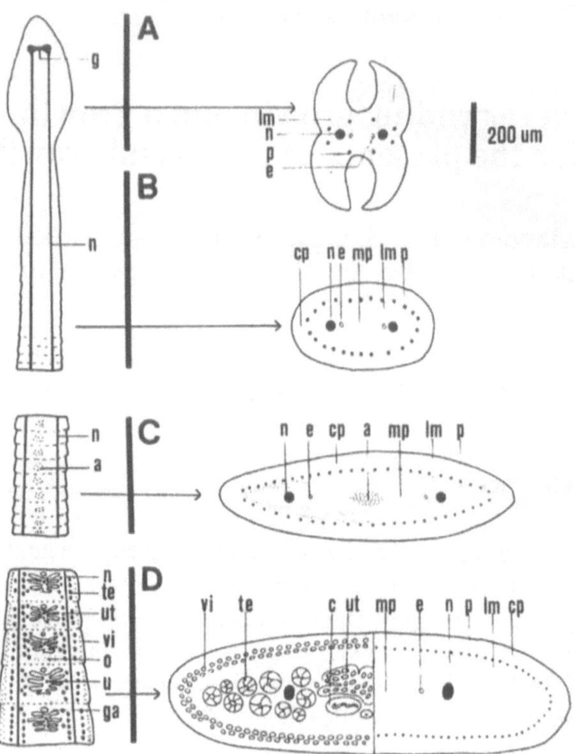

Fig. 1. Diagram showing the histology of adult *Diphyllobothrium dendriticum*. At left a longitudinal section through scolex (head) (A), neck region (B), region with primary genital anlage (C), and mature proglottid (D). At right cross sections of corresponding parts. Ganglionic commissure (*g*), main nerve cord (*n*), peripheral nerve bundle (*p*), longitudinal muscle layer (*lm*), main excretory duct (*e*), medullary parenchyma (*mp*), cortical parenchyma (*cp*), primary genital anlage (*a*), testicular follicle (*te*), uterus (*ut*), uterine pore (*u*), vitelline gland (*vi*), ovary (*o*), male copulatory organ (*c*), genital atrium (*ga*). (from Gustafsson *et al.*, 1986)

wall of the testicular follicles (Fig. 7) and 8. close to the vitelline glands.

Discussion

A growth factor is a substance which coordinates different aspects of cell growth, proliferation, differentiation, and morpho-functional maintenance by inducing replicative DNA synthesis and cell division (Rozengurt, 1986). EGF induces these effects on ectodermal and mesodermal cells. Thirty years ago EGF was discovered by Stanley Cohen (1962) in mice. EGF is synthesized as a precursor of 1217 amino acids, which includes at least seven repeat amino acid sequences, homologous to the original 53 amino acid EGF mito-

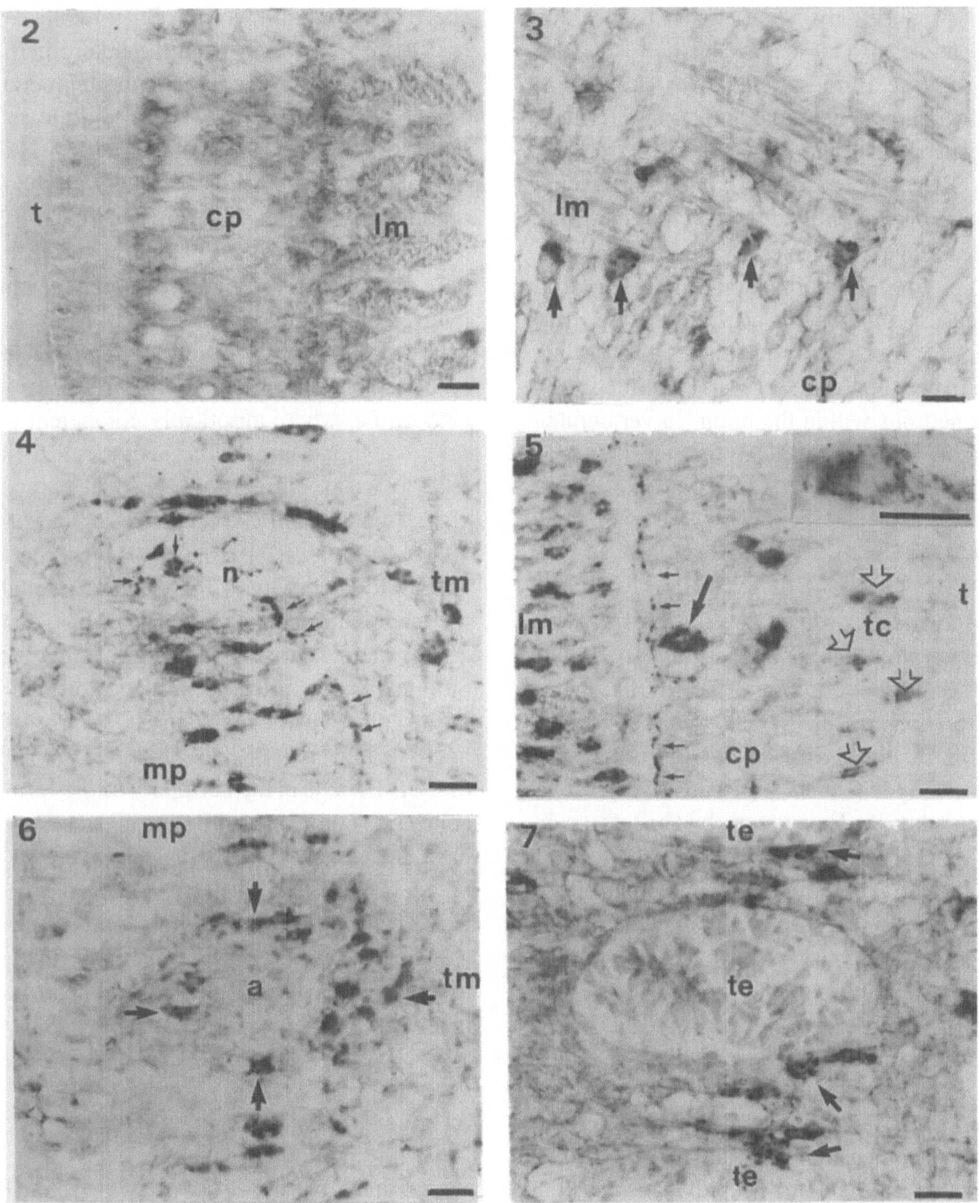

Figs 2–7. Immunocytochemical reactivity (IR) to anti-EGF in the gull-tapeworm *Diphyllobothrium dendriticum*, peroxidase-anti-peroxidase staining, Nomarski optics. Scale = 25 μm in all figures. (Fig. 2) Micrograph of control for anti-EGF-IR. Section of surface tegument (*t*), cortical parenchyma (*cp*) and longitudinal muscle layer (*ml*) stained with anti-EGF absorbed with EGF peptide (BT-102). No EGF-IR occurs. (Fig. 3) EGF-IR nerve cells (arrows) along outer border of longitudinal muscle layer (*lm*) in plercercoid larva cultivated for 10 h at 38 °C. Note granular EGF-IR. Cortical parenchyma (*cp*). (Fig. 4) EGF-IR in nerve cells surrounding the main nerve cord (*n*) and in the medullary parenchyma (*mp*) and transverve muscle layer (*tm*) of adult worm. Note EGF-IR nerve fibres (arrows) inside and reaching out from nerve cord. (Fig. 5) EGF-IR in peripheral nerve fibre (small arrows) in cortical parenchyma (*cp*) of adult worm. Note EGF-IR nerve cell (large arrow) associated with peripheral nerve fibre and EGF-IR projections (open arrows) towards growing layer of tegumental cells (*tc*). Longitudinal muscle layer (*lm*) with EGF-IR nerve cells. Inset. Bipolar EGF-IR neuron with EGF positive granules. Surface tegument (*t*). (Fig. 6) EGF-IR cells (arrows) in association with primary genital anlage (*a*) in medullary parenchyma (*mp*). The EGF-IR cells are located peripherally, where mitotic divisions take place. In the centre of the anlage the cells are differentiated and do no longer divide. EGF–IR cell bodies also occur in medullary parenchyma (*mp*) and the transverse muscle layer (*tm*), where mitoses take place. (Fig. 7) EGF-IR cells (arrows) close to the testicular follicles (*te*). Note the granular content of the EGF-IR cells.

gen. EGF is produced by a variety of tissues, of non-neuronal and neuronal origin and in foetal and adult organisms (see Burgess, 1989; Plata-Salamán, 1991). EGF-IR neurons, neuronal fibres and terminals are widely distributed in the mammalian central nervous system (CNS), and EGF mRNA and EGF precursor mRNA have been identified there. Thus synthesis of EGF in the CNS is suggested by ICC studies and demonstrated by *in situ* hybridization techniques. This suggests a neuromodulatory or neurotransmitter role of EGF in the CNS (see Plata-Salamán, 1991). It has also been suggested that EGF serves as a trophic factor for particular neurons within the brain. In vertebrates EGF thus occurs both in neuronal and non-neuronal tissues.

EGF-IR in D. dendriticum

The few data available on the occurrence of growth factors in flatworms point specifically to a neuronal localization (see Introduction) and the results of the present investigation on EGF-IR cells support this. The cytomorphology of the EGF-IR cells conforms with that of the paraldehyde fuchsin (PAF) positive neurons described from adult *D. dendriticum*, by Gustafsson & Wikgren (1981a). These neurons are distributed along the CNS and the PNS, predominantly in the periphery (Fig. 1), as are the EGF-IR cells. As flatworms lack a circulatory system, the need for a dispersed population of secretory neurons is great. The neurosecretory material is released either synaptically or paracrinally (Gustafsson & Wikgren, 1981c; Gustafsson, 1984).

D. dendriticum larvae live encysted on the stomach wall of whitefish and grow slowly (Bylund, 1969). Adult worms live in the small intestine of sea-gulls (or in laboratory hamsters) and grow rapidly, at first even exponentially (Bråten, 1966). The results of this study show that EGF-IR occurs in the actively growing adult tapeworm but not in the 'non-growing' larva. Anti-bFGF, however, was detected with ICC and Western blot techniques in the CNS of both larval and adult *D. dendriticum* by Gustafsson & Eriksson (1992).

During the change of host from cold fish to warm bird *D. dendriticum* goes through great and rapid alterations: metabolic and growth rates and mobility, all increase. The PAF positive neurons become activated (Gustafsson & Wikgren, 1981b), release from neuronal terminals containing large dense vesicles commences (Gustafsson & Wikgren, 1981c), and cell proliferation increases (Gustafsson, 1968). The mitotic index rises steeply from close to 0.0 in the larvae to 1.0 in the neck region of adult worms (Gustafsson, 1990). When larvae are transferred from the heterothermic host to an *in vitro* culture medium at the temperature of the homeothermic host (38 °C), the mitotic activity rises rapidly (Wikgren & Gustafsson, 1967). An increase in EGF-IR takes place in parallel with the increase in mitotic activity after short time *in vitro* cultivation. The results of the *in vitro* cultivation of larvae thus support the existence of a correlation between the mitotic activity/growth rate and the presence of EGF-IR cells.

The EGF-IR cells in adult *D. dendriticum* are located in regions which are characterized by very active growth, *i.e.* where mitotically competent germinative cells are found and where active cell differentiation takes place, such as in the cortical parenchyma and close to the main nerve cords (Wikgren & Knuts, 1970; Gustafsson, 1976, 1990). Another centre for high mitotic activity is the primary genital anlage in the medullary parenchyma. The anlage consists of an aggregation of actively dividing germinative cells. The mitotic divisions are especially located at the periphery of the anlage, while the centre contains differentiated, non-dividing cells (Wikgren *et al.*, 1971). The finding of EGF-IR cells in the periphery of the forming genital anlage is significant. Although 17 neuronal mediators have been localized in *D. dendriticum* with ICC so far (Gustafsson, 1992; Gustafsson *et al.*, 1993), this is the first time a specific immunoreaction has been found associated with the forming genital anlage. In order to verify the role/s of the growth factors in the gull-tapeworm further investigations have to be made.

Acknowledgments

We are thankful for the support of the Finnish Academy of Sciences, the Research Institute of the Åbo Akademi Foundation and Jubileumsfonden at Åbo Akademi University.

References

Baguñà, J., E. Saló & R. Romero, 1989. Effects of activators and antagonists of neuropeptides substance P and substance K on cell division in planarians. Int. J. Dev. Biol. 33: 261–264.

Baguñà, J., R. Romero, E. Saló, J. Collet, C. Auladell, M. Ribas, M. Riutort, J. García-Fernandez, F. Burgaya & D. Bueno, 1990. Growth, degrowth and regeneration as developmental phenomena in adult freshwater planarians. In H. Merthy (ed.), Experimental Embryology in Aquatic Plants and Animals. Plenum Press, New York: 129–162.

Bråten, T., 1966. Studies of the helminth fauna of Norway. VII. Growth, fecundity, and fertility of Diphyllobothrium norvegicum Vik, (Cestoda) in the golden hamster. Nytt Mag. Zool. (Oslo) 13: 39–51.

Burgess, A. W., 1989. Epidermal growth factor and transforming growth factor α. In W. D. Waterfield (ed.), Growth Factors. Br Medical Bull. 45: 401–424.

Bylund, G., 1969. Experimentell undersökning av Diphyllobothrium dendriticum (= D. norvegicum) från Norra Finland. Information 10: 3–17.

Cohen, S., 1962. Isolation of a mouse submaxillary gland protein accelerating incisor eruption and eyelid opening in the new born animal. J. Biol. Chem. 237: 1555–1562.

Gustafsson, M. K. S., 1968. Effect of colchicine on DNA synthesis in plerocercoids of Diphyllobothrium dendriticum (Cestoda). Exp. Cell Res. 50: 1–8.

Gustafsson, M. K. S., 1976. Observations on the histogenesis of nervous tissue in Diphyllobothrium dendriticum Nitzsch, 1824 (Cestoda, Pseudophyllidea). Z. Parasitenkd. 50: 313–321.

Gustafsson, M. K. S., 1984. Synapses in Diphyllobothrium dendriticum (Cestoda). An electron microscopical study. Ann. Zool. Fenn. 21: 167–175.

Gustafsson, M. K. S., 1990. The cells of a cestode. Diphyllobothrium dendriticum as a model in cell biology. In M. K. S. Gustafsson & M. Reuter (eds), The Early Brain. Acta Acad Aboensis Ser B 50: 13–44.

Gustafsson, M. K. S., 1992. The neuroanatomy of parasitic flatworms. Adv. Neuro immunol. 2: 267–286.

Gustafsson, M. K. S. & K. Eriksson, 1992. Never ending growth and a growth factor. I. Immunocytochemical evidence for the presence of basic fibroblast growth factor in a tapeworm. Growth Factors 7: 327–334.

Gustafsson, M. K. S. & M. C. Wikgren, 1981a. Peptidergic and aminergic neurons in adult Diphyllobothrium dendriticum Nitzsch, 1824 (Cestoda Pseudophyllidea). Z. Parasitenkd. 64: 121–134.

Gustafsson, M. K. S. & M. C. Wikgren, 1981b. Activation of the peptidergic neurosecretory system in Diphyllobothrium dendriticum (Cestoda, Pseudophyllidea). Parasitology 83: 243–247.

Gustafsson M. K. S. & M. C. Wikgren, 1981c. Release of neurosecretory material by protrusions of bounding membranes extending through the axolemma, in Diphyllobothrium dendriticum (Cestoda). Cell Tissue Res. 220: 473–479.

Gustafsson, M. K.S., M. A. I. Lehtonen & F. Sundler, 1986. Immunocytochemical evidence for the presence of 'mammalian' neurohormonal peptides in neurones of the tapeworm Diphyllobothrium dendriticum. Cell Tiss Res. 243: 41–49.

Gustafsson, M. K. S., D. Nässel & A. Kuusisto, 1993. Immunocytochemical evidence for the presence of Substance P-like peptide in Diphyllobothrium dendriticum. Parasitology 106: 83–89.

Plata-Salamán, C. R., 1991. Epidermal growth factor and the nervous system. Peptides 12: 653–663.

Rozengurt, E., 1986. Early signals in the mitogenic response. Science 234: 161–166.

Reuter, M., 1994. Substance-P like positive cells innervating sensory structures and pharyngeal nervous system of Stenostomum leucops (Catenulida) and Microstomum lineare (Macrostomida). Cell Tissue Res., 276: 173–180.

Reuter, M. & A. Kuusisto, 1992. Growth factors in asexually reproducing Catenulida and Macrostomida (Platyhelminthes)? A confocal, immunocytochemical and experimental study. Zoomorphology 112: 155–166.

Sternberger, L. A., 1974. Immunocytochemistry. In A. Oster & L. Weiss (eds), Foundation of Immunology Series. Prentice Hall Inc, Englewood Cliffs, New Jersey.

Wikgren, B.-J.P. & M. K. S. Gustafsson, 1967. Duration of the cell cycle of germinative cells in plerocercoids of Diphyllobothrium dendriticum. Z. Parasitenkd. 29: 275–281.

Wikgren, B.-J.P. & G. Knuts, 1970. Growth of the subtegumental tissue in cestodes by cell migration. Acta Acad. Aboensis (B) 30: 1–6.

Wikgren, B.-J.P., M. K. S. Gustafsson & G. M. Knuts, 1971. Primary anlage formation in Diphyllobothriid tapeworms. Z. Parasitenkd. 36: 131–139.

Wikgren, B.-J.P., G. M. Knuts & M. K. S. Gustafsson, 1970. Circadian rhythm of mitotic activity in the adult gull-tapeworm, Diphyllobothrium dendriticum (Cestoda). Z. Parasitenkd. 34: 3242–250.

Hydrobiologia **305**: 235–240, 1995.
L.R.G. Cannon (ed.), Biology of Turbellaria and some Related Flatworms.

TCEN-49, a monoclonal antibody that identifies a central body antigen in the planarian *Dugesia (Girardia) tigrina*. Implications for pattern formation and positional signalling mechanisms

David Bueno*, Lluis Espinosa, Marc Aureli Soriano, Eduard Batlle, Jaume Baguñà & Rafael Romero
Departament de Genètica, Facultat de Biologia, Universitat de Barcelona, Av. Diagonal 645, Barcelona 08071, Catalonia, Spain (author for correspondence)*

Key words: monoclonal antibodies, regional antigen, regeneration, pattern formation, immunohistochemistry, *Dugesia (G.) tigrina*

Abstract

We have produced monoclonal antibodies (mAbs) against antigens of the freshwater planarian *Dugesia (G.) tigrina* (Girard) using standard protocols. One of these mAbs, TCEN-49, detects an antigen (TCEN-49Ag) present in most cells of the central area of the body, including the pharynx. Labelled cells seem more related by position than by lineage, suggesting that TCEN-49Ag is involved somehow in the expression of central body positional identity. The spatial and temporal changes in TCEN-49Ag expression during growth/degrowth and regeneration have been monitored and the implications of these results are discussed.

Introduction

Freshwater planarians (Platyhelminthes, Turbellaria, Tricladida) are remarkable for their great power of regeneration and for their capacity of sustained growth and degrowth (see Brønsted, 1969; and Baguñà *et al.*, 1990, for general reviews). Both processes are based on the presence of a class of undifferentiated cells, called neoblasts (Baguñà, 1981). Regeneration and growth/degrowth processes are fairly well understood at the tissue and cell levels. However, some basic questions still remain unanswered. First, what is the origin and the lineage of cells involved in regeneration? Furthermore, how is the pattern of structures determined during regeneration, and how do the regional shifts that take place during growth and degrowth occur? Secondly, how is the antero-posterior polarity maintained? Finally, the molecular and genetic mechanisms at the base of both phenomena are unknown (Gremigni, 1988; Baguñà *et al.*, 1990).

The introduction of hybridoma techniques to produce monoclonal antibodies (mAbs) by Kohler & Milstein (1975), represented a key method to obtain molecular markers of cell types and their subsets and to detect antigens that are spatially and temporally restricted during developmental processes. Due to their high degree of sensitivity and specificity, mAbs have been employed mainly to monitor the early stages of pattern formation.

In this paper, we describe the mAb TCEN-49 that detects an antigen, TCEN-49Ag, present in most cells of the central area of the body in the planarian *Dugesia (G.) tigrina*. Its spatial and temporal changes of expression (detected by immunohistochemical procedures) during regeneration and growth/degrowth have been monitored and the implications to unravel the mechanisms of pattern formation are discussed.

Materials and methods

Species and culture conditions

Organisms used in this study belong to an asexual race (class A; Ribas *et al.*, 1989) of *Dugesia (G.) tigrina* collected near Barcelona. They were maintained in

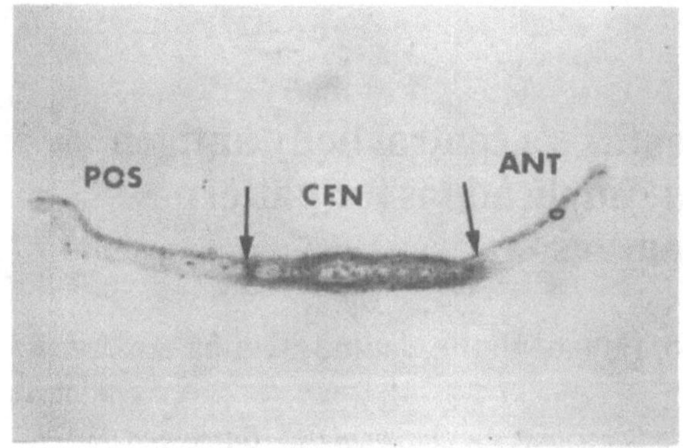

Fig. 1. *Dugesia (G.) tigrina* immunostained with the mAb TCEN-49. Saggital section. ABC method (50×). ANT, anterior region; CEN, central region; POS, posterior region; arrows, boundaries between regions.

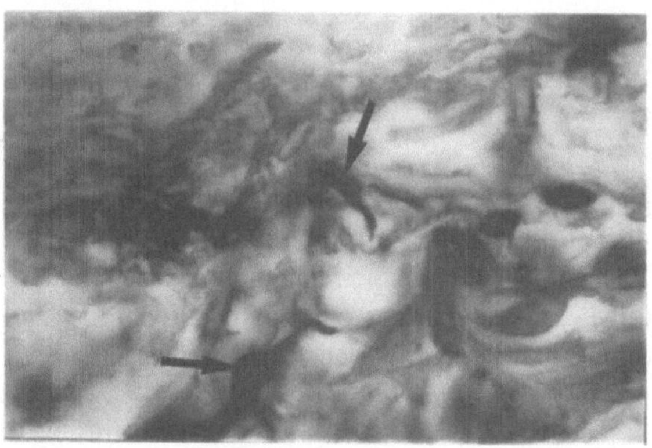

Fig. 2. *Dugesia (G.) tigrina* immunostained with the mAb TCEN-49. Saggital section. ABC method (500 ×). Arrows, immunostained cytoplasm of TCEN-49Ag positive cells.

planarian saline solution (Saló, 1984) in the dark at 4 °C and fed once a month with beef liver. Specimens chosen for the immunization and immunocytochemical procedures were starved for 15 days before use. All regeneration experiments were done at 22 °C.

Production of the mAb TCEN-49.

We followed a modified protocol from that described by Harlow & Lane (1988) (Romero *et al.*, 1991). For immunization, we used enriched fractions of live neoblasts obtained as described by Baguñà *et al.* (1989). 1–2 million live neoblast (80% purity) in 400 ml PBS (Phosphate-buffered saline 100 mM) were injected into 8 weeks old BALB/C female mice following the immunization and fusion procedures described by Romero *et al.* (1991). Hybridomas were screened by incubating cell-culture supernatant on ELISA plates in the presence of PO-conjugated goat anti mouse IgG, on paraffin sections of 7 mm long organisms and on macerated cells.

Immunocytochemical procedures.

(a) Paraffin sections.
Immunostained paraffin sections were obtained as described by Romero *et al.* (1991). Organisms were fixed in 3% paraformaldehyde – 0.02% glutaraldehyde in PBS, embedded in 51–53 °C paraffin, cut at 10 μm thick saggital sections, and mounted on gelatin-coated slides. Specimens were permeabilized in 0.4% pepsine (Sigma) in 0.1N HCl pH 5, blocked, and incubated in TCEN-49 – PBS solution. The presence of TCEN-49 linked antibody was detected with the Avidin-Biotin Complex method (ABC, PO conjugated, Vector).

(b) Macerated cells.
Planarians were macerated following a modified protocol from that described by Baguñà & Romero (1981). Organisms were macerated for 12 h at 4 °C in 1 ml of macerated solution (acetic acid/glycerol/PBS, 1:1:13), and fixed for 1 h with paraformaldehyde 4%, Tween-20 0.12% in PBS. Cell suspensions were placed on a slide and dried overnight. Slides were then rinsed in PBS and permeabilized in 0.1% pepsin in 0.1N HCl pH 5 (Espinosa, 1993). The rest of the protocol is the same as that for immunostaining of paraffin sections.

Results

The mAb TCEN-49 (IgG) detects an antigen (TCEN-49Ag) present in most cells of the central area of the body, including the pharynx (Fig. 1). Boundaries of expression of TCEN-49Ag are very sharp and located within the prepharyngeal and postpharyngeal regions.

TCEN-49 Ag is present mainly on the cell membrane of most cell types described in planarians (e.g. nerve cells, muscle cells, gastrodermal cells, parenchyma cells, cyanophilic secretory cells and other secretory cell types, neoblasts), as well as in the pericytoplasm of a few cells (Figs 2 & 3) and in the extracellular matrix. The cell type most clearly expressing this antigen in its cytoplasm seems to be the cyanophilic secretory cells (Fig. 3d).

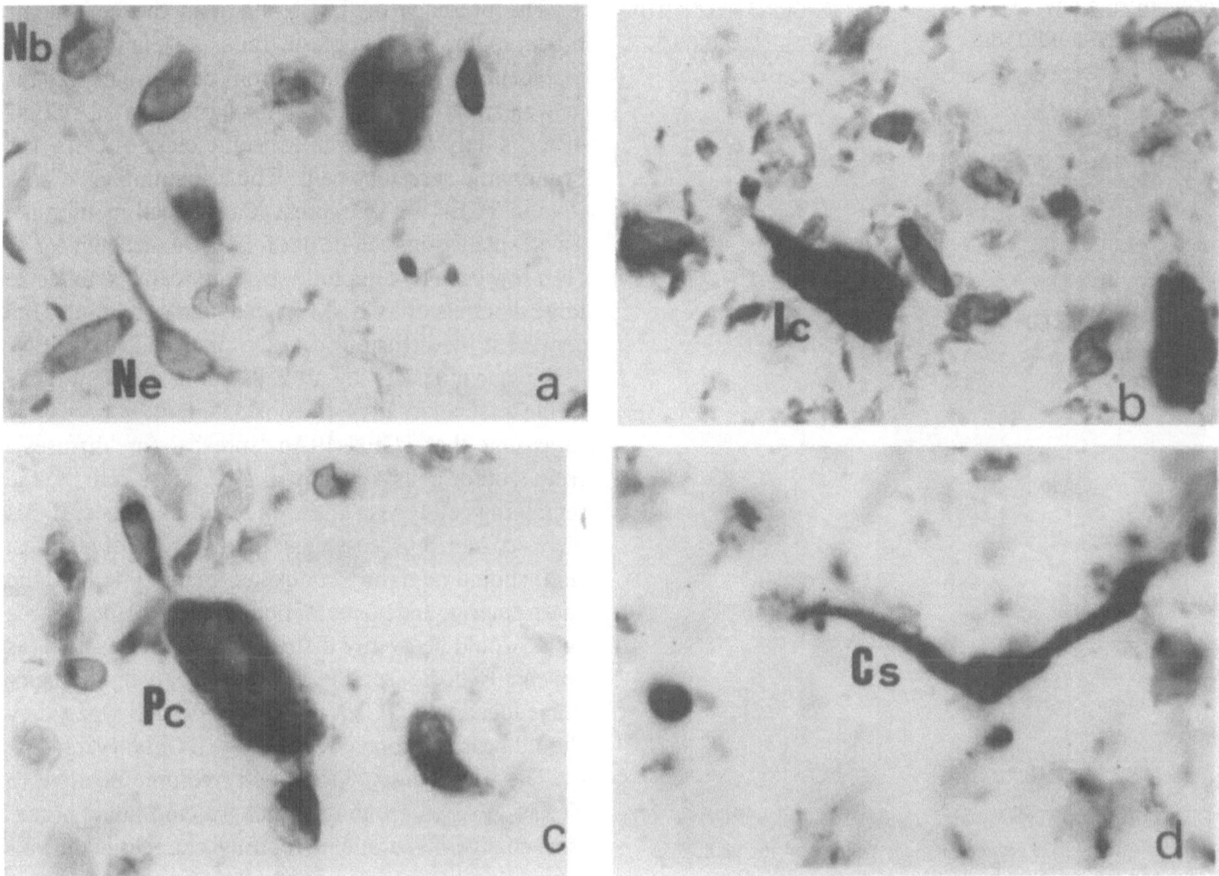

Fig. 3. Macerated cells immunostained with TCEN-49. a, b, c and d are different fields of a slide of macerated cells. Cs, cyanophilic secretory cell; Ic, intestinal cell; Nb, neoblast; Ne, nerve cell; Pc, parenchyma cell; (500×).

The body of planarians have been usually divided into five regions according to broad morphological criteria (Fig. 4): cephalic, prepharyngeal, pharyngeal, postpharyngeal and caudal (Brønsted, 1969). On the basis of the presence or absence of TCEN-49Ag immunostained with the mAb TCEN-49, three regions can be differentiated (Fig. 1 & 4): anterior, central and posterior. The percentage in volume of these regions as defined by TCEN-49Ag boundaries, decreases and increases (measured through serial sections) during growth and degrowth, respectively (Fig. 5). Thus, the posterior region has a positive allometry, whereas the anterior and the central regions have negative allometries.

During regeneration the spatial domain of TCEN-49Ag changes, often extensively, according to the level of body cut. When the cut is done in front or behind the boundaries of the central body area defined by TCEN-49Ag, the regenerating head or tail must rebuild this region. Thus, if the cut level is behind the postpharyngeal region (level E; Saló 1984) (Fig. 6), the antigen appears '*de novo*' following an antero-posterior sequence. First, it appears in the cytoplasm of a few round cells at the second day of regeneration (at 22 °C), where the pharynx is to be formed from the 3rd day on. These cells increase in number and extend several processes in the antero-posterior direction. Finally, the antigen acquires its regional distribution at the sixth/seventh day of regeneration.

When the regenerant includes some part of the central region (e.g. when the cut is made through the central region), the TCEN-49Ag boundaries shift anteriorly or posteriorly according to the level of the cut. At a prepharyngeal level, the boundary shifts anteriorly, whereas at a postpharyngeal level it shifts posteriorly (not shown). These changes occur over short periods of time, usually in less than 24 h. Later, the antigen expands following a proximo-distal sequence (as

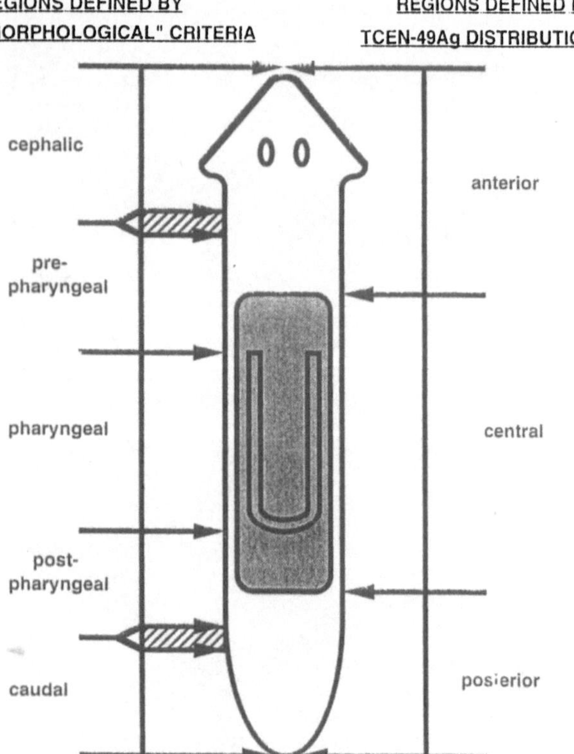

Fig. 4. Comparison between regions defined by morphological criteria and those defined by TCEN-49Ag expression. Dotted area represents the region expressing TCEN-49 Ag; arrows, sharp boundaries defined by TCEN-49Ag expression; double arrows, undistinct boundaries between regions according to morphological criteria.

referred to the regeneration blastema) until the new central region is defined.

Discussion

The antigen recognized by the mAb TCEN-49 is the first regional molecular marker so far reported in planarians. Cells expressing this antigen (TCEN-49Ag) belong to most cell types so far described (Baguñà & Romero, 1981). These cells are not related by lineage, but all are positioned in the central area of the body. This suggests that TCEN-49Ag could be involved somehow in the expression of central body positional identity. The regional expression of TCEN-49Ag has a striking parallelism to that of an homeobox containing gene in the nematode *Caenorhabditis elegans* which is also expressed in most cells of the central body region (Clark *et al.*, 1993).

The presence of TCEN-49Ag in the cell membrane and/or the cytoplasm (and presumably in the extracellular matrix) varies from cell to cell. As far as we can tell, the cytoplasmic expression of TCEN-49Ag is only detected in a specific class of cells, the cyanophilic secretory cells. The remaining cells positive to TCEN-49Ag express it at the cell membrane. Two explanations can be put forward to account for it: (1) a few cells (cyanophilic secretory cells) synthesize large amounts of TCEN-49Ag and the remaining cells acquire it by diffusion from the first; and (2) TCEN-49Ag is synthesized by all cells but at different rates. If the first alternative is considered, it is necessary to assume that TCEN-49Ag 'spreads' (or 'diffuses') from 'source cells' (cyanophilic secretory cells) to the remaining cell types *via* the extracellular space. If this were so, such a process must involve several features: (a) it should be tightly controlled spatially to hold the sharp anterior and posterior boundaries; (b) the 'source cells' would be evenly distributed within this area (as appears to occur, see Fig. 2); and (c) given the membrane location of TCEN-49Ag in most cell types, they should bear membrane receptors to TCEN-49Ag.

The percentage of total body volume positive to TCEN-49Ag in growing and degrowing intact organisms changes in parallel to reported changes in the volume of central body regions (including the pharynx) obtained by integrated serial sections using conventional histological techniques (Baguñà, 1976). In addition, anterior and posterior boundaries of the TCEN-49Ag positive region shift anteriorly during growth and posteriorly during degrowth in agreement with similar changes in equivalent areas measured by histological methods. A clue to both features is the higher rate of cell proliferation at postpharyngeal and caudal regions as compared to more anterior regions (Saló & Baguñà, 1985). This means that the boundaries of TCEN-49Ag are very sensitive to and may be a reflection of changes in positional values along the A-P axis that should occur during growth and degrowth. This makes of TCEN-49 a very useful marker to monitor changes in the positional identity of the central body regions.

The very fast changes in the spatial expression of TCEN-49Ag seen during regeneration and its relation to the level of the body cut, makes the mAb TCEN-49 a valuable marker to monitor changes in positional value and to detect the early determination of central body areas. The most pressing question to be asked is how the changes in TCEN-49Ag are brought about. To answer it, the biochemical nature and the precise

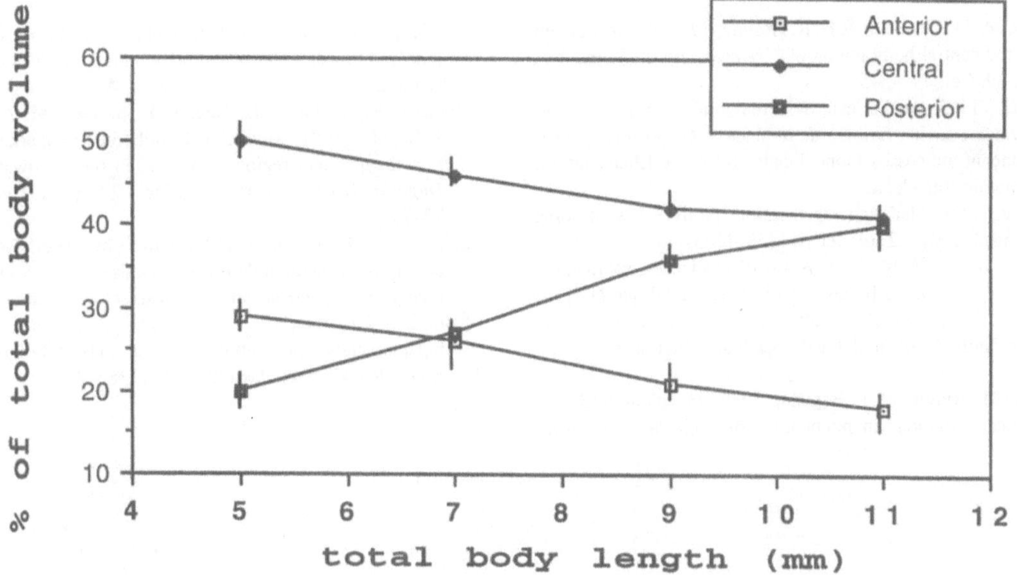

Fig. 5. Allometric changes of anterior, central and posterior regions during growth and degrowth.

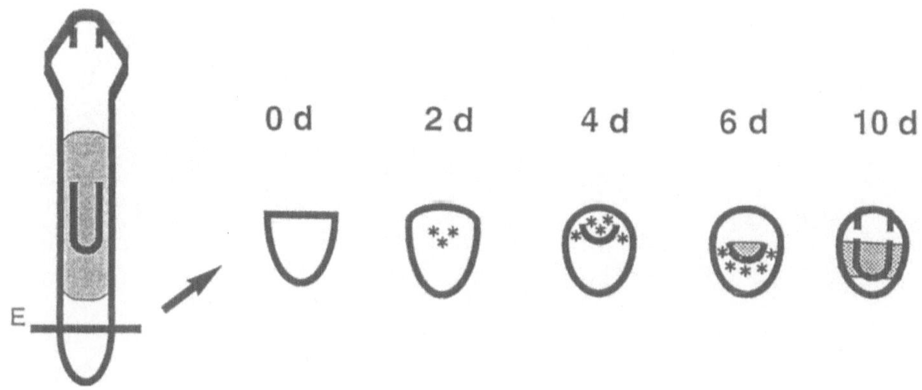

Fig. 6. Diagrammatic representation on TCEN-49 immunostaining during anterior regeneration over 10 days of caudal pieces. Dotted area represents the region expressing TCEN-49Ag; stars represent single cells expressing TCEN-49Ag.

cellular location of TCEN49Ag as well as a more thorough study, at the cell level, of its changes during regeneration are the next steps to be tackled.

Acknowledgment

This work was supported by the grant PB 89- 0249 (DGICYT, Ministerio de Educaciôn y Ciencia).

References

Baguñà, J., 1976. Mitosis in the intact and regenerating planarian *Dugesia mediterranea* n. sp. J. exp. Zool. 195: 53–64.

Baguñà, J., 1981. Planarian neoblasts. Nature 290: 14–15.

Baguñà, J. & R. Romero, 1981. Quantitative analysis of cell types during growth, degrowth and regeneration in the planarians *Dugesia (S) mediterrania* and *Dugesia (G.) tigrina*. Hydrobiologia 84: 181–194.

Baguñà, J., E. Saló & M. C. Auladell, 1989. Regeneration and pattern formation in planarians. III. Evidence that neoblast are totipotent stem cells and the source of blastema cells. Development 107: 77–86.

Baguñà, J., R. Romero, E. Saló, J. Collet, M. C. Auladell, M. Ribas, M. Riutort, J. Garcia-Fernandez, F. Burgaya & D. Bueno, 1990. Growth, degrowth and regeneration as developmental phenomena in adult fresh-water planarians. In H. J. Marthy (ed.), Experimental Embryology of Aquatic Plant and Animal Organisms. NATO-ASI Series, Plenum Press. New York: 129–162.

Brønsted, H. V., 1969. Planarian Regeneration. Pergamon Press, London.

Clark, S. G., A. D. Chisholm & H. R. Horvitz, 1993. Control of cell fates in the central body region of *C. elegans* by the homeobox gene lin-39. Cell 74: 43–55.

Espinosa, Ll., 1993. Anàlisi espacial i temporal de la regeneració de la faringe i cavitat faringia de la planària *Dugesia (G.) tigrina* mitjançant anticossos monoclonals. Tesina de Llicenciatura, Universitat de Barcelona.

Gremigni, V., 1988. Planarian regeneration: an overview of some cellular mechanisms. Zool. Sci. 5: 1153–1163.

Harlow, E. & D. Lane (eds), 1988. Antibodies: a laboratory manual. Cold Spring Harbour Laboratory, Cold Spring Harbour, NY.

Kohler, G. & C. Milstein, 1975. Continuous cultures of fused cells secreting antibody of predefined specificity. Nature 256: 495–497.

Ribas, M., M. Riutort & J. Baguñà, 1989. Morphological and biochemical variations in populations of *Dugesia (G.) tigrina* (Turbellaria, Tricladida, Paludicola) from the western mediterranean: biogeographical and taxonomical implications. J. Zool., Lond. 218: 609–626.

Romero, R., F. Fibla, D. Bueno, L. Sumoy, M. A. Soriano & J. Baguñà, 1991. Monoclonal antibodies as markers of specific cell types and regional antigens in the freshwater planarian *Dugesia (G.) tigrina*. Hydrobiologia 227 (Dev. Hydrobiol. 69): 73–79.

Saló, E., 1984. Formació del blastema i re-especificació del patró durant la regeneració de les planàries *Dugesia (S.) mediterrània* i *Dugesia (G.) tigrina*. Ph. D. Thesis, Universitat de Barcelona.

Saló,, E. & J. Baguñà, 1985. Proximal and distal transformation during intercalary regeneration in the planarian *Dugesia (S.) mediterranea*. Roux's Arch. Dev. Biol. 194: 364–386.

Hydrobiologia **305**: 241–246, 1995.
L.R.G. Cannon (ed.), *Biology of Turbellaria and some Related Flatworms*.

Localization of the antigen for the monoclonal antibody ② 9F11-B-E_4 and interspecific cross-reaction in the freshwater triclads

Sachiko Ishida, Mikiko Iida, Hiromi Takahashi & Yasuhiko Sato
Department of Biology, Faculty of Science, Hirosaki University, Hirosaki 036, Japan

Key words: monoclonal antibody, antigen, testis, copulatory apparatus, interspecific cross-reaction

Abstract

The localization of the antigen for monoclonal antibody ② 9F11-B-E_4 was clarified by immuno-electron microscopy. The antigens were localized on the mitochondria and Golgi bodies in the male germ cells and on the secretory granules of various gland cells in the penis bulb and subepidermal parenchymal tissue of *Phagocata vivida*. The results of the interspecific cross-reaction tests with seven other freshwater triclads showed that these secretory granules are species-specific. A positive interspecific reaction was showed with *Dugesia* (family Dugesiidae), but not with *Polycelis* within the same family Planariidae which suggests the position of *Phagocata* within the Planariidae needs to be reassesed.

Introduction

Monoclonal antibody (mAb) specific to male germ cells and epidermis of *Phagocata vivida* (Planariidae) has been produced (Shirakawa *et al.*, 1991), but localization of the specific antigen has not been clarified yet. We identified the localization of this antigen by immuno-electron microscopy.

Furthermore, we examined the interspecific cross-reaction of this monoclonal antibody with seven other freshwater triclads.

Materials and methods

Animals. Specimens of triclads were collected by hand from freshwater streams in Japan: *Phagocata vivida* (Ijima & Kaburaki) and *Polycelis (Seidlia) auriculata* Ijima & Kaburaki (Mt. Iwaki, near Hirosaki), *Dugesia japonica* Ichikawa & Kawakatsu, *Phagocata teshirogii* Ichikawa & Kawakatsu, *Bdellocephala brunnea* Ijima & Kaburaki (Hirosaki), *Phagocata kawakatsui* (Nagano), *Polycelis sapporo* (Ijima & Kaburaki) (Aomori), *Dendrocoelopsis lactea* Ichikawa & Okugawa (Kuroishi).

The monoclonal antibody was produced by injecting separate cells of *Ph. vivida* as antigens into mice and screened by immuno-histostaining using the ABC (avidin-biotin peroxidase complex) method (Shirakawa *et al.*, 1991). *Ph. vivida* was used to survey the localization of specific antigen and seven other species were used to test the interspecific cross-reactions.

Immunohistochemistry. More than 10 specimens of *Ph. vivida* were fixed with 4% formalin in one-quarter strength PBS overnight at 4 °C. The interspecific cross-reaction tests were examined by the ABC method (Hsu *et al.*, 1981). The monoclonal antibody (② 9F11-B-E_4) produced already by Shirakawa *et al.* (1991) was used as the primary antibody.

Immuno-electron microscopy. To clarify the localization of the specific antigen, we used pre-embedding staining (Nakane & Pierce, 1966). Specimens were fixed in 4% PLP fixative (McLean & Nakane, 1974) for 16 h at 4 °C. After being washed with cold one-quarter strength PBS, they were frozen with hexane and sectioned 6 μm. The immunoreaction was carried out according to the ABC method. Then the samples were post-fixed with 2% osmic acid in phosphate buffer for 2 h, dehydrated in graded ethanols and embedded in epoxy resin. Ultrathin sections were stained for 1 min with lead citrate.

Post-embedding staining using colloidal gold followed Faulk & Taylor (1971). Specimens were fixed with 4% PLP fixative or 2.5% glutalaldehyde in 0.1 M phosphate buffer for 2 h at 4 °C. After being washed in buffer, they were dehydrated in graded ethanols and embedded in LR-White. Thin sections on nickel grids were rinsed with PBS and 1% BSA for 30 min. The sections were incubated for 3–4 h with primary antibody, washed with PBS, then incubated for 2 h with gold-labelled goat anti-mouse antibody IgG+M 15 nm (Zymed) diluted 1:20 in 0.1% BSA-PBS. The sections washed with PBS and distilled water were finally stained with uranyl acetate and lead citrate and observed with a Hitachi H600 electron microscope.

In negative controls, the primary antibody was replaced by mouse myeloma ascites diluted 1:1000 in PBS.

Results

Though Shirakawa *et al.* (1991) reported that this antibody was specific to male germ cells and epidermal cells, we found that the cells in the penis and subepidermal parenchymal tissue also showed a positive reaction. The copulatory bursa, bursa stalk and vagina often showed a positive reaction due to the sperm mass stored inside following copulation, but this reaction was not consistently seen. First, we tried to examine the immunocytochemical localization of the antigen in male germ cells. The positive reaction seen in the cytoplasm suggested that the antigens were distributed in the cytoplasm, not in the nucleus. Fig. 1 shows immuno-electron micrographs prepared by the pre-embedding staining method. An early spermatid in the testis shows the nucleus is surrounded by many mitochondria, and a Golgi body is observed below the mitochondria (Fig. 1A). Spermatids in spermiogenesis are seen in Figs. 1B and 1C. The mitochondria, Golgi body and secretory vacuoles originating from the Golgi body showed a positive reaction. Secretory vacuoles were seen adhering to mitochondria of the spermatid (Fig. 1B). Spermatozoa in a seminiferous tubule are seen in Figs. 1D and 1E. Mitochondria showed a positive reaction. These results show that the antigens in the male germ cells localize on the mitochondria and Golgi bodies.

The post-embedding staining method showed colloidal golds localized on the secretory granules in gland cells of the penis bulb. We calculated the non-specific adhesion of gold was from 0 to 2 gold particles per granule (Fig. 2F). There were various gland cells, i.e., with the secretory granules having a core, with the secretory granules of medium electron density and with the secretory granules of high electron density, in the parenchymal tissue of the penis bulb. Many colloidal gold particles were observed on the granules of medium electron density (Fig. 2A,B,E) and on the granules having a core of various electron densities (Fig. 2C,D). But no gold particles were found on the cores of high electron density (Fig. 2D). Some granules contained fewer gold particles than their neighbors (Fig. 2B). Few gold particles were found on the granules of high electron density (Fig. 2E). We could find no gold particles on the mitochondria in these gland cells and other cells such as muscle cells (Fig. 2A,C).

Furthermore, we found that the cells which showed a positive reaction in the sub-epidermal parenchymal tissue were the same gland cells as we found abundantly in the penis bulb. In other words, the distribution of the gland cells which reacted with this antibody were revealed also. Though these gland cells were abundant in the penis bulb, they were scattered in the sub-epidermal parenchymal tissue.

The antibody reaction with the male germ cells and secretory granules of other species is summarized in Table 1. Two species of genus *Phagocata* and *Dugesia japonica* showed a positive reaction in only the male germ cells. In these species, the seminiferous tubule, ejaculatory duct and copulatory bursa often showed a positive reaction due to the sperm mass going through the ducts or stored inside the copulatory bursa following copulation. Table 1 shows that the positive reactions in the secretory granules and epidermis were specific to *Ph. vivida*.

Figure 4 shows schematically the cross section of the penis papilla of *Ph. vivida* based on light and electron microscopical observations. The wall of the penis papilla consists of six or seven layers: 1, a layer of outer epithelium with microvilli lies adjacent to the male genital antrum; 2, an outer circular muscle layer; 3, an outer longitudinal muscle layer; 4, then follows parenchymal tissue mixed with several kinds of gland cells, nerve cells, muscle cells and parenchymal cells; 5, an inner longitudinal muscle layer (ill-defined); 6, an inner circular muscle layer; and 7, an inner epithelial layer with cilia and microvilli. This revealed the gland cells with antibody-positive secretory granules reached the ejaculatory duct.

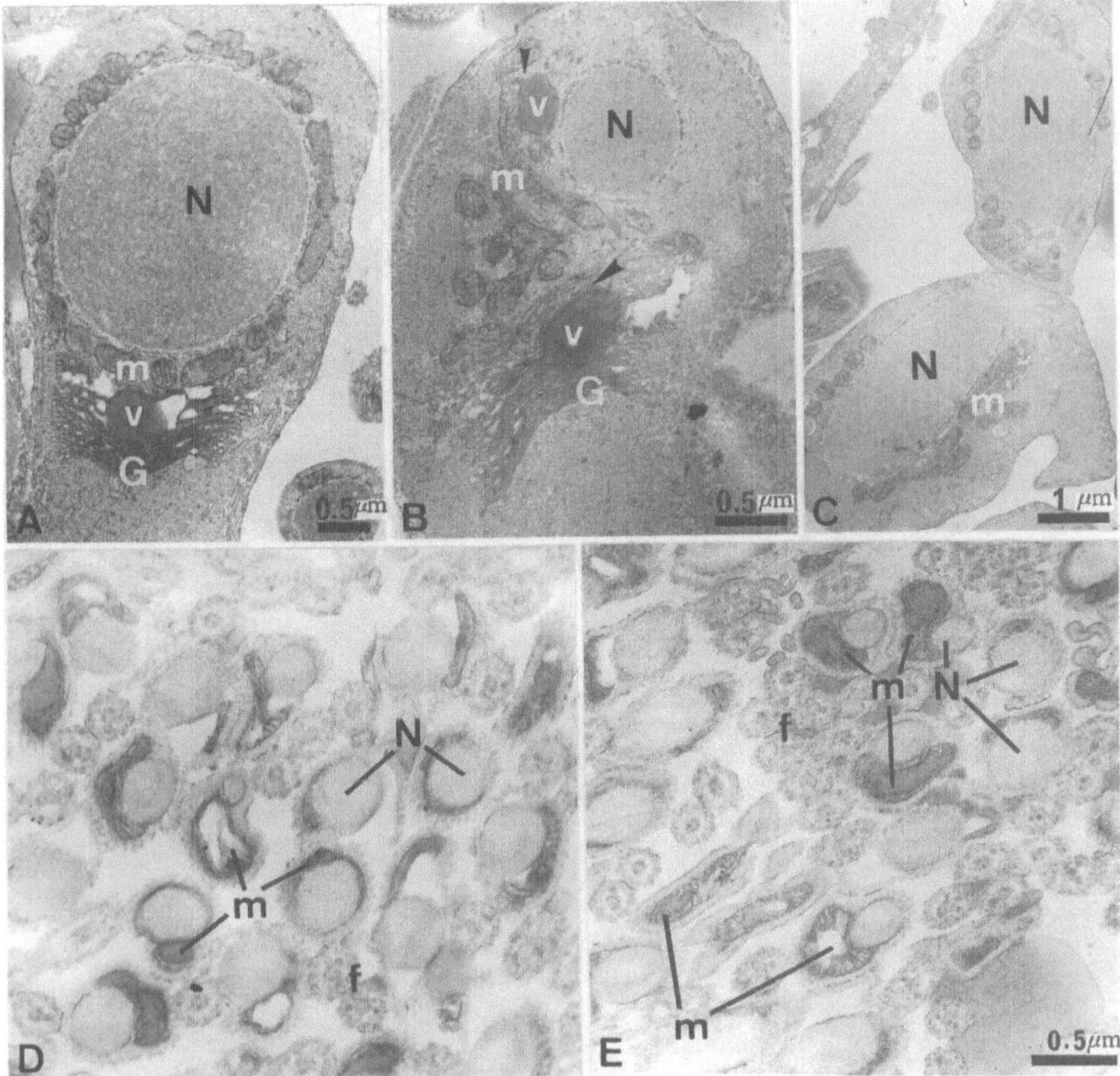

Fig. 1. Immuno-electron micrographs produced by the ABC method of spermatids in the testis (A-C) and spermatozoa in a seminiferous tubule (D-E). Mitochondria (m), Golgi body (G) and secretory vacuoles (v) originating from the Golgi body are stained. Arrows show secretory vacuoles adhering to mitochondria; N, nucleus. D and E at the same magnification.

Discussion

In addition to the observation by Shirakawa *et al.* (1991), that this antibody was specific to male germ cells and epidermal cells of *Ph. vivida*, we found that the secretory granules in the gland cells distributed in the sub-epidermal parenchymal tissue and penis also showed a positive reaction. Although the copulatory bursa, bursa stalk and vagina often showed positive reactions, we consider the antigen was not localized there, but the reaction was due to the bundles of spermatozoa and the secretion derived from the penis gland cells of the partner following copulation. These results show that male genital organs (testis and penis) have the antigens for this mAb, but female genital organs (ovary, copulatory bursa, bursa stalk and vagina) do not.

In the spermatozoa of freshwater triclads, an acrosomal vesicle similar to that found in other metazoan spermatozoa has not been observed yet (see Ishida

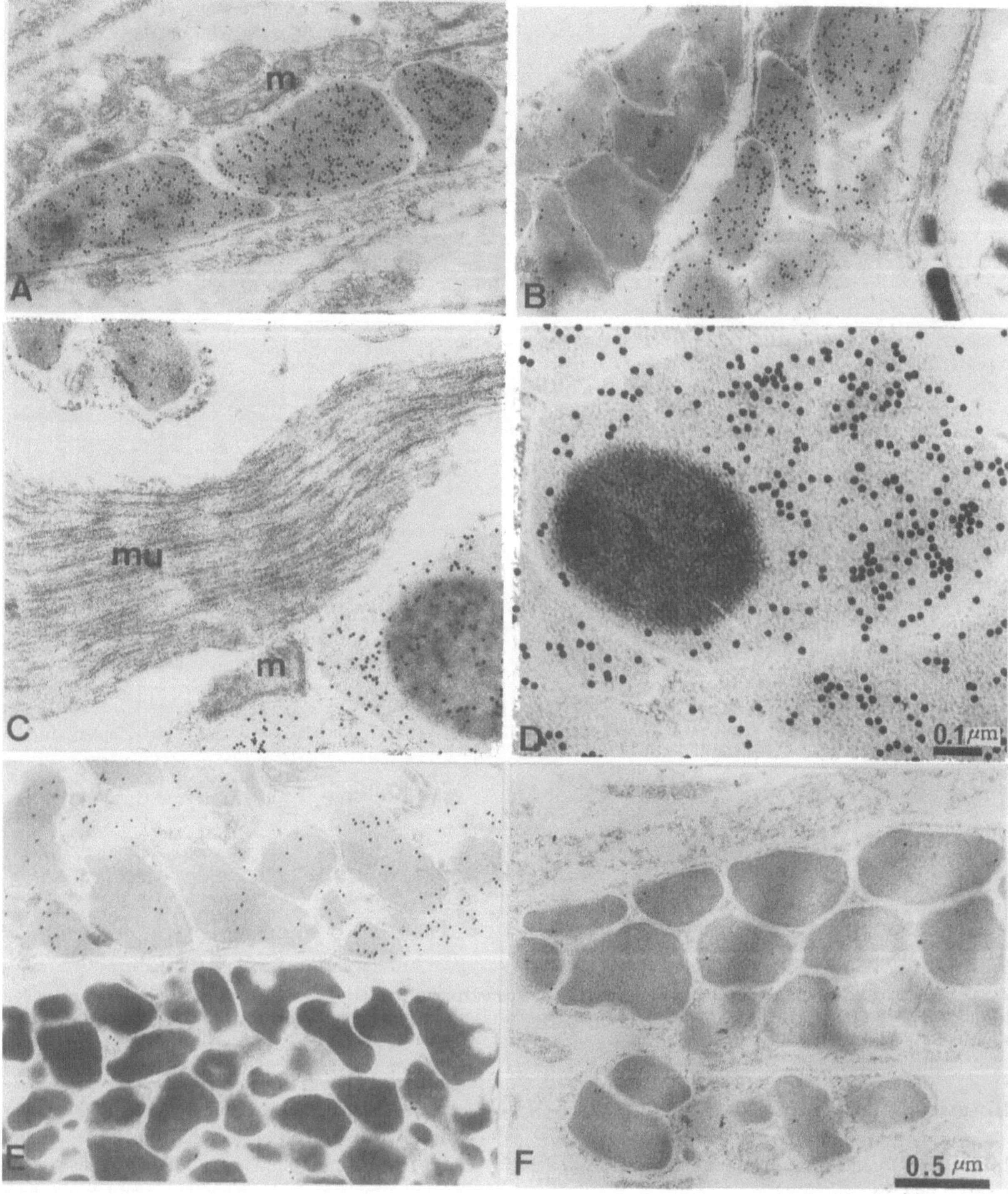

Fig. 2. Immuno-electron micrographs by colloidal gold labeling. The secretory granules in the penis gland cells show positive reactions (A-E). Secretory granules (E) and core (D) of high electron density show no reaction. Mitochondria (m) in the gland cells and muscle cells (mu) also show no reaction. F is a negative control. A, B, C, D, E, and F at the same magnification.

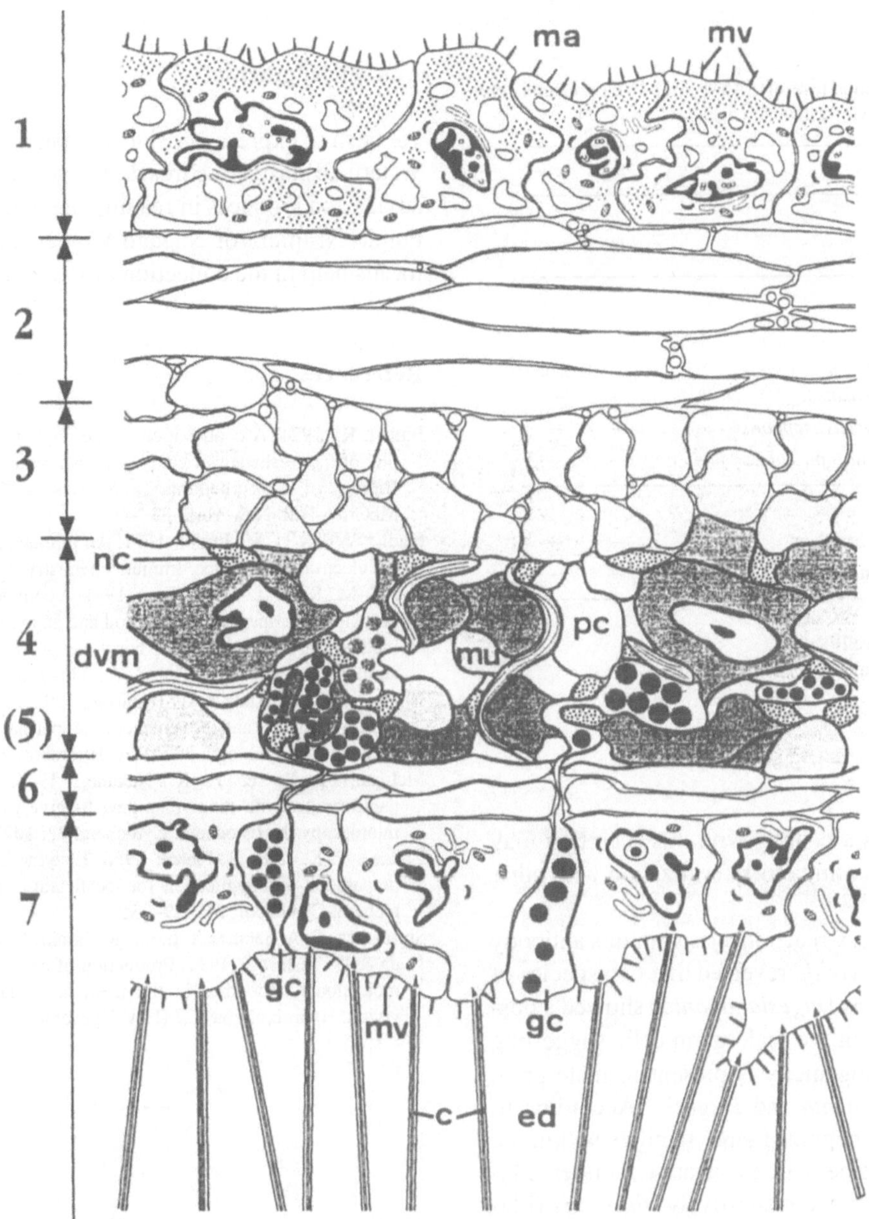

Fig. 3. Scheme showing the cross section of the penis papilla of *Ph. vivida*. The wall of the penis papilla consists of six or seven layers (1-7). 1, outer epithelium layer with microvilli (mv); 2, outer circular muscle layer; 3, outer longitudinal muscle layer; 4, parenchymal tissue mixed with several kinds of gland cells (gc), nerve cells (nc), muscle cells (mu) and parenchymal cells (pc); 5, inner longitudinal muscle layer (ill-defined); 6, inner circular muscle layer; 7, inner epithelial layer with cilia (c) and microvilli (mv). dvm, dorsoventral muscle; ed, ejaculatory duct; ma, male genital antrum.

et al., 1991). It is still not clear what role the developed Golgi bodies in spermatids perform. In the male germ cells, mitochondria and Golgi bodies showed a positive reaction and the secretory vacuoles often adhered to mitochondria. Perhaps the antibody-positive secretions originating from Golgi bodies are transported to mitochondria.

In the penis, the antigens were localized on the secretory granules in some gland cells, but not on the secretory granules of high electron density and cores of high electron density. Since antibody-positive gland cells in the penis discharge to the ejaculatory duct, the secretion would coat the passing spermatozoa. The antigenicity of these secretory granules was specific

Table 1. Cross reaction tests of mAb ② 9F11-B-E_4 prepared against *Ph. vivida* for freshwater triclads.

Species	m.g.c.	s.g.	ep.
Dugesiidae			
Dugesia japonica	+	–	–
Planariidae			
Phagocata vivida	+	+	+
Phagocata teshirogii	+	–	–
Phagocata kawakatsui	+	–	–
Polycelis (Polycelis) sapporo	–	–	–
Polycelis (Seidlia) auriculata	–	–	–
Dendrocoelidae			
Bdellocephala brunnea	–	–	–
Dendrocoelopsis lactea	–	–	–

ep., epidermis
m.g.c., male germ cell
s.g., secretory granule

to *Ph. vivida*, thus a coating with this secretion may play a role in preventing cross-fertilization with other species.

Interspecific cross-reaction tests with this antibody produced from *Ph. vivida* revealed that two species of *Phagocata* as well as *Dugesia japonica* showed a positive reaction only in the male germ cells suggesting that a common antigenicity is present in male germ cells of both *Phagocata* and *Dugesia*. According to Ball (1974) who recognized three families within the Paludicola, these genera are in separate families, Planariidae and Dugesiidae respectively. *Polycelis* (Planariidae) showed no reaction, however. We think it is necessary to examine further the other species of the family Dugesiidae and their phylogenetic relationships.

Acknowledgments

We wish to express our sincere thanks to President Dr Wataru Teshirogi of Hirosaki University for his critical advice and kindness in reading the manuscript and Dr Fumio Ni-imura of Nagano Women's Junior College for his help in the collection of *Ph. kawakatsui*.

References

Ball, I. R., 1974. A contribution to the phylogeny and biogeography of the freshwater triclads (Platyhelminthes: Turbellaria). In Biology of the Turbellaria. N. W. Riser & M. P. Morse (eds), McGraw-Hill, New York: 339–401.

Faulk, W. P. & G. M. Taylor, 1971. An immunocolloid method for the electron microscope. Immunochemistry 8: 1081–1083.

Hsu, S.-M., Raine, L. & H. Fanger, 1981. A comparative study of the peroxidase-antiperoxidase method and an avidin-biotin complex method for studying peptide hormones with radioimmunoassay antibodies. Am. J. Clin. Pathol. 75: 734–738.

Ishida, S., Y. Yamashita & W. Teshirogi, 1991. Analytical studies of the ultrastructure and movement of the spermatozoa of freshwater triclads. Hydrobiologia 227 (Dev. Hydrobiol. 69): 95–104.

McLean, I. W. & P. K. Nakane, 1974. Periodate-lysine-paraformaldehyde fixative: a new fixative for immunoelectron microscopy. J. Histochem. Cytochem. 22: 1077–1083.

Nakane, P. K. & G. B. Pierce, 1966. Enzyme-labeled antibodies: preparation and application for localization of antigens. J. Histochem. Cytochem. 14: 929–931.

Shirakawa, T., A. Sakurai, T. Inoue, K. Sasaki, Y. Nishimura, S. Ishida & W. Teshirogi, 1991. Production of cell- and tissue-specific monoclonal antibodies in the freshwater planarian Phagocata vivida. Hydrobiologia 227 (Dev. Hydrobiol. 69): 81–91.

Hydrobiologia **305**: 247–253, 1995.
L.R.G. Cannon (ed.), *Biology of Turbellaria and some Related Flatworms.*

Regulation factor for planarian regeneration

Takao Shinozawa*, Syuichi Shiozaki, Masanobu Ezaki, Hideki Fujino, Takeshi Tanaka & Toshihiko Saheki
*Department of Biological and Chemical Engineering, Faculty of Engineering, Gunma University, Kiryu, Gunma, 376, Japan (*author for correspondence)*

Key words: planarian, regeneration, regeneration factor

Abstract

Planarian head extract was fractionated and the fractions were assayed for their effect on cultured cells and planarian regeneration. One fraction (molecular weight larger than 10 000; unadsorbable by DEAE-Sephadex, CM-Sephadex and Con A-Sepharose; and precipitable by ammonium sulfate) inhibited the growth of both Neuro 2a and PC-12 cell lines as well as planarian head- regeneration. This effect was specific for head-regeneration (regeneration of tails was not influenced), trypsin sensitive, reversible and stable after heat-treatment at 80°C for 30 min.

Introduction

Planarians are interesting and unique animals with respect to their extremely high regenerability. The process of regeneration includes several fundamental problems of developmental biology. The determination of the number and position of heads during the regeneration is one of those problems.

Since 1977 (Lender, 1955, 1956, 1960; Steele & Lange, 1977), no progress has been made concerning this problem, especially the kind of substance(s) participating in this process. An attempt has been made to detect and characterize a regulation factor for planarian regeneration by assaying the effect of a planarian head extract on cultured cells as well as on planarian regeneration. We isolated a fraction that could inhibit planarian head-regeneration as well as the growth of Neuro 2a and PC-12 cells. This inhibition was reversible. The activity was heat stable and trypsin sensitive.

Materials and methods

Reagents: 3-(4,5-Dimethyl-2-thiazolyl)-2,5-diphenyl-2H-tetrazolium bromide (MTT) was obtained from Sigma. Eagle's Minimum Essential Medium with Hank's Salt (MEMH) was from Hazelton. RPMI 1640 medium was from JHR Biosciences. Foetal bovine serum (FBS) and Horse serum were from Gemini Bio-Products, Inc. Non-Essential Amino Acid Mixture (NEAAM) was obtained from BIO-PRODUCTS, INC. Cat. No. 13-11 4A Lot# 0M0098. *Dulbecco's phosphate-buffered saline (PBS):* PBS contained 137 mM NaCl, 3 mM KCl, 8 mM Na_2HPO_4 and 1.5 mM KH_2PO_4 (pH 7.4). *Dialysis tubes and resins:* Dialysis tubes with a molecular weight cut off of 1000 (Spectrum No. 530–10161) or 10 000 (Spectrum No. 536–17081) were obtained from Spectrum Medical Industries Inc. Diethylaminoethyl (DEAE)-Sephadex A-50 caroboxymethyl (CM)-Sephadex C-50, concanavalin (Con A)-Sepharose 4B and Phenyl-Sepharose were purchased from Pharmacia.

Planarians

Specimens of an asexual strain of *Dugesia japonica japonica* were collected from Kiryu river near Gunma University and were selected for normal morphology (average length, 2.1 cm). Planarians were maintained in dechlorinated water (boiled city water: 'breeding water') and fed once a week on fresh raw chicken liver. Before use in the experiments, they were kept fasting for more than three weeks.

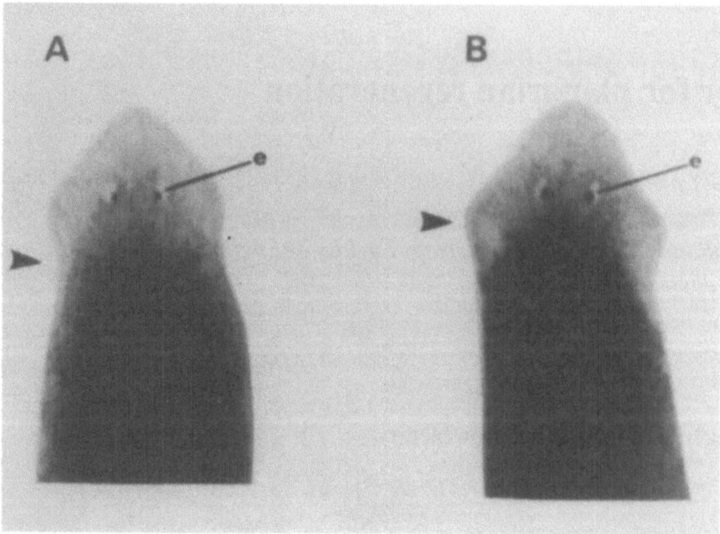

Fig. 1. Effect of the addition of crude homogenate on planarian head-regeneration. The decapitated planarians were kept at 16°C for 14 days in the absence (A) or in the presence (B) of the crude homogenate (final protein content: 80 μg ml^{-1}). Arrow head shows the position of the decapitation; *e*: eye

Preparation and fractionation of the planarian homogenates

The anterior 2–3 mm (heads, 2.5 g wet weight) was cut from about two thousand planarians with a razor blade; these were homogenized in 5 ml PBS using a Teflon-glass homogenizer, and then the homogenate was sonicated (30 sec × 3 times. Tomy Seiko Co. LTD; TOMY UR-200P). This suspension was centrifuged at 12000 × ***g*** for 30 min and the supernatant was used as a 'crude (unfractionated) homogenate'. The crude homogenate was dialyzed against PBS (100 times the volume of the homogenate) three times, each time 12 h, using a dialyse tube which cuts off the molecules with Mw less than 10000. The outside solution of this dialysis was applied to a DEAE-Sephadex column equilibrated with PBS and the adsorbed proteins were eluted with PBS containing 2.0 M NaCl. The eluted solution was dialyzed against PBS using a cut-off 1000 dialysis tube. The inside solution of this dialysis was used as a fraction of 'molecular weight (Mw) less than 10000 and DEAE-Sephadex bound'. The inside solution (Mw>10000) of the first dialysis was treated similarly and used as a fraction of 'Mw larger than 10000 and DEAE-Sephadex bound'. The solution (Mw >10000) which passed through the DEAE-Sephadex column was applied to a CM-Sephadex column equilibrated with PBS.

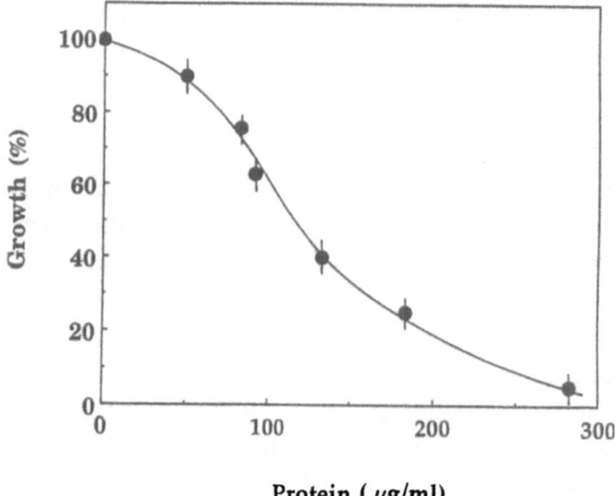

Fig. 2. Dose dependence of the Neuro 2a cell-growth with respect to the ASP fraction. Various amounts of ASP fraction were added to 1×10^4 cells of the line Neuro 2a and the cell-growth was measured after 72 h. Vertical bars indicate standard deviations in 6 wells assays.

Then, the fraction with 'Mw larger than 10000 and CM-Sephadex bound' was prepared in the same way as the 'Mw larger than 10000 and DEAE-Sephadex bound' fraction.

The solution which passed through the CM-Sephadex column was supplemented with 1 mM $MnCl_2$ and 1 mM $CaCl_2$ and applied to a Con A-Sepharose column equilibrated with PBS containing 1 mM $MnCl_2$ and 1 mM $CaCl_2$. The proteins which

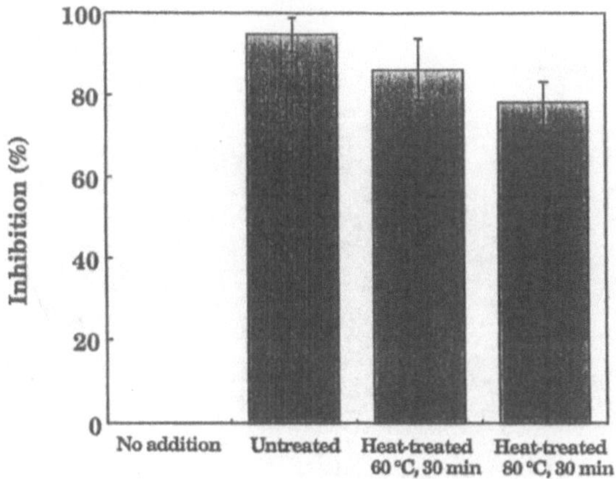

Fig. 3. Effect of heat-treatment on the inhibitory activity of the ASP fraction. Thirty μg of non-treated or heat-treated (60° C for 30 min or 80° C for 30 min) the ASP fraction were added to 10^4 Neuro 2a cells and the cell-growth measured. Vertical bars indicate standard deviations in 6 wells assay.

bound to the Con A-Sepharose were eluted with PBS containing 1 mM $MnCl_2$, 1 mM $CaCl_2$ and 100 mM α-Methyl-D-mannoside. The eluted proteins were dialyzed using a dialysis tube of Mw cut-off 1000 and the inside solution was used as 'Mw larger than 10 000 and Con A-Sepharose bound fraction'. The solution which passed through the Con A-Sepharose column was supplemented with 40% (1.7 M final concentration) of ammonium sulfate. The precipitated proteins were obtained by centrifugation at 12 000 × ***g*** for 30 min. After the dialysis against PBS using a dialysis tube of Mw cut-off 1000, the precipitated proteins were used as 'Ammonium Sulfate Precipitated Fraction (ASP fraction)'.

The ammonium sulfate supernatant fraction was applied to a Phenyl- Sepharose column equilibrated with PBS containing 40% (1.7 M) ammonium sulfate. The solution which passed through Phenyl- Sepharose was dialyzed against PBS using a dialysis tube of Mw cut-off 1000. The inside solution of this dialysis was used as 'Mw larger than 10 000 and Phenyl Sepharose unbound fraction'. The protein which bound to Phenyl-Sepharose was eluted with PBS containing 6.0 M urea. By a dialysis as described above, a fraction of 'Mw larger than 10 000 and Phenyl-Sepharose bound' was obtained. All the preparation procedures were carried out under 4°C. For the assay of cell-growth, all the solutions or fractions were sterilized by passing them through filters (pore size 0.2 μm, Corning).

Cell culture and cell-growth assay

Neuro 2a cells (from Dr T. Horiuchi, Institute for Molecular and Cellular Regulation, Gunma University) were cultured in MEMH supplemented with 10% (V/V) FBS and 1% (V/V) non-essential amino acid mixture (NEAAM). PC-12 cells (from Dr K. Inoue, Institute for Molecular and Cellular Regulation, Gunma University) were cultured in RPMI 1640 medium supplemented with 5% (V/V) FBS and 10% (V/V) Horse serum (HS). Incubation was carried out at 37°C, 5% CO_2 and 90% humidity. For the cell-growth assay, 100 μl of the culture medium containing 1×10^4 Neuro 2a cells or 5×10^3 PC-12 cells were added to each well of a 96 wells culture dish. After 24 h cultivation, 20μl of testing sample, or PBS as a control, were added to each well. The Neuro 2a or PC-12 cells were further incubated for 72 h or 48 h, respectively. The cell-growth was measured by the MTT method described by Mosmann (1983), measuring the $OD_{570-630}$ using the plate reader (Toso, Type MPR-A4).

Assay of planarian regeneration

For the regeneration assay, the above breeding water was autoclaved. For the head-regeneration assay, worms were decapitated at a level midway between head and pharynx (separating an anterior piece about 5 mm in length) with a razor blade. With the tail-regeneration assay, about 5 mm of the tail was cut off. The decapitated or tailless worms were kept in autoclaved breeding water overnight. For each experimental condition, twelve decapitated or tailless worms were selected.The regeneration of each worm was observed in one ml breeding water either supplemented or not supplemented with 20μl of each testing sample. Each worm was kept in a well with a diameter of 1.6 cm. The dishes were maintained at 17°C and solutions changed every two days.

Determination of protein

The amount of protein was determined by Lowry's method (Lowry *et al.*, 1951) using bovine serum albumin (BSA) as a standard. Wherever reported, amount and concentrations of fractions refer to protein content determined in this way.

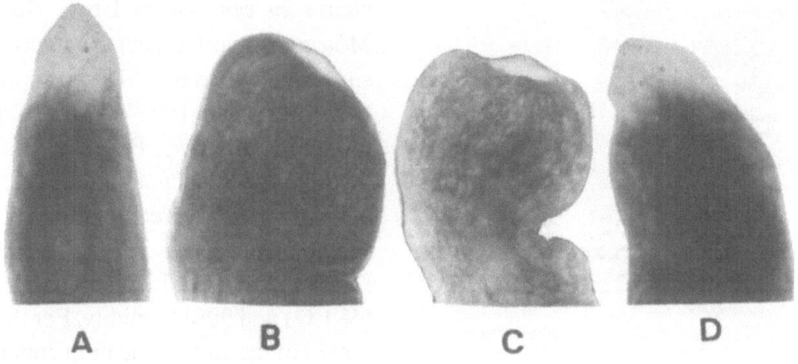

Fig. 4. Effect of the addition of ASP fraction on the planarian head-regeneration. Various amounts (A: no addition, B: 30 $\mu g\ ml^{-1}$, C: 15 $\mu g\ ml^{-1}$, D: 10 $\mu g\ ml^{-1}$) of the ASP fraction were added to the decapitated planarians. The pictures show the regeneration achieved by the 9th day. Of 12 decapitated planarians used with each concentration at least 8 specimens regenerated to the respective stage above.

Trypsin treatment of the ASP fraction

The ASP fraction (490 $\mu g\ ml^{-1}$) was treated with 200 $\mu g\ ml^{-1}$ of trypsin at 37°C for 60 min. This solution was heated at 80°C for 20 min to denature the supplemented trypsin. The trypsin solution itself (200 $\mu g\ ml^{-1}$) was also heat-treated (at 80 °C for 20 min) as a control.

Results

Effect of the crude homogenate on planarian regeneration

The addition of the crude planarian head homogenate, prepared as described, to the decapitated planarians resulted in deformed (short-necked) heads (Fig. 1). Observation for more than 14 days was necessary: of 12 decapitated planarians, six survived and regenerated to deformed planarians. Normal regeneration was observed in every planarian in the absence of homogenate.

Inhibition of the growth of cultured cells by the fractionated planarian homogenate

To speed up the experiment and to intensify the effect(s) by concentrating the active substance(s), fractionation of the planarian head homogenate was combined with a test assay using cultured cells. The cell lines Neuro 2a and PC-12, having characteristics of neural cells, were used. This idea originated from the fact that the neural regeneration (or development) precedes the other organ's regeneration (Reuter, 1988; Kurabuchi & Kishida, 1988; Palmberg, 1991; Reuter & Eriksson, 1991). As shown in Table 1, there were two fractions which inhibited the cell-growth of Neuro

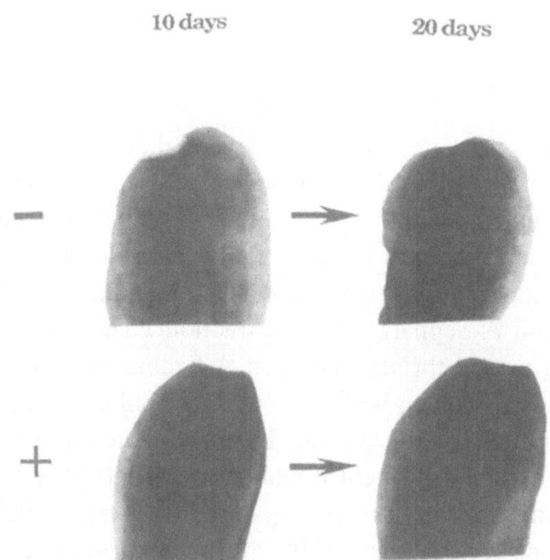

Fig. 5. Effect of the non-treated and heat-treated ASP fractions on planarian head-regeneration. Forty $\mu g\ ml^{-1}$ of non- treated (–) or heat treated (+, 80°C for 30 min) ASP fraction were added to the decapitated planarians (12 for '–' and '+' each). The regeneration was observed for 20 days. At least 8 of 12 decapitated specimens survived under each condition and behaved as shown in the respective photograph.

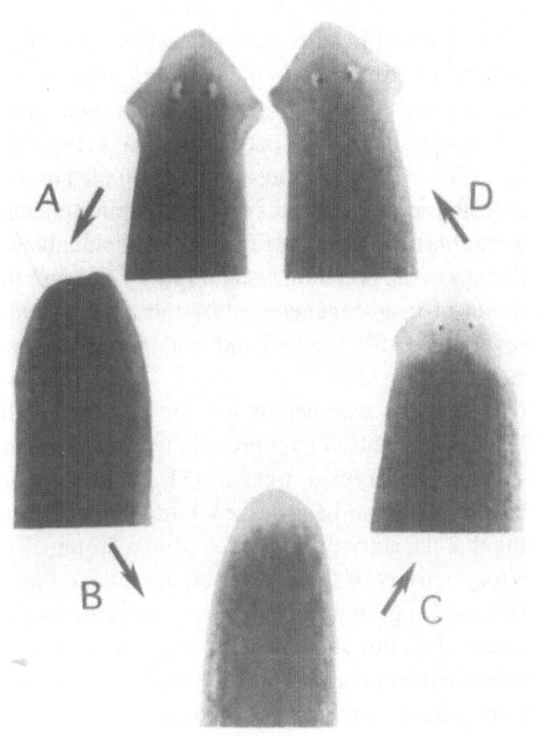

Fig. 6. Reversibility of the inhibition of planarian head-regeneration by the ASP fraction. Planarians were decapitated and incubated for 20 days in the presence of 40 μg ml^{-1} of the ASP fraction (A). Then, the specimens were returned to the culture water (which did not contain the ASP fraction) and incubated for 7 days (B), 17 days (C) and 27 days (D). Of 12 decapitated planarians 10 regenerated.

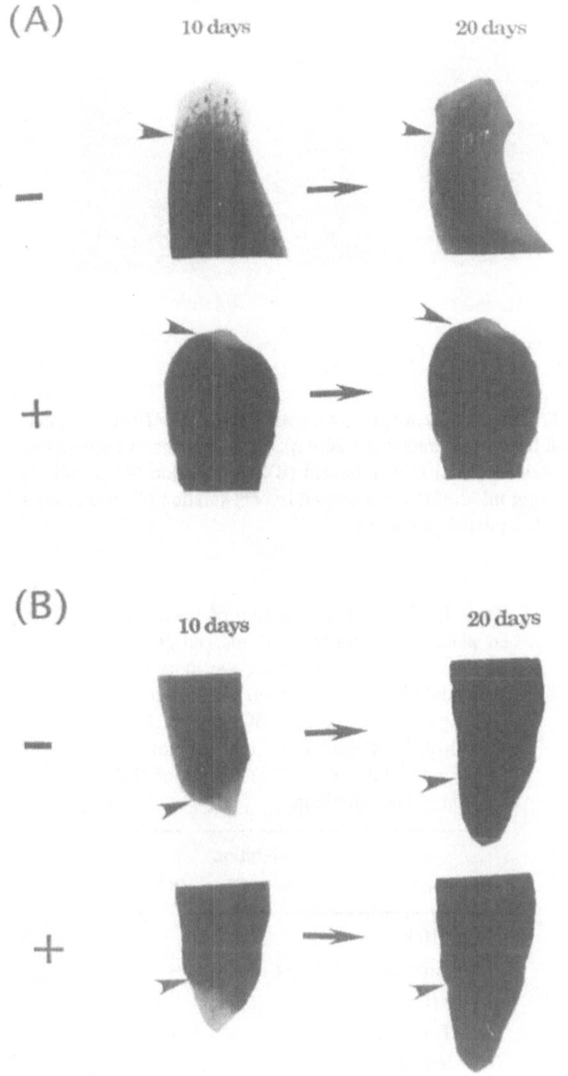

(A) 10 days 20 days

(B) 10 days 20 days

Fig. 7. Effect of the addition of ASP fraction on planarian head- and tail- regeneration. Decapitated (A) or posteriorly cut (B) planarians were incubated in the absence (−) or presence (+) of 40 μg ml^{-1} of the ASP fraction. Head-regeneration (A) or tail-regeneration (B) was observed for 20 days. Arrows show amputation levels. More than 7 of 12 decapitated planarians showed these regeneration patterns. Of 12 planarians with their tails amputated, all regenerated.

2a by more than 40%. One of these was a fraction of 'Mw less than 10 000 and bound to DEAE-Sephadex'. This fraction did not inhibit planarian regeneration. The other was the ammonium sulfate precipitated (ASP) fraction which also inhibited PC-12 cell-growth and head-regeneration (see below). Motivated by these characteristics, a further analysis of the latter fraction was done.

Figure 2 shows the dose effect of the ASP fraction on the Neuro 2a cell growth. Fifty percent inhibition was obtained by the addition of 120 μg ml^{-1} of this fraction. As a control, 83–333 μg ml^{-1} BSA were added to the cells without any inhibition being observed (data not shown). The decrease of inhibitory activity of this fraction on the Neuro 2a cells by heat treatment at 60°C for 30 min or 80°C for 30 min was small (Fig. 3).

Inhibition of the planarian head-regeneration by the fractionated planarian homogenate

An attempt was made to check the effect of the addition of the ASP fraction on planarian regeneration. Figure 4 shows that the addition of 30 μg ml^{-1} ASP fraction completely inhibited the head-regeneration (Fig. 4B). When the amount of ASP fraction was decreased to 10 μg ml^{-1}, the regeneration recovered and a short-

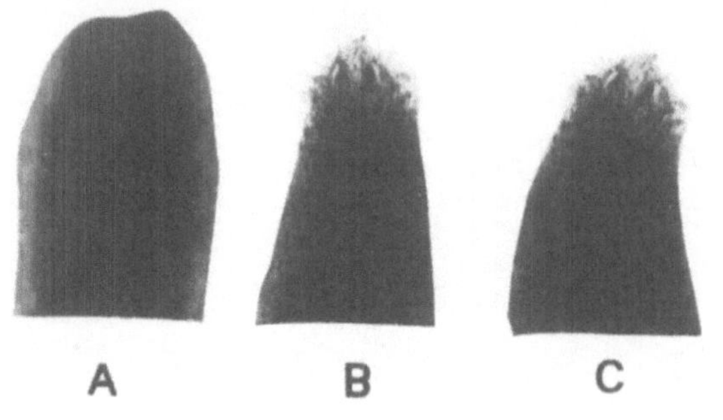

Fig. 8. Effect of the addition of trypsin-treated ASP fraction on planarian head-regeneration. At 20 days (12 planarians in each series) non- treated (A) or trypsin-treated (B) ASP fraction (40 μg ml^{-1}) and 16 μg ml^{-1} of the heat-treated trypsin solution (C) were added to the decapitated planarians.

Table 1. Effect of the addition of fractionated planarian homogenate on the cell-growth of Neuro 2a and PC-12 cells. ++: more than 40% inhibition by the addition of 83 μg ml^{-1} (10 μg well^{-1}) protein of that fraction. +: 10–40% inhibition by the addition of 83 μg ml^{-1} (10 μg well^{-1}) protein of that fraction. –: no inhibition.

Fraction	Inhibition Neuro 2a	PC-12
Mw<10 k		
DEAE-bound	++	–
Mw>10 k		
DEAE-bound	–	–
CM-bound	–	–
Con A-bound	+	+
Ammonium sulfate precipitate	++	++
Phenyl-Sepharose bound	+	+
Phenyl-Sepharose unbound	–	–

necked head was formed (Fig. 4D) as observed in the case of crude homogenate (Fig. 1). This inhibitory activity was stable after heat-treatment at 80°C for 30 min (Fig. 5).

In order to estimate the physiological significance of the inhibition of planarian regeneration by the ASP fraction, the reversibility (Fig. 6) and the specificity for the head- regeneration (Fig. 7) were analyzed. The decapitated planarians were incubated for 20 days in the presence of the ASP fraction resulting in the complete inhibition of head-regeneration (Fig. 6A). These planarians were returned to the breeding water which did not contain the ASP fraction. Head-regeneration recovered after incubation periods of 7 days (Fig. 6B), 17 days (Fig. 6C) and 27 days (Fig. 6D) in the absence of the ASP fraction. Thus, the non-regenerated planarians were able to completely regenerate after incubation for 27 days in the breeding water. Figure 7A shows the inhibition of head-regeneration by the ASP fraction. However, the ASP fraction did not inhibit the tail-regeneration (Fig. 7B).

To determine whether or not the activity of the ASP fraction is caused by a protein, trypsin treatment was carried out. Trypsin- treated ASP fraction did not inhibit the planarian head- regeneration (Fig. 8) suggesting that the activity originated from a protein. The inhibitory activity of the ASP fraction on the cultured cell (Neuro 2a) growth was not sensitive to the trypsin treatment (data not shown). This suggests that these activities on the planarians and cultured cells originate from the different molecules or moieties.

Discussion

Fractionation of planarian head homogenate and assays for the inhibition of the growth of cultured cells and also for planarian regeneration were carried out. The ASP fraction (Mw larger than 10 000, unadsorbable to the DEAE-Sephadex, CM-Sephadex, Con A- Sepharose, precipitable by ammonium sulfate) inhibited the growth of both Neuro 2a and PC-12 cells and the planarian regeneration. The inhibition of planarian regeneration by this fraction was reversible and also specific for head-regeneration. This activity on planarian regeneration was resistant to heat-treatment and trypsin sensitive. Lender (1955, 1956, 1960) reported the presence of a head-regeneration inhibitory factor. This appears in the supernatant upon centrifugation at 10 000× *g* of the planarian head homogenate and is stable after heat-treatment at 60 °C. Steele & Lange (1977) also described an inhibitory substance for planarian regeneration. This substance appeared in the supernatant of centrifugation at 10 000× *g* and was destroyed by pronase treatment. The characteristics of our ASP fraction is consistent with their results and the hypothesis that some unknown protein or peptide factor, produced in the planarian head-region, inhibits the formation of a second head. It was reported (Palmberg & Reuter, 1983; Palmberg, 1986; Kurabuchi & Kishida, 1988) that brain regeneration precedes the regeneration of other organs in planarians. The occurrence of fission in regions far from the head and the formation of neural structures just behind the fission plane were reported by Reuter's group for various orders of microturbellarians (Reuter, 1988; Reuter & Eriksson, 1991). These findings support the above hypothesis indicating the importance of the brain regeneration (including its position) followed by fission and the regeneration of other organs. We have reported the inhibition of planarian regeneration by melatonin (Yoshizawa *et al.*, 1991). However, this inhibition was not head-specific. We suppose that melatonin inhibits neural regeneration, resulting in the inhibition of both head- and tail-regeneration. The addition of crude planarian homogenate or low concentration of ASP fraction to decapitated planarians resulted in the formation of short- necked heads. Since regeneration of the brain precedes that of the other organs, it is reasonable to assume that the same might be the case with inhibition: due to the inhibitory effect, only a small brain is regenerated, resulting in the development of a short-necked head. The purification of the inhibitory factor (probably a protein) of planarian head-regeneration and the analysis of the distribution of this factor in the planarian body are the next targets of investigation.

Acknowledgments

We wish to thank Dr S. A. Salehi of Kansai Medical University for his critical reading of the manuscript. This work was supported by a Grant-in-Aid from Ministry of Education, Science and Culture of Japan (No. 03804059).

References

Kurabuchi, S. & Y. Kishida, 1988. The role of nervous system in planarian regeneration. In S. Inoue *et al.* (eds), Proc. 6th M. Singer Symposium: 99–110.

Lender, T., 1955. Sur l'inhibition de la régénération du cerveau de la planaire *Polycelis nigra*. C. R. Acad. Sci., Paris 241: 1863–1865.

Lender, T., 1956. Analyse des phénomènes d'induction et d'inhibition dans la régénération des planaires. Ann. Biol. 32: 457–471.

Lender, T., 1960. L'inhibition spécifique de la différenciation du cerveau des planaires d'eau douce en régénération. J. Embryol. exp. Morph. 8: 291–301.

Lowry, O. H., N. J. Rosebrough, A. L. Farr & R. J. Randall, 1951. Protein measurement with folin phenol reagent. J. Biol. Chem. 193: 265–275.

Mosmann, T., 1983. Rapid colorimetric assay for cellular growth and survival. J. Immunol. Meth. 65: 55–63.

Palmberg, I. & M. Reuter, 1983. Asexual reproduction in Microstomum lineare (Turbellaria). I. An autoradiographic and ultrastructural study. Int. J. invert. Reprod. 6: 197–206.

Palmberg, I., 1986. Cell migration and differentiation during wound healing and regeneration in Microstomum lineare (Turbellaria). Hydrobiologia 132 (Dev. Hydrobiol. 32): 181–188.

Palmberg, I., 1991. Differentiation during asexual reproduction and regeneration in a microturbellarian. Hydrobiologia 227 (Dev. Hydrobiol. 69): 1–10.

Reuter, M., 1988. Development and organization of nervous systems visualized by immunocytochemistry in three flatworm species. Progr. Zool. 36: 181–184.

Reuter, M. & K. Eriksson, 1991. Catecholamines demonstrated by glyoxylic-acid-induced fluorescence and HPLC in some microturbellarians. Hydrobiologia 227 (Dev. Hydrobiol. 69): 209–219.

Steele, V. E. & C. S. Lange, 1977. Characterization of an organ-specific differentiator substance in the planarian *Dugesia etrusca*. J. Embryol. exp. Morph. 37: 159–172.

Yoshizawa, Y., K. Wakabayashi & T. Shinozawa, 1991. Inhibition of planarian regeneration by melatonin. Hydrobiologia 227 (Dev. Hydrobiol. 69): 31–40.

Hydrobiologia **305**: 255–257, 1995.
L.R.G. Cannon (ed.), *Biology of Turbellaria and some Related Flatworms.*

Preparation of monoclonal antibodies against planarian organs and the effect of fixatives

Takao Shinozawa[1,*], Hajime Kawarada[1], Kazuhisa Takezaki[1], Hideaki Tanaka[2] & Kinji Inoue[3]
[1]*Department of Biological and Chemical Engineering, Faculty of Engineering, Gunma University, Kiryu, Gunma 376, Japan (*author for correspondence)*
[2]*Department of Pharmacology, Faculty of Medicine, Gunma University, Maebashi, Gunma 371*
[3]*Institute for Molecular and Cellular Regulation, Gunma University, Maebashi, Gunma 371*

Key words: planarian, monoclonal antibody, fixing reagent

Abstract

Monoclonal antibodies were obtained against all part of the body of the planarian *Dugesia japonica japonica* except the nerve cord, though parts of the enteric canal were also non reactive. The effect of a variety of fixatives (acetone, Bouin, formalin) used on planarian tissues prior to screening with antibodies, was assessed.

Introduction

Monoclonal antibodies against planarian organs will allow their corresponding antigens to function as differentiating markers during regeneration (Romero *et al.*, 1991, Shirakawa *et al.*, 1991). We describe the monoclonal antibodies against epidermis, testis, gland cells and part of the enteric canal as well as all other parts of the body except the nerve cord and some of the enteric canal. The effect of fixatives on monoclonal antibody screening was also investigated.

Materials and methods

Animals and Reagents: Planarians (*Dugesia japonica japonica*) and BALB/C mice were obtained from Kiryu river and Gunma University's Animal Center, respectively. Rabbit anti- serotonin antibody was from Sigma; fluoroscein isothiocyanate (FITC)-conjugated Goat anti-Rabbit IgG(H+L) and FITC-conjugated goat anti-BALB/C mouse IgG(H+L) from Cappel. Other reagents were of special grade.

Fixation: Formalin fixation; decapitated planarians were fixed in 10% formalin at 4 °C over night, dehydrated by graded ethanol series-1 (50%, 70%, 80%, 90%, 95%, 99.5%, 100%), cleared by xylene, and then

Table 1. Effect of fixatives on the monoclonal antibody screening. Antigens and the monoclonal antibodies obtained against them, are shown in the first column. The results of screening after the use of formalin, Bouin's or acetone as fixatives are shown. The percentages calculated from the ratio of the number of wells containing antibody against each specific antigen to the number of total screened wells are shown. –: negative reaction, /: not registered. ca: common antigen; ca (-ec): common antigen except for enteric canal; ca (-nc): common antigen expect for nerve cord; ca (-ec, ep): common antigen except for enteric canal and epidermis; ec: enteric canal; ep: epidermis; t: testis; gc: gland cell; bm: basement membrane; lm: longitudinal muscle.

Antigen	Formalin fixation	Bouin fixation	Acetone fixation
ca	16.2 (%)	18.2 (%)	/ (%)
ca (-ec)	2.7	3.4	/
ca (-nc)	2.5	3.2	/
ca (-ec,ep)	4.5	5.0	/
ec	3.4	5.1	/
ec,ep	5.3	3.5	/
t	–	–	0.9
gc	–	1.1	0.5
bm,lm	–	1.8	3.0
other organs	–	0.7	0.9
total	37.6	44.0	5.3

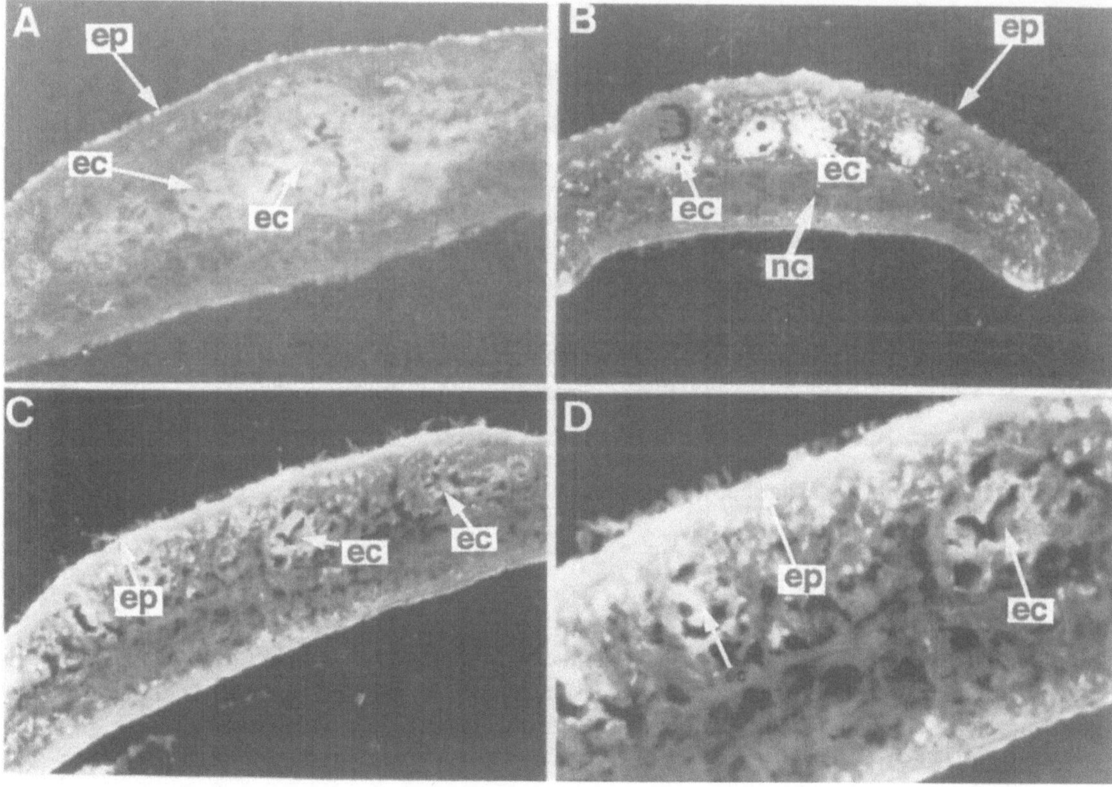

Fig. 1. Immunofluorescent staining of Transverse sections of enteric canal (*ec*) and epidermis (*ep*) with monoclonal antibodies. *A*) Formalin fixation. Control using Dulbecco's Phosphate-Buffered Saline (137 mM NaCl, 3 mM KCl, 7 mM Na_2HPO_4 and 1 mM KH_2PO_4) instead of hybridoma supernatant. *B*) Formalin fixation using hybridoma supernatant. Specific staining of enteric canal (*ec*). *C*) fixed as in B, enteric canal (*ec*) and epidermis (*ep*). *D*) fixed as in B, high magnification of the epidermis (*ep*) *nc*: nerve cord.

embedded in paraffin (mp 42–44, 56– 58 °C). Bouin's fixation followed a similar procedure. Acetone fixation; decapitated planarians were fixed and dehydrated by acetone at −80 °C for a week, cleared by xylene, and then embedded in paraffin.

Identification of (formalin fixated) planarian organs: Pharynx, ventral nerve cord, epidermis, testis, enteric canal, basement membrane and longitudinal muscle were distinguished by hematoxylin and eosin yellow staining. Ventral nerve cords, nerve cords, and dorsal-ventral nerve cord were identified by anti- serotonin antibody staining as reported by Reuter (1988). *Immunization of mice and preparation of monoclonal antibodies*: Immunization of mice with the planarian homogenate and preparation of monoclonal antibodies were carried out according to Fujita *et al.* (1982). Ten series of cell- fusion, 2000 wells screening in each series, have been carried out.

Results and discussion

Monoclonal antibodies on the enteric canal (ec) and epidermis (ep) are shown in Fig. 1. The part of the enteric canal (ec) that reacted positively to FITC is easily distinguished from the non- reacted parts (Fig. 1B). Staining with monoclonal antibody against enteric canal (ec) and epidermis (ep) is shown in Figure 1C & 1D. Specific staining with monoclonal antibodies of testis (t) (Fig. 2A), gland cells (gc) (Fig. 2B), common antigen except for enteric canal (ec) and epidermis (ep) (Fig. 2C), and common antigen except for nerve cord (nc) and enteric canal (ec) (Fig. 2D) are shown. Monoclonal antibodies to the basement membrane or longitudinal muscle were also isolated (data not shown). Results of screening using various fixatives are shown in Table 1. In the case of formalin and Bouin's fixation, antibodies to the enteric canal, enteric canal and epidermis, and other common antigens (except enteric canal, except nerve cord, except for enteric canal

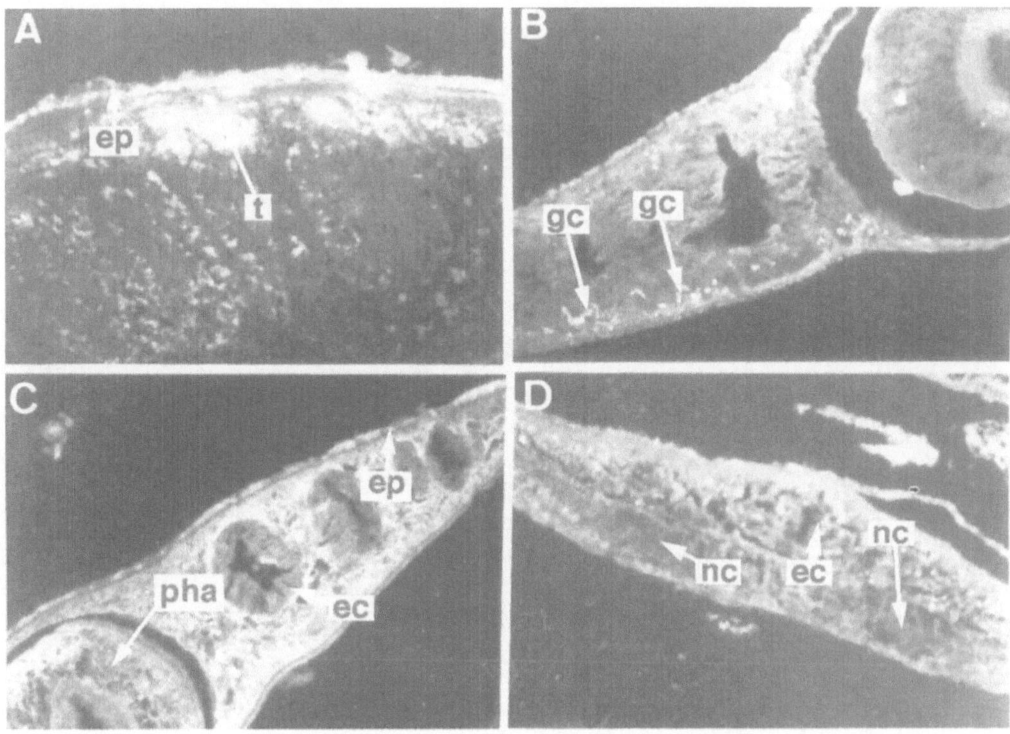

Fig. 2. Immunofluorescent staining of planarian organs with monoclonal antibodies. *A*) acetone: specific staining of testis (*t*). *B*) Bovin's: gland cells (gc). *C*) Bovin's: common antigen except for enteric canal (*ec*) and epidermis (*ep*). *D*) formalin: common antigen except for nerve cord (*nc*) and enteric canal (*ec*): *pha*: pharynx.

and epidermis) were isolated. Several other antibodies to the gland cells (gc), basement membrane (bm) and longitudinal muscle (lm) were also detected by Bouin's fixation. By acetone fixation, several antibodies to testis, gland cells (gc) and basement membrane (bm) and longitudinal muscle (lm) were detected. No antibodies reacting to the antigen in testis were detected by formalin or Bouin's fixation. These findings suggest that the method of fixation is very important in screening for specific monoclonal antibodies.

References

Fujita, S., S. L. Zipursky, S. Benzer, A. Ferrus & S. L. Shotwell, 1982. Monoclonal antibodies against the Drosophila nervous system. Proc. natn. Acad. Sci. USA 79: 7929–7933.

Reuter, M., 1988. Development and organization of nervous systems visualized by immunocytochemistry in three flatworm species. Fortschr. Zool. 36: 181–184.

Romero, R., J. Fibla, D. Bueno, L. Sumoy, M. A. Soriano & J. Baguña, 1991. Monoclonal antibodies as markers of specific cell types and regional antigens in the freshwater planarian Dugesia (G.) tigrina. Hydrobiologia 227 (Dev. Hydrobiol. 69): 73–79.

Shirakawa, T., A. Sakurai, T. Inoue, K. Sasaki, Y. Nishimura, S. Ishida & W. Teshirogi, 1991. Production of cell- and tissue-specific monoclonal antibodies in the freshwater planarian Phagocata vivida. Hydrobiologia 227 (Dev. Hydrobiol. 69): 81–91.

Hydrobiologia **305**: 259–260, 1995.
L.R.G. Cannon (ed.), *Biology of Turbellaria and some Related Flatworms.*

Abnormality in the fissioning control of a strain of asexual *Dugesia dorotocephala* found near Fort Collins, Colorado

Jay Boyd Best
Center for Environmental Toxicology, Colorado State University, Fort Collins, Colorado 80523, USA

Key words: asexual reproduction, *Dugesia dorotocephala*, fissioning control, group size, planarians, population density, Tricladida, Turbellaria

Abstract

Asexual *Dugesia dorotocephala*, collected from a new site several miles northwest of Fort Collins, Colorado, appeared similar, at the time of collection, to ones collected from other sites in the same general locale. However, these exhibited marked differences in the more crowded conditions of laboratory culture: abnormally long, 30–50 mm specimens, seemingly unable to fission, developed. Since previous studies of asexual *D. dorotocephala* had revealed an inhibitory effect of cohorts on fissioning (Best *et al.*,1969, 1974, 1975; Pigon *et al.*, 1974), experiments were conducted with this new strain to ascertain the effect of population density (group size) on fissioning incidence. Results are described which show that this inhibitory effect of cohorts on fissioning is exaggerated in this new strain.

Introduction

Previously published studies on fissioning of asexual *Dugesia dorotocephala* revealed that: its incidence is suppressed in a graduated manner by cohorts; it is more likely in longer than shorter specimens (Ss); it is modulated by photoperiod; it occurs with a high incidence in decapitated Ss, irrespective of cohorts or photoperiod (Best *et al.*, 1969, 1974, 1975; Pigon *et al.*, 1974; Morita & Best, 1984). All of these previous studies were done with laboratory cultures of Ss originally collected from a pond and stream fed from outflow from the Soldier Canyon Dam (SCD) approximately 6 miles northwest of Fort Collins, Colorado. More recently, Ss have been collected from a small stream generated by outflow from the Bellevue Fish Hatchery (BFH). The two sites are approximately 2 miles apart and both approximately 6 miles northwest of Fort Collins, but with no direct waterway linking them. At collection, the BFH strain appeared similar to, but slightly larger than, the SCD strain. Other differences only became apparent following 2–3 months of laboratory culture. Instead of fissioning, many of the BFH strain grew abnormally long in the laboratory colony pans, giving them an unusual serpentine appearance. While Ss in the SCD colony pans typically ranged 12–25 mm, those in the BFH colony pans ranged 25–60 mm in length. One unnatural aspect of our laboratory culture was the relatively high population density in the colony pans. However, the SCD strain appeared to thrive under those conditions. Was the BFH strain more sensitive to the inhibitory effect of cohorts on fissioning than the SCD strain in which the effect had been originally demonstrated? The following experiment was designed to answer this question.

Materials and methods

Subject specimens (Ss) for the study were 242 asexual *Dugesia dorotocephala* of the SCD strain 18–25 mm long and 242 of the BFH strain 25–40 mm long, selected from the laboratory colony pans 3 days before experimental time zero. Conditions of maintenance of the laboratory cultures and selection criteria have been described previously (Best *et al.*, 1974). The Ss were last fed 3 days prior to selection. Following selection, all Ss were maintained in cylindrical glass bowls with 105 mm inside diameter. The study was performed in two replicate experiments, each utilizing 121 SCD strain Ss and 121 BFH Ss tested concurrently. Imme-

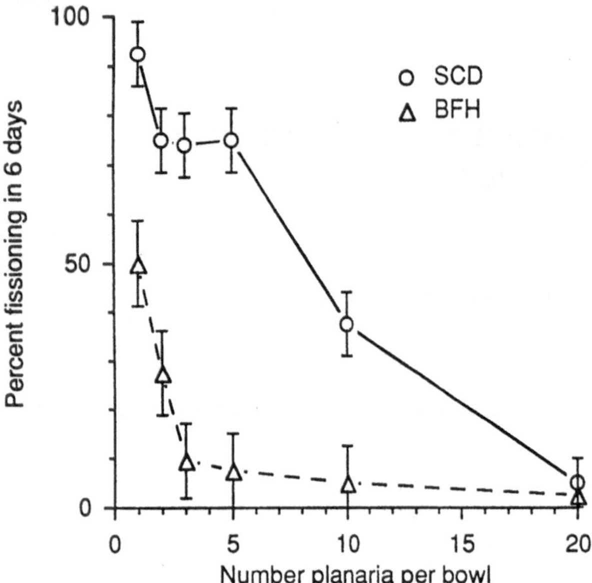

Fig. 1. Fissioning incidence as a function of cohort numbers for two different asexual strains (SCD & BFH) of the fresh water planarian, *Dugesia dorotocephala.*

diately following selection, the Ss were maintained for acclimation and to suppress fissioning for 3 days in bowls containing 70 Ss and 200 ml of culture water per bowl (b). Water was changed daily. At the end of this adaptation period, i.e. at experimental time zero, 121 Ss of each strain were segregated into the following groups: 20×1S/b, 10×2S s/b, 7×3S s/b, 4×5S s/b, 2×10S s/b, 1×20S s/b, each bowl containing 50 ml of culture water. Observations were made, tail fragments removed and the water changed daily at the same time of day.

Results

The percent of Ss fissioning within the first six days for each of the various experimental groups is shown in Fig. 1. Vertical bars represent the standard deviations of the means of such fissioning incidences. Although 92% of the isolates of the SCD strain fissioned in the first 6 days, only 50% of the BFH isolates did so within the same time period.

Increasing the group size relative to Ss/b decreased SCD fissioning to 75% while that of the BFH strain was decreased to only 7%. With group sizes of 20/b, fissioning is reduced to essentially zero in both strains (no significant difference between zero and the fissioning incidences observed for the SCD and BFH strains).

Discussion

Although the 6 day fissioning incidence of the BFH strain isolates (50%) is significantly less than that of the SCD isolates (92%), the ratio of these two incidences (1.84) is dramatically enhanced (to 10.7) when the number of cohorts is increased to 4 (group size of 5/b).

In general it can be anticipated that the average length of asexual planarians will be the resultant of growth rate and fissioning incidence. For a fixed growth rate, increased fissioning incidences should yield more, but smaller Ss. Such considerations generally accord with prior observations and present results: prior studies with the SCD strain showed suppression of fissioning with increased population density; the present study demonstrates the same effect, in exaggerated form, in the BFH strain, i.e. the BFH strain's fissioning incidence is more severely depressed by cohorts than that of the SCD strain. The present study also demonstrates a slight, but significant, difference in fissioning incidence of the isolates for the two strains; the lower fissioning incidence of the BFH isolates may explain their slightly larger size than the SCD strain at the time of collection. With population densities approximating those of our laboratory colony pans, present results indicate fissioning incidences for the BFH strain may be only a tenth that of the SCD strain, which could explain the extremely elongated Ss previously observed in the BFH strain laboratory cultures.

Determining which of the physiological components of the fissioning system previously elucidated in the SCD Ss (Best *et al.*; 1969, 1974, 1975; Pigon *et al.*, 1974; Morita & Best, 1984) have been altered in the BFH Ss will require further study.

References

Best, J. B., A. B. Goodman & A. Pigon, 1969. Fissioning in planarians: Control by the brain. Science 164: 565–566.

Best, J. B., W. Howell, V. Riegel & M. Abelein, 1974. Cephalic mechanism for social control of fissioning in planarians. I. Feedback cue and switching characteristics. J. Neurobiol. 5: 421–442.

Best, J. B., M. Abelein, E. Kreutzer & A. Pigon, 1975. Cephalic mechanism for social control of fissioning in planarians. III. Central nervous system centers of facilitation and inhibition. J. comp. physiol. Psych. 89: 923–932.

Morita, M. & J. B. Best, 1984. Effects of photoperiod and melatonin on planarian asexual reproduction. J. exp. Zool. 231: 273–282.

Pigon, A., M. Morita & J. B. Best, 1974. Cephalic mechanism for social control of fissioning in planarians. II. Localization and identification of the receptors by electron micrographic and ablation studies. J. Neurobiol. 5: 443–462.

Hydrobiologia **305**: 261, 1995.
L.R.G. Cannon (ed.), *Biology of Turbellaria and some Related Flatworms.*

Regenerative and reproductive capacities of the fissiparous planarian *Dugesia tahitiensis*

Roland Peter
Department of Genetics and General Biology, University of Salzburg, Hellbrunnerstr. 34, A-5020 Salzburg, Austria

Key words: architomy, asexual reproduction, *Dugesia tahitiensis*, fission, head frequency, photoperiod, planarians, regeneration, Tricladida, Turbellaria

Abstract

A pilot study was performed to assess the regenerative capacities of *Dugesia tahitiensis* Gourbault, 1977, an exclusively fissiparous planarian species. Animals measuring 9.5–12.5 mm in length were used. Head regeneration rate determined by the appearance of eye spots (Brøndsted, 1969: 29–46) was extremely high: at 23 °C, it took 43–59 h to regenerate clearly discernible eye spots in 28 specimens. For comparison, 3.8 days were reported for the regeneration of eye spots in 11.9–12.7 mm long *D. tigrina* (Girard) at 24 °C (Mead, 1985). As all posterior fragments regenerated a head, irrespective of the cutting level, *D. tahitiensis* seems to match the *Phagocata velata* (Stinger) type with a head frequency of 100% at every level (Teshirogi *et al.*, 1977; cf. also Brøndsted, 1969:30).

Fissioning (architomy) occurs postpharyngeally. It cuts off the hindmost fifth to third of the body. A constriction of variable width precedes fission, the process itself being executed by mechanical tearing, apparently due to reduced coordination between head and tail: see Morita & Best (1984) for *D. dorotocephala* (Woodworth). Fission rate depends on population density and/or group size. A maximal population growth, by a factor of 7 within 22 days, was achieved when animals were kept singly (1 specimen per 44.2 cm^2 of surface area, 10 ml of culture water) with a photoperiod of 12 h per day (700–1,400 lx). They were fed *Tubifex* every fourth day. Fission rate declined by 22% for 30 specimens kept together (1 animal per 6.8 cm^2). Fission occurred even in groups of 60 animals, with 1 planarian per 1.3 cm^2. *D. dorotocephala* exhibits a much more pronounced dependence of fission on population density (Best *et al.*, 1974; see there for further references and cf. Davison, 1973, for *D. tigrina*).

References

Best, J. B., W. Howell, V. Riegel & M. Abelein, 1974. Cephalic mechanism for social control of fissioning in planarians. I. Feedback cue and switching characteristics. J. Neurobiol. 5: 421–442.

Brøndsted, H. V., 1969. Planarian Regeneration. Pergamon Press, Oxford.

Davison, J., 1973. Population growth in planaria *Dugesia tigrina* (Girard) – Regulation by the absolute number in the population. J. Gen. Physiol. 61: 767–785.

Mead, R. W., 1985. Proportioning and regeneration in fissioned and unfissioned individuals of the planarian *Dugesia tigrina*. J. exp. Zool. 235: 45–54.

Morita, M. & J. B. Best, 1984. Effects of photoperiods and melatonin on planarian asexual reproduction. J. exp. Zool. 231: 273–282.

Teshirogi, W., S. Ishida & H. Yamazaki, 1977. Regenerative capacities of transverse pieces in the two species of freshwater planarian, *Dendrocoelopsis lactea* and *Polycelis sapporo*. Sci. Rep. Hirosaki Univ. 24: 55–72.

Hydrobiologia **305**: 263–264, 1995.
L.R.G. Cannon (ed.), Biology of Turbellaria and some Related Flatworms.

TCAV-1, a monoclonal antibody specific to epithelial pharyngeal cells in the planarian *Dugesia (Girardia) tigrina*. Application to pattern formation of the pharynx during regeneration

David Bueno*, Eduard Batlle, Marc Aureli Soriano, Lluis Espinosa, Jaume Baguñà & Rafael Romero
Departament de Genètica, Facultat de Biologia, Universitat de Barcelona, Av. Diagonal 645, Barcelona 08071, Catalonia, Spain
(author for correspondence)*

Key words: Monoclonal antibodies, regeneration, immunohistochemistry, pharyngeal epithelial cells, differentiation, *Dugesia (G.) tigrina*

Abstract

During regeneration in planarians, anterior (head and prepharyngeal) and posterior (postpharyngeal and tail) fragments rebuild one of the most peculiar structures of planarians: the pharynx and the pharynx cavity. Previous studies (see Brønsted, 1969, for a general review, and Asai, 1990, 1991, for anterior regeneration) have shown that within postpharyngeal pieces both structures appear in the old stump from clusters of undifferentiated cells. However, the lineage and differentiation of their elements (inner and outer epithelial cells, muscle layers, gland cells, nerve rings) and the overall pattern of growth and differentiation is not clear.

To improve our understanding of the processes of regeneration of the pharynx and the pharynx cavity and how their patterns unfold, molecular markers for the pharynx were used to monitor the early stages of both processes. Monoclonal antibodies (mAbs) were used during development and regeneration to detect antigens expressed in spatially and/or temporally restricted areas. During planarian regeneration, mAbs were extremely useful in monitoring when and where pharyngeal markers appear. We have screened panels of mAbs raised again extracts of intact *Dugesia (G.) tigrina* to detect pharynx specific mAbs. One of them, TCAV-1 (code letters for mAb specificity: T for the species, *D. tigrina*, CAV for the structure, pharynx cavity; code number for the mAb order), has been used to study the heterogeneity of epithelial cells in the intact pharynx and the appearance of TCAV-1 positive cells during regeneration.

With that aim, organisms belonging to an asexual race of *Dugesia (G.) tigrina*, starved for two weeks, were cut at the postpharyngeal level. Tail pieces, kept at 22 °C, were tested at 2.5, 3, 3.5, 4 and 5 days of regeneration for the earliest appearance of cells posi-

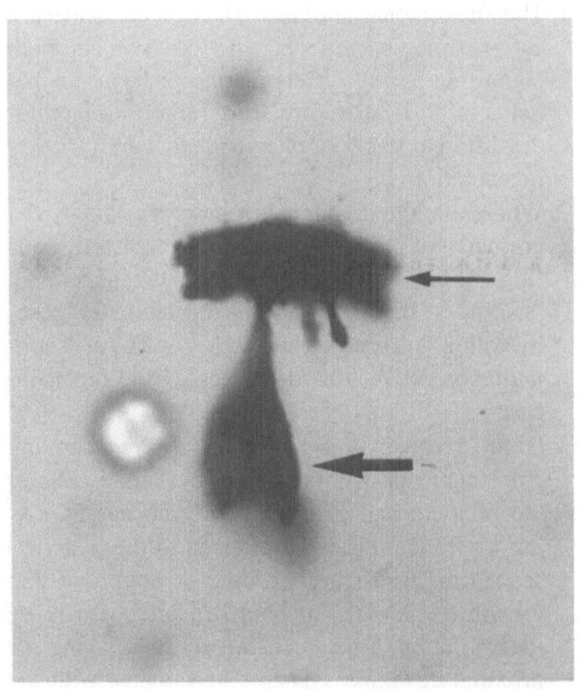

Fig. 1. Macerated cell immunostained with TCAV-1. ABC method. (1000 ×). Thin arrow, ciliated region; thick arrow, nuclear region.

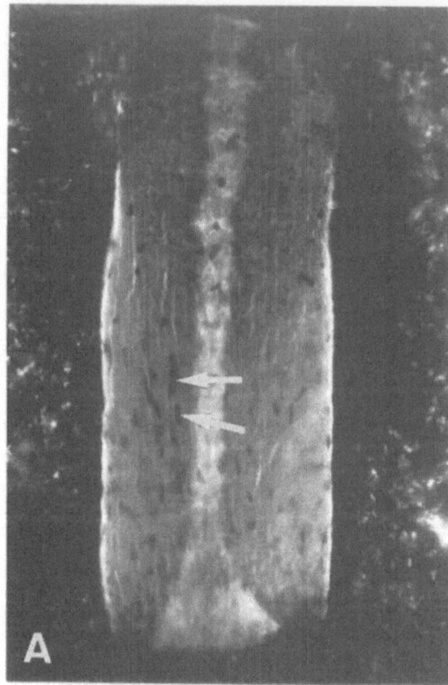

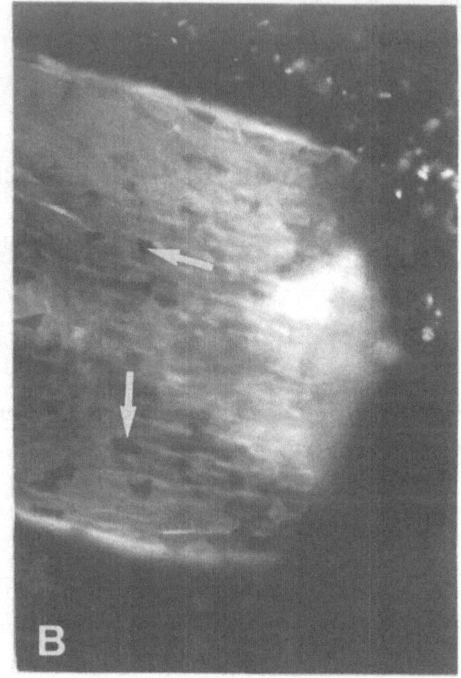

Fig. 2. Pharynx immunostained with the mAb TCAV-1, showing also TCAV-1 non-reactive cells (arrows). Whole mount procedure. FITC method. (A) 250 ×, (B) 500 ×.

tive to TCAV-1 using standard immunohistochemical techniques on sagittal paraffin sections (Romero *et al.*, 1991), macerated cells (Espinosa, 1993) and whole mounts (Burgaya & Baguñà, 1992).

The results obtained and the suggestions that stem from these results are as follows:

1. The monoclonal antibody TCAV-1 recognizes a cytoplasmic antigen (Fig. 1, arrows), in both mature and early differentiating pharynx epithelial cells. It is, therefore, a useful tool to monitor the different stages of pharynx and pharynx cavity regeneration.
2. Whole mount immunocytochemistry shows the presence of TCAV-1 non-immunoreactive cells (Fig. 2, arrows) in both intact and regenerating pharynges, disclosing an unexpected heterogeneity within this organ. The number of TCAV-1 non-immunoreactive cells decreases during regeneration.
3. During regeneration of anteriorly regenerating tail fragments, TCAV-1 immunostained cells at the pharynx and pharynx cavity appear following a proximo-distal (as referred to the pharynx) sequence. This suggests, though do not prove, that growth (expansion) of both structures occurs by incorporation of new cells at any point along the proximo-distal axis. This is at variance with the mechanism shown by Assai (1991) to occur during the growth (expansion) of the pharynx in anteriorly regenerating head fragments: the disto-proximal fusion of transitory cavities or slits. These differences could be related to mechanical constraints imposed upon the pharynx anlage by its relative proximity to the wound epithelium.

References

Asai, E., 1990. The behavior of pharyngeal outer epithelial cells during regeneration of the planarian *Dugesia japonica japonica*. J. Morph. 206: 313–325.

Asai, E., 1991. Regeneration of the pharynx in a fresh-water planarian: An electron microscopic study with special reference to the formation of the pharyngeal cavity and pharyngeal lumen. Zool. Sci. 8: 775–784.

Brønsted, H. V., 1969. Planarian Regeneration. Pergamon Press, London.

Burgaya, F. & J. Baguñà, 1992. Seguiment per mètodes immunohistoquímics de la regeneració del sistema nerviós de planàries processades 'in toto'. Biologia del Desenvolupament SCB 10: 201–210.

Espinosa, Ll., 1993. Anàlisi espacial i temporal de la regeneració de la faringe i cavitat faringia de la planària *Dugesia (G.) tigrina* mitjançant anticossos monoclonals. Tesina de Llicenciatura, Universitat de Barcelona.

Romero, R., F. Fibla, D. Bueno, L. Sumoy, M. A. Soriano & J. Baguñà, 1991. Monoclonal antibodies as markers of specific cell types and regional antigens in the freshwater planarian *Dugesia (G.) tigrina*. Hydrobiologia 227 (Dev. Hydrobiol. 69): 73–79.

Hydrobiologia **305**: 265, 1995.
L.R.G. Cannon (ed.), Biology of Turbellaria and some Related Flatworms.

Restitution bodies from planarian cell suspensions: formation and survival in an isotonic culture medium

Roland Peter
Department of Genetics and General Biology, University of Salzburg, Hellbrunnerstr. 34, A-5020 Salzburg, Austria.

Key words: cell aggregation, *Dugesia polychroa, Dugesia tahitiensis*, planarian cells, pseudogastrulae, restitution bodies, Tricladida, Turbellaria

Abstract

Restitution bodies are more or less spherical cell aggregates covered with an epithelial layer and ranging from 10 to 500 μm in diameter. They were first observed by Freisling & Reisinger (1958) who pressed planarians through fine mesh gauze and suspended the resulting cell mass in a relatively small amount of hypotonic medium (50% Holtfreter's solution). To optimize and standardize conditions, planarians were disintegrated into their cells using a Dounce homogenizer with a tolerance of 50 μm (Franquinet, 1981; Schürmann & Peter, 1988). This was done in an isotonic medium (126 mOsmol l^{-1} pH 7.4; Schürmann, 1993) based on a composition published by Teshirogi & Tohya (1988). A 40- to 200-fold excess of medium was applied under sterile conditions. Restitution bodies originated immediately after disintegration, *Dugesia polychroa* (Schmidt) and *Dugesia tahitiensis* Gourbault being equally efficient. This spontaneous cell aggregation apparently depends critically on the ciliary action of epithelial cells. The presence of free calcium ions (1 mM) is a prerequisite for ciliary activity as well as for the formation of restitution bodies.

Morphology, histology and sizes of the restitution bodies were similar to the original description, with varying cellular composition and different forms. Spheres prevailed, a central cavity was either filled with different cell types, or contained a vacuolized cell mass, or was empty. Double spheres and 'pseudogastrulae' with varied invaginations were observed. Light microscopy suggests the wide-spread occurrence of syncytia; these findings must, however, be reexamined by ultrastructural studies. The restitution bodies survived up to one week at 20 °C without a change of medium. When larger aggregates were separated by filtration through a 40 μm nylon gauze 12 h after tissue disintegration and the medium was changed every two days, they survived for two weeks.

References

Franquinet, R., 1981. Synthèse d'ADN dans les cellules de planaires cultivées *in vitro*. Rôle de la sérotonin. Biol. Cell. 34: 71–76.

Freisling, M. & E. Reisinger, 1958. Zur Genese und Physiologie von Restitutionskörpern aus Planarien-Gewebsbrei. Roux' Arch. Entwickl. Mech. Org. 150: 581–606.

Schürmann, W., 1993. Isolierung von Neoblasten aus Planarien. Doctoral thesis submitted to the Faculty of Sciences at the University of Salzburg.

Schürmann, W. & R. Peter, 1988. Isolation and cultivation of planarian neoblasts by a novel combination of methods. Fortschr. Zool. 36: 111–114.

Teshirogi, W. & K. Tohya, 1988. Primary tissue culture of freshwater planarians in a newly devised medium. Fortschr. Zool. 36: 91–96.

Hydrobiologia **305**: 267, 1995.
L.R.G. Cannon (ed.), Biology of Turbellaria and some Related Flatworms.

Separation of planarian neoblasts based on density gradient centrifugation

Wolfgang Schürmann & Roland Peter
Department of Genetics and General Biology, University of Salzburg, Hellbrunnerstr. 34, A-5020 Salzburg, Austria

Key words: cell culture, density gradients, *Dugesia polychroa*, neoblast separation, planarians, Tricladida, Turbellaria

Abstract

For highly purified preparations of neoblasts, density gradient centrifugation in Percoll solutions (Pertoft *et al.*, 1978) was applied to cell suspensions obtained by disintegrating *Dugesia polychroa* (Schmidt) in culture medium contained in a Dounce homogenizer (tolerance: 50 μm; one animal 12 mm in length per ml). To reduce the high viscosity caused by mucus, 0.00063% (w/v) of dithiothreitol was added during disintegration and purification. Based on previous experiments (Schürmann & Peter, 1988), five media were compared.

For prepurification, four washing steps (differential centrifugations at $500 \times \mathbf{g}$ for 5 min each) were followed by subsequent filtration through a series of nylon gauzes (40, 30, 20 and 15 μm mesh size) and a final washing step. The resulting cell suspensions were then fractionated by isopycnic centrifugation ($500 \times \mathbf{g}$, 45 min) in one continuous (1.018–1.121 g ml^{-1}) or one of seven different discontinuous Percoll gradients (Schürmann, 1993). The best yield and highest purity of neoblasts in one fraction was obtained with a four step gradient (1.03–1.09 g ml^{-1}): the neoblasts (purity: 91%) were concentrated in one sharp band at the boundary between the densities 1.05 and 1.07 g ml^{-1}. The spherical cells (diameter from 10 to 13 μm *in vivo*) stained as typical neoblasts (Pedersen, 1959).

Primary cultures were obtained with all media. The medium developed by Teshirogi and Tohya (1988) and its isotonic modification (Schürmann, 1993) proved best, resulting in 86% of viable cells without signs of differentiation after 17 days of culture at 18 °C, with still 46% being left after 31 days. Earlier reports state that isolated neoblasts only survive for 4 days (Betchaku, 1967) and total planarian cell suspensions only 2–3 weeks (Teshirogi & Tohya, 1988).

References

Betchaku, T., 1967. Isolation of planarian neoblasts and their behavior *in vitro* with some aspects of the mechanism of the formation of the regeneration blastema. J. exp. Zool. 164: 407–434.

Pedersen, K. J., 1959. Cytological studies on planarian neoblasts. Z. Zellforsch. 50: 799–817.

Pertoft, H., T. C. Laurent, T. Laas & L. Kagedal, 1978. Density gradients prepared from colloidal silicon particles coated by polyvinylpyrrolidone (Percoll). Anal. Biochem. 88: 271–282.

Schürmann, W., 1993. Isolierung von Neoblasten aus Planarien. Doctoral thesis submitted to the Faculty of Sciences at the University of Salzburg.

Schürmann, W. & R. Peter, 1988. Isolation and cultivation of planarian neoblasts by a novel combination of methods. Fortschr. Zool. 36: 111–114.

Teshirogi, W. & K. Tohya, 1988. Primary tissue culture of freshwater planarians in a newly devised medium. Fortschr. Zool. 36: 91–96.

Hydrobiologia **305**: 269–275, 1995.
L.R.G. Cannon (ed.), *Biology of Turbellaria and some Related Flatworms.*

The freshwater planarian *Dugesia (G.) tigrina* contains a great diversity of homeobox genes

Emili Saló*, Ana Maria Muñoz-Mármol, José Ramon Bayascas-Ramirez, Jordi Garcia-Fernàndez, Agusti Miralles, Andreu Casali, Montserrat Corominas & Jaume Baguñá
Departament de Genètica, Facultat de Biologia, Universitat de Barcelona, Diagonal 645, 08071 Barcelona, Spain (author for correspondence)*

Key words: Turbellaria, *Dugesia*, homeobox, gene expression

Abstract

To identify potential pattern control and cell determination and/or differentiation genes in the freshwater planarian *Dugesial (G.) tigrina*, we searched for homeobox genes of different types in the genome of this primitive metazoan. We applied two basic approaches: 1) Screening the cDNA library with degenerate oligonucleotides corresponding to the most conserved amino acid sequence from helix-3 of the homeodomain of each family; and 2) PCR amplification of genomic DNA or cDNA, using two sets of degenerated oligonucleotides corresponding to helices 1 and 3 of the homeodomain or two specific domains of the POU family. Using the first strategy we have identified and characterized two tissue-specific cell determination and/or differentiation NK-type homeobox genes. Using the second strategy we have identified several homeobox genes that belong to the HOM/Hox, paired (prd) or POU families.

Introduction

The homeobox is a 180 bp DNA sequence that codifies for 60 residues, the homeodomain, with a specific motif the helix-turn-helix that binds DNA. The homeobox-containing genes are transcription factors involved in the control of pattern formation and cell determination and differentiation and they are widely distributed on the phylogenetic tree, from plants to vertebrates (Gehring, 1987).

Freshwater planarians belong to the phylum Platyhelminthes, an old monophyletic group which branched from the rest of metazoans at an early stage. Actually, this group can be considered the sister group of the other bilateral triploblastic organisms (Riutort *et al.*, 1992; Adoutte & Philippe, 1993).

Freshwater planarians show great morphological plasticity. An adult organism can grow or shrink continuously depending on food availability, and a whole organism can be regenerated from a small piece of the body. Both phenomena have a common cellular basis: the presence of a population of small, undifferentiated, self-renewing cells (neoblasts) which give rise to all differentiated cell types in the intact organisms and to blastema cells during regeneration (Baguñà *et al.*, 1990). Thus in planarians, the morphogenetic mechanisms are always active in order to maintain body proportions during changes in body size or regeneration. Although regeneration in planarians is fairly well understood at the tissue and cell levels, no genetic data have so far been reported. In order to make the transition from cellular to molecular level, as has been done in mouse, and some other metazoans without any genetic tradition, we decided to isolate regulatory genes using conservative sequences like the homeobox. Moreover, the key phyletic position of the Platyhelminthes may be of help in tracing the origin and evolution of the homeobox-containing genes and explore their role within development and evolution.

In previous reports, we described the cloning, DNA sequence determination and expression analysis of the first two Platyhelminth homeobox genes, *Dth*-1 and *Dth*-2 from *Dugesia (G.) tigrina* (Garcia-Fernandez *et al.*, 1991; 1993). Here we review and extend our analyses of these homeobox-containing genes in freshwater planarians, and report the isolation of new ones. We also discuss the relevance of these results to the molecular control of spatial patterning in adult and regenerating freshwater planarians.

Materials and methods

The planarians used in this study belong to an asexual population of the species *Dugesia (G.) tigrina* and were collected near Barcelona. DNA was extracted from adults (Garcia-Fernandez *et al.*, 1993) and used for Polymerase Chain Reaction (PCR). DNA was also purified from an aliquot of intact, early and late regenerating planarian cDNA libraries (Muñoz-Mármol *et al.*, unpublished) and used as a cDNA template for PCR. Two pairs of primers were used for PCR amplification. The first were 'universal' homeobox primers, based on the degenerate primers described by Forhman *et al.* (1990). They are complementary to regions encoding the alpha helices 1 and 3, conserved in the homeodomain type *Antennapedia*. The second pair designed to amplify POU homeobox sequences, were as described by Billin *et al.* (1991) and Bürglin *et al.*, 1989) for the upstream and downstream primers respectively.

The first pair of primers were mixed with genomic DNA or cDNA and denatured (5 min, 95 °C), and amplified by 40 PCR cycles of 94 °C, 1 min, 50 °C, 1 min, 72 °C, 1 min, followed by a final 5 min extension at 72 °C. The specific amplification of POU type genes was performed in the same conditions, except that the annealing was at 55 °C and the elongation time was increased to 3 min. Recombinant DNA cloning of PCR products was performed by standard methods. DNA sequencing was used to identify different homeobox genes.

To analyze the spatial distribution of gene transcripts, *in situ* hybridization using ^{35}S-labelled riboprobes was used, as previously described (Garcia-Fernández *et al.*, 1993).

Results and discussion

Intact and regenerating freshwater planarian morphology

Freshwater planarians are unsegmented and bilaterally symmetrical. They lack circulatory, respiratory and skeletal structures. The digestive system is formed by a pharynx and a blind gut lacking an anus. The mesenchyme or parenchyma is a solid mass of tissue that fills the space between the epidermis and the gut and surrounds the internal organs. The parenchyma consists of several non-proliferating differentiated cell types and a particular class of undifferentiated mitotic cells called neoblasts (Baguñà *et al.*, 1990).

When a freshwater planarian is cut, the wound rapidly contracts and a thin layer of old epidermal cells covers it. During the early regenerative stage (0–2 days) an accumulation of neoblasts takes place by continuous addition of new cells formed by cell division in the underlying parenchyma (Saló & Baguñà, 1984; 1989). During this early period the determination of new structures occurs along the antero-posterior axis (Saló, 1984; Baguñà *et al.*, 1990). In the following days this accumulation of undifferentiated cells (blastema) grows exponentially and at 5–7 days of regeneration the differentiation of new structures initiates within the blastema; the regeneration is completed after 3–4 weeks at 17 °C.

Diversity of freshwater planarian homeobox genes

Homeobox genes comprise diverse and phylogenetically widespread types of gene. Their sequence comparisons allow recognition and classification in many families (for example HOM/Hox, NK, Prd, POU; Bürglin, 1993). We were interested in determining whether the freshwater planarian genome has different families of homeobox genes and how many genes are present for each class.

Planarian cDNA libraries were screened with degenerate oligonucleotides from the helix-3 of the homeodomain. Two homeobox genes, called *Dth*-1 and *Dth*-2, and belonging to the NK-type, were isolated (Garcia-Fernàndez *et al.*, 1991). We also employed the polymerase chain reaction (PCR) to assess the diversity and complexity of homeobox gene families in the freshwater planarian genome. Using a set of degenerate oligonucleotides representing two conserved regions in the POU-domain we amplified a portion of the POU-box from freshwater planarian genomic DNA and cDNA. Three POU-boxes were amplified (Table 1), from which the nucleotide sequences and the corresponding predicted amino acid sequences were obtained (Fig. 1). Clone DtPOU-1 codes for a putative polypeptide, identical over the entire POU-domain to the 'DjPOU-1' from the planarian *Dugesia japonica* (Orii *et al.*, 1993). The predicted amino acid sequence of clone DtPOU-2 shows 91% and 62% similarity to the DtPOU-1 clone in the POU-specific and homeotic domains respectively, while the linker region is completely divergent. DtPOU-1 and DtPOU-2 show the typical features of the POU-III class (He *et al.*, 1989). The last clone, DtPOU-3, shows low similarity with the first two, (61-63% to POU specific and 40% to POU-homeodomain). DtPOU-3 is very similar to the IV class (He *et al.*, 1989), including three characteristic additional amino acid residues in the POU-specific domain.

Using these three POU amplified fragments we screened the cDNA and genomic libraries and we isolated the corresponding DtPOU-1 cDNA clone, and the DtPOU-3 genomic clone (work in progress). Moreover, genomic Southern blot analysis with the same probes (results not shown) confirms that these three POU genes are of freshwater planarian origin.

Finally, using homeobox primers capable of amplifying most HOM/Hox genes plus some other families we amplified freshwater planarian genomic DNA and cDNA and cloned these into plasmid vectors. Four novel homeobox genes were identified (Table 1). Three of them, named DthoxB, DthoxF and DthoxD, belong to the HOM/Hox family, and one, called Dtprd-1, to the paired family. DtoxB shows the highest similarity (67%) with the proboscipedia class genes, DtoxF corresponds to the Antennapedia class (93%), and DtoxD could be considered, homologous to the Abdominal-A class (85%). Although Dtprd-1 shows a low similarity to the paired family (44%), additional sequences from the cDNA and genomic clones isolated confirms that it belongs to that family. Planarians like some other lower metazoa, have an AT-rich genome and A or T is more frequently found at the third position of the codon, whereas *Drosophila* and vertebrate species more frequently display G or C (Garcia-Fernàndez *et al.*, 1991). The sequences from these four novel homeobox genes showed an A/T rich sequence and a clear codon bias. Moreover, the genomic or cDNA clones of some of these PCR amplified fragments were isolated from the planarian libraries (work in progress). From these results these four genes appear to be of freshwater planarian origin.

Expression of freshwater planarian homeobox genes.

To investigate the pattern of spatial expression of Dth-1 and Dth-2 in adult freshwater planarians by in situ hybridization, a 1100 and a 550 nucleotide specific single-stranded antisense RNA probe, from Dth-1 and Dth-2 cDNA clones respectively, were used. Dth-1 is detected specifically in cells of the gastrodermis (Fig. 2A,C), but it is important to note that additional expression has been observed in some clusters of cells surrounding the gut (Fig. 2A). These clusters could be neoblasts committed or determined to become intestinal cells in order to maintain the cell turnover of this tissue. Dth-2 expression shows a clear expression in the peripheral parenchyma (Fig. 2B), with higher values in dorsal and lateral regions than in ventral regions and a nonhomogeneous silver grain distribution (Fig. 2D). Since specific organs precursors are not known in these regions, this pattern of expression could be correlated with a non-homogeneous distribution of a specific peripheral parenchymal cell type or types. During regeneration, the expression of both genes is not enhanced at early stages, though a slight increase, linked to cell differentiation, is seen at later stages (Garcia-Fernàndez *et al.*, 1993). These results strongly suggest that Dth-1 and Dth-2 are tissue-specific homeobox genes involved in processes of determination and/or differentiation of specific cell types, but do not indicate any role in early pattern formation during regeneration.

In situ hybridization experiments using a DtPOU-1 specific cDNA fragment of 450 nucleotides did not give clear results. It should be pointed out, however, that northern blot and reverse transcription and PCR (RT-PCR) analysis with the homologous POU gene of Dugesia japonica, DjPOU-1, showed a weak expression with higher values in the head regions than in the tail regions (Orii *et al.*, 1993). These results suggest a specific head expression of DjPOU-1 within nerve cells of head ganglia.

The other homeobox genes have not yet been studied (work in progress), but the number of positive clones detected in the cDNA library screening suggests that transcripts are not abundant, even during regeneration.

Homeobox genes and pattern control in adult and regenerating planarians

The mechanisms of pattern formation and maintenance in planarians are still unsolved. To establish how the lost pattern is built again during regeneration and how it is maintained in intact adults, we have searched for HOM/Hox homeobox gene clusters. These genes are known to have a key role in higher organisms in setting up the antero-posterior polarity during early develop-

```
           1                                                                      60
DtPOU-1    LGYTQADVGL ALGTLYGNVF ...SQTTICR FEALQLSFKN MCKLRPLLQK WLHEADSSSE
DtPOU-2    LGFTQADVGL ALGNLYGSVF ...SQTTICR FEALQLSFKN MCKLRPLLGK WLLEADSTGC
DtPOU-3    LGVTQADVGK ALGNLKLSGV GSLSQSTICR FESLTLSHNN MVALKPILQA WLEEAERQAA

           61                                                                    117
DtPOU-1    SPTNFDKISA QS.RKRKKRT SIEANVKSIL ESSFMKLSKP SAQDISSLAE KLSLEKE
DtPOU-2    APSGIDKVTA A.GRKRKKRT SIEVGVKNAL ESHFSRQSKP TAQIITQLAD NLGLEKE
DtPOU-3    ERLKNPDIYN DSLDRKRKRT SITDAEKRSL EAYFAVQPRP SSEKIAQIAD KLNLKKN
```

Fig. 1. Partial predicted protein sequences of DtPOU-1, DtPOU-2 and DtPOU-3 deduced from PCR amplified planarian genomic DNA. Partial POU-specific domain and POU-homeodomain correspond respectively to positions 1–56 and 74–117. Dots indicate gaps inserted to maintain sequence alignment.

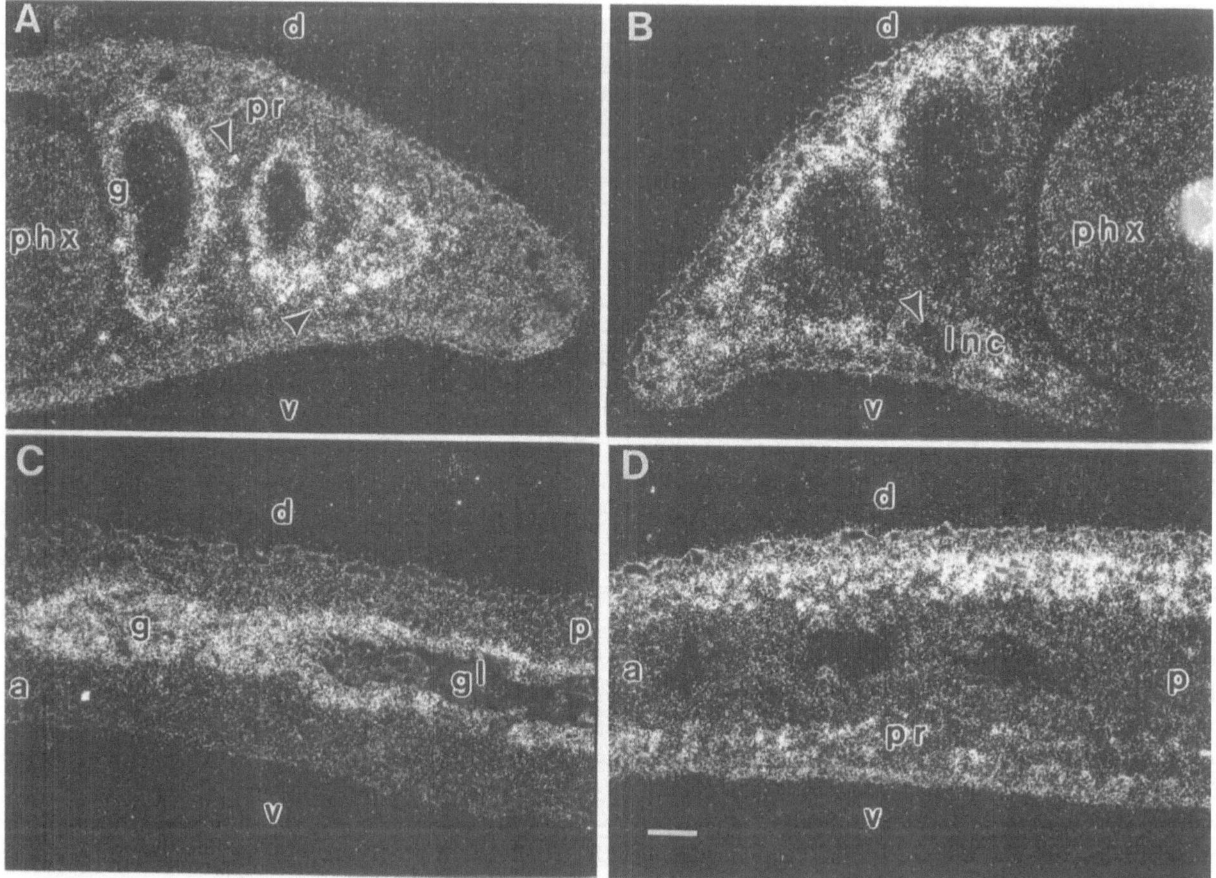

Fig. 2. Expression of Dth-1 (A, C) and Dth-2 (B, D) homeobox genes in a freshwater planarian *Dugesia (G.) tigrina*, as revealed by *in situ* hybridization. (A) Transverse section hybridized to Dth-1 antisense probe. Clusters of hybridization in areas surrounding the gastrodermis are indicated by arrows. (B) Transverse section hybridized to Dth-2 antisense probe. Higher values of expression could be seen in dorsal and lateral areas than in the ventral regions of peripheral parenchymal cells. Lack of hybridization of Dth-2 probe to the longitudinal nerve cord is indicated by an arrow. (C) Sagittal sections hybridized to Dth-1 antisense probe. Stronger expression in the intestinal cells can be observed. (D) Sagittal sections hybridized to Dth-2 antisense probe. Their expression is scattered in small clusters of cells from the peripheral parenchyma. *a-p*, anteroposterior axis; *d-v*, dorsoventral axis; *phx*, pharnyx; *pr*, parenchyma; *g*, gastrodermis; *gl*, gut lumen; *lnc*, longitudinal nerve cord. (Scale = 50 μm.)

ment (Holland, 1992; McGinnis & Krumlauf, 1992). Their ubiquitous distribution from Cnidarians to Vertebrates and their remarkable similarity in genomic organization and function have been taken to represent apomorphic features defining the whole animal kingdom (Slack *et al.*, 1993). Although, in Cnidarians

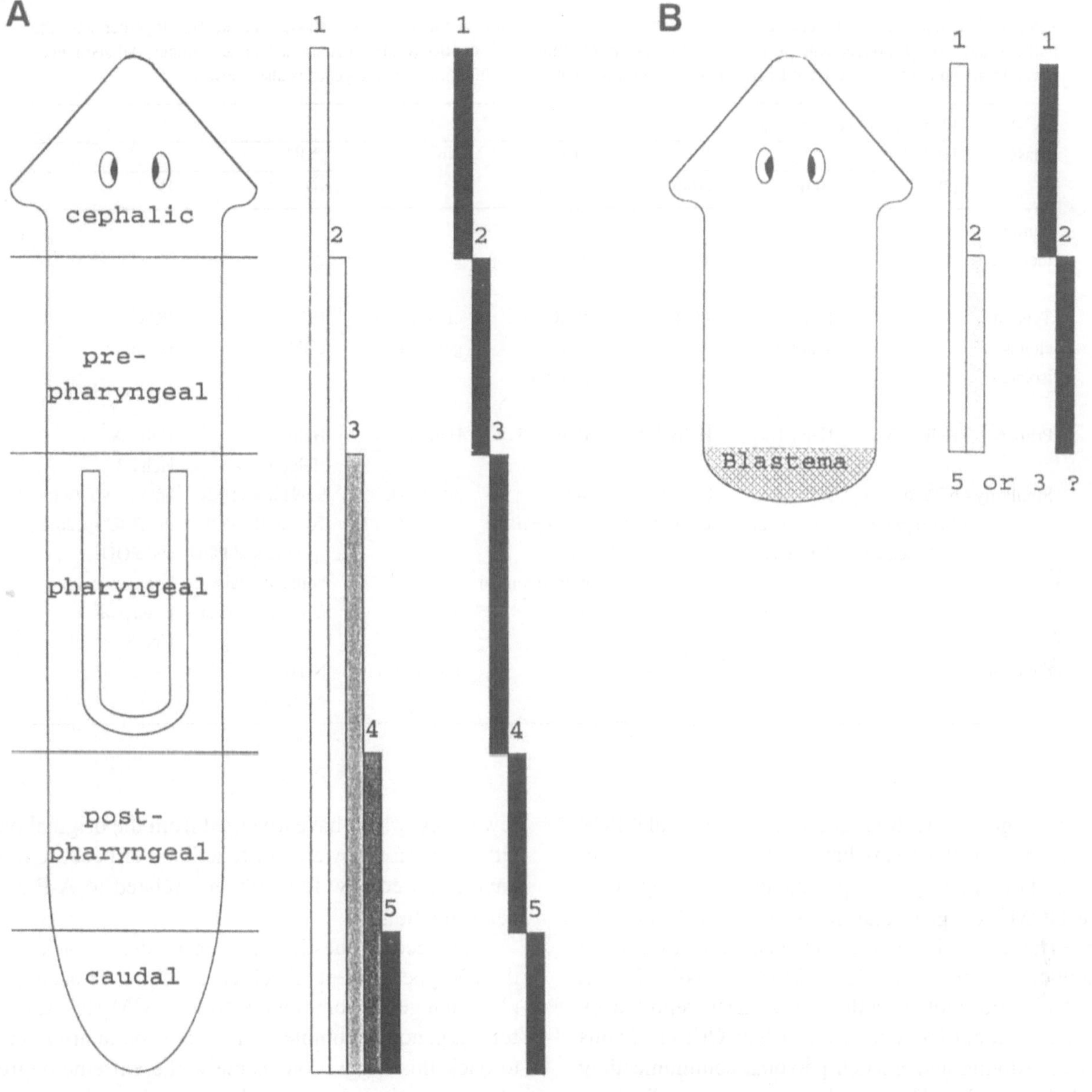

Fig. 3. (A) Scheme showing two possible models of pattern expression of five hypothetical freshwater planarian HOM/Hox genes. Inside, regions defined by morphological criteria; right, pattern of expression of five different HOM/Hox hypothetical genes. Different shadowed vertical bars represent the nested model and black vertical bars represent the region-specific model. (B) Hypothetical HOM/Hox gene expression during posterior regeneration according to a disto-proximal or a proximo-distal mechanism of pattern restitution.

has been demonstrated a linkage between even-skipped and Antennapedia class genes (Miler & Miles, 1993), almost nothing is known on the genomic organization and pattern expression of HOM/Hox genes in lower invertebrates.

Are the HOM/Hox genes found in planarians homologous to the genes within the HOM/Hox gene clusters of higher organisms?. To answer this question, some previous problems must be tackled: (1) How many HOM/Hox genes could be expected to be present in planarians?; (2) are they organized in clusters?; and (3) do they regulate the antero-posterior polarity in the intact and regenerating organism?

Table 1. The number of different genes from each homeobox class thus far identified in freshwater planarians, together with the number and type of independent clones isolated and level of amino acid identity to the most similar homeodomain. All sequences for comparison are taken from the EMBL and SwissProt databases. The function of some genes is also shown.

Class	Planarian homeobox genes						
	HOM/Hox			Prd	NK	POU	
	pb	Antp	Antp			III	IV
Number identified	1	1	1	1	2	2	
Type of clones isolated	PCR	PCR genomic	PCR	PCR cDNA genomic	cDNA genomic	PCR cDNA	PCR genomic
Primers	Helix 1,3	Helix 1,3	Helix 1,3	Helix 1,3	Helix 3	Helix A Helix 3	Helix A Helix 3
Similarity	67% to *Drosophila* proboscipedia	93% to *Drosophila* Antennapedia	85% to *Drosophila* Abdominal A	44% to *Drosophila* paired homeodomain	80-82% to *Drosophila* NK	93-91% and 72-76% to the POU specific and POU homeodomain of *Drosophila* Cfla	86% and 67% to the POU specific and POU homeodomain of *C. elegans* Unc-86
Function					Tissue specific	Nerve cell expression?	

As regards the first question, large scale PCR screenings in planarians have revealed a large number of homeobox genes, presumably homologous to the HOM/Hox gene cluster Six of these HOM/Hox genes have been introduced in the GenBank (accession numbers: z22600, z22601, z22539, z22540, z22541, z22542) and another sixteen have been reported by several authors (Bartels *et al.*, 1993; Orii *et al.* this issue; Adoutte and Ehrlich personal communication; and this work). This number is unexpectedly high considering the rather simple body plan of planarians. Indeed, the more complex *Drosophila* has two clusters (ANTC and BX-C) that comprise only eight HOM/Hox genes. On the other hand, a thorough comparative analysis of many homeobox sequences postulates that at the diploblast/triploblast divergence, three classes of homeobox genes must have been present (Schubert *et al.*, 1993). So, a reasonable guess for the number of HOM/Hox genes in planarians would be between three and eight. However, most planarian species screened so far are polyploids or diploids derived from polyploid ancestors. Therefore, it can be speculated that the large number of HOM/Hox genes found in planarians may include duplicated copies of a few genes, which have diverged from the original one either because they are not functional or because they have acquired new functions not related to A-P axis determination.

The second question is even harder to tackle. If the zootype concept (Slack *et al.*, 1993) is accepted, planarian genes homologous to the HOM/Hox should form a genomic complex. There are sound arguments to back this suggestion; namely, the extreme degree of conservation along the phylogenetic scale of these genes in clusters. To check this we are planning to analyze the genomic organization of planarians using three different approaches: a) chromosome walking in the genomic library; b) Southern blots of pulse field electrophoresis of large fragments of genomic DNA; and c) in situ hybridization on chromosome preparations.

Finally, to determine whether or not these genes are involved in the control of A-P polarity in both intact and regenerating organisms, it is important to obtain, first, some data on its spatial and temporal expression. Northern blots of different A-P regions of the organism, and *in situ* hybridization on sagittal paraffin sections seem the most straightforward approaches.

Later on, some methods can be devised to alter their expression and test their functional role in a more direct way. Which pattern of expression should be expected for such genes in intact and regenerating organisms? According to the patterns found in *Drosophila*, some vertebrates (McGinnis & Krumlauf, 1992), and in one Cnidarian homeobox gene (Shenk *et al.*, 1993), we could expect either a nested expression or a region-specific expression of five genes that would determine the five different regions drawn according to morphological criteria (Fig. 3A). Using these genes as regional markers and analyzing their order of appearance during regeneration we could answer long-standing problems such as the proximo-distal or distoproximal polarity of pattern restitution (Fig 3B).

Acknowledgments

This work was supported by grants from the Dirección General de Investigación Cientifica y Técnica (DGI-CYT) (Ministerio de Educación y Ciencia, España; PB89-0249) and the Comissió Interdepartemental de Recerca i Innovació Tecnológica (Generalitat de Catalunya; AR-90; AR91). A.M.M.-M. and J.R.B.-R. are recipient of a FPI fellowship.

References

Addoutte,A. & H. Philippe, 1993. The major lines of metazoan evolution: Summary of traditional evidence and lessons from ribosomal RNA sequence analysis. In. Y. Pichon (ed.), Comparative Molecular Neurobiology. Birhäuser Verlag. Basel: Switzerland: 1–34.

Baguñà, J., R. Romero, E. Saló, J. Collet, C. Auladell, M. Ribas, M. Riutort, J. Garcia-Fernàndez, F. Burgaya & D. Bueno, 1990. Growth, degrowth and regenerations as developmental phenomena in adult freshwater planarians. In: Experimental Embryology in Aquatic Plants and Animals. (ed. H. J. Marthy.): New York: Plenum Press. pp.: 129–162.

Bartels, J. L., M. T. Murtha, & F. H. Ruddle, 1993. Multiple Hox/HOM-class homeoboxes in Platyhelminthes. Mol. Phyl. and Evol. 2 No 2: 143–151.

Billin, A. N., K. A. Cockerill, & S. J. Poole (1991). Isolation of a family of Prosophila POU domain genes expressed in early development. Mech Dev. 34: 75–84.

Bürglin, T. R., M. Finney, A. Coulson, & G. Ruvkun, 1989. *Caenorhabditis elegans* has scores of homeobox-containing genes. Nature 341: 239-243.

Bürglin, T. R. 1993. The homeodomain phylum. In A guidebook for homeobox genes. D. Duboule (ed.). Oxford: Oxford University Press. In press.

Forhman, M. A., M. Boyle, & G. R. Martin, 1990. Isolation of the mouse *Hox-2.9* gene; analysis of embryonic expression suggests that positional information along the anterior-posterior axis is specified by mesoderm. Development 110: 589–607.

Garcia-Fernàndez, J., J. Baguñà, & E. Saló, 1991. Planarian homeobox genes: Cloning, sequence analysis, and expression. Proc. Natl. Acad. Sci. USA 88: 7338–7342.

Garcia-Fernàndez, J., J. Baguñà, & E. Saló, 1993. Genomic organization and expression of the planarian homeobox genes Dth-1 and Dth-2. Development 11 8: 241–253.

Gehring, W. J. 1987. Homeoboxes in the study of development. Science 236: 1245–1252.

He, X., M. N. Treacy, D. M. Simmons, H. A. Ingraham, L. W. Swanson, & M. G. Rosenfeld, 1989 Nature 340: 35–42.

Holland, P. W. H., 1992. Homeobox genes in vertebrate evolution. Bio Essays 14: 267–273.

McGinnis, W. & Krumlauf, R. 1992. Homeobox genes and axial patterning. Cell 68: 283–302.

Miller, D. J. & A. Miles, 1993. Homeobox genes and the zootype. Nature 365: 215–216.

Orii, H., K. Agata, & K. Watanabe, 1993. POU-domain genes in planarian *Dugesia japonica*: The structure and expression. Biochem. Bioph. Res. Commun. 192: 1395–1402.

Riutort, M., K. G. Field, J. M. Turbeville, R. A. Raff, & J. Baguñà, 1992. Enzyme electrophoresis, ^{18}S rRNA sequences, and levels of phylogenetic resolution among several species of freshwater planarians (Platyhelminthes, Tricladida, Paludicola). Can. J. Zool., 70: 1425–1439.

Saló, E. 1984. Formació del blastema i re-especificació del patró durant la regeneració de les planàries Dugesia (S) mediterranea i Dugesia (G.) tigrina: anàlisi morfologica, cel.lular i bioquímica, Ph. D. Thesis, University of Barcelona.

Saló, E. & J. Baguñà, 1984. Regeneration and pattern formation in planarians. I. The pattern of mitosis in anterior and posterior regeneration in *Dugesia (G) tigrina*, and a new proposal for blastema formation. J. Embryol. Exp. Morph. 83: 63–80.

Saló, E. & J. Baguñà, 1989. Regeneration and pattern formation in planarians. II. Local origin and role of cell movements in blastema formation. Development 107: 69–76.

Schubert, F. R., K. Nieselt-Struwe & P. Gruss, 1993. The Antennapediatype homeobox genes have evolved from three precursors separated early in metazoan evolution. Proc. Natl. Acad. Sci. USA. 90: 143–147.

Shenk, M. A., H. Bode & R. E. Steele, 1993. Expression of Cnox-2, a HOM/HOX homeobox gene in hydra, is correlated with axial pattern formation. Development 117: 657–667.

Slack, J. M. W., P. W. H. Holland & C. F. Graham, 1993. The zootype and the phylotypic stage. Nature 361: 490–492.

Hydrobiologia **305**: 277–279, 1995.
L.R.G. Cannon (ed.), Biology of Turbellaria and some Related Flatworms.

cDNA cloning and partial sequencing of homeobox genes in *Dugesia japonica*

Hidefumi Orii*, Takuya Miyamoto, Kiyokazu Agata & Kenji Watanabe
Laboratory of Regeneration Biology, Department of Life Science, Faculty of Science, Himeji Institute of Technology, Harima Science Park City, Akou, Hyogo 678-12, Japan (author for correspondence)*

Key words: planarian, homeobox gene, regeneration, cell differentiation

Abstract

In the freshwater planarian *Dugesia japonica*, four types of cDNAs of homeobox-containing genes have been isolated by screening a cDNA library using a homeobox guessmer. Partial sequencing analysis of two types of cDNAs revealed that one was a homolog of *Dth2* which is a homeobox gene in *Dugesia tigrina* and another was similar to *Distal-less* gene in *Drosophila*. This suggests that planarians have many homeobox genes.

Introduction

Planarians are well-known for their high regeneration potentiality and have been attractive animals for developmental biologists. However, our knowledge of planarian regeneration processes at molecular level is very poor. For example, we have no idea of molecular events occurring in the regenerating blastema, when, where and what types of cells differentiate and what genes control cell differentiation in the blastema.

Recently, homeobox genes have been isolated and analysed in various animals (for example; Scott *et al.*, 1989). The products of homeobox genes which contain a conserved amino acid sequence (61 residues) called homeodomain, function as transcription factors to regulate gene expression. Many homeobox genes are thought to have key roles in development and cell differentiation. For example, in *Drosophila* each homeobox gene in the *Antennapedia* and *Bithorax* complexes specifies the identity of a particular region of the body along the antero-posterior axis.

To understand the mechanism of planarian regeneration, we focused on the planarian homeobox genes, some of which were expected to be involved in the pattern formation of regeneration. We report here the cloning and partial sequencing of several cDNAs of homeobox genes in the freshwater planarian *Dugesia japonica*.

Materials and methods

A cDNA library was prepared in lambda ZAP from poly A^+ RNA of the anterior part of the planarian *Dugesia japonica* and was screened to isolate homeobox genes with a guessmer corresponding to the amino acid sequence KIWFQNRR in the homeobox (Scott *et al.*, 1989; Orii *et al.*, 1993). The condition for hybridization and washing in the screening was described by Bürglin *et al.* (1989). Several positive phage clones have been isolated and subcloned into the pBluescriptSK$^-$ plasmid vector (USB) by *in vivo* excision. These clones have been categorized by restriction mapping. DNA sequence analysis was performed by the method of Sanger *et al.* (1977).

Results and discussion

Using the homeobox guessmer, we have isolated 10 positive clones out of 5×10^5 independent phage clones. The clones were classified into four types by restriction mapping (Fig. 1). Partial sequencing analysis on clone 11 revealed that it was a homolog to *Dth2* which is a homeobox gene in *Dugesia tigrina* (Fig. 2, Garcia-Fernàndez *et al.*, 1991). Only one different amino acid residue between clone 11 and *Dth2* was observed in the amino acid sequence of the amino terminal region which has been sequenced. Therefore we named the clone 11 *Djh2* (*Dugesia japonica*

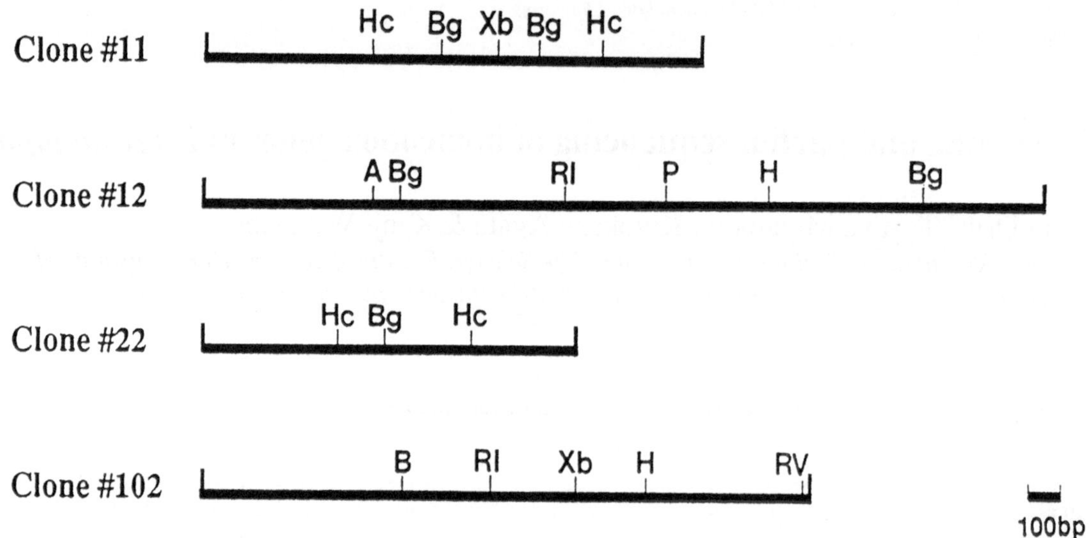

Fig. 1. Restriction maps of isolated cDNA clones of homeobox genes. A, *Acc*I; B, *Bam*HI; Bg, *Bgl*II; H, *Hin*dIII; Hc, *Hin*cII; P, *Pst*I; RI, *Eco*RI; RV, *Eco*RV; Xb, *Xba*I; Xh, *Xho*I.

```
#11   CAATCATCTATTTTAATAACACTTAACACTCTGAGAACCTAAGTTGGTATTAGGAGAAGT

#11   ACTACTTGAAGACTACTCAATCTGGTTATACACCCATCATTCAAAAAGCTACTATGAGTG
                                                             M  S
Dth2                                                         M  S

#11   GTATACCAGGGCTCTCTGGAATGTCCTCAATGTCTCCGTATTCTGCATACG
      G  I  P  G  L  S  G  M  S  S  M  S  P  Y  S  A  Y
Dth2  G  I  P  G  L  A  G  M  S  S  M  S  P  Y  S  A  Y
                     -
```

Fig. 2. Partial sequence of clone 11 and comparison with *Dth2*. Predicted amino acid sequence of clone 11 is shown under nucleotide sequence. *Dth2* amino acid sequences are from Garcia– Fernàndez *et al.* (1991). A variant amino acid is underlined.

homeobox 2). In *Dugesia tigrina*, *Dth2* is expressed in the peripheral parenchyma and may be involved in specific cell or tissue differentiation (Garcia-Fernàndez *et al.*, 1993). Northern blot analysis of clone 11 showed that it was the same as *Dth2* in the transcript size and expression pattern during regeneration. This suggests that clone 11 (*Djh2*) in *D. japonica* has the same function(s) as *Dth2* in *D. tigrina*.

We have also analysed the nucleotide sequences of the clone 22. Partial sequence data showed that clone 22 was a new type of planarian homeobox gene. It has a high similarity in the homeobox with *Distal-less* (*Dll*) gene in *Drosophila* (59%), *cnox3* gene in hydra (51%) and muscle segment homeobox genes in *Ciona* (51%) and *Drosophila* (51%) (Fig. 3). It is interesting that in the hydra the accumulation of *cnox3* mRNA are observed from 0 to 48 hours after amputation (Schummer *et al.*, 1992). The *Dll* is expressed in the leg primordia of the thoracic segments of the embryo and is required in these cells for the development of larval and adult leg structure (Cohen, 1990). Recently, it was revealed the *Dll* gene was repressed in abdominal segments of the *Drosophila* embryo by the homeobox genes of the Bithorax complex and so that limb development in the abdomen was repressed (Vachon *et al.*, 1992).

In addition to two homeobox genes characterized in this paper, we have already isolated and sequenced *Djh3* and *Djh4* cDNAs which were similar to the *abdominal-A* and *Deformed* of *Drosophila*, respectively. By the polymerase chain reaction (PCR) technique, it has been revealed that *Dugesia japonica* has at least 19 homeobox genes including *Djh1* (homolog to *Dth1*), *Djh3* and *Djh4* (Orii *et al.*, unpublished data). Considering that the nematode *Caenorhabditis elegans* has more than 50 homeobox genes (Bürglin *et al.*, 1989),

	clone 22	Dll	DLX-2	cnox3	bee H17	Dromsh	Ciomsh
clone 22	-						
Dll	**59**	-					
DLX-2	**61**	84	-				
cnox3	**51**	54	56	-			
bee H17	**52**	51	54	44	-		
Dromsh	**51**	51	52	46	97	-	
Ciomsh	**51**	52	54	48	92	93	-

Fig. 3. Percentage similarities between clone 22 and several homeoboxes. References are as follows; *Dll* (Vachon *et al.*, 1992), DLX-2 (Selski *et al.*, GENBANK accession no. L07919), *cnox3* (Schummer *et al.*, 1992), bee H17 (Walldorf *et al.*, 1989), Dromsh (*Drosophila* muscle segment homeobox gene) and Ciomsh (*Ciona* muscle segment homeobox gene) (Holland, 1991).

we guess that the planarian might also have many homeobox genes. Some of these homeobox genes may be present as a cluster on the planarian genome, as well as the homeobox gene complex (HOM-C) in higher animals (Scott, 1992), and we expect our cloned *Djh3* and *Djh4* to be members of HOM-C. It has been suggested that homeobox genes, especially HOM-C genes control body planning in the development of phylogenetically higher animals. We expect that the research on these homeobox genes will reveal molecular mechanism of planarian regeneration.

Acknowledgments

We thank E. Saló and Y. Umesono for helpful discussion. This work was supported partly by Grants-in-Aid from the Japanese Ministry of Education, Science and Culture to H.O (no. 05780552), to K.A (no. 04304007) and to K.W (no. 05277217) and from Hyogo Science and Technology Association to H.O (5H16).

References

Bürglin, T. R., M. Finney, A. Coulson & G. Ruvkun, 1989. Caenorhabditis elegans has scores of homeobox-containing genes. Nature 341: 239–243.

Cohen, S. M., 1990. Specification of limb development in the Drosophila embryo by positional cues from segmentation genes. Nature 343: 173–177.

Garcia-Fernàndez, J., J. Bagunà & E. Saló, 1991. Planarian homeobox genes: Cloning, sequence analysis, and expression. Proc. nat. Acad. Sci. USA 88: 7338–7342.

Garcia-Fernàndez, J., J. Bagunà & E. Saló, 1993. Genomic organization and expression of the planarian homeobox genes Dth-1 and Dth-2. Development 118: 241–253.

Holland, P. W., 1991. Cloning and evolutionary analysis of msh-like homeobox genes from mouse, zebrafish and ascidian. Gene 98: 253–257.

Orii, H., K. Agata & K. Watanabe, 1993. POU-domain genes in planarian Dugesia japonica: The structure and expression. Biochem. Biophys. Res. Comm. 192: 1395–1402.

Sanger, F., S. Nicklen & A. R. Coulson, 1977. DNA sequencing with chain terminating inhibitors. Proc. nat. Acad. Sci. USA 74: 5463–5467.

Schummer, M., I. Scheurlen, C. Schaller & B. Galliot, 1992. HOM/HOX homeobox genes are present in hydra (Chlorohydra viridissima) and are differentially expressed during regeneration. The EMBO J. 11: 1815–1823.

Scott, M., 1992. Vertebrate homeobox gene nomenclature. Cell 71: 551–552.

Scott, M. P, J. W. Tamkun & G. W. Hartzell, 1989. The structure and function of the homeodomain. Biochim. Biophys. Acta Rev. Cancer 989: 25–48.

Vachon, G., B. Cohen, C. Pfeifle, M. E. McGuffin, J. Botas & S. M. Cohen, 1992. Homeotic genes of the Bithorax complex repress limb development in the abdomen of the Drosophila embryo through the target gene Distal-less. Cell 71: 437–450.

Walldorf, U., R. Fleig & W. J. Gehring, 1989. Comparison of homeobox-containing genes of the honeybee and Drosophila. Proc. nat. Acad. Sci. USA 86: 9971–9975.

Hydrobiologia **305**: 281, 1995.
L.R.G. Cannon (ed.), Biology of Turbellaria and some Related Flatworms.

Sequence analysis and expression studies of a collagen cDNA in *Diphyllobothrium dendriticum* (Cestoda)

Kaj A. Karlstedt, Monica H. Wahlberg, Anne J. West & Gun I. L. Paatero
Åbo Akademi University, Department of Biology, BioCity, Artellerigatan 6, FIN-20520 Åbo, Finland

Key words: invertebrate collagen, Northern blot, *in situ* hybridization, *Diphyllobothrium dendriticum*, Cestoda

Abstract

Collagens are a group of evolutionary related structural proteins that are found in extracellular matrices. The vertebrate collagens have been divided into fibril-forming and nonfibril-forming collagens based on the structure of the proteins as well as the organization of the corresponding genes. In invertebrates, both fibrillar and nonfibrillar collagen genes that resemble their vertebrate counterparts have been identified (see Vuorio & Crombrugghe, 1990). The present study is the first characterization of a collagen gene in flatworms. A cDNA library (Uni-Zap XR, Stratagene) of the flatworm *Diphyllobothrium dendriticum* was screened with a murine pro-α1 type II collagen cDNA probe (a gift from Prof. Eero Vuorio). Several positive clones have been isolated and one of the cDNA inserts, named DidC1, has been further characterized. DidC1 covers 2631 bp from the 3′-end of an approximately 4.5 kb transcript. The deduced amino acid sequence of DidC1 was determined, and it represents a protein that shows structural similarities to vertebrate fibril-forming collagens. The DidC1 polypeptide includes a C-propeptide that contains 7 conserved cysteine residues and a region of Gly-X-Y triplets with one Gly-X-Y-Z imperfection. Northern blot hybridizations indicate that DidC1 is expressed approximately 20 times more in the scolex and neck region of the adult flatworm than in the posterior part of the worm. *In situ* hybridization reveals that DidC1 is expressed in all developmental stages of the adult worm and that the expression is localized to the longitudinal, transverse and dorso-ventral muscles. Cells in association with the developing genital organs also express DidC1, but only until the organs are developed fully.

Reference

Vuorio, E. & B. de Crombrugghe, 1990. The family of collagen genes. Annu. Rev. Biochem. 59: 837-72.

Hydrobiologia **305**: 283, 1995.
L.R.G. Cannon (ed.), *Biology of Turbellaria and some Related Flatworms.*

Sequence analysis of an actin isoform in the flatworm *Diphyllobothrium dendriticum* (Cestoda)

Monica H. Wahlberg, Kaj A. Karlstedt & Gun I. L. Paatero
Åbo Akademi University, Department of Biology, BioCity, Artillerigatan 6, FIN-20520 Åbo, Finland

Key words: Actin cDNA, multigene family, *Diphyllobothrium dendriticum*, Cestoda

Abstract

We have examined actin cDNA of the flatworm *Diphyllobothrium dendriticum* (Cestoda). Actin is a contractile protein that has been implicated in a variety of developmental and cellular processes. It is highly conserved and present in all eukaryotic cells. It is of particular interest to analyze evolutionary preserved genes in flatworms, because ancestral flatworms are regarded to play a central role in the evolution of the metazoans (Barnes *et al.*, 1988). Screening a cDNA library of *D. dendriticum* (UniZap XR, Stratagene) with a human γ-actin probe resulted in several positive clones. One of the cDNA inserts, *Didact1*, consisting of 1392 bp was completely sequenced. The established nucleotide sequence revealed a 5′ untranslated region of 33 bp, the entire open reading frame of 1128 bp and a 3′ untranslated region of 231 bp which ends in a stretch of 21 A residues. The potential polyadenylation signal (AATAAA) is located 14 bp upstream of the poly(A) tail. The deduced amino acid sequence of *Didact1* is 376 amino acids long. It is a typical invertebrate actin (Fyrberg *et al.*, 1981) resembling more the cytoplasmic than the muscular isoforms of vertebrate actins. *Didact1* is for example 96% homologous to human cytoplasmic γ-actin but only 92.6% identical with human smooth muscle α-actin. The actin proteins are generally encoded by a multigene family which differs in size from species to species. Most organisms have four to eight genes coding for actin in their genome, but the number of actin genes can also be over 20 (Hamelin *et al.*, 1988). Sequence comparisons of *Didact1* and the partly sequenced cDNA clones indicate that *D. dendriticum* has at least four different genes coding for actin in its genome.

Acknowledgment

This work was supported by the Victoria foundation.

References

Barnes, R. S. K., P. Calow & P. J. W. Olive, 1988. The invertebrates: a new synthesis. Blackwell Scientific Publications, 582 pp.

Fyrberg, E. A., B. J. Bond, N. D. Hershey, K. S. Mixter & N. Davidson, 1981. The actin genes of Drosophila: Protein coding regions are highly conserved but intron positions are not. Cell 24: 107–116.

Hamelin, M., L. Adam, G. Lemieux & D. Pallotta, 1988. Expression of the three unlinked isocoding actin genes of Physarum polycephalum. DNA 7: 317–328.

Hydrobiologia **305**: 285–289, 1995.
L.R.G. Cannon (ed.), Biology of Turbellaria and some Related Flatworms.

GABA in the nervous system of the planarian *Polycelis nigra*

Krister Eriksson, Pertti Panula & Maria Reuter
Department of Biology, Åbo Akademi University, Biocity, SF-20520 Åbo, Finland

Abstract

Gamma-aminobutyric acid (GABA) is an important inhibitory neurotransmitter in vertebrates and it has a similar inhibitory role in several invertebrate taxa. The transmitters serotonin, octopamine, catecholamines and histamine are present in flatworms while evidence for GABA is still lacking. Therefore, we have studied the occurrence of GABA-like immunoreactivity (IR) in the planarian nervous system. Specimens of *Polycelis nigra* were fixed in 4% 1-ethyl-3-(3-dimethyl-aminopropyl) carbodiimide with 2% paraformaldehyde. The GABA-antiserum was raised in rabbits against GABA conjugated to keyhole limpet hemocyanin. Preabsorption with GABA-ovalbumin conjugate abolished all IR. The results were further confirmed with an monoclonal antibody and high pressure liquid chromatography (HPLC). In *P. nigra* GABA-like IR was seen as long, often varicose, sparsely distributed fibers in the ventral longitudinal nerve cords. IR was also located in a few cell somata in the brain and in the neuropil of the brain. The IR was restricted to the central nervous system and was absent in peripheral nerves and plexuses. The HPLC analysis supported the presence of GABA.

Our results suggest that GABA is an interneuronal transmitter in *P. nigra*. The results also suggest a phylogenetically old origin of GABAergic neurotransmission.

Introduction

The flatworms have the simplest nervous systems that exhibit centralization and cephalization. Their frontal ganglion is generally referred to as a brain. It is a common assumption that all higher animals have evolved from a flatworm-like ancestor (e.g. Barnes *et al.*, 1988) and this makes flatworms an obvious target for research. It is likely that many features of the ancestral flatworm are retained in present day flatworms.

In the evolution of nervous systems, the appearance of the various types of chemical neurotransmission is of much theoretical interest. This has given rise to many studies on potential neurotransmitters in the nervous system of several flatworm taxa. In the free-living turbellarians the catecholamines dopamine (DA) and noradrenaline (NA) and the indoleamine serotonin (5-HT) have been detected histochemically. Furthermore, immunoreactions (IR) to 5-HT and to histamine have been detected (Franquinet & Catania, 1979; Hauser & Koopowitz, 1987; Reuter & Palmberg, 1989; Wikgren *et al.*, 1990). Chromatographical methods have verified the presence of DA, NA, 5-HT (Welsh & King, 1970; Algeri *et al.*, 1983) and melatonin (Morita *et al.*, 1987) in planarians of the genus *Dugesia* and DA and its precursor 3,4-dihydroxyphenylalanine (dopa) in the microturbellarian *Stenostomum leucops* (Reuter & Eriksson, 1991). Studies on DA and 5-HT have provided strong evidence for an actual neurotransmitter role in flatworms for these two substances (Venturini *et al.*, 1989; Kabotyanski *et al.*, 1991; Webb, 1988).

Antisera raised against many neuropeptides show IR in flatworms (for review, see Gustafsson & Reuter, 1992), and recently a flatworm neuropeptide was isolated and sequenced (Maule *et al.*, 1991). Different forms of this peptide, neuropeptide F, have this far been isolated and characterized in two cestodes, *Moniezia expansa* (Maule *et al.*, 1991) and *Proteocephalus pollanicola* (see Marks *et al.*, 1993), and in the terrestrial planarian *Artioposthia triangulata* (see Curry *et al.*, 1992).

The amino acid γ-aminobutyric acid (GABA) is the most important inhibitory interneuronal transmitter in vertebrates and it is present in high concentrations in the brain (Roberts, 1986). In invertebrates it is an interneuronal transmitter in all studied phyla (Walker & Holden-Dye, 1991). It is an inhibitory transmitter in neuromuscular synapses in several phyla. In the nematode *Ascaris* it is present in inhibitory motorneurons (Johnson & Stretton, 1987) while it is restricted to the central nervous system (CNS) in the mollusc *Limax maximus* (see Cooke & Gelperin, 1988). In crustaceans it is present in inhibitory motorneurons in the legs and in interneurons in the thoracic ganglion (Homberg *et al.*, 1993). In insects GABA is present in the CNS and inhibitory motomeurons (Watson, 1986) and there is evidence that GABA is the inhibitory transmitter in insect neuromuscular junctions (Usherwood & Grundfest, 1965).

Despite this widespread occurrence of GABA in the animal kingdom, no reports of GABA in flatworms have to date been published. In marine polyclad flatworms treatment with GABA or glycine at concentrations down to 1 μM decrease the electrical activity in the ventral nerve cords. This inhibitory effect can be prevented by treatment with picrotoxin, bicuculline and strychnine (Keenan *et al.*, 1979). This indicates that GABA-binding receptors are present in flatworms.

In an attempt to ascertain whether this response to GABA reflects an actual presence of the substance in flatworms, we have studied the occurrence of GABA in the planarian *Polycelis nigra* with immunocytochemical (ICC) methods and high pressure liquid chromatography (HPLC).

Materials and methods

Animals

The planarians were from an old laboratory culture of *P. nigra* that had been kept in tap water and fed weekly with raw beef live for several years. Starved specimens ranging from about 4 to 12 mm in length were used.

Fixation

The animals were put on a damp filter paper in a 5 cm petri dish on ice prior to fixation. Only about 500 μl of cold fixative was added during the first 20 min. This amount saturated the filter paper and enabled the fixative to penetrate from the ventral side while the worms were unable to move or curl. When the worms were dead they were immersed in fixative. The fixatives were either 4% 1-ethyl-3(3-dimethyl-aminopropyl)-carbodiimide (EDAC, Sigma) and 2% formaldehyde (FA, Merck) in 0.1 M phosphate buffer, pH 7.4, with 0.25% Triton X-100 (PBS-T) for 3–12 h at 4 °C (Panula *et al.*, 1988) or 4% FA and 1% glutaraldehyde (GA, Merck) in PBS-T for 2–3 h in 4 °C. After fixation the worms were rinsed in PBS and kept in 20% sucrose in PBS in 4 °C for at least 3 h. They were then cut on a cryostat into 20 μm thick sections which were collected on coated glass slides. The EDAC/FA-fixed sections were treated with 0.5% sodium borohydride (Merck) for 30 min at room temperature (RT) to reduce free aldehydes that could cause excess background staining (Airaksinen *et al.*, 1992).

Antisera

Most of the stainings were performed with an anti-GABA antiserum (#1H) raised in rabbits against GABA conjugated with EDCDI to the carrier-protein keyhole limpet hemocyanin (Airaksinen & Panula, 1990). Controls for the specificity of the antiserum for GABA were done by incubation of the diluted antiserum with GABA-ovalbumin conjugate (25 μg ml^{-1}). To verify the results a monoclonal mouse anti-GABA antibody (mAb) was used. The specificity for GABA of this mAb (#115AD5) has been thoroughly described (Szabat *et al.*, 1991).

Immunocytochemistry

The antiserum was diluted 1:500 in PBS-T with 1% normal swine serum (NSS) and applied to the EDCDI/FA-fixed sections overnight at 4 °C. The mAb was diluted 1:10 in PBS-T with 1% NSS and applied to the FA/GA-fixed sections either overnight in 4 °C or 2 h in RT. After the incubation with anti-GABA serum or mAb the slides were washed for 45 min in 3 changes of PBS-T. The procedure continued with the indirect immunofluorescence method (Coons, 1958). The secondary antisera were either fluorescein-labelled swine antirabbit IgG (DAKO) for the anti-GABA serum or fluorescein-labelled rabbit antimouse IgG (DAKO) for the mAb. In both cases the secondary antiserum was diluted 1:40 in 1% NSS and applied for 45–60 min in RT. After 3 washes comprising 45 min the slides were coverslipped with 50% glycerol in PBS.

High pressure liquid chromatography

To further verify the presence of GABA in *P. nigra* reverse phase HPLC with fluorescence detection was performed. This was done according to the method described in Tamura *et al.* (1990) with minor modifications. Briefly, the amino acids were derivatized in 10 mM *o*-phthalaldehyde (OPA, Fluka) in the presence of 4.5 μl ml^{-1} 2-mercaptoethanol (Sigma) in 0.4 M

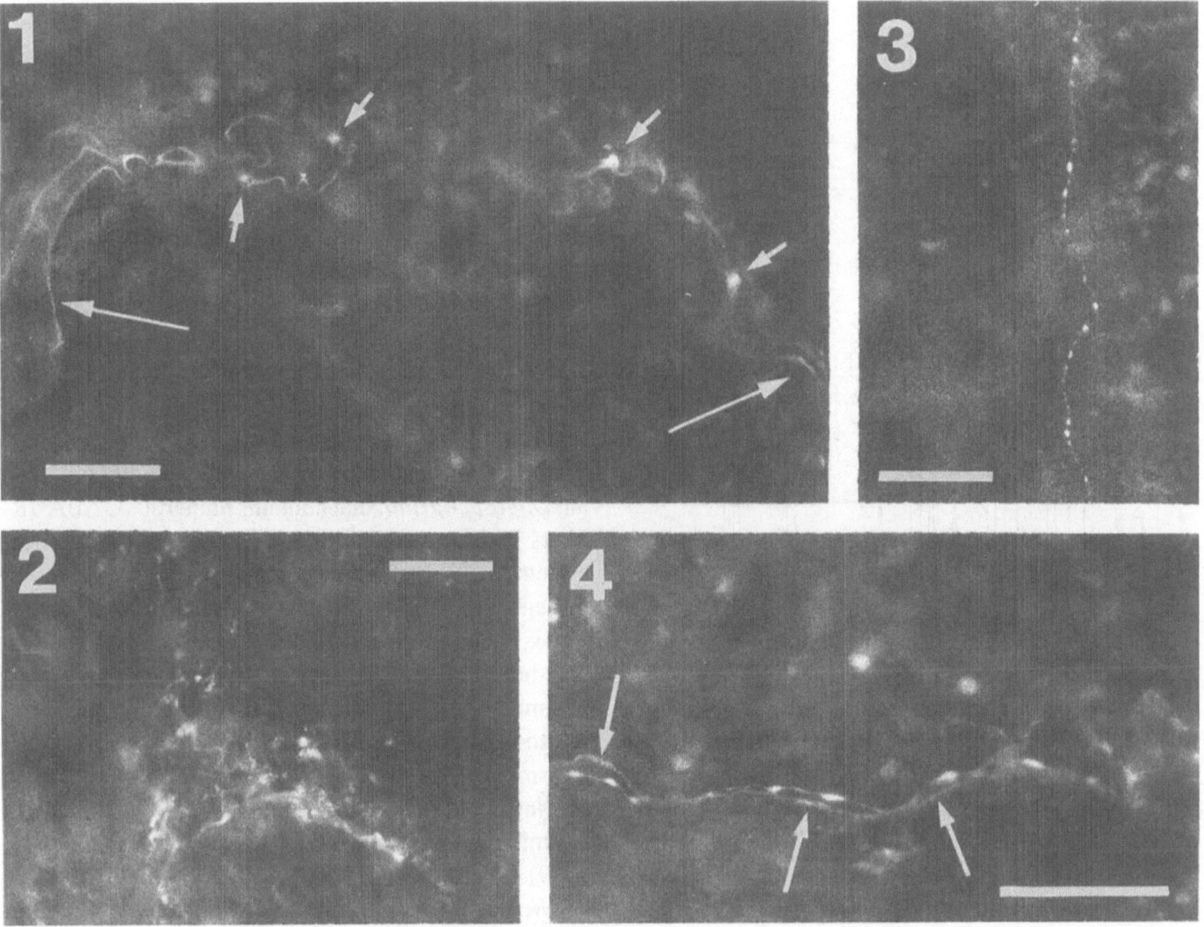

Figs 1–4. Frontal sections of *P. nigra* showing GABA-IR. *1*) GABA-IR in the brain. The frontal direction is up in the picture. Fibers which run into the longitudinal nerve cords (LNC) leave the brain (long arrows). The short arrows indicate cell somata. *2*) A part of the GABA-IR in the neuropil of the brain. Most or all of the IR are seen in fibers. Cell somata can not be identified with certainty. *3*) A part of a single very long varicose fiber in one of the LNC. This stretch is about 100 μm long. *4*) A fiber without varicosities (arrows) that runs in parallel with a varicose fiber in a LNC. The varicose fiber show stronger IR. In the mesenchyme surrounding the LNC autofluorescent tissue elements can be seen. Scale = 20 μm.

sodium borate buffer (pH 10.2) during 2 min at 12 °C. The OPA-derivatives of amino acids were separated on a TSKgel ODS-80 TM (4.6 × 150 mm, 5 μm, Tosoh) reverse phase column in two different 30 min non-linear gradients from 25% to 80% methanol in 0.1 M potassium acetate buffer, pH 5.1, at a flow rate of 0.8–1.0 ml min^{-1}. The excitation wave length was 360 nm and the fluorescence was measured at 455 nm.

Results

The results obtained with all three methods supported each other and they all indicated the presence of GABA in *P. nigra*. The localizations of IR to the anti-GABA antiserum and IR to the monoclonal antibody were similar. All GABA-IR were restricted to the CNS, consisting of the brain and the longitudinal nerve cords (LNC). IR were detected in fibers running in the LNC and in the neuropil (NP) of the brain. A few cell somata located in the periphery of the brain showed IR (Fig. 1). It was difficult to identify cell somata because of their small size. The IR in the NP appeared as irregular patches of IR surrounded by non-immunoreactive tissue (Fig. 2). No cell somata could be identified in the NP. The positive fibers in the LNC were few, sometimes only one and seldom more than two running in parallel. They were often several hundred μm long

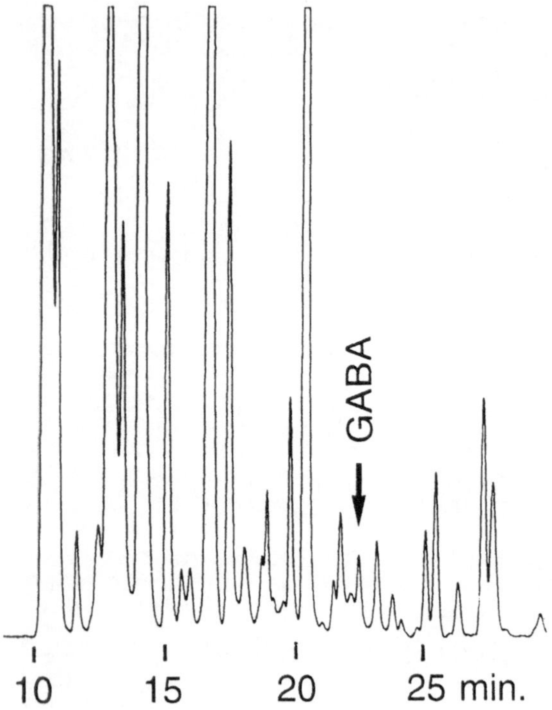

Figs 5. Chromatogram showing the separation of OPA-derivatives of amino acids as described in the text. GABA is indicated with an arrow. Several high peaks, corresponding to free amino acids, can be seen. The numbers refer to minutes after sample injection.

and varicose (Fig. 3). A common situation was one thin fiber with varicosities running in parallel with a thicker fiber without varicosities (Fig. 4). The fibers got more abundant closer to the brain and no GABA-IR fibers were seen in the posterior half of the worm. No IR fibers leave the brain in any other direction than into the LNC. No branching from the LNC into lateral nerves or into the mesenchyme was seen. Neither peripheral nerve plexuses (subepidermal, submuscular, gastrodermal and pharyngeal) nor structures containing sense receptors (e.g. eyes and auricles) showed GABA-IR. Incubation of the antiserum with GABA-ovalbumin conjugate completely abolished all staining.

Due to a shortage of animals, only two HPLC runs were done on this species. Both runs, performed in different gradients, confirmed the presence of GABA. In one of the runs the concentration of GABA was calculated to a value between 100 and 240 picomole per mg protein (Fig. 5).

Discussion

Our results convincingly show that GABA is present in the CNS of *P. nigra*. The results from the two ICC methods support each other well and the HPLC analysis further verifies the presence of GABA. The presence in the nervous system together with its known reversible inhibitory action at low concentrations that can be blocked by GABA-antagonists (Keenan *et al.*, 1979) suggest a function in inhibitory synapses in flatworms.

We could only detect GABA in the CNS of *P. nigra* and this suggests that if GABA really acts as a transmitter, it has an interneuronal function in this species. This is in contrast with our results on another planarian, *Dugesia tigrina*, that contains numerous GABA-IR fibers both in the CNS and in the peripheral nervous system (PNS), namely in lateral nerves, fibers in the mesenchyme and in the subepidermal plexus (Eriksson & Panula, 1994). In other invertebrate taxa GABA acts both as an interneuronal and an neuromuscular transmitter. A similar dual role seems likely in planarians too. It seems likely that *P. nigra* uses some other transmitter instead of GABA in its PNS. Especially glycine ought to be studied, since the effect of glycine is similar to that of GABA in polyclads (Keenan *et al.*, 1979).

We had difficulties in finding cell somata showing GABA-IR. One reason for this is that planarian nerve cell somata often are rather small (Welsh & Williams, 1970) which makes them difficult to identify. Another reason is that the brain in *P. nigra* is very poorly demarcated from the mesenchyme and this, in combination with the high number of autofluorescent tissue elements, further hampers the identification of cell somata.

GABA is present in all studied phyla, and our results show that it is present also in flatworms. This suggests a phylogenetically old origin of GABAergic neurons and that the appearance of GABAergic neurotransmission may have occurred already in an flatworm-like ancestor and thereby precedes the evolutionary radiation into higher phyla.

Acknowledgment

This work was supported by the Magnus Ehrnrooth Foundation.

References

Airaksinen, M. S., S. Alanen, E. Szabat, T. J. Visser & P. Panula, 1992. Multiple neurotransmitters in the tuberomammillary nucleus: comparison of rat, mouse and guinea pig. J. Comp. Neurol. 323: 103–116.

Airaksinen, M. S. & P. Panula, 1990. Comparative neuroanatomy of the histaminergic system in the brain of the frog *Xenopus laevis*. J. Comp. Neurol. 292: 412–423.

Algeri, S., A. Carolei, P. Ferretti, C. Gallone, G. Palladini & G. Venturini, 1983. Effects of dopaminergic agents on mono amine levels and motor behavior in planaria *Dugesia gonocephala*. Comp. Biochem. Physiol. 74C: 27–30.

Barnes, R. S. K., P. Calow, P. J. W. Olive, 1988. The invertebrates: a new synthesis. Blackwell Scientific, Oxford, 582 pp.

Cooke, I. R. C. & A. Gelperin, 1988. Distribution of GABA-like immunoreactive neurons in the slug *Limax maximus*. Cell Tissue Res. 253: 77–81.

Coons, A. H., 1958. Fluorescent antibody methods. In J. F. Danelli (ed.), General cytochemical methods. Academic Press, New York: 399–442.

Curry, W. J., C. Shaw, C. F. Johnston, L. Thim & K. D. Buchanan, 1992. Neuropeptide F: Primary structure from the turbellarian, *Artioposthia triangulata*. Comp. Biochem. Physiol. 101C: 269–274.

Eriksson, K. S. & P. Panula, 1994. Gamma-aminobutyric acid in the nervous system of a planarian. J. Comp. Neurol. 345: 528–536.

Franquinet, R. & R. Catania, 1979. Localization histofluorimetrique et tude microspectrofluorimetrique de la sérotonine et des catécholamines chez une planaire entire et en cours de régénération. C. Hebd. Acad. Sci. Ser. D. Sci. Nat. 289: 339–342.

Gustafsson, M. K. S. & M. Reuter, 1992. The map of neuronal signal substances in flatworms. In R. N. Singh (ed.), Nervous systems. Principles of design and function. Wiley Eastern Limited, New Delhi: 165–188.

Hauser, M. & H. Koopowitz, 1987. Age-dependent changes in fluorescent neurons in the brain of *Notoplana acticola*, a polyclad flatworm. J. exp. Zool. 241: 217–255.

Homberg, U., A. Bleick & W. Rathmayer, 1993. Immunocytochemistry of GABA and glutamic acid decarboxylase in the thoracic ganglion of the crab *Eriphia spinifrons*. Cell Tissue Res. 271: 279–288.

Johnson, C. D. & A. O. W. Stretton, 1987. GABA-like immunoreactivity in inhibitory motor neurones of the nematode, *Ascaris*. J. Neurosci. 7: 223–235.

Kabotyanski, E. A., L. P. Nezlin & D. A. Sakharov, 1991. Serotonin neurones in the planarian pharynx. In D. A. Sakharov & W. Winlow (eds), Simpler Nervous Systems. Studies in Neuroscience 13. Manchester University Press, Manchester, New York: 138–152.

Keenan, L., H. Koopowitz & K. Bernardo, 1979. Primitive nervous systems. Action of aminergic drugs and blocking agents on activity in the ventral nerve cord of the flatworm *Notoplana acticola*. J. Neurobiol. 10: 397–408.

Marks, N. J., A. G. Maule, D. W. Halton, C. Shaw, C. F. Johnston, 1993. Distribution and immunochemical characteristics of neuropeptide-F (NPF) (*Moniezia expansa*) immunoreactivity in *Proteocephalus pollanicola* (Cestoda, Proteocephalidea). Comp. Biochem. Physiol. 104C: 381–386.

Maule, A. G., C. Shaw, D. W. Halton, T. Thim, C. F. Johnston, I. Fairweather & K. D. Buchanan, 1991. Neuropeptide F: a novel parasitic flatworm regulatory peptide from *Moniezia expansa* (Cestoda: Cyclophyllidea). Parasitology 102: 309–316.

Morita, M., F. Hall, J. B. Best & W. Gern, 1987. Photoperiodic modulation of cephalic melatonin in planarians. J. exp. Zool. 241: 383–388.

Panula, P., O. Häppölä, M. S. Airaksinen, S. Auvinen & A. Virkamäki, 1988. Carbodiimide as a tissue fixative in histamine immunohistochemistry and its application to developmental neurobiology. J. Histochem. Cytochem. 36: 259–270.

Reuter, M. & K. Eriksson, 1991. Catecholamines demonstrated by glyoxylic-acid-induced fluorescence and HPLC in some microturbellarians. Hydrobiologia 227 (Dev. Hydrobiol. 69): 209–219.

Reuter, M. & I. Palmberg, 1989. Development and differentiation of neuronal subsets in asexually reproducing *Microstomum lineare* – immunocytochemistry of 5-HT, RF-amide and SCP_B. Histochemistry 91: 123–131.

Roberts, E., 1986. GABA: the road to neurotransmitter status. In R. W. Olsen & J. C. Venter (eds), Benzodiazepine/GABA receptors and chloride channels. Structural and functional properties. Alan R Liss, New York: 1–39.

Szabat, E., S. Soinila, O. Häppölä, A. Linnala & I. Virtanen, 1992. A new monoclonal antibody against the GABA-protein conjugate shows immunoreactivity in sensory neurons of the rat. Neuroscience 47: 409–420.

Tamura, H., T. P. Hicks, Y. Hata, T. Tsumoto & A. Yamatodani, 1990. Release of glutamate and aspartate from the visual cortex of the cat following activation of afferent pathways. Exp. Brain Res. 80: 447–455.

Usherwood, P. N. R. & H. Grundfest, 1965. Peripheral inhibition in skeletal muscle of insects. J. Neurophysiol. 28: 497–518.

Venturini, G., F. Stocchi, V. Margotta, S. Ruggieri, D. Bravi, P. Bella & G. Palladini, 1989. A pharmacological study of dopaminergic receptors in planaria. Neuropharmacology 28: 1377–1382.

Walker, R. J. & L. Holden-Dye, 1991. Evolutionary aspects of transmitter molecules, their receptors and channels. Parasitology 102: S7–S29.

Watson, A. D. H., 1986. The distribution of GABA-like immunoreactivity in the thoracic nervous system of the locust *Schistocerca gregaria*. Cell Tissue Res. 246: 331–341.

Webb, R. A., 1988. Endocrinology of acoelomates. In H. Laufer & R. G. H. Downer (eds), Endocrinology of selected invertebrate types. Allan R. Liss, New York: 31–62.

Welsh, J. H. & E. C. King, 1970. Catecholamines in planarians. Comp. Biochem. Physiol. 36: 683–688.

Welsh, J. H. & L. D. Williams, 1970. Monoamine-containing neurons in planaria. J. Comp. Neurol. 138: 103–116.

Wikgren, M., M. Reuter, M. Gustafsson & P. Lindroos, 1990. Immunocytochemical localization of histamine in flatworms. Cell Tissue Res. 260: 479–484.

Hydrobiologia **305**: 291–295, 1995.
L.R.G. Cannon (ed.), *Biology of Turbellaria and some Related Flatworms.*

An immunocytochemical method for histamine: application to the planarians

Pertti Panula, Krister Eriksson, Margaretha Gustafsson & Maria Reuter
Department of Biology, Åbo Akademi University Biocity, Artillerigatan 6A, 20520 Åbo, Finland

Key words: carbodiimide, planarians, *Dugesia*, *Polycelis*, neurotransmitter

Abstract

Histamine-immunoreactivity was investigated in the planarians *Dugesia tigrina* and *Polycelis nigra*. Specific antisera against a histamine-protein conjugate were used, and 1-ethyl-3 (3-dimethyl– aminopropyl) carbodiimide was used both as coupling agent to prepare the antigen and as a tissue fixative. In *D. tigrina*, histamine-immunoreactivity was restricted to photoreceptor cells in the cerebral eye. In *P. nigra*, nerve fibers were found in the ventral nerve cord and nerves running laterally from these. The epidermal eyes did not display histamine-immunoreactivity. The results suggest that histamine may be a transmitter in some of the most primitive animals. They also suggest that the distribution of histamine may differ in planarians.

Introduction

Histamine (β-aminoethylimidazole, HA) is synthesized in many species by a specific enzyme, L-histidine decarboxylase (HDC; Watanabe *et al.*, 1991). Although histamine was found in animal tissues as early as 1927 (Best *et al.*, 1927), the exact cellular localization of the amine has been difficult due to methodological problems. The fluorescence histochemical method based on ortho-phthalaldehyde (OPT; Juhlin & Shelley, 1966) detects at least two fluorophores, and at least in some tissues the fluorophore is not histamine (Håkanson, 1970). Antibodies against rat HDC have been produced (Taguchi *et al.*, 1984) and applied in immunocytochemistry (Watanabe *et al.*, 1984), but they are species-specific and of limited use in invertebrate studies. Antisera against histamine itself can be produced if the amine is coupled to a carrier protein (Panula *et al.*, 1984; Steinbusch & Mulder, 1984). With this method, HA-containing neurons in the mammalian brain (Panula *et al.*, 1984; Airaksinen & Panula, 1988; Airaksinen *et al.*, 1989; 1991), mast cells in various tissues and endocrine cells in mammalian stomach (Panula *et al.*, 1985) have been demonstrated. Application of 1-ethyl-3(3-dimethyl-aminopropyl) carbodiimide (EDAC) to both HA-protein conjugate preparation and tissue fixation has led to development of a highly sensitive and specific immunohistochemical method for HA (Panula *et al.*, 1988; 1990).

The role of HA in the mammalian brain is well established (Haas, 1985; Prell & Green, 1986; Panula, 1987; Schwartz *et al.*, 1991; Yamatodani *et al.*, 1991), and both electrophysiology (McCaman & Weinreich, 1985) and immunocytochemistry (Soinila *et al.*, 1990; Elste *et al.*, 1990) have indicated that HA is a neurotransmitter in the marine mollusc *Aplysia*. In the insect eye, HA functions as a photoreceptor cell transmitter (Hardie, 1987; Pirvola *et al.*, 1988; Nässel *et al.*, 1988). Very little is known about the synthesis, distribution and function of HA in other invertebrates. This study was undertaken to localize HA in two different, commonly used turbellarians to investigate the phylogeny and function of HA in primitive animals.

Materials and methods

Production and specificity of the antibodies

Keyhole limpet hemocyanin (KLH), ovalbumin and bovine serum albumin (BSA); all from Sigma, St. Louis, USA; were succinylated as described earlier (Panula *et al.*, 1984; 1988). Histamine or L-histidine (L-His) was coupled to these carrier proteins with 1-ethyl-3(3-dimethyl-aminopropyl)carbodiimide (EDAC, Sigma, St. Louis, USA) as described earlier (Panula *et al.*, 1984). Rabbits were immunized with the HA-KLH conjugate. The first intradermal injection consisted of 1 ml of emulsion containing 500 μg of the conjugate in 0.9% saline and 500 μl of Freund's complete adjuvant. Four injection sites in the back of each rabbit were used. Booster injections (300 μg of conjugate in Freund's incomplete adjuvant) were given 40 days after the first injection. Blood was taken for testing every 10 days after the booster injection. The sera were tested with dot-blot tests and immunocytochemical incubations on rat stomach, a rich source of HA. In dot-blot tests, the antisera detected all conjugates that contained HA, whereas the conjugates containing L-His were not detected. The reaction for HA was blocked totally when the antiserum was preabsorbed with HA-protein conjugate, whereas the L-His-protein conjugate had no effect (Panula *et al.*, 1988; 1990). In cryostat sections of rat stomach, the antisera detected bright immunofluorescence for HA, which was abolished by the HA-protein conjugate but not by the L-His protein conjugate.

Animals and fixation

Animals from laboratory cultures of *Dugesia tigrina* and *Polycelis nigra* were used. They were fixed by immersion in 4% EDAC in 0.1 M sodium phosphate buffer (pH 7.0) with 0.25% Triton X-100 for 3–12 h at 4 °C. For some animals, 2% paraformaldehyde (PFA, Merck, Darmstadt, Germany) was added to the fixative. After fixation the worms were rinsed in phosphate buffered saline (PBS) and kept in 20% sucrose-PBS at 4 °C for at least 3 h. Cryostat sections (thickness 20 μm) were collected on gelatin-coated glass slides. Sections fixed with a mixture of EDAC and PFA were treated with 0.5% sodium borohydride (Merck) for 30 min at room temperature to reduce free aldehyde groups.

Immunocytochemistry

The specific HA antiserum was diluted 1:1000 in phosphate-buffered saline containing 0.25% Triton X-100 (PBS-T). Normal swine serum (1%) was added to reduce non-specific binding of the secondary antiserum. The sections were incubated overnight at 4 °C. They were then washed three times for 45 min in PBS-T. The secondary antiserum was fluorescein-labeled swine anti-rabbit IgG (DAKOPATTS, Copenhagen, Denmark), diluted 1:40 in PBS-T. Incubation was carried out at room temperature for 45–60 min. After three washes for a total of 45 min in PBS without Triton-X-100, the sections were mounted with 50% glycerol in PBS and examined under a Leitz Aristoplan fluorescence microscope equipped for epi-illumination. Control sections were incubated with the specific histamine antiserum preabsorbed with HA-ovalbumin conjugate (100 μg ml^{-1}), prepared using EDAC as coupling agent. All specific green fluorescence was abolished in these samples.

Results

In *D. tigrina*, the only histamine-immunoreactive (HA-ir) structures were seen in the cerebral eyes, where the photoreceptors exhibited strong HA-immunoreactivity. This HA-ir was restricted to the 'cup' of the eye, and no fibres in association with the eyes showed immunoreactivity. The HA-ir cells were seen only in the eyes of animals sacrificed during the night, and the immunoreaction was very intense (Fig. 1a). The auricles were examined carefully, because other sense receptors could also be expected to contain HA, but no HA-ir was found. The auricles contained strongly autofluorescent greenish yellow structures which may cover specific immunofluorescence. In *P. nigra*, HA-ir nerves were detected in the anterior end of the ventral longitudinal nerve cords (LNC; Fig. 1b) slightly posterior to the brain, and in a few nerve fibres running laterally from these (Figs 1c–d). The immunoreactivity in the LNC was restricted to the vicinity of the brain, and connection of these nerves with the epidermal eyes in the peripheral parts could not be ruled out. The eyes displayed no immunoreactivity.

Discussion

Although the neurotransmitter nature of HA has been discovered relatively late, it is obvious that HA plays a crucial role in many animals. Histaminergic neurons mediate both excitatory and inhibitory functions on follower cells in *Aplysia* (see McCaman & Weinreich, 1985) and HA hyperpolarizes *Onchidium* neurons (Gotow *et al.*, 1980). In insects (Hardie *et al.*, 1987, Pirvola *et al.*, 1988) and giant barnacle *Balanus nubilus* (see Callaway & Stuart, 1989), HA acts as a photoreceptor transmitter. In the spiny lobster *Panulirus*

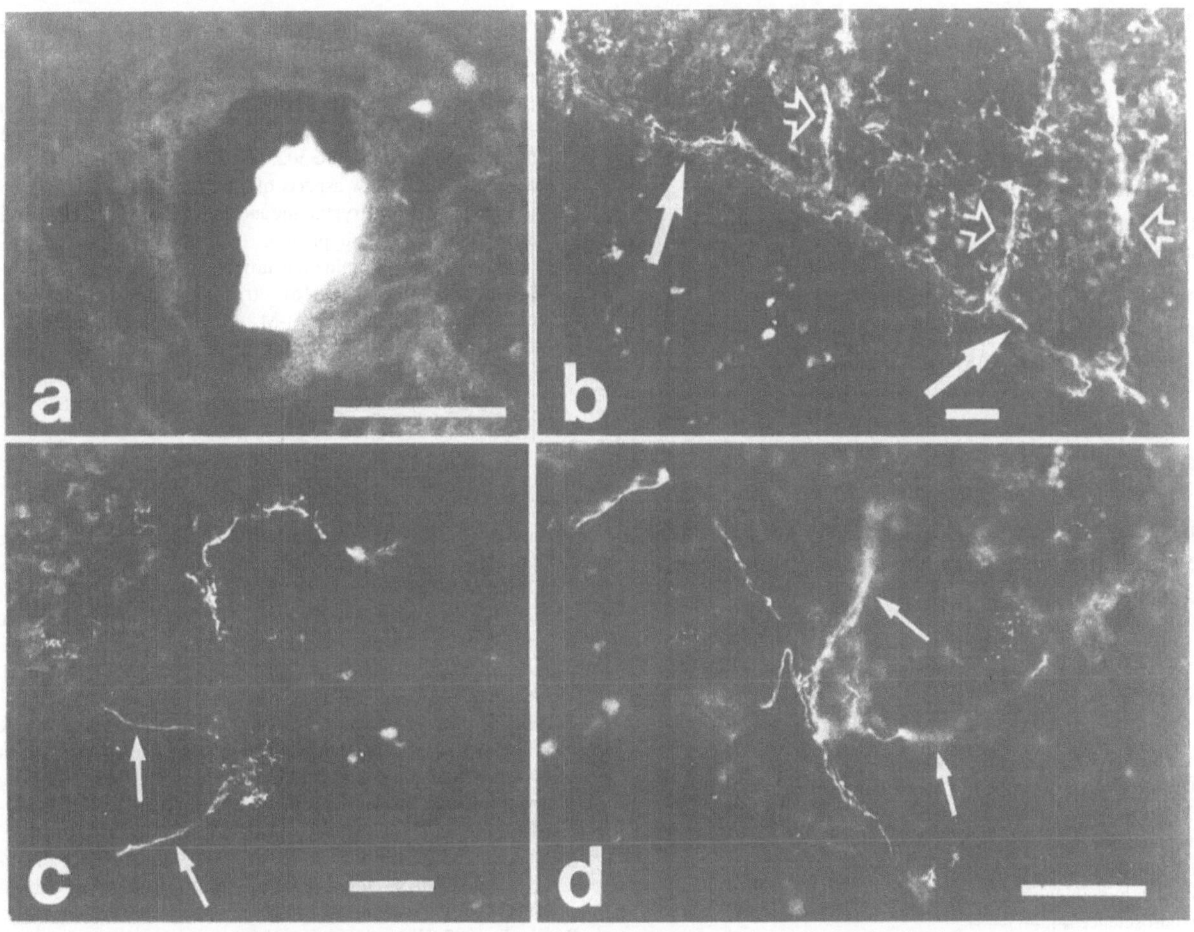

Fig. 1. Histamine-immunoreactivity in planarians. (*a*) Brigthly fluorescent cells in the eye of *D. tigrina*. Immunoreactive cells are partly surrounded by the pigmented eyecup. (*b*) HA-ir nerves in the longitudinal nerve cord (arrows) and lateral branches (open arrows) in *P. nigra*. (*c–d*) HA-ir nerve fibres associated with the brain of *P. nigra*. The arrows indicate two lateral nerves running to the left (*c*) and to the right (*d*). Scale = 20 μm.

interruptus HA also mediates inhibitory responses in the stomatogastric ganglion (Claiborne & Selverston, 1984). In mammals, HA can switch thalamic rhythmic burst firing to single-spike activity (McCormick & Williamson, 1991) and increase the excitability of hippocampal neurons (Haas & Konnerth, 1983; Greene & Haas, 1990), thereby providing a modulatory mechanism for targeted actions of other transmitters. Surprisingly little is known about the biology of HA in flatworms. In a parasitic flatworm *Diphyllobothrium dendriticum* numerous HA-ir nerve cells and fibres are found laterally to the main nerve cord (Wikgren *et al.*, 1990). In the microturbellarian *Microstomum lineare* two symmetrically placed pear-shaped HA-ir neurons are located on each side of the both excretory pores close to the presumptive photoreceptor cells (Palmberg *et al.*, 1980; Wikgren *et al.*, 1990).

The finding of HA-immunoreactivity in the eyes of *D. tigrina* is in agreement with the presence and neurotransmitter function of HA in the eyes of insects (Hardie, 1987; Pirvola *et al.*, 1988) and barnacles (Callaway & Stuart, 1989). The reason for our failure to detect HA-ir nerves in other parts of the nervous system in *D. tigrina* is not clear. There is either a true difference in the nervous function between *D. tigrina* and *P. nigra*, or the HA concentrations in the nerves differ so that our method is not sensitive enough to detect immunoreactivity in all structures. The possible different evolutionary origin of the epidermal and cerebral eyes may explain the differences in the distribution of HA in the two species (Eakin, 1979). Our preliminary studies with high pressure liquid chromatography coupled to fluorometry confirm the presence of authentic HA in *D. tigrina*. We are currently testing if there are also functional differences between these species. L-Histidine decarboxylase, the enzyme that synthesizes histamine, has been cloned and sequenced from the rat (Joseph *et al.*, 1990) and human (Zahnow *et al.*, 1991). This has allowed studies to detect the sites of histamine synthesis in mammalian tissues (Castren & Panula, 1991). Such studies are needed in planarians to find out where HA is synthesized.

Acknowledgments

This study was supported by the Academy of Finland, the Sigrid Juselius Foundation, and the Magnus Ehrnrooth Foundation.

References

Airaksinen, M. S., G. Flugge, E. Fuchs & P. Panula, 1989. Histaminergic system in the tree shrew brain. J. Comp. Neurol. 286: 289–310.

Airaksinen, M. S., A. Paetau, L. Paljärvi, K. Reinikainen, P. Riekkinen, R. Suomalainen & P. Panula, 1991. Histamine neurons in human hypothalalmus in normal and alzheimer diseased brains. Neuroscience 44: 465–481.

Airaksinen, M. S. & P. Panula, 1988. The histaminergic system in the guinea pig central nervous system: An immunocytochemical mapping study using an antiserum against histamine. J. Comp. Neurol. 273: 163–186.

Best, C. H., H. H. Dale, H. W. Dudley & W. V. Thorpe, 1927. The nature of the vasodilator constituents of certain tissue extracts. J. Physiol. 62: 397–417.

Callaway, J. & A. E. Stuart, 1989. Biochemical and physiological evidence that histamine is the transmitter of barnacle photoreceptors. Visual Neurosci. 3: 311–325.

Castren, E. & Panula, P., 1991. The distribution of histidine decarboxylase mRNA in the rat brain: an *in situ* hybridization study using synthetic oligonucleotide probes. Neurosci. Lett. 120: 113–116.

Claiborne B. J. & A. I. Selverston, 1984. Histamine as a neurotransmitter in the stomatogastric nervous system of the spiny lobster. J. Neurosci. 4: 708–721.

Eakin, R., 1979. Evolutionary significance of photoreceptors: In retrospect. Am. Zool. 19: 647–653.

Elste, A., J. Koester, E. Shapiro, P. Panula & J. H. Schwartz, 1990. Identification of histaminergic neurons in Aplysia. J. Neurophysiol. 64: 736–744.

Gotow, T., C. T. Kirkpatrick & T. Tomita, 1980. Excitatory and inhibitory effects of histamine on molluscan neurons. Brain Res. 196: 169–182.

Greene, R. W. & H. L. Haas, 1990. Effects of histamine on dentate granule cells *in vitro*. Neuroscience 34: 299–303.

Haas, H., 1985. Histamine. In M. A. Rogawski & J. L. Barker, Neurotransmitter actions in vertebrate nervous system. Plenum Press, New York: 321–337.

Haas, H. L. & A. Konnerth, 1983. Histamine and noradrenaline decrease calcium-activated potassium conductance in hippocampal pyramidal cells. Nature 302: 432–434.

Håkanson, R., 1970. New aspects of the formation and function of histamine, 5-hydroxytryptamine and dopamine in gastric mucosa. Acta Physiol. Scand. Suppl. 340.

Hardie, R., 1987. Is histamine a neurotransmitter in insect photoreceptors? J. Comp. Physiol. 161: 201–213.

Joseph, D. R., P. M. Sullivan, Y.-M. Wang, C. Kozak, D. A. Fenstermacher, M. E. Behrendsen & C. A. Zahnow, 1990. Characterization and expression of the complementary DNA encoding rat histidine decarboxylase. Proc. Natn. Acad. Sci. USA 87: 733–737.

Juhlin, L. & W. Shelley, 1966. Detection of histamine by a new fluorescent o-phthalaldehyde stain. J. Histochem. Cytochem. 14: 525–528.

McCaman, R. E. & D. Weinreich, 1985. Histaminergic synaptic transmission in the cerebral ganglion of Aplysia. J. Neurophysiol. 53: 1016–1037.

McCormick, D. A. & A. Williamson, 1991. Modulation of neuronal firing mode in cat and guinea pig LGNd by histamine: possible cellular mechanisms of histaminergic control of arousal. J. Neurosci. 11: 3188–3199.

Nässel, D., M. H. Holmqvist, R. C. Hardie, R. Håkanson & F. Sundler, 1988. Histamine-like immunoreactivity in photoreceptors of the compound eyes and ocelli of the flies Calliphora eryhtrocephala and Musca domestica. Cell Tissue Res. 253: 639–646.

Palmberg, I., M. Reuter & M. Wikgren, 1980. Ultrastructure of epidermal eyespots of Microstomum lineare (Turbellaria, Macrostomida). Cell Tissue Res. 210: 21–32.

Panula, P., 1987. Histaminergic mechanisms in the brain. Med. Biol. 65: 9–11.

Panula, P., M. S. Airaksinen, U. Pirvola & E. Kotilainen, 1990. A histamine-containing neuronal system in the human brain. Neuroscience 34: 127–132.

Panula, P., O. Häppölä, M. S. Airaksinen, S. Auvinen & A. Virkamäki, 1988. Carbodiimide as a tissue fixative in histamine immunohistochemistry and its application in developmental neurobiology. J. Histochem. Cytochem. 36: 259–269.

Panula, P., M. Kaartinen, M. Mäcklin & E. Costa, 1985. Histamine-containing peripheral neuronal and endocrine systems. J. Histochem. Cytochem. 33: 933–941.

Panula, P., H.-Y.T. Yang & E. Costa, 1984. Histamine-containing neurons in the rat hypothalamus. Proc. Natl. Acad. Sci. USA 81: 2572–2576.

Pirvola, U., A. Yamatodani, L. Tuomisto & P. Panula, 1988. Distribution of histamine in the cockroach brain and visual system: an immunocytochemical and biochemical study. J. Comp. Neurol. 276: 514–526.

Prell, G. D. & J. P. Green, 1986. Histamine as a neuroregulator. Ann. Rev. Neurosci. 9: 209–254.

Schwartz, J.-C., J.-M. Arrang, M. Garbarg, H. Pollard & M. Ruat, 1991. Histaminergic transmission in the mammalian brain. Physiol. Rev. 71: 1–51.

Soinila, S., G. J. Mpitsos & P. Panula, 1990. Comparative study of histamine immunoreactivity in nervous systems of Aplysia and Pleurobranchaea. J. Comp. Neurol. 298: 83–96.

Steinbusch, H. W. M. & A. Mulder, 1984. Immunohistochemical localization of histamine in neurons and mast cells in the rat brain. In A. Björklund, T. Hökfelt & M. J. Kuhar, Handbook of Chemical Neuroanatomy, Vol. 3. Elsevier, Amsterdam: 126–140.

Taguchi, Y., T. Watanabe, H. Kubota, H. Hayashi & H. Wada, 1984. Purification of histidine decarboxylase from the liver of fetal rats and its immunochemical and immunohistochemical characterization. J. Biol. Chem. 259: 5214–5221.

Watanabe, T., Y. Taguchi, K. Maeyama & H. Wada, 1991. Formation of histamine: Histidine decarboxylase. In B. Uvnäs Histamine and Histamine Antagonists, Handbook of Experimental Pharmacology, Vol. 97. Springer, Heidelberg: 145–163.

Watanabe, T., Y. Taguchi, S. Shiosaka, J. Tanaka, H. Kubota, Y. Terano, H. Hayashi, M. Tohyama & H. Wada, 1984. Distribution of the histaminergic neuron system in the central nervous system of rats: a fluorescent immunohistochemical analysis with histidine decarboxylase as a marker. Brain Res. 295: 13–25.

Wikgren, M., M. Reuter, M. K. S. Gustafsson & P. Lindroos, 1990. Immunocytochemical localization of histamine in flatworms. Cell Tissue Res. 260: 479–484.

Yamatodani, A., N. Inagaki, P. Panula, N. Itowi, T. Watanabe & H. Wada, 1991. Structure and functions of the histaminergic neurone system. In B. Uvnäs, Histamine and Histamine Antagonists, Handbook of Experimental Pharmacology, Vol. 97. Springer, Berlin: 243–283.

Zahnow, C. A., H.-F. Yi, O. W. McBride & D. R. Joseph, 1991. Cloning of the cDNA encoding human histidine decarboxylase from an erythroleukemia cell line and mapping of the gene locus to chromosome 15. J. DNA Seq. Mapp. 1: 395–400.

Hydrobiologia **305**: 297–303, 1995.
L.R.G. Cannon (ed.), *Biology of Turbellaria and some Related Flatworms.*

Neuropeptide F: a ubiquitous invertebrate neuromediator?

Aaron G. Maule[1], David W. Halton[2] & Chris Shaw[1]
Comparative Neuroendocrinology Research Group, School of [1]Clinical Medicine and [2]Biology & Biochemistry, The Queen's University of Belfast, Belfast BT7 1NN, Northern Ireland, UK

Key words: invertebrate neuropeptides, platyhelmith neuropeptides, neuropeptide Y superfamily, neuropeptide F

Abstract

Using specific antisera, neuropeptide F (NPF)-related peptides have been identified immunocytochemically as widespread and abundant in the nervous systems of all invertebrate taxa examined so far. To date, four NPFs have been isolated and sequenced: from the cestode, *Moniezia expansa* and the turbellarian, *Artioposthia triangulata*, and from the molluscs, *Helix aspersa* and *Aplysia californica*; a related nonapeptide has been sequenced also from *Loligo vulgaris*. These peptides all display structural characteristics of the vertebrate NPY superfamily of peptides and appear, therefore, to represent invertebrate members of this superfamily. In this respect, invertebrate NPFs most likely represent the precursors of the vertebrate NPY superfamily. Homologies between the gene structure of human NPY and molluscan NPF (*A. californica*) support the view that the NPY/NPF gene is of ancient lineage. Although NPF (*A. californica*) has been found to inhibit the activity of the abdominal ganglia in *Aplysia*, its widespread expression in this mollusc would suggest multiple functions; the physiological role(s) of NPFs in other invertebrates awaits examination. The abundance and apparent ubiquitous nature of NPF-related peptides establishes them as evolutionarily-ancient molecules that likely serve important physiological functions in invertebrate neurobiology.

Introduction

During the 1980s, there were numerous documented accounts of the occurrence of extensive immunostaining to the mammalian regulatory peptide, pancreatic polypeptide (PP) in invertebrate nervous systems (see Maule *et al.*, 1992a; Fairweather & Halton, 1991; Rajpara *et al.*, 1992; Schoofs *et al.*, 1988); however, nothing was known of the structural characteristics of this immunoreactivity. PP is one of the structurally-related peptides which constitute a family of vertebrate peptides that have been termed the PP family or the neuropeptide Y (NPY) superfamily of regulatory peptides. In this article, family members are considered to belong to the NPY superfamily of peptides and include PP, NPY and peptide YY (PYY). Within vertebrates, PP occurs exclusively in pancreatic endocrine cells; NPY is the most abundant brain neuropeptide and is synthesised by chromaffin cells in the adrenal medulla and by neurones situated within the central and peripheral nervous systems; and PYY is localised to mucosal endocrine-like L cells of the distal intestine. All family members consist of 36 amino acid residues (except chicken PYY which contains 37 residues) and are C-terminally amidated. Although NPY superfamily peptides are diverse in structure, certain key residues have been conserved throughout its members, including Pro^{5}, Pro^{8}, Ala^{12}, Tyr^{27}, Arg^{33} and Arg^{35} (see *Abbreviations* Table 1).

Initially, research within our laboratory concentrated on the screening and identification of regulatory peptide-immunoreactivities within the nervous systems of parasitic platyhelminths, with a view to identifying novel targets for anthelmintics that could disrupt possible peptide-controlled neuromuscular functioning. The employment of C-terminally-directed NPY superfamily antisera in immunocytochemical techniques has identified widespread immunostaining throughout helminth nervous systems (see Fairweather & Halton, 1991; Halton *et al.*, 1992). The immunostaining patterns delineated during these investigations indicated the presence of extensive and well developed peptidergic nervous systems within these primitive invertebrates.

Table 1. Primary sequences of known neuropeptide Fs (NPFs)

Moniezia expansa	PDKDFIVNPSDLVLDNKAALRDYLRQINEYFAIIGRPRF NH_2
Artioposthia triangulata	KVVHLRPRSSFSSEDEYQI***NVSK*IQLY***** NH_2
Helix aspersa	STQMLSPPERPREFRHPNE**Q**KEL***Y**M**T** NH_2
Aplysia californica	DNSEMLAPP*RPEEFTSAQQ**Q**AAL***YS*M***** NH_2
Loligo vulgaris	Y**VA* *** NH_2
Homologies in NPFs	————————YL—Y—R-RF NH_2
Vertebrate NPY superfamily homologies	—-P–P—A————Y—R-R- NH_2

* Residue homologous with NPF (*M. expansa*).
It should be noted that all NPY superfamily peptides possess either a C-terminal tyrosinamide or a C-terminal phenylalaninamide.
Abbreviations used for each amino acid as a three-letter and single-letter equivalent are as follows: alanine Ala A; arginine Arg R; asparagine Asn N; aspartic acid Asp D; cysteine Cys C; glutamine Gln Q; glutamic acid Glu E; glycine Gly G; histidine His H; isoleucine Ile I; leucine Leu L; lysine Lys K; methionine Met M; phenylalanine Phe F; proline Pro P; serine Ser S; threonine Thr T; tryptophan Trp W; tyrosine Tyr Y; valine Val V.

Table 2. Invertebrates in which NPF has been demonstrated*

PHYLUM	SPECIES
CNIDARIA	- *Hydra vulgaris*
PLATYHELMINTHES	- Turbellaria: *Artioposthia triangulata, Microstomum lineare, Stenostomurn leucops.*
	- Cestoda: *Caryophyllaeus laticeps, Diphyllobothrium dendriticum, Grillotia erinaceus, Hymenolepis diminuta, Ligula intestinalis, Mesocestoides corti, Moniezia expansa, Proteocephalus pollanicola, Taenia crassiceps.*
	- Trematoda - Monogenea: *Diclidophora merlangi, Entobdella soleae*; - Digenea: *Bucephaloides gracilescens, Cercaria emasculans, Cotylurus erraticus, Cryptocotyle lingula, Diplostomum spathaceum, Fasciola hepatica, Haplometra cylindracea, Himasthla* spp., *Microphallus similis, Plagioporus varius, Schistosoma mansoni.*
NEMATODA	- *Ascaris suum*
ANNELIDA	- *Hirudo medicinalis, Lumbricus terrestris, Nereis virens.*
ARTHROPODA	- *Calliphora vomitoria, Neobellieria bullata, Periplaneta americana.*
MOLLUSCA	- *Aplysia californica**, *Helix aspersa, Loligo vulgaris.*

* It should be noted that NPF-immunoreactivity has been recorded either immunocytochemically or radioimmunometrically in all of the above species using antiserum 792.3, which was raised against the C-terminal decapeptide of NPF (*M. expansa*) (Maule *et al.*, 1992b) except *A. californica*. Immunostaining was demonstrated in the latter using an antiserum raised against apNPY ie. NPF (*A. californica*) (Rajpara *et al.*, 1992).

NPY superfamily peptide-immunoreactivities in parasitic helminths

The most intense immunostaining of parasitic platyhelminth nervous systems was obtained using the C-terminally directed PP antiserum, PP221, although some C-terminally-directed PYY and NPY antisera were also cross-reactive with native helminth neuropeptides (see Maule *et al.*, 1990a, 1990b). PP-immunostaining was abundant in both the central and the peripheral nervous systems of those parasites examined. Interestingly, PP-immunoreactivity was often associated with the innervation of muscular organs, such as suckers, holdfasts and copulatory organs, and with the female reproductive system, most notably the egg-forming apparatus or ootype/Mehlis' gland complex (Halton *et al.*, 1991; Magee *et al.*, 1989; Maule *et al.*, 1990b). At the ultrastructural level, PP-immunostaining, as revealed by immunogold labelling, was localised exclusively to large dense-cored secretory vesicles occupying the majority of axons in the central and peripheral nervous systems of helminths, such as the monogenean gill-fluke, *Dicli-*

dophora merlangi and the cestode, *Moniezia expansa* (see Brennan *et al.*, 1993a, b).

Immunostaining with antisera to NPY superfamily peptides has not only been documented in flatworm parasites but also in nematodes. In this respect, PP-, PYY- and NPY-immunoreactivities have been recorded in the nervous system of the parasitic nematode *Ascaris suum* by Brownlee *et al.*(1993). There is one other report of NPY- -immunostaining in *A. suum*, but in this case the antiserum employed, which was designated NPY-1, was actually raised against PP (Sithigorngul *et al.*, 1990).

NPY superfamily peptides or FMRFamide related peptides

It soon became evident that in many invertebrate species the immunostaining obtained with PP antisera overlapped with that obtained with antisera to the invertebrate cardioexcitatory peptide, FMRFamide, such that there appeared to be co-localisation of these immunoreactivities (Brennan *et al.*, 1993a; Maule *et al.*, 1990a, 1990b). In some cases, this overlap in immunostaining could be accounted for by the non-specific cross-reactivity of some PP antisera with FMRFamide-related peptides (FaRPs) (Dockray & Williams, 1983). However, C-terminally-directed PP antisera which did not cross-react with FMRFamide (*i.e.* PP221) identified extensive immunostaining patterns in the central and peripheral nervous systems of helminth parasites (Halton *et al.*, 1992; Brownlee *et al.*, 1993). Using PP and FMRFamide antisera with double immunogold-labelling techniques indicated consistently that both immunoreactivities resided in the same large, dense-cored secretory vesicles, that have been found throughout the nervous systems of those parasitic platyhelminths examined (Brennan *et al.*, 1993a). This suggested either the co-localisation of PP and FMRFamide neuronal mediators or the cross-reactivity of both antisera with a single native parasite neuropeptide.

Characterisation of helminth NPY superfamily peptides

Characterisation of the PP-immunoreactive peptide from the fish-gill parasite, *D. merlangi*, identified a single form which had a molecular weight similar to that of PP (*i.e.* 4200–4500 Da.) and, therefore, much larger than any known FaRP (Maule *et al.*, 1992a). The PP-like neuropeptide identified in *D. merlangi* was fully cross-reactive with antiserum PP221, but did not cross-react with a range of NPY antisera or with an N-terminally-directed PP antiserum. This indicated that the *D. merlangi* peptide was more analogous to PP than to NPY in its C-terminus and that its N-terminus was different from that of mammalian PPs. Also, since antiserum PP221 did not cross-react with PP free acid (*i.e.* unamidated PP), but fully cross-reacted with the PP-immunoreactive peptide from *D. merlangi*, it appeared that the parasite peptide was C-terminally amidated.

Chemical characterisation of NPY superfamily peptides within nematodes indicated the presence of both PP- and NPY-immunoreactive peptides in tissue extracts of *A. suum* (Smart *et al.*, 1992a). However, sequence data for these immunoreactive peptides is required to establish whether or not *A. suum* possesses members of this peptide family in its nervous system.

Neuropeptide F (NPF)

NPF-identification and characterisation

In 1991, a 39-residue C-terminally amidated peptide which, fortuitously, cross-reacted under radioimmunometrical conditions with the antiserum PP221 was isolated and sequenced from the parasitic platyhelminth, *Moniezia expansa* by Maule *et al.* (1991). The peptide, which terminated C-terminally in RPRFamide, was significantly larger than previously identified FaRPs and displayed distinct structural homology with vertebrate members of the NPY superfamily. Indeed, the C-terminal tetrapeptide amide of NPF was found to be identical to that of known amphibian and reptilian PPs (*i.e.* RPRFamide). This neuropeptide, designated neuropeptide F (NPF), was the first identified platyhelminth neuropeptide and did not appear to be related to any previously identified invertebrate neuropeptide. However, it did display C-terminal structural homology with vertebrate members of the NPY superfamily; hence, its cross-reactivity with antisera raised to this region of vertebrate NPY superfamily members.

NPF-related peptides in invertebrates

Since the full structural characterisation of NPF (*M. expansa*), analogous peptides have been isolated and sequenced from the free-living turbellarian, *Artioposthia triangulata*, by Curry *et al.*, 1992), the gastropod mollusc, *Helix aspersa* by Leung *et al.* (1992), and the sea hare, *Aplysia californica*, by Rajpara *et al.* (1992) (see Table 1). A related nonapeptide, designated peptide tyrosine phenylalanine (PYF), has been isolated and sequenced from nervous tissue extracts of the cephalopod mollusc, *Loligo vulgaris*, by Smart *et al.* (1992b); however, two peaks of PP-immunoreactivity were resolved, a small molecular weight form (PYF) and a larger form (possibly squid NPF). Smart *et al.* (1992b) suggested that PYF may have been (a) generated by non-specific endopeptidase activity prior to extraction, (b) a C-terminal fragment of squid NPF which was generated by specific endopeptidase activity, or (c) a highly truncated, receptor-active variant of NPF. Elucidation of the primary structure of the large PP-immunoreactive peptide from squid neural tissues, coupled with physiological examinations of the roles of these peptides are required to determine which of these possibilities is correct.

NPF-immunoreactivity in invertebrates

Whole-molecule and fragments of NPF (*M. expansa*) have been synthesised and region-specific NPF antisera have been generated that did not cross-react with mammalian PPs or with FMRFamide (Maule *et al.*, 1992b, 1992c). One of these antisera, namely 792.3, which was C-terminally directed, identified NPF-immunoreactivity immunocytochemically in representative members from the major invertebrate phyla examined to date, including Cnidaria, Platyhelminthes, Nematoda, Annelida, Arthropoda and Mollusca (see Table 2). These findings identified NPF-immunoreactivity as being widespread amongst invertebrate phyla and indicated the existence of either a novel family of invertebrate neuropeptides or the existence within invertebrate nervous systems of the precursors of vertebrate NPY superfamily peptides.

The C-terminally-directed NPF (*M. expansa*) antiserum 792.3 cross-reacts with all known NPFs and with PPs which terminate in RPRFamide (*i.e.* known amphibian and reptilian forms). Immunocytochemical observation and radioimmunometric quantification of the NPF-immunoreactivity in a range of invertebrate species, using the specific NPF antisera, indicated that NPF was the most widespread and abundant neuropeptide-immunoreactivity demonstrable (unpublished observations by the authors). It should be noted that the vertebrate analogue of NPF, namely NPY, as well as being ubiquitous in vertebrate nervous systems, is the most abundant neuropeptide in the vertebrate brain (Adrian *et al.*, 1983; McDonald, 1988).

All but one (Rajpara *et al.*, 1992) of the documented examinations of the distribution of NPF-immunoreactivity in invertebrates, which employed specific NPF antisera, were carried out on parasitic platyhelminths (Marks *et al.*, 1993; Maule *et al.*, 1992b, 1992c, 1993a). The results of these studies revealed that NPF-immunoreactivity was identical in distribution in platyhelminth nervous systems to that obtained using NPY superfamily peptide antisera. Preliminary examinations have indicated that this is true for the other major invertebrate phyla. Characterisation of the NPF-immunoreactivity in the proteocephalidean cestode, *Proteocephalus pollanicola* from Lough Neagh pollan (*Coregonus autumnalis*) indicated the presence of two molecular forms of NPF, both of which had molecular weights similar to those of known NPFs (4400–4900 Da.) (Marks *et al.*, 1993). However, the chromatographic resolution of two immunoreactive peptides was thought to be due to the presence of an oxidisable methionine residue within NPF (*P. pollanicola*).

NPF cross-reactivity with FMRFamide antisera

NPF-related peptides were found to cross-react with a number of FMRFamide antisera under immunocytochemical conditions. This is not surprising since the C-terminal dipeptide-amides of NPFs and FaRPs are identical (*i.e.* RFamide). Preabsorption studies revealed that NPFs could be used to block not only PP- and PYY-immunostaining in parasitic platyhelminths but also FMRFamide-immunoreactivity (Maule *et al.*, 1992b, 1992c). It was evident, therefore, that reports of the co-localisation of PP- and FMRFamide-immunoreactivities, which, in some cases, were accounted for by the non-specific binding of PP-antisera with FMRFamide, were likely to be due to the non-specific binding of FMRFamide antisera with NPF-related peptides. The cross-reactivity of FMRFamide antisera with NPF-related peptides calls into question previous immunocytochemical descriptions of the distribution of FMRFamides in invertebrates, most of which were carried out before NPFs were known to exist.

The distribution of NPF-immunoreactivity in the nervous systems of parasitic platyhelminths has been found to mirror that identified using, not only C-terminally directed NPY superfamily peptide antisera, but also FMRFamide antisera at both the light and the ultrastructural levels (Maule *et al.*, 1992b, 1992c), thus providing further evidence that much of the previously documented PP- and FMRFamide-immunoreactivities in parasitic platyhelminths was due to NPF-related peptides. This situation has been complicated further by the identification and sequencing of the first platyhelminth FaRP (GNFFRFamide) from the cestode, *M. expansa* by Maule *et al.* (1993b). While this hexapeptide was found not to cross-react with the NPF antiserum 792.3, it does not preclude as yet unidentified FaRPs in other invertebrate species from cross-reacting with such antisera. In view of these findings, future studies of RFamide-related peptides warrant particular caution.

NPF evolution

The widespread distribution of NPF-related peptides within the platyhelminths, namely, the earliest organisms to display significant neurocephalisation, indicates an early evolutionary origin for NPF. The primary structures of the known NPFs indicate significant C-terminal structural homologies; indeed, of the C-terminal 20 residues in NPF (*H. aspersa*), 13 are identical to those in NPF (*M. expansa*), and differences in the remaining 7 amino acids can be accounted for by single-base substitutions in the genetic code (see Table 1). This indicates strong structural conservation in the carboxy-terminus of NPFs between the flatworms and the molluscs, a fact which is perhaps not surprising since molluscs are believed to have evolved from flatworm-like ancestors.

It should be noted that the 3 C-terminal residues which occur in all identified members of the NPY superfamily (Tyr^{27}, Arg^{33}, Arg^{35}) are present in analogous positions in known NPFs (eg. in *M. expansa*, Tyr^{30}, Arg^{36}, Arg^{38}), and may, therefore, have remained essential for receptor binding or biological activity over vast evolutionary timescales. Also, the polyproline helix in the amino-terminal region of vertebrate NPY superfamily peptides is present in the 2 known molluscan NPFs; thus, the conserved residues Pro^5 and Pro^8 of vertebrate NPY superfamily peptides are present in analogous positions in *H. aspersa* (Pro^8 and Pro^{11}) and in *A. californica* (Pro^9 and Pro^{12}). This would indicate that the polyproline amino-terminus of higher vertebrate NPY superfamily peptides had been established very early in the evolution of this peptide family. However, although N-terminal prolines are present in the two known platyhelminth NPFs they would not appear to be in analogous positions. Clearly, strong structural constraints would seem to have been placed on members of the NPY superfamily of peptides approximately 1400 million years ago, such that key amino- and carboxy-terminal residues, probably essential for receptor binding and/or biological function, are present in both primitive invertebrate and higher vertebrate homologues.

Interestingly, in common with the vertebrate homologues, the molluscan NPF gene transcribes a precursor which is amongst the smallest of those encoding regulatory peptides (Rajpara *et al.*, 1992). The N-terminus is generated by direct signal-peptide cleavage and the C-terminus is flanked by a Gly-Lys-Arg (GKR)-processing site, followed by a short C-terminal extension peptide (see Fig. 1). These data would support the view that the NPF/NPY gene is of ancient lineage and may have been condensed early in the course of metazoan evolution.

NPF function

The role of NPF within the nervous systems of parasitic helminths is unknown. However, its abundance and distribution throughout the central and peripheral nervous systems of parasitic helminths and other invertebrate groups suggests that NPF plays an important basic role in neuronal communication within these organisms. Its prevalence in nerves which innervate muscle, for example, the suckers, holdfasts and copulatory organs of flatworms, points to an involvement in neuromuscular function. Probably the most notable feature of NPF-immunostaining patterns in parasitic flatworms is the peptide's association with the reproductive system, especially the egg-forming apparatus of those flatworm parasites examined (Halton *et al.*, 1993). This would indicate a potentially vital role for the peptide in egg-formation, and clearly earmarks NPF-controlled mechanisms as potential anthelmintic targets in control strategies aimed at disrupting parasite reproductive function.

In studies where the peptidergic, cholinergic and aminergic components of flatworm parasite central nervous systems were compared, it was evident that the peptidergic component more closely resembled the cholinergic than the aminergic system (Maule *et al.*, 1990a, 1990b, 1993a). This would indicate the possible co-localisation and, therefore, co-release of cholinergic and peptidergic neuronal mediators within the nervous systems of these organisms. In such cases, the peptidergic messengers (eg. NPF) may influence the cholinergic signalling.

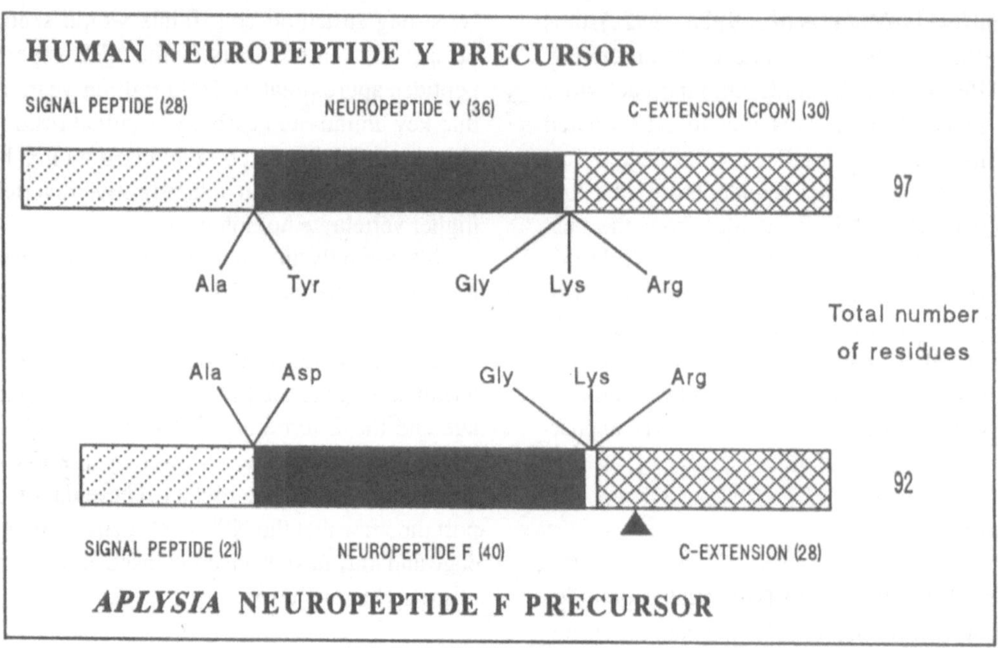

Fig. 1. A schematic drawing comparing the structures of the precursor peptides encoded by the human NPY gene and the *Aplysia* (molluscan) NPF gene. The number of amino acid residues that comprise the signal peptides, NPY/NPF peptides, and the C-terminal extensions are shown in parenthesis. The N-terminus of each peptide is generated by signal-peptide cleavage (at Ala-Tyr for NPY; Ala-Asp for NPF), and the Gly and Lys-Arg residues mark the sites for amidation (NH2) and cleavage, respectively, at the C-terminus of NPY and NPF. Note that an additional potential tribasic cleavage site is present within the predicted C-terminal extension peptide of *Aplysia* NPF (arrow). CPON, *C*arboxy-terminal *p*eptide *o*f *N*PY.

The only examination of the physiological role of NPF in the invertebrate nervous system was carried out on *Aplysia californica*. The addition of either synthetic or natural NPF (*A. californica*), referred to by Rajpara *et al.*, (1992) as *Aplysia* NPY (apNPY), to abdominal ganglia caused a prolonged hyperpolarisation of the cell membranes and a reduction of the spontaneous contractility of these cells. These responses were concentration-dependent and were thought to be consistent with those of a co-transmitter released from the bag-cell neurones at the initiation of egg-laying. Its prolonged actions led Rajpara *et al.* (1992) to suggest that it played a role in the control of visceromotor function during egg-laying. It should be noted, however, that the prolonged effects of its inhibition were obtained in the presence of a number of protease inhibitors. The expression of NPF was not confined to a single population of neurones in *A. californica*, therefore, it was intimated that NPF (*A. californica*) fulfilled multiple roles. Since NPF-immunoreactivity is widespread in other invertebrate phyla, this assumption would, at this stage, seem to apply equally to the NPF-related peptides identified in them.

Conclusions

Evidence to date indicates that NPF-related peptides are invertebrate members of the NPY superfamily and, as such, the precursors from which this important regulatory peptide family has evolved. Its occurrence and abundance in neural tissues in representatives of all of the major invertebrte phyla suggest that NPF is a ubiquitous neuronal mediator, and therefore much like its vertebrate analogue, NPY, in vertebrate nervous systems. To date, the only known physiological role for this neuropeptide is as an inhibitor of abdominal ganglia activity in *Aplysia*. Clearly, it is imperative that its physiological functions be determined throughout the invertebrate phyla.

Acknowledgments

The authors thank the following for kindly providing unpublished data for inclusion in Table 2: W. J. Curry, I. Fairweather, M. S. Gustafsson, G. Hrckova, N. J. Marks, J.-Z. Pan, S. Quinn, M. Reuter, P. J. Skuce, D. Smart & P. Verhaert.

References

Adrian, T. E., J. M. Allen, S. R. Bloom, M. A. Ghatei, M. N. Rossor, G. W. Roberts, T. J. Crow, K. Tatemoto & J. M. Polak, 1983. Neuropeptide Y distribution human brain. Nature 306: 584–586.

Brennan, G. P., D. W. Halton, A. G. Maule, C. Shaw, C. F. Johnston, S. Moore & I. Fairweather, 1993a. Immunoelectron microscopical studies of regulatory peptides in the nervous system of the monogenean parasite, *Diclidophora merlangi*. Parasitology 106: 171–176.

Brennan, G. P., D. W. Halton, A. G. Maule & C. Shaw, 1993b. Electron immunogold labelling of regulatory peptide immunoreactivities in the nervous system of Moniezia expansa (Cestoda: Cyclophyllidea). Parasit. Res. 79(5): 409–415.

Brownlee, D J. A., I. Fairweather, C. F. Johnston, D. Smart, C. Shaw & D. W. Halton, 1993. Immunocytochemical demonstration of neuropeptides in the central nervous system of the roundworm, *Ascaris suum* (Nematoda: Ascaroidea). Parasitology 106: 305–316.

Curry, W. J., C. Shaw, C. F. Johnston, L. Thim & K. D. Buchanan, 1992. Neuropeptide F: primary structure from the turbellarian, Artioposthia triangulata. Comp. Biochem. Physiol. 101C: 269–274.

Dockray, G. J. & R. G. Williams, 1983. FMRFamide-like immunoreactivity in rat brain: Development of a radioimmunoassay and its application in studies of distribution and chromatographic properties. Brain Res. 266: 295–303.

Fairweather, I. & D. W. Halton, 1991. Neuropeptides in platyhelmints. Parasitology 102: S77–S92.

Halton, D. W., G. I. P. Brennan, A. G. Maule, C. Shaw, C. F. Johnston & I. Fairweather, 1991. The ultrastructure and immunogold labelling of pancreatic polypeptide-immunoreactive cells associated with the egg-forming apparatus of a monogenean parasite, *Diclidophora merlangi*. Parasitology 102: 429–436.

Halton, D. W., C. Shaw, A. G. Maule, C. F. Johnston & I. Fairweather, 1992. Peptidergic messengers: a new perspective of the nervous system of parasitic platyhelminths. J. Parasitol. 78: 179–193.

Halton, D. W., A. G. Maule & C. Shaw, 1993. Neuronal mediators in monogenean parasites. Bull. fr. Piscic. 328: 82–104.

Leung, P. S., C. Shaw, A. G. Maule, L. Thim, C. F. Johnston & G.B. Irvine, 1992. The primary structure of neuropeptide F (NPF) from the garden snail, *Helix aspersa*. Regul. Pept. 41: 71–81.

Magee, R. M., I. Fairweather, C. F. Johnston, D. W. Halton & C. Shaw, 1989. Immunocytochemical demonstration of neuropeptides in the nervous system of the liver fluke, *Fasciola hepatica*. Parasitology 98: 227–238.

Marks, N. J., A. G. Maule, D. W. Halton, C. Shaw & C. F. Johnston, 1993. Distribution and immunochemical characteristics of neuropeptide F (NPF) (*Moniezia expansa*)-immunoreactivity in *Proteocephalus pollanicola* (Cestoda: Proteocephalidea). Comp. Biochem. Physiol. 104C: 381–386.

Maule, A. G., D. W. Halton, C. F. Johnston, C. Shaw & I. Fairweather, 1990a. The serotoninergic, cholinergic and peptidergic components of the nervous system in the monogenean parasite, *Diclidophora merlangi*: a cytochemical study. Parasitology 100: 255–273.

Maule, A. G., D. W. Halton, C. F. Johnston, C. Shaw & I. Fairweather, 1990b. A cytochemical study of the serotoninergic, cholinergic and peptidergic components of the reproductive system in the monogenean parasite, *Diclidophora merlangi*. Parasitol. Res. 76: 409–419.

Maule, A. G., C. Shaw, D. W. Halton, L. Thim, C. F. Johnston, I. Fairweather & K D. Buchanan, 1991. Neuropeptide F: a novel parasitic flatworm regulatory peptide from *Moniezia expansa* (Cestoda: Cyclophyllidea). Parasitology 102: 309–316.

Maule, A. G., C. Shaw, D. W. Halton, C. F. Johnston & I. Fairweather, 1992a. Immunochemical and chromatographic analyses of a neuropeptide from the monogenean parasite, *Diclidophora merlangi*: evolutionary aspects of the neuropeptide Y superfamily. Comp. Biochem. Physiol. 102C: 517–522.

Maule, A. G., C. Shaw, D. W. Halton, G. P. Brennan, C. F. Johnston & S. Moore, 1992b. Neuropeptide F (*Moniezia expansa*): localisation and characterisation using specific antisera. Parasitology 105: 505–512.

Maule A. G., G. P. Brennan, D. W. Halton, C. Shaw, C. F Johnston & S. Moore, 1992c. Neuropeptide F-immunoreactivity in the monogenean parasite *Diclidophora merlangi*. Parasitol. Res. 78: 655–660.

Maule, A. G., D. W. Halton, C. Shaw & I. Fairweather, 1993a. The cholinergic, serotoninergic and peptidergic components of the nervous system of *Moniezia expansa* (Cestoda, Cyclophyllidea). Parasitology 106: 429–440.

Maule, A. G., C. Shaw, D. W. Halton & L. Thim, 1993b. GNFFRFamide: a novel FMRFamide-immunoreactive peptide isolated from the sheep tapeworm, *Moniezia expansa*. Biochem. Biophys. Res. Comm. 193: 1054–1060.

McDonald, J K., 1988. Neuropeptide Y and related substances. In J. Nelson (ed.), CRC Critical Reviews in Neurobiology, CRC Press, Boca Raton, Florida. 4: 97–135.

Rajpara, S. M., P. D. Garcia, R. Roberts, J. C. Eliassen, D. F. Owens, D. Maltby, R. I. Myers & E. Mayeri, 1992. Identification and molecular cloning of a neuropeptide Y homolog that produces prolonged inhibition in *Aplysia* neurons. Neuron 9: 505–513.

Schoofs, L., J. M. Danger, S. Jegou, G. Pelletier, R. Huybrechts, H. Vaudry & A. De Loof, 1988. NPY-like peptides occur in the nervous system of the migratory locust, Locusta migratoria and in the brain of the grey fleshfly, *Sarcophaga bullata*. Peptides 9: 1027–1036.

Sithigorngul, P., A. O. W. Stretton & C. Cowden, 1990. Neuropeptide diversity in *Ascaris*: An immunocytochemical study. J. Comp. Neurol. 294: 362–376.

Smart, D., C. Shaw, C. F. Johnston, D. W. Halton, I. Fairweather & K. D. Buchanan, 1992a. Chromatographic and immunochemical characterisation of neuropeptide Y-like and pancreatic polypeptide-like peptides from the nematode *Ascaris suum*. Comp. Biochem. Physiol. 102C: 477–481.

Smart, D., C. Shaw, C. Johnston, L. Thim, D. Halton & K. Buchanan, 1992b. Peptide tyrosine phenylalanine: a novel neuropeptide F-related nonapeptide from the brain of the squid, *Loligo vulgaris*. Biochem. Biophys. Res Comm. 186: 1616–1623.

Hydrobiologia **305**: 305–306, 1995.
L.R.G. Cannon (ed.), Biology of Turbellaria and some Related Flatworms.

The pattern of neuropeptide F and RF-amide in two tapeworms

I. Elo[1], A. G. Maule, M. Grahn[1], C. Shaw[2], M. K. S. Gustafsson[1] & D. W. Halton[2]
[1]*Department of Biology, Åbo Akademi University, FIN- 20 520 Åbo, Finland*
[2]*Comparative Neuroendocrinology Research Group, School of Biology and Biochemistry, The Queen's University of Belfast, Belfast BT7 1NN, UK*

Key words: Cestode, nervous system, neuropeptide F, FMRF, immunocytochemistry

Abstract

Two years ago the first platyhelminth regulatory peptide, neuropeptide F (NPF), was isolated from the tapeworm *Moniezia expansa* by Maule *et al.* (1991). NPF is a 39 amino acid peptide with a C terminal phenylalaninamide. NPF is the first platyhelminth neuropeptide to be sequenced fully. Preabsorption with NPF quenches the immunostaining with anti-FMRF-amide and anti-bovine PP (Halton *et al.* 1992). As the first authentic flatworm neuropeptide, the occurrence and distribution of NPF along the whole flatworm line are under investigation. Both free-living and parasitic flatworms are being studied. So far NPF-immunoreactivity has been reported from three free- living flatworms (see Grahn *et al.*, 1995) and from four parasitic flatworms (Marks *et al.*, 1993).

FMRF- and RF-amide immunoreactive (IR) nerve cells and fibres are common in the gull-tapeworm *Diphyllobothrium dendriticum*. In order to test whether the patterns for NPF- and RF-immunoreactivity co- localize in the gull-tapeworm, immunostaining with anti-NPF and anti-RF were performed. To broaden the study, adult *Proteochepalus exiguus* from the intestine of whitefish were included in the experiment.

The study was performed on whole mounts of skinned worms (Gustafsson, 1991). Anti-NPF was used in concentrations 1:500 and 1:1000. Controls included liquid phase absorption with the homologous antigen (1000 ng ml^{-1}).

In *D. dendriticum* NPF-immunoreactivity occurs in nerve cells and varicose nerve fibres of larval and adult worms. The NPF-IR cell bodies are more common in the peripheral nerve cords than along the main nerve cords, which contain nerve fibres with large varicosities. The cell bodies in the PNS are often triangular in shape. Immediately beneath the tegumental surface a thin NPF-IR nerve fibre is observed. As to the co-localization of NPF and RF nothing definite can be said but the general pattern seems to be the same. In the brain commissure of *D. dendriticum* one large ganglion cell stains with both antisera, indicating coexistence.

In *P. exiguus* NPF- and RF-immunoreactivity was observed in the two main nerve cords situated laterally and in the pairs of thin dorsal and ventral longitudinal nerve cords. Numerous transverse commissures connect the longitudinal cords forming an orthogonal pattern. The cell bodies along the nerve cords are multipolar. Thin projections extend from the main nerve cords to the surface of the worm. The main nerve cords are lined with NPF-and RF-IR cell bodies. The general staining patterns of NPF and RF are very similar.

References

Grahn, M., A. G. Maule, I. Elo, C. Shaw, M. Reuter & D. W. Halton, 1995. Antigenicity to neuropeptide F (NPF) in *Stenostomum leucops* and *Microstomum lineare*. Hydrobiologia 305 (Dev. Hydrobiol. 108): 307–308.

Gustafsson, M. K. S., 1991. Skin your tapeworms before you stain their nervous system! A new method for whole-mount immunocytochemistry. Parasitol. Res. 77: 509–516.

Halton, D. W., C. Shaw, A. G. Maule, C. F. Johnston & I. Fairweather, 1992. Peptidergic messengers: a new perspective of the nervous system of parasitic platyhelminths. J. Parasitol. 78: 179–193.

Marks, N. J., A. G. Maule, D. W. Halton, C. Shaw & C. F. Johnston, 1993. Distribution and immunochemical characteristics of neuropeptide F (NPF) (*Moniezia expansa*) — immunoreactivity in *Proteocephalus pollanicola* (Cestoda: Proteocephalidae). Comp. Biochem. Physiol. 104: 381–386.

Maule, A. G., C. Shaw, D. W. Halton, L. Thim, C. F. Johnston, I. Fairweather & K. D. Buchanan 1991. Neuropeptide F: a novel parasitic flatworm regulatory peptide from *Moniezia expansa* (Cestoda: Cyclophyllidae). Parasitology 12: 309–316.

Hydrobiologia **305**: 307–308, 1995.
L.R.G. Cannon (ed.), *Biology of Turbellaria and some Related Flatworms.*

Antigenicity to neuropeptide F (NPF) in *Stenostomum leucops* and *Microstomum lineare*

M. Grahn[1], A. G. Maule[2], I. Elo[1], C. Shaw[2], M. Reuter[1] & D. W. Halton[2]
[1]*Department of Biology, Åbo Akademi University, Åbo, Finland*
[2]*Comparative Neuroendocrinology Research Group, School of Biology and Biochemistry, The Queen's University of Belfast, Belfast BT7 1NN, Northern Ireland, UK*

Key words: Plathelminthes, Platyhelminthes, nervous system, immunocytochemistry, neuropeptide F, FMRF

Abstract

The first 'native' flatworm regulatory peptide, neuropeptide F (NPF) has recently been isolated and sequenced from the cestode *Moniezia expansa* (see Maule *et al.*, 1991) and the turbellarian *Artioposthia triangulata*, (see Curry *et al.*, 1992). NPF belongs to the neuropeptide Y (NPY) superfamily and the antiserum is known to show cross-reactivity to the vertebrate neuropeptides of the NPY superfamily. It terminates in RFamide, like the invertebrate neuropeptides FMRFamide and RFamide, and may cross-react with neuropeptides of the FMRFamide family. Strong immunoreactivity (IR) to FMRF- and RF-amide has been demonstrated in members of most flatworm groups. In the present study, IR to NPF (diluted 1:1000) is demonstrated in *Stenostomum leucops* (Catenulida) and *Microstomum lineare* (Macrostomida). The controls included: omitting primary antibody, using non-immune serum and liquid-phase absorption with the homologous antigen (1000 ng ml^{-1}). The NPF IR pattern was compared to the FMRF and RF-amide IR patterns in order to reveal differences or co-localization. In addition, the sequential appearance of NPF-positive cells in developing zooids was followed and double staining with a-5-HT made to complete the study.

The antigenicity to NPF, obtained in *Stenostomum leucops* and *M. lineare*, strongly indicates the presence of NPF in flatworms of the taxa Catenulida and Macrostomida, which occupy positions near the base of the flatworm phylogenetic tree. The distribution of IR to NPF in the central and peripheral nervous system (NS) in the worms corresponds in a high degree to the FMRF-amide/RF-amide IR pattern. In *S. leucops*, the main difference is the weak IR in the pharyngeal NS compared to the very strong FMRF-amide/RF-amide IR. In *M. lineare*, the use of confocal microscopy revealed NPF–positive cells connected by a network of immunoreactive nerve fibres meandering in the intestinal wall. The occurrence of neuron-like cells in the intestinal wall has been postulated on the basis of ultrastructural studies, but the NPF IR in cells of the intestinal nerve net is the first demonstration of a regulatory peptide in these cells.

The pattern of positive cells associated with the brain and the pharyngeal nerve ring, and of immunoreactive fibres in the nerve cords and intestinal wall, corresponds to that observed in FMRF-amide/RF-amide immunostaining. A positive reaction in the intestinal granular club cells was also observed. Similar results have previously been obtained with antisera to FMRF-amide and to bovine pancreatic peptide (BPP). The discussion of cross-reactivity or co-localization has to wait for further investigations.

The study of the sequential appearance of NPF-positive cells and fibres in the new NS of developing zooids of *M. lineare* revealed new details. IR to both NPF and RF-amide was demonstrated in a cell pair in the rear end of mature zooid chains. In addition, a strongly positive pair of cells associated with the lateral cords and connected by a thin circumferential fibre was sometimes observed very close to the division zone in developing zooids. No 5-HT IR, indicating the initiation of the development of the brain of a new zooid, could be demonstrated in this area by double staining. The appearance of the cell pair apparently marks the finished development of a new zooid. It seems to correspond to the cell pair observed in the rear end of mature zooid chains. Variability of the number of neurons in regenerating turbellarians has been proposed. The present observation, however, points in the direction of constancy in organization and number of neurons in this worm with high regenerative capacity.

References

Curry, W. J., C. Shaw, C. F. Johnston, L. Thim & K. D. Buchanan, 1992. Neuropeptide F: primary structure from the turbellarian, *Artioposthia triangulata*. Comp. Biochem. Physiol. 101C: 269–274.

Maule, A. G., C. Shaw, D. W. Halton, L. Thim, C. F. Johnston, I. Fairweather and K. D. Buchanan, 1991. Neuropeptide F: a novel parsitic flatworm regulatory peptide from *Moneizia expansa* (Cestoda: Cyclophyllidea). Parasitology 102: 309–316.

Hydrobiologia **305**: 309–316, 1995.
L.R.G. Cannon (ed.), *Biology of Turbellaria and some Related Flatworms.*

Flatworm neuropeptides – present status, future directions

I. Fairweather & P. J. Skuce
School of Biology and Biochemistry, The Queen's University of Belfast, Belfast BT7 1NN, Northern Ireland, UK

Key words: platyhelminths, neuropeptides, localisation, isolation, molecular studies, physiological roles

What are regulatory peptides?

Peptides represent a highly-diverse group of multi-purpose chemical messengers produced by the nervous, endocrine and immune systems. In the nervous system, peptides function as neurotransmitters or neuromodulators, modifying the action of other transmitter substances. Neuronal peptides can also serve as blood-borne neurohormones. In the endocrine system, peptides function as classical blood-borne hormones or as localised paracrine mediators. In the immune system, peptides serve immune functions. The bidirectional communication between the immune and neuroendocrine systems allows immunologic peptides to play hormonal roles and neuroendocrine peptides to serve immunoregulatory functions. The fact that the three systems share the same peptide molecules gives rise to highly versatile and complex mechanisms that are involved in the maintenance of homeostasis. A more detailed discussion of regulatory peptides is beyond the scope of this brief review. For further information, the reader is referred to the many excellent reviews and texts in this field, including Blalock (1989), Holmgren (1989), Polak (1989) and Turner (1987).

Significance of the peptidergic (neurosecretory) system in platyhelminths

Why are peptides of particular importance to flatworms? Perhaps that question can best be answered in the following way. Flatworms lack a circulatory system and endocrine glands, and so are unable to produce classical, blood-borne hormones. Consequently, in these organisms, the nervous system takes over an endocrine role by its production of peptidergic (neurosecretory) molecules – after all, it is the only system capable of producing and transporting hormone-like substances. The peptidergic system, then, will be responsible for controlling and co-ordinating many aspects of flatworm physiology, especially growth and development which are processes normally under the control of hormones in higher organisms. Therefore, the platyhelminth nervous system serves the dual role of nervous system and endocrine system.

Given the significance of the peptidergic system, it is not surprising that the main focus of attention in flatworm neurobiology has shifted towards it. The aim is to gain a greater insight into the range of peptide molecules present, their chemical identities and, most importantly, their roles. This review will attempt to assess what has been achieved to date and highlight new areas for research, to expand our understanding of this major component of the flatworm nervous system.

Immunocytochemical localisation of neuropeptides

The presence of peptidergic nerve cells in flatworms has been known for some time, through the use of special staining techniques for neurosecretory substances and from observation of dense-cored (neurosecretory) vesicles in electron microscope studies on the nervous system (see reviews by Fairweather & Halton, 1991; Lender, 1974, 1980; Webb, 1988). However, it has been the application of immunocytochemical techniques, using well-characterised antisera raised against a wide range of invertebrate and vertebrate peptides that has been the main impetus behind studies on neuropeptides in flatworms. These studies have been greatly aided by the application of confocal scanning laser microscopy. The ability to optically section through intact whole-mount specimens and generate three-dimensional images of neural elements has revolutionised neuroanatomical studies (Johnston *et al.*, 1990). Immunoreactivities to more than 30 vertebrate and invertebrate peptides have been demonstrated to date and the number is increasing all the time. However, there is still a long way to go to reach the recent estimate of a few hundred biologically active peptides in each species (De Loof & Schoofs, 1990). Collectively, the immunocytochemical studies have reinforced the view that the peptidergic nervous system predominates over the cholinergic and aminergic components. Only the major findings will be highlighted here; for more detailed accounts the reader is referred to the reviews of Fairweather & Halton (1991, 1992), Fairweather *et al.* (1992), Gustafsson (1992), Halton *et al.* (1992) and Reuter & Gustafsson (1989).

Peptides have been localised throughout the central nervous system of those species examined, although the distribution may vary depending on the peptide antiserum used. The main immunoreactivities demonstrated have been to members of the neuropeptide Y (NPY) superfamily (pancreatic polypeptide [PP], peptide YY [PYY] and NPY), tachykinins such as substance P and the molluscan peptide, FMRFamide. Many of the immunocytochemical studies have hinted at the possibility of co-localisation between different peptides within the same neuron. However, only one study involving a double-labelling method has confirmed this. In this study, co-localisation between RFamide and the molluscan peptide, small cardiac peptide B was demonstrated in the cestode, *Diphyllobothrium dendriticum*, by Gustafsson & Wikgren (1989). Peptides are also widely distributed in the nerve plexuses that constitute the peripheral nervous system, that is, the plexuses innervating the main body musculature, the musculature of attachment organs such as the suckers, the outer muscle coat of the gut and reproductive ducts and the wall of the excretory ducts. Innervation of sensory receptors has also been documented in some species, suggesting a role for peptides in sensory perception and transduction.

A major role for peptides in the reproductive biology of flatworms has been indicated by the innervation of the gonads in a number of parasitic species. However, a finding of perhaps greater potential significance has been the association between peptidergic nerve elements and the egg-producing apparatus (or ootype/Mehlis' gland complex) in trematode parasites. Peptides may control the release of ova and vitelline cells and their movements along the oviduct, vitelline duct and ovovitelline duct; the entry of ova and vitelline cells into, and the exit of the newly-formed egg from, the ootype; the motility of the ootype; and exert a paracrine-like influence over the secretory activity of the Mehlis' gland cells. High reproductive capacity, in the form of egg production, is essential to the survival of parasites, to enable them to overcome the hazards of transmission from one host to the next in the often complex life cycles displayed by these organisms. In view of this, it is not surprising to find a key role in egg production being taken by neuropeptides; this aspect of peptide functioning has been reviewed by Fairweather & Halton (1992).

Ontogeny of the peptidergic system

Peptidergic nerve elements are present in all life cycle stages of parasitic flatworms. In trematodes, for example, they have been demonstrated in the miracidium and cercaria of the human blood fluke, *Schistosoma mansoni*, by Skuce *et al.* (1990). Amongst cestodes, peptidergic nerve cells have been demonstrated in the procercoid and plerocercoid stages of the pseudophyllidean, *D. dendriticum* (see Gustafsson *et al.* 1985, 1986; Wikgren, 1986; Wikgren *et al.*, 1986); in the plerocercoid of the pseudophyllidean, *Schistocephalus solidus*, by Wikgren *et al.* (1986) and the tetraphyllidean, *Trilocularia acanthiaevulgaris*, by Fairweather *et al.* (1990), and in the cysticercoid larva of the cyclophyllidean, *Hymenolepis diminuta*, by Fairweather *et al.* (1988).

Confocal microscopy has made it possible to trace the neuronal circuitry and ontogeny of the nervous system in what are often very small organisms. This has been carried out for the liver fluke, *Fasciola hepatica*, by Magee (1990). In the miracidium larva, which is only 60 μm in length, the nervous system takes the form of a single, central anterior nerve mass, from which emanate six nerve tracts. Some centralisation of the nerve elements into paired anterior ganglia occurs in the redia, but the predominant component of the nervous system takes the form of a peripheral nerve net. The nervous system in the cercaria more closely resembles that of the adult. Paired anterior ganglia are joined by a commissure and the three pairs of nerve cords are linked by transverse connectives that form an orthogonal pattern typical of platyhelminths. Peptidergic nerve elements have been demonstrated in each larval stage, the number of immunoreactive nerve cells increasing in each successive larval stage as the nervous system develops and becomes more complex (Magee, 1990).

Ultrastructural localisation of neuropeptides –immunogold labelling studies

Few immunoelectron microscopical studies have been carried out on peptides in flatworms (Reuter *et al.* 1990; Palmberg, 1991; Brennan *et al.* 1993a, b). They have demonstrated labelling of dense-cored vesicles resembling typical neurosecretory (peptidergic) vesicles in nerve cell bodies and nerve fibres, thus confirming the neuronal localisation of peptide immunoreactivities obtained at the light microscope level. Labelled vesicles have been localised at classic synapses within the nervous system and at neuromuscular junctions, supporting the supposition that peptides act as neurotransmitters in flatworms (Reuter *et al.* 1990). Double and triple labelling studies have been carried out, demonstrating that co-localisation of peptide molecules may occur within the same neuronal vesicle. However, they involved the use of antisera to PP, FMRFamide and NPF and blocking experiments indicated that at least some, and perhaps all, of the PP and FMRFamide immunolabelling may be due to an NPF-like peptide (Brennan *et al.*, 1993a, b). However, the technique used did not permit discrimination between these closely-related epitopes. Consequently, the true situation regarding the potential co-localisation of unrelated peptide molecules within the same neuron remains to be evaluated.

Isolation of 'native' flatworm peptides

Radioimmunometrical studies have been carried out on a number of trematode and cestode parasites, to quantify the levels of peptide immunoreactivity in these organisms (McKay *et al.*, 1990, 1991; Magee *et al.*, 1991; see also Halton *et al.*, 1992). The predominant peptide immunoreactivity has been shown to be due to a PP-like molecule, confirming the immunocytochemical data. The same C-terminally-directed PP antiserum used in the localisation and assay studies has also been used for the isolation and complete amino acid sequencing of a peptide from the sheep tapeworm, *Moniezia expansa*; this peptide has been designated neuropeptide F (NPF) (Maule *et al.* 1991). Analogous peptides have subsequently been isolated from another flatworm, the predatory turbellarian, *Artioposthia triangulata*, by Curry *et al.* (1992) and the molluscs *Helix aspersa* by Leung *et al.* (1992) and *Aplysia californica* by Rajpara *et al.* (1992). For a more detailed account of the NPF story, see Maule *et al.* (1995).

Molecular approaches

Isolation of parasite peptides by conventional chromatographic techniques is feasible with parasites of the size of *M. expansa*, which is >5 m in length, and contains a high level of PP immunoreactivity (193 ng g^{-1} wet weight). This is an exceptional case, however, because most other parasites are small organisms and display lower levels of PP immunoreactivity (for examples, see Halton *et al.* 1992). The human blood fluke, *S. mansoni*, for example, is very small – a worm pair weighs only 0.5 mg – and the level of PP immunoreactivity is extremely low: $\simeq$25 ng g^{-1} wet weight. Consequently, the amount of tissue required for isolation purposes would be prohibitively large and so, in this case, an alternative, molecular approach has been adopted. The aim of this approach is to clone and sequence the gene coding for the PP-(presumed NPF) like peptide from cDNA expression-vector libraries of the worm. Such an approach has proved successful for other invertebrate groups, including molluscs, insects, nematodes and coelenterates (Linacre *et al.*, 1990; Taussig & Scheller, 1986; Schneider & Taghert, 1988; Rosoff *et al.*, 1993; Grimmelikhuijzen *et al.*, 1992).

Peptide hormones are typically encoded by one or more genes and synthesised in the form of long precursor proteins (preprohormones) from which the bioactive peptide(s) are enzymatically excised and post-translationally modified. In the case of neuropeptides, this modification usually takes the form of C-terminal amidation, which is essential in eliciting biological activity (Bradbury & Smyth, 1991). An absolute prerequisite for successful immunological screening is an antiserum which will recognise the likely precursor of NPF, since it is the sequence of the precursor and not the mature peptide which will be represented in a cDNA library. Moreover, the antiserum must be able to recognise this molecule in its non-native, unprocessed form, because fusion proteins are not correctly processed, that is, cleaved and amidated, in a bacterial-based expression-vector system. Unfortunately, the available C-terminally-directed PP antisera used initially to localise, quantify and characterise the schistosome peptide did not meet these criteria and were unsuitable for screening purposes. This was mainly due to the high proportion of antibodies directed against the extremely immunogenic C-terminal α-amide structure. Antisera to the unprocessed C-terminus and the glycine-extended C-terminal of NPF are being raised to target the precursor in future screening experiments.

In view of the problems associated with the immunological screening experiments, an alternative approach has been adopted to screen the same schistosome libraries with oligonucleotide probes. In the absence of schistosome-specific amino acid sequence information, the probes were designed on the basis of regions of homology in available PP/NPF sequences from other species. The probes are by necessity highly degenerate, representing all possible nucleotide sequences capable of encoding specific tracts of amino acids, bearing in mind the typical codon usage of schistosomes (Meadows & Simpson, 1989). Prior to embarking on screening experiments, the various probes were evaluated in hybridisation experiments with schistosome DNA and RNA. One degenerate antisense probe, a 17-mer directed against the likely C-terminal connecting sequence of the PP/NPF-precursor, recognises a small (≃500 bp) transcript in Northern blots of total RNA that is comparable in size to that of the precursor proteins of the NPY superfamily (Minth *et al.*, 1984) and NPF (Rajpara *et al.*, 1992) (Skuce, unpublished observations).

Another strategy, which would actually obviate the need to clone the NPF gene, is to employ the polymerase chain reaction (PCR) with opposing sense and antisense primers in order to synthesise the schistosome NPF homologue *in vitro*. PCR primers are usually unique oligonucleotides designed from known DNA sequences. However, in the case of an uncloned gene, such as that for NPF, it is possible to employ highly degenerate primers designed from known (or inferred!) amino acid sequence information. The DNA amplified in this way then represents the gene of interest. However, because the NPF molecule is small and all the sequence homology resides in the mid-to-C-terminal region, the expected PCR products will also be relatively small (≃75 bp). Nevertheless, a 75 bp PCR product has been generated from genomic DNA, using the aforementioned C-terminal primers, which when isolated and sequenced proved to encode a tract of amino acids terminating in the heptapeptide YYGRPRF, which is identical to the C-terminus of planarian NPF (Skuce, unpublished observations). This product is currently being evaluated by Southern analysis of genomic DNA prior to its use as a schistosome-specific probe for further library screening, in order to identify the complementary full-length cDNA molecule(s). Moreover, as yet unspecified PCR products of the predicted size have been amplified directly from the cDNA libraries using a degenerate oligonucleotide primer in combination with a vector-specific primer in an 'anchored PCR' strategy.

Once a schistosome PP/NPF clone has been isolated, it should then be possible to derive the complete amino acid sequence of the schistosome peptide precursor and to identify the mature peptide and its likely biosynthetic processing sites, such as the amidation-cleavage signal and possible dibasic/monobasic cleavage sites. As previously stated, neuropeptides are not correctly processed, that is, cleaved and amidated, in a bacterial-based prokaryotic expression-vector system. However, positive cDNA clones identified by screening such systems can be introduced into a eukaryotic expression-vector system by cloning into a virus infective to eukaryotic cells. Vaccinia virus has been used successfully to express amidated neuropeptides in a variety of vertebrate cell lines (Hruby & Thomas, 1987). It would then be possible to investigate the biosynthetic processing of the peptide by site-directed mutagenesis and induce the large-scale production of the encoded peptide for evaluation in physiological studies.

While progress with the molecular approach has been frustratingly slow, the potential rewards from its successful accomplishment are considerable. Identification of the amino acid sequence of the peptide and its precursor, together with an understanding of the biosynthetic processing of the precursor, have already been alluded to. In comparison with what is known about the precursors of other known invertebrate peptides, it is likely that sequences of additional peptides will be obtained once the precursor has been identified. It will also be possible to investigate the expression of the gene during the life cycle of the parasite.

Physiological roles

Little progress has been made towards elucidating the functions of peptides in flatworms but, of what information is available, this is one area in which more is known about turbellarian than parasitic species. It has been known for more than 20 years that a cycle of neurosecretory activity is associated with the scissiparity cycle in freshwater planarians (see reviews by Lender, 1974, 1980; Webb, 1988). A cycle of RNA synthesis parallels the neurosecretory cycle, although it is slightly displaced. This led to the suggestion that the neurosecretory material may activate the totipotent neoblast cells during the second half of the scissiparity cycle, inducing them to synthesize RNA and so prepare the worm for the regeneration that follows the next fission (Lender, 1974, 1980; Webb, 1988). Neurosecretions have also been implicated in the control of regeneration and sexual reproduction in planarians; in the latter, neurosecretions may stimulate the maturation and differentiation of gametes from neoblast cells (Lender, 1974, 1980; Webb, 1988).

A more substantial role for neurosecretions in planarian regeneration was provided by a classic set of experiments by Webb & Friedel (Webb & Friedel, 1978, 1979; Friedel & Webb, 1979). In these experiments, a peptidergic fraction was isolated from *Dugesia tigrina* and was shown to be localised in neurosecretory cells in the brain and ventral nerve cords. Addition of this fraction to post-pharyngeal pieces of worm stimulated protein and RNA synthesis, also mitotic activity. Thus, the experiments provided evidence for a morphogenetic role of an uncharacterised peptide in planarians.

More recent work on specific peptides, namely, the tachykinins substance P and neurokinin A, has shown that they stimulate mitosis and cellular differentiation in intact and regenerating planarians (Baguñà, 1986, 1987; Baguñà *et al.* 1989; Saló & Baguñà, 1986). In contrast, it has been suggested that somatostatin plays an inhibitory role in regeneration (Bautz & Schilt, 1986).

In parasitic species, neurosecretions have been implicated in the control of sexual reproduction in the trematode species *Leucochloridiomorpha constantiae* and *F. hepatica* (see Harris & Cheng, 1972; Grasso & Quaglia, 1972). Also, they may regulate morphogenesis in the cestode, *D. dendriticum* (see Gustafsson & Wikgren, 1981; Gustafsson *et al.* 1983). A number of vertebrate peptides (cholecystokinin, motilin, caerulein, somatostatin) affect the motility of *F. hepatica* and *H. diminuta in vitro* (Sukhdeo *et al.* 1984; Sukhdeo & Sukhdeo, 1989). Of these, only cholecystokinin-immunoreactivity has been demonstrated in one of the parasites (*F. hepatica*: Magee, 1990), so the significance of these results is difficult to assess. Intravenous administration of secretin, cholecystokinin and insulin influences the diurnal migrations of *H. diminuta* within the rat gut, indicating that the worm possesses receptors that interact with the gut hormones and so help it to determine its position within the gut (Cho, 1984; Mettrick & Podesta, 1982).

In addition to more neurally-directed roles, potential morphogenetic roles have been examined in the trematodes, *F. hepatica* and *S. mansoni*. FMRFamide has been shown to stimulate protein and nucleic acid synthesis in *F. hepatica*, the liver fluke (Fairweather, unpublished observations). In contrast, NPF exerts an inhibitory effect over protein and nucleic acid synthesis in *S. mansoni*, the blood fluke (Fairweather, unpublished observations). Such an inhibitory influence is not surprising given that *Aplysia* NPF causes inhibition of nerve cell activity in the abdominal ganglion of this mollusc, the response being characterised by hyperpolarization of the cell membrane and reduction of spike activity (Rajpara *et al.* 1992). The NPF-like peptides are believed to be invertebrate members of the NPY superfamily; NPY itself is one of the most abundant neuropeptides in mammals and is known to be a potent inhibitory transmitter (Allen & Bloom, 1986; McDonald, 1988).

A number of studies have investigated the possibility that growth factors are present in flatworms and may play a role in morphogenesis. Immunostaining for epidermal growth factor (EGF), epidermal growth factor receptor (EGF-r) and basic fibroblast growth factor (bFGF) has been demonstrated in the nervous system of the microturbellarians, *Stenostomum leucops* and *Microstomum lineare* and the tapeworm, *D. dendriticum* (see Gustafsson & Eriksson, 1992; Gustafsson *et al.*, 1995; Reuter & Kuusisto, 1992). Antibodies to the growth factors blocked asexual reproduction in the two turbellarian species (Reuter & Kuusisto, 1992). In *D. dendriticum*, immunoreactivity for bFGF was obtained in the larval and adult stages, although there was no correlation between the presence of bFGF-immunoreactive nerve cells and the general growth rate of the worm (Gustafsson & Eriksson, 1992). EGF has been shown to act as a potent mitogen in the planarian, *D. tigrina*, stimulating mitotic activity in intact and regenerating specimens (Baguñà *et al.*, 1989). Finally, nerve growth factor stimulates neuronal hypertrophy in *Dugesia gonocephala* (see Palladini *et al.*, 1988).

The lack of supporting physiological data has not prevented speculation as to the possible roles that peptides might play in flatworms; the ideas are based largely on the results of localisation studies. Immunolabelling, particularly of dense-cored vesicles in synapses, points strongly to a role in neurotransmission and/or neuromodulation. Innervation of sensory receptors suggests that peptides may be involved in sensory perception and transduction. A major role in gametogenesis and egg production is indicated by the innervation of the gonads and reproductive ducts, particularly the ootype/Mehlis' gland complex. Finally, innervation of the excretory ducts in certain species suggests a possible role in fluid transport. However, all these roles await experimental verification.

Conclusions

A considerable number of immunocytochemical studies have now been carried out on flatworms. They have established the peptidergic system as the main component of the nervous system, containing a greater variety of potential transmitter substances than the cholinergic and aminergic systems. Moreover, they have moved the peptidergic system centre stage in terms of flatworm neurobiology and revived interest in this neglected area of flatworm physiology. While much of the initial screening for peptide immunoreactivities has been completed, immunocytochemical studies are likely to continue for some time to come, as new peptide antisera become available. Neuronal localisation of peptides has been confirmed by immunolabelling at the ultrastructural level, but only a limited number of these studies has been carried out to date. Experiments to quantify and characterise peptide immunoreactivities have been performed for a few species and this has led to the isolation of NPF. More peptides will be isolated in future, but for many flatworms this will involve an alternative, molecular approach; that breakthrough has yet to happen. So, some progress has been made towards understanding what peptide molecules are present and their distribution. However, given that peptides are likely to be involved in both neural and 'hormonal' activities, relatively little is known about their roles in these two areas. It is perhaps in these two key areas that the main thrust of work will focus in the near future, both in terms of their gross roles and the intracellular signalling systems.

Acknowledgments

The schistosome work included in this review was supported by the UNDP/World Bank/WHO Special Programme for Research and Training in Tropical Diseases, Project No. 900336.

References

Allen, J. M. & S. R. Bloom, 1986. Neuropeptide Y: a putative neurotransmitter. Neurochem. Int. 8: 1–8.

Baguñà, J., 1986. Efecte de neuropèptids i factors de creixement sobre la proliferació cel.lular a planàries. Biol. Desenv. 4: 71–78.

Baguñà, J., 1987. Efectes des estimuladors i antagonistes dels neuropéptids Substància i Substància K sobre la proliferació cel.lular a planàries. Biol. Desenv. 5: 274–283.

Baguñà, J., E. Saló & R. Romero, 1989. Effects of activators and antagonists of the neuropeptides substance P and substance K on cell proliferation in planarians. Int. J. Dev. Biol. 33: 261–264.

Bautz, A. & J. Schilt, 1986. Somatostatin-like peptide and regeneration capacities in planarians. Gen. Comp. Endocrinol. 64: 267–272.

Blalock, J. E., 1989. A molecular basis for bidirectional communication between the immune and neuroendocrine systems. Physiol. Rev. 69: 1–32.

Bradbury, A. F. & D. G. Smyth, 1991. Peptide amidation. TIBS 16: 112–115.

Brennan, G. P., D. W. Halton, A. G. Maule, C. Shaw, C. F. Johnston, S. Moore & I. Fairweather, 1993a. Immunoelectron microscopical studies of regulatory peptides in the nervous system of the monogenean parasite, Diclidophora merlangi. Parasitology 106: 171–176.

Brennan, G. P., D. W. Halton, A. G. Maule & C. Shaw, 1993b. Electron immunogold labeling of regulatory peptide immunoreactivity in the nervous system of Moniezia expansa (Cestoda: Cyclophyllidea). Parasitol. Res. 79: 409–415.

Cho, C. H., 1984. Study of the effects of insulin on the migration of Hymenolepis diminuta in rats. J. Helminthol. 58: 291–293.

Curry, W. J., C. Shaw, C. F. Johnston, L. Thim & K. D. Buchanan, 1992. Neuropeptide F: primary structure from the turbellarian, Artioposthia triangulata. Comp. Biochem. Physiol. 101C: 269–274.

De Loof, A. & L. Schoofs, 1990. Homologies between the amino acid sequences of some vertebrate peptide hormones and peptides isolated from invertebrate sources. Comp. Biochem. Physiol. 95B: 459–468.

Fairweather, I. & D. W. Halton, 1991. Neuropeptides in platyhelminths. Parasitology 102: S77–S92.

Fairweather, I. & D. W. Halton, 1992. Regulatory peptide involvement in the reproductive biology of flatworm parasites. Invert. Reprod. Devel. 22: 117–125.

Fairweather, I., D. W. Halton & C. Shaw, 1992. Regulatory peptides in host-parasite interactions: characterisation and roles in pathophysiology and immune responses. Adv. Neuroimmunol. 2: 249–265.

Fairweather, I., G. A. Macartney, C. F. Johnston, D. W. Halton & K. D. Buchanan, 1988. Immunocytochemical demonstration of 5-hydroxytryptamine (serotonin) and vertebrate neuropeptides in the nervous system of excysted cysticercoid larvae of the rat tapeworm, Hymenolepis diminuta (Cestoda, Cyclophyllidea). Parasitol. Res. 74: 371–379.

Fairweather, I., S. Mahendrasingam, C. F. Johnston, D. W. Halton & C. Shaw, 1990. Peptidergic nerve elements in three developmental stages of the tetraphyllidean tapeworm Trilocularia acanthiaevulgaris. Parasitol. Res. 76: 497–508.

Friedel, T. & R. A. Webb, 1979. Stimulation of mitosis in Dugesia tigrina by a neurosecretory fraction. Can. J. Zool. 57: 1818–1819.

Grasso, M. & A. Quaglia, 1972. Ultrastructural studies in neurosecretion and gamete ripening in platyhelminthes. Gen. Comp. Endocrinol. 18: 593–594.

Grimmelikhuijzen, C. J. P., K. Carstensen, D. Darmer, A. Moosler, H.-P. Nothacker, R. K. Reinscheid, C. Schmutzler, H. Vollert, I. McFarlane & K. L. Rinehart, 1992. Coelenterate neuropeptides: structure, action and biosynthesis. Am. Zool. 32: 1–12.

Gustafsson, M. K. S., 1992. The neuroanatomy of parasitic flatworms. Adv. Neuroimmunol. 2: 267–286.

Gustafsson, M. K. S. & K. Eriksson, 1992. Never ending growth and a growth factor. I. Immunocytochemical evidence for the presence of basic fibroblast growth factor in a tapeworm. Growth Factors 7: 327–334.

Gustafsson, M. K. S., K. Eriksson & A. Hydén, 1995. Never ending growth and a growth factor. II. Immunocytochemical evidence for the presence of epidermal growth factor in a tapeworm. Hydrobiologia 305 (Dev. Hydrobiol. 108): 229–233.

Gustafsson, M. K. S., A. C. Jukanen & M. C. Wikgren, 1983. Activation of the peptidergic neurosecretory system in Diphyllobothrium dendriticum (Cestoda) at suboptimal temperatures. Z. Parasitenk. 69: 279–282.

Gustafsson, M. K. S., M. A. I. Lehtonen & F. Sundler, 1986. Immunocytochemical evidence for the presence of 'mammalian' neurohormonal peptides in neurones of the tapeworm Diphyllobothrium dendriticum. Cell Tissue Res. 243: 41–49.

Gustafsson, M. K. S. & M. C. Wikgren, 1981. Activation of the peptidergic neurosecretory system in Diphyllobothrium dendriticum (Cestoda: Pseudophyllidea). Parasitology 83: 243–247.

Gustafsson, M. K. S. & M. C. Wikgren, 1989. Development of immunoreactivity to the invertebrate neuropeptide small cardiac peptide B in the tapeworm Diphyllobothrium dendriticum. Parasitol. Res. 75: 396–400.

Gustafsson, M. K. S., M. C. Wikgren, T. J. Karhi & L. P. C. Schot, 1985. Immunocytochemical demonstration of neuropeptides and serotonin in the tapeworm Diphyllobothrium dendriticum. Cell Tissue Res. 240: 255–260.

Halton, D. W., C. Shaw, A. G. Maule, C. F. Johnston & I. Fairweather, 1992. Peptidergic messengers: a new perspective of the nervous system of parasitic platyhelminths. J. Parasitol. 78: 179–193.

Harris, K. R. & T. C. Cheng, 1972. Presumptive neurosecretion in Leucochloridiomorpha constantiae (Trematoda) and its possible role in governing maturation. Int. J. Parasitol. 2: 361–367.

Holmgren, S., 1989. The comparative physiology of regulatory peptides. Chapman and Hall Ltd., London.

Hruby, D. E. & G. Thomas, 1987. Use of vaccinia virus to express biopharmaceutical products. Pharmaceutical Research 4: 92–97.

Johnston, C. F., C. Shaw, D. W. Halton & I. Fairweather, 1990. Confocal scanning laser microscopy and helminth neuroanatomy. Parasitol. Today 6: 305–308.

Lender, T., 1974. The role of neurosecretion in freshwater planarians. In N. W. Riser & M. P. Morse (eds), Biology of the Turbellaria. McGraw-Hill, New York: 460–475.

Lender, T., 1980. Endocrinologie des planaires. Bull. Soc. zool. Fr. 105: 173–191.

Leung, P. S., C. Shaw, A. G. Maule, L. Thim, C. F. Johnston & G. B. Irvine, 1992. The primary structure of neuropeptide F (NPF) from the garden snail, Helix aspersa. Regul. Pept. 41: 71–81.

Linacre, A., E. Kellett, S. Saunders, K. Bright, P. R. Benjamin & J. F. Burke, 1990. Cardioactive neuropeptide Phe-Met-Arg-Phe-NH_2 (FMRFamide) and novel related peptides are encoded in multiple copies by a single gene in the snail Lymnaea stagnalis. J. Neurosci. 10: 412–419.

McDonald, J. K., 1988. NPY and related substances. CRC Crit. Rev. Neurobiol. 4: 97–135.

McKay, D. M., C. Shaw, D. W. Halton, C. F. Johnston, I. Fairweather & K. D. Buchanan, 1990. Mammalian regulatory peptide immunoreactivity in the trematode parasite Haplometra cylindracea and the lung of its frog host, Rana temporaria: comparative chromatographic characterisation using reverse-phase high-performance liquid chromatography. Comp. Biochem. Physiol. 96C: 345–351.

McKay, D. M., I. Fairweather, C. F. Johnston, C. Shaw & D. W. Halton, 1991. Immunocytochemical and radioimmunometrical demonstration of serotonin- and neuropeptide- immunoreactivities in the adult rat tapeworm, Hymenolepis diminuta (Cestoda, Cyclophyllidea). Parasitology 103: 275–289.

Magee, R. M., 1990. An ontogenetic study of neuropeptides and 5-hydroxytryptamine (serotonin) in the liver fluke Fasciola hepatica. PhD Thesis, The Queen's University of Belfast.

Magee, R. M., I. Fairweather, C. Shaw, J. M. McKillop, W. I. Montgomery, C. F. Johnston & D. W. Halton, 1991. Quantification and partial characterisation of regulatory peptides in the liver fluke, Fasciola hepatica, from different mammalian hosts. Comp. Biochem. Physiol. 99C: 201–207.

Maule, A. G., D. W. Halton & C. Shaw, 1995. Neuropeptide F: a ubiquitous invertebrate neuromediator? Hydrobiologia 305 (Dev. Hydrobiol. 108): 295–301.

Maule, A. G., C. Shaw, D. W. Halton, L. Thim, C. F. Johnston, I. Fairweather & K. D. Buchanan, 1991. Neuropeptide F: a novel parasite flatworm regulatory peptide from Moniezia expansa (Cestoda: Cyclophyllidea). Parasitology 102: 309–316.

Meadows, H. M. & A. J. G. Simpson, 1989. Codon usage in Schistosoma. Mol. Biochem. Parasitol. 36: 291–299.

Mettrick, D. F. & R. B. Podesta, 1982. Effect of gastrointestinal hormones and amines on intestinal motility and the migration of Hymenolepis diminuta in the rat small intestine. Int. J. Parasitol. 12: 151–154.

Minth, C. D., S. R. Bloom, J. M. Polak & J. E. Dixon, 1984. Cloning, characterisation and DNA sequence of a human cDNA encoding neuropeptide tyrosine. Proc. Natn. Acad. Sci. USA 81: 4577–4581.

Palladini, G., L. Medolago-Albani, V. P. Gallo, G. M. Lauro, G. Diana, D. Scorsini, L. Alfei & V. Margotta, 1988. The answer of the planarian Dugesia gonocephala neurons to nerve growth factor. Cell Mol. Biol. 34: 53–63.

Palmberg, I., 1991. Differentiation during asexual reproduction and regeneration in a microturbellarian. Hydrobiologia 227 (Dev. Hydrobiol. 69): 1–10.

Polak, J. M., 1989. Regulatory Peptides. Birkhäuser Verlag, Basel.

Rajpara, S. M., P. D. Garcia, R. Roberts, J. C. Eliassen, D. F. Owens, D. Maltby, R. M. Myers & E. Mayeri, 1992. Identification and molecular cloning of a neuropeptide Y homolog that produces prolonged inhibition in Aplysia neurons. Neuron 9: 505–513.

Reuter, M. & M. Gustafsson, 1989. 'Neuroendocrine cells' in flatworms – progenitors to metazoan neurons? Arch. Histol. Cytol. 52: 253–263.

Reuter, M., M. K. S. Gustafsson, J. Lang & C. J. P. Grimmelikhuijzen, 1990. The release sites and targets of nerve cells immunoreactive to RFamide – an ultrastructural study of Microstomum lineare and Diphyllobothrium dendriticum (Plathelminthes). Zoomorphology 109: 303–308.

Reuter, M. & A. Kuusisto, 1992. Growth factors in asexually reproducing Catenulida and Macrostomida (Plathelminthes)? A confocal, immunocytochemical and experimental study. Zoomorphology 112: 155–166.

Rosoff, M. L., K. E. Doble, D. A. Price & C. Li, 1993. The *flp*-1 propeptide is processed into multiple, highly similar FMRFamide-like peptides in Caenorhabditis elegans. Peptides 14: 331–338.

Saló, E. & J. Baguñà, 1986. Stimulation of cellular proliferation and differentiation in the intact and regenerating planarian Dugesia (G) tigrina by the neuropeptide substance P. J. exp. Zool. 237: 129–135.

Schneider, L. E. & P. H. Taghert, 1988. Isolation and characterization of a Drosophila gene that enoodes multiple neuropeptides related to Phe-Met-Arg-Phe-NH_2 (FMRFamide). Proc. Natn. Acad. Sci. USA 85: 1993–1997.

Skuce, P. J., C. F. Johnston, I. Fairweather, D. W. Halton & C. Shaw, 1990. A confocal scanning laser microscope study of the peptidergic and serotoninergic components of the nervous system in larval Schistosoma mansoni. Parasitology 101: 227–234.

Sukhdeo, M. V. K., S. C. Hsu, C. S. Thompson & D. F. Mettrick, 1984. Hymenolepis diminuta: behavioural effects of 5-hydroxytryptamine, acetylcholine, histamine and somatostatin. J. Parasitol. 70: 682–688.

Sukhdeo, M. V. K. & S. C. Sukhdeo, 1989. Gastrointestinal hormones: environmental cues for Fasciola hepatica? Parasitology 98: 239–243.

Taussig, R. & R. H. Scheller, 1986. The Aplysia FMRFamide gene encodes sequences related to mammalian brain peptides. DNA 5: 453–461.

Turner, A. J., 1987. Neuropeptides and their peptidases. Ellis Horwood Ltd., Chichester.

Webb, R. A., 1988. Endocrinology of Acoelomates. In R. G. H. Downer & H. Laufer (eds), Invertebrate Endocrinology, vol. 2, Endocrinology of Selected Invertebrate Types. Alan R. Liss, New York: 31–62.

Webb, R. A. & T. Friedel, 1978. Localization of a putative regeneration hormone from Dugesia (Turbellaria). An immunoenzyme immunocytochemical study. Bull. Can. microsc. Soc. 6: 21–22.

Webb, R. A. & T. Friedel, 1979. Isolation of a neurosecretory substance which stimulates RNA synthesis in regenerating planarians. Experientia 35: 657–658.

Wikgren, M. C., 1986. The nervous system of early larval stages of the cestode Diphyllobothrium dendriticum. Acta Zool. 67: 155–163.

Wikgren, M., M. Reuter & M. Gustafsson, 1986. Neuropeptides in free-living and parasitic flatworms (Platyhelminthes). An immunocytochemical study. Hydrobiologia 132 (Dev. Hydrobiol. 32): 93–99.